Microbial Physiology: A Practical Approach

Microbial Physiology: A Practical Approach

Editor: Elsa Cooper

R CALLISTO
REFERENCE

www.callistoreference.com

Callisto Reference,
118-35 Queens Blvd., Suite 400,
Forest Hills, NY 11375, USA

Visit us on the World Wide Web at:
www.callistoreference.com

ISBN: 978-1-63239-985-4 (Hardback)

Cataloging-in-Publication Data

Microbial physiology : a practical approach / edited by Elsa Cooper.
    p. cm.
Includes bibliographical references and index.
ISBN 978-1-63239-985-4
1. Microorganisms--Physiology. I. Cooper, Elsa.
QR84 .M53 2018
576.11--dc23

Table of Contents

Preface

Every book is a source of knowledge and this one is no exception. The idea that led to the conceptualization of this book was the fact that the world is advancing rapidly; which makes it crucial to document the progress in every field. I am aware that a lot of data is already available, yet, there is a lot more to learn. Hence, I accepted the responsibility of editing this book and contributing my knowledge to the community.

Microbial physiology studies the cell structure, metabolism and genetic structure of microorganisms. The fundamental concern of this subject is to understand the characteristics of complex organisms by studying single-celled organisms. Microorganisms are vital in many environmental and industrial processes such as digestion, bioremediation, decomposition, fermentation etc. This book explores all the important aspects of microbial physiology in the present day scenario. Those in search of information to further their knowledge will be greatly assisted by this book.

While editing this book, I had multiple visions for it. Then I finally narrowed down to make every chapter a sole standing text explaining a particular topic, so that they can be used independently. However, the umbrella subject sinews them into a common theme. This makes the book a unique platform of knowledge.

I would like to give the major credit of this book to the experts from every corner of the world, who took the time to share their expertise with us. Also, I owe the completion of this book to the never-ending support of my family, who supported me throughout the project.

Editor

Role of TLR5 and Flagella in *Bacillus* Intraocular Infection

Salai Madhumathi Parkunan[1], Roger Astley[2], Michelle C. Callegan[1,2,3]*

1 Departments of Microbiology and Immunology, University of Oklahoma Health Sciences Center, Oklahoma City, Oklahoma, United States of America, 2 Department of Ophthalmology, University of Oklahoma Health Sciences Center, Oklahoma City, Oklahoma, United States of America, 3 Dean A. McGee Eye Institute, Oklahoma City, Oklahoma, United States of America

Abstract

B. cereus possesses flagella which allow the organism to migrate within the eye during a blinding form of intraocular infection called endophthalmitis. Because flagella is a ligand for Toll-like receptor 5 (TLR5), we hypothesized that TLR5 contributed to endophthalmitis pathogenesis. Endophthalmitis was induced in C57BL/6J and TLR5−/− mice by injecting 100 CFU of *B. cereus* into the mid-vitreous. Eyes were analyzed for intraocular bacterial growth, retinal function, and inflammation by published methods. Purified *B. cereus* flagellin was also injected into the mid-vitreous of wild type C57BL/6J mice and inflammation was analyzed. TLR5 activation by *B. cereus* flagellin was also analyzed *in vitro*. *B. cereus* grew rapidly and at similar rates in infected eyes of C57BL/6J and TLR5−/− mice. A significant loss in retinal function in both groups of mice was observed at 8 and 12 hours postinfection. Retinal architecture disruption and acute inflammation (neutrophil infiltration and proinflammatory cytokine concentrations) increased and were significant at 8 and 12 hours postinfection. Acute inflammation was comparable in TLR5−/− and C57BL/6J mice. Physiological concentrations of purified *B. cereus* flagellin caused significant inflammation in C57BL/6J mouse eyes, but not to the extent of that observed during active infection. Purified *B. cereus* flagellin was a weak agonist for TLR5 *in vitro*. These results demonstrated that the absence of TLR5 did not have a significant effect on the evolution of *B. cereus* endophthalmitis. This disparity may be due to sequence differences in important TLR5 binding domains in *B. cereus* flagellin or the lack of flagellin monomers in the eye to activate TLR5 during infection. Taken together, these results suggest a limited role for flagellin/TLR5 interactions in *B. cereus* endophthalmitis. Based on this and previous data, the importance of flagella in this disease lies in its contribution to the motility of the organism within the eye during infection.

Editor: Fu-Shin Yu, Wayne State University, United States of America

Funding: This study was supported by National Institutes of Health (NIH) Grant R01EY012985 (to MCC, http://www.nih.gov). The authors' research is also supported in part by NIH CORE Grant P30EY012191 (Robert E. Anderson, OUHSC), National Center for Research Resources COBRE Grant P20RR017703 (Robert E. Anderson, OUHSC), and an unrestricted grant to the Dean A. McGee Eye Institute from Research to Prevent Blindness. The funders had no role in study design, data collection and analysis, decision to publish, or preparation of the manuscript.

Competing Interests: The authors have declared that no competing interests exist.

* Email: michelle-callegan@ouhsc.edu

Introduction

B. cereus is a Gram-positive, sporulating bacterium that is more commonly recognized for causing food-borne illnesses, chronic skin infections, and systemic diseases such as meningitis and pneumonia [1]. Nosocomial infection pseudo-outbreaks caused by *B. cereus* have been reported in the last decade and have been attributed to contaminated disinfecting agents like ethyl alcohol [2] and alcohol swabs [3], or contaminated equipment like airflow sensors, intravenous catheters [1,4], and ventilator and filtration units [1,5,6]. A recent nosocomial outbreak identified *B. cereus* in contaminated alcohol Prep Pads [7]. *B. cereus* is also highly associated with a blinding ocular infection termed endophthalmitis. Endophthalmitis is characterized by intraocular inflammation and damage to the retina, resulting in partial or complete loss of vision. Microbes can enter the posterior segment following an ocular injury (post-traumatic), surgery (post-operative) or from another site of infection (endogenous) [8,9]. While cases of post-operative endophthalmitis generally respond positively to treatment, cases of post-traumatic and endogenous endophthalmitis caused by *B. cereus* have a significantly greater failure rate, necessitating the search for better strategies to combat the disease.

The pathogenicity of *B. cereus* in endophthalmitis is associated with the inflammogenicity of its cell wall and the production of secreted toxins and proteases [10–14]. Previous studies have shown that *B. cereus* endophthalmitis develops faster and is more virulent than endophthalmitis caused by other Gram-positive ocular pathogens such as *Staphylococcus aureus* [15,16], *Enterococcus faecalis* [17,18], or *Streptococcus pneumoniae* [19,20]. The explosive nature of *B. cereus* endophthalmitis dictates the need for immediate and aggressive therapy to stop the progression of the disease. Currently, there is no universal therapeutic regimen which prevents vision loss that occurs during severe forms of endophthalmitis. The use of anti-inflammatory agents in addition to antibiotics has not proven effective [21–25]. In addition, current therapies ignore toxins which are proven to contribute to pathogen virulence in the eye [10,12–19].

Innate immune mechanisms drive inflammation by the recognition of distinguishing molecules on the surface of the invading bacterium via a class of pattern recognition receptors called Toll-like receptors (TLRs) expressed on host cells. TLRs are expressed in ocular surface, retinal, iris, and corneal epithelial cells [26–28]. In the context of intraocular infections, TLRs have been found to be important in inflammation in *S. aureus* [29] and *B.*

cereus [30] endophthalmitis. For experimental *B. cereus* endophthalmitis, the absence of TLR2 resulted in a diminished inflammatory environment when compared to controls [30], but there was still some degree of inflammation in *B. cereus*-infected TLR2−/− eyes. This suggests that other TLRs and/or components of innate immunity are involved in intraocular inflammation during *B. cereus* endophthalmitis.

When *B. cereus* infects the eye, the organism migrates rapidly throughout all parts of the eye, from the initial site of injection in the vitreous into the anterior segment within 6 to 12 hours [31]. This ability of *B. cereus* to migrate throughout the eye contributes to endophthalmitis pathogenesis [12,32,33]. The absence of motility affects toxin production and hence non-motile *Bacillus* caused less severe disease pathogenesis [12,32,33]. *B. cereus* use peritrichous flagella [34] as motility appendages which render the bacterium capable of movement throughout the eye. Moreover, flagella may impact the inflammatory response mounted against *B. cereus* since flagellin, the monomer which comprises full-length flagella, is a natural ligand for TLR5 [35]. Since *B. cereus* is a flagellated bacterium, we hypothesized that *B. cereus* flagella contributed to the pathogenesis during endophthalmitis by activating the ocular inflammatory response via TLR5. This hypothesis was tested by analyzing the immune response against *B. cereus* flagellin *in vitro* and *in vivo*, and by comparing the pathogenesis of *B. cereus* infection in an experimental model of endophthalmitis in wild type control and TLR5−/− mice.

Methods

Ethics Statement

These experiments involved the use of mice. All procedures were carried out in strict accordance with the recommendations in the Guide for Use of Laboratory Animals of the National Institutes of Health, institutional guidelines set forth by the University of Oklahoma Health Sciences Center IACUC, and the Association for Research in Vision and Ophthalmology Statement for the Use of Animals in Ophthalmic and Vision Research. The OUHSC IACUC approved these studies under protocols 11–068 and 11–090.

Experimental *B. cereus* endophthalmitis

Wild type C57BL/6J mice were purchased from commercially available colonies (Stock No. 000664, Jackson Labs, Bar Harbor ME). An original breeding pair of TLR5−/− mice on the C57BL/6 background was a kind gift from Dr. Richard A. Flavell (Yale University, New Haven CT). Following rederivation, TLR5−/− mice were bred on the C57BL/6J background and maintained in-house on a 12 hour on/12 hour off light cycle under barrier facility conditions. All animals were acclimated to conventional housing after arrival/weaning for at least 2 weeks and were used in experiments at 8–10 weeks of age.

Experimental endophthalmitis was induced by injecting 100 CFU *B. cereus* strain ATCC 14579 into the mid-vitreous using a sterile capillary needle as previously described [30,36–38]. At different time points postinfection, quantitation of intraocular bacterial growth, proinflammatory cytokines and chemokines, myeloperoxidase (MPO, to estimate PMN infiltration), and retinal function were performed, as described below.

Intraocular Bacterial growth

Bacteria were quantified by harvesting infected eyes at 0, 4, 8, and 12 hours postinfection. The eyes were homogenized with 1 mm sterile glass beads (Biospec products, Inc., Bartlesville OK) in 400 μl PBS. Bacteria were then track diluted 10-fold onto brain-heart infusion (BHI) agar and quantified [30,37,38]. Values represent the mean ± standard deviation (SD) for N≥4 eyes per time point.

Electroretinography

Retinal function was analyzed in wild type and TLR5−/− mice by electroretinography (ERG) as previously described [30,37,38]. ERGs were performed at 8 and 12 hours postinfection (Espion E2, Diagnosys LLC, Lowell MA). After dark adaptation for at least 6 hours, eyes were exposed to a transient flash of light. Bright flashes resulted in a response which consisted of an A wave initial negative amplitude followed by a B wave positive deflection. A-wave provides a direct measure of photoreceptor activity, while B-wave represents the action of Muller cells, bipolar cells, and second order neurons. A- and B-wave amplitudes were recorded for each infected eye and compared with the uninfected eye. The percentage of retinal function retained was then calculated using the formula 100 − {[1 − (experimental A-wave amplitude/control A-wave amplitude)] ×100} or 100 − {[1 − (experimental B-wave amplitude/control B-wave amplitude)] ×100} [38]. Values represent the mean ±SD for N≥4 eyes per time point.

Histology

Whole eyes were harvested at 0, 4, 8, and 12 hours after infection and incubated in buffered zinc formalin fixative for 24 hours at room temperature [30,37,38]. Globes were then transferred to 70% ethanol and embedded in paraffin, sectioned, and stained with hematoxylin and eosin. Images are representative of 4 eyes/time point.

Inflammatory Cell Influx

PMN influx into the eye was estimated by quantifying MPO levels in whole eye homogenates by sandwich ELISA (Mouse MPO ELISA Kit, Hycult Biotech, Plymouth Meeting PA), as previously described [30]. Eyes were harvested and analyzed for MPO activity at 0, 4, 8 and 12 h postinfection. Harvested eyes were suspended in PBS containing protease inhibitor cocktail (Roche Applied Science, Indianapolis IN) and homogenized. Homogenates of uninfected eyes served as negative controls. The lower limit of detection for this assay was 2 ng/ml. Results are reported as mean ±SD for N≥4 eyes per group per time point.

Inflammatory Mediator Expression

Ocular proinflammatory cytokine and chemokine expression was quantified as previously described [30,36,37]. Eyes were harvested at 0, 4, 8 and 12 hours postinfection. Harvested eyes were suspended in PBS containing protease inhibitor cocktail and homogenized. Concentrations of IL6, TNFα, IL1β, and KC were quantified in harvested wild type and TLR5−/− eyes using commercial enzyme linked immunosorbent assay (ELISA) kits (Quantikine, R&D Systems, Minneapolis MN). The lower limits of detection for each assay were: TNFα, 2 pg/ml; KC, 2 pg/ml; IL6, 2 pg/ml; IL1β, 2 pg/ml. Values represent mean ±SD for N≥4 eyes/time point.

Purification of *B. cereus* Flagellin

Flagellin preparations from *B. cereus* were generated based on a previously described method [39,40]. Motile *B. cereus* was grown overnight in 1L Luria Bertani (LB) media with minimal rotary shaking (80 rpm) to avoid damage to intact flagella. Bacteria were harvested by centrifugation at 4000×g for 30 min, the pellet was resuspended in 15 ml of PBS containing protease inhibitor cocktail, and the suspension was vigorously mechanically shaken to remove the flagella. Two cycles of differential centrifugation were done at 15,000×g for 30 min to remove bacterial debris and again at 78,000×g for 2 h to sediment flagella. Purified flagella was then resuspended overnight in 1ml PBS with protease inhibitor and stored at 4°C. Purity of flagellar monomers was analyzed by SDS-PAGE. A single band of approximately 29 kD was identified, extracted from the gel, and sequenced (LC/MS/MS; OUHSC Laboratory for Molecular Biology and Cytometry Research, Oklahoma City OK). The purified flagellin protein sequence matched those of *B. cereus* ATCC 14579 flagellin (Accession No. gi|30019803) and *B. thuringiensis* flagellin A1 (Accession No. gi|189164115).

Flagella-Induced Inflammation in the Eye

Purified *B. cereus* flagellin suspensions in PBS were injected into C57BL/6J or TLR5−/− mouse eyes as described above. 0.5 ng/0.5 μL flagellin were injected into each eye. ERGs were performed and eyes were harvested for histology at 0, 8, and 12 hours postinfection, as described above. ERG values represent the mean ±SD for N≥2 eyes/time point.

Flagellin Activation of TLR5

Purified flagellin was tested for its ability to activate TLR5 in a TLR5 reporter cell line which expresses human TLR5 and secreted alkaline phosphatase reporter gene under the transcriptional control of NFκB (IML-105, TLR5/SEAPorter HEK 293 cells, Imgenex, San Diego CA). The positive control for this assay was purified flagellin FliC from *Salmonella typhimurium* (IMG-2205, Imgenex) tested at equal concentrations (0.1, 0.5, 1.0, 5.0, and 10.0 ng/mL). Results were analyzed by reading absorbance at 405 nm. Values represent mean ±SD for 2 replicates per concentration.

Statistics

Results represent the arithmetic means ± standard deviations (SD) for all samples from each experimental group. A two-tailed, two-sample Student t test assuming equal variance was used to compare the statistical significance of the experimental groups. Statistical significance was determined at P≤0.05.

Sequence Analysis

Flagellin sequences for *B. cereus* (NP_831435.1), *B. anthracis* (WP_001222388.1), *B. thuringiensis* (ABD33778.1), and *S. enterica* serovar *typhimurium* (S07276) were aligned, displayed, and analyzed with ClustalW (European Bioinformatics Institute, Cambridgeshire UK) [41].

Results

Intraocular Growth of *B. cereus*

Bacterial growth in wild type C57BL/6J and TLR5−/− eyes is shown in Figure 1. The rates of bacterial growth followed a similar pattern but were statistically different at 4 hours (P = 0.002) and 12 hours (P = 0.01) postinfection. *B. cereus* reached maximum concentrations of approximately 6.5 (TLR5−/−) and 7.3 (C57BL/6J) $\log_{10}$ CFU/eye by 12 hours postinfection. This result suggested that the absence of TLR5 did not greatly affect the overall rate of *B. cereus* growth in the eye.

Retinal Function

Retinal function analysis of *B. cereus* endophthalmitis in wild type C57BL/6J and TLR5−/− mice is summarized in Figure 2. Amplitudes of A-wave and B-wave declined significantly both in

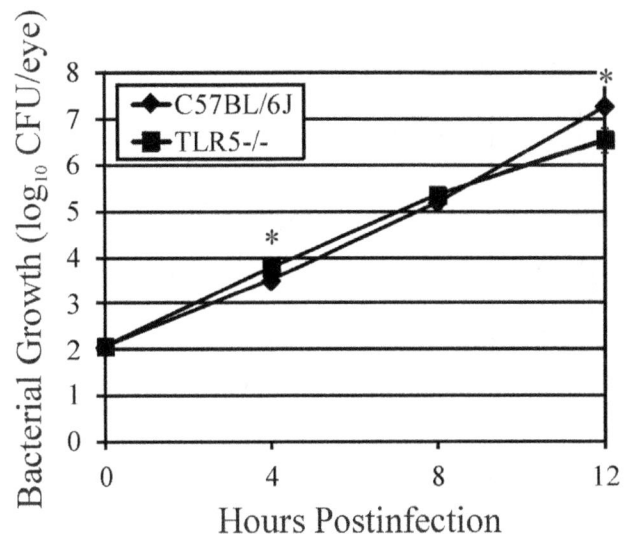

Figure 1. Influence of TLR5 on bacterial growth during experimental *B. cereus* endophthalmitis. C57BL/6J wild type and TLR5−/− mouse eyes were injected with 100 CFU *B. cereus*. Eyes were harvested, homogenized, and analyzed for bacterial growth. Overall, *B. cereus* grew to similar concentrations in infected eyes of TLR5−/− and C57BL/6J mice, suggesting that the absence of TLR5 did not influence the overall growth of *B. cereus* in the eye. Values represent the mean $\log_{10}$ CFU±SD of N≥4 eyes per time point for at least 2 separate experiments. *P≤0.05.

wild type C57BL/6J and TLR5−/− eyes at 8 and 12 hours following infection with *B. cereus*. The A-wave amplitudes in C57BL/6J infected eyes was similar at 8 h postinfection (P = 0.07), but slightly less at 12 h postinfection (P = 0.02). B-wave amplitudes retained were greater in TLR5−/− infected eyes at 8 h postinfection (P = 0.008), but both groups had similar ERG values at 12 h postinfection (P = 0.1). By 12 hours postinfection, both A-wave and B-wave amplitudes retained decreased to approximately 20% or less in infected eyes, indicating significant and comparable retinal function loss in both groups of mice.

Intraocular Inflammation

Histology of uninfected (control) and *B. cereus*-infected globes in wild type C57BL/6J and TLR5−/− mice is depicted in Figure 3. At 4 hours postinfection, wild type C57BL/6J and TLR5−/− mice had similar levels of fibrin deposition in the anterior segment and minimal fibrin and polymorphonuclear leukocyte (PMN) infiltration in the posterior segment. At this time, retinas were intact in eyes of both groups. At 8 hours postinfection, eyes of both groups had significant fibrin deposition in the anterior chamber and in the posterior segment, corneas were edematous, and significant numbers of PMN were present in the vitreous. In C57BL/6J and TLR5−/− mouse eyes, retinal layers were intact but retinal detachments were present. At 12 hour postinfection, whole globe inflammation was significant and retinal layers were indistinguishable in both groups of mice.

PMN infiltration in whole eyes following *B. cereus* infection is depicted in Figure 4. Myeloperoxidase (MPO) levels increased significantly after 4 hours postinfection in C57BL/6J and TLR5−/− infected eyes. MPO levels were similar in these groups at all time points postinfection (P≥0.17). These results suggest that the absence of TLR5 did not alter the PMN response during infection, supporting the histology data.

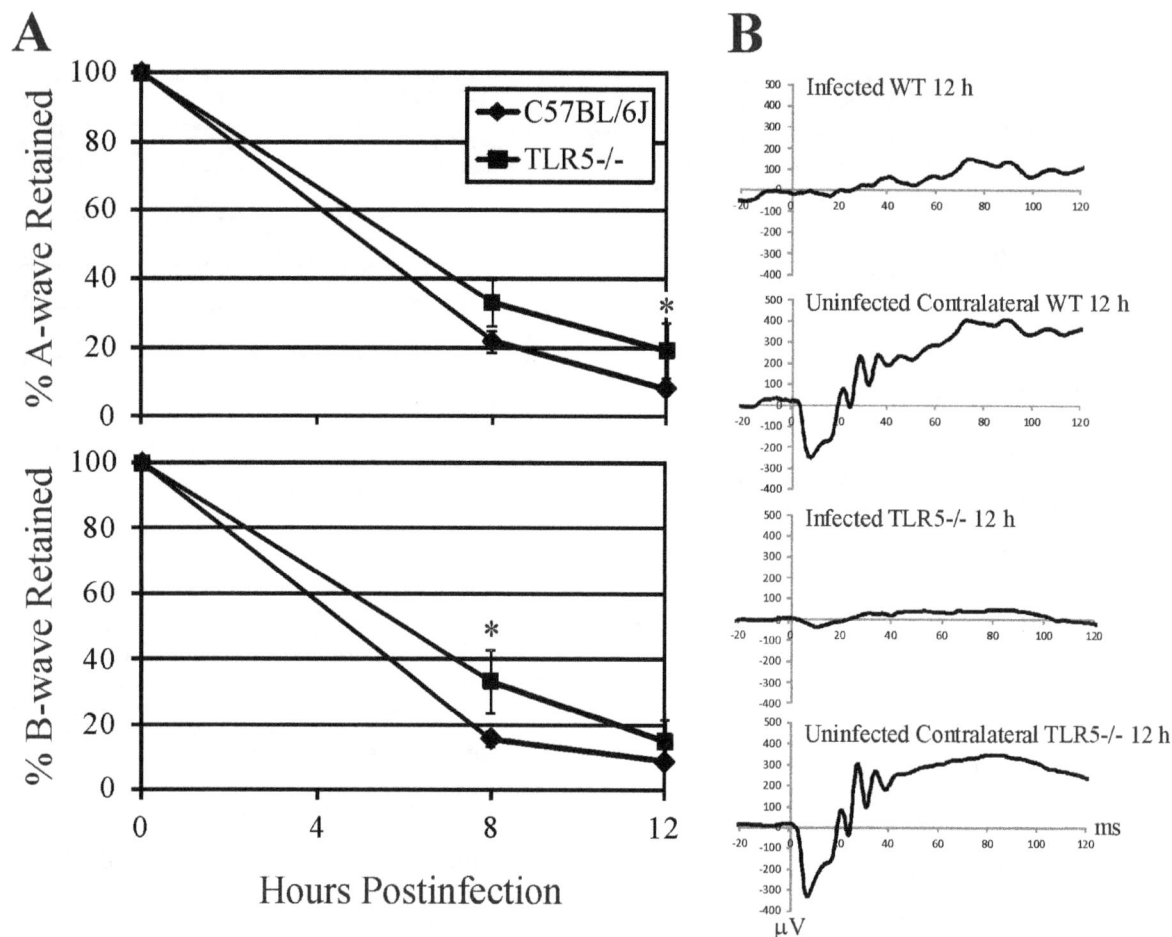

Figure 2. Influence of TLR5 on retinal function during experimental *B. cereus* endophthalmitis. C57BL/6J wild type and TLR5−/− mouse eyes were injected with 100 CFU *B. cereus*. Retinal function was assessed by electroretinography (ERG). **A**) A-wave amplitudes were slightly greater in C57BL/6J infected eyes at 12 h postinfection (P = 0.02), while B-wave amplitudes were greater in TLR5−/− infected eyes at 8 h postinfection (P = 0.008). By 12 hours postinfection, A-wave and B-wave amplitudes retained in both groups decreased to approximately 20% or less in infected eyes, indicating significant retinal function loss in both groups of mice regardless of the presence of TLR5. Values represent the mean ±SD of N≥4 eyes per time point for at least 2 separate experiments. *P≤0.05. **B**) Representative averaged waveforms from wild type (WT) and TLR5−/− mice at 12 h postinfection, with one eye infected and the contralateral eye serving as the uninfected control. Representative of N≥4 eyes per time point.

The presence of proinflammatory cytokines and chemokines in the eye during infection is depicted in Figure 5. In general, all cytokines and chemokines tested increased significantly in both groups of mice during experimental endophthalmitis. TNFα levels were similar in C57BL/6J and TLR5−/− eyes at 8 and 12 hours postinfection (P≥0.1). TNFα levels increased approximately 12-fold in both groups between 4 and 12 hours postinfection. KC levels were similar at all time points postinfection (P≥0.05), with an approximate 7-fold increase in KC in both groups between 4 and 12 hours postinfection. IL6 levels were similar in C57BL/6J and TLR5−/− eyes at 8 and 12 hours postinfection (P≥0.57). IL6 levels increased an average of 30-fold in both groups between 4 and 12 hours postinfection. IL1β levels were similar in C57BL/6J and TLR5−/− mice, except at 8 hours postinfection when IL1β levels were slightly but significantly greater in infected C57BL/6J eyes (P=0.001). IL1β levels increased 8-fold and 13-fold in C57BL/6J and TLR5−/− mouse eyes between 8 and 12 hours postinfection (P≤ 0.0001). Despite a few time points where proinflammatory mediators were slightly but significantly greater in C57BL/6J eyes compared to that of TLR5−/− eyes, the data suggest that overall, the cytokine and chemokine response

to *B. cereus* infection was not altered by the absence of TLR5. These results coincided with the histology and MPO data, indicating that TLR5 did not significantly contribute to the inflammatory response in experimental *B.cereus* endophthalmitis.

Intraocular Effects of Flagellin

It has been reported that flagellin monomers elicit a TLR5-mediated response because the flagellar TLR5-binding domain is exposed in monomers, but not in polymerized flagellin [42,43]. These findings were reported for *Salmonella* and *Pseudomonas* flagellin, but, to our knowledge, these effects have not been analyzed for *B. cereus* flagellin. To determine whether flagellin alone could cause intraocular inflammation in the absence of *B. cereus* organisms, flagellin monomers (Figure 6A) were purified and intravitreally injected into mouse eyes as described in the Methods. Based on a report of an average of 11 flagella per *B. cereus* cell [44], an estimated 20,000 flagellin subunits per filament [45], and a calculated molecular mass of 29 kD [46], we estimated that injecting 0.5 ng flagellin into an eye would equate to that quantity of flagellin found in 4.72×10^5 CFU *B. cereus*. Extrapolation of the CFU data in Figure 1 suggests that this concentration of *B. cereus*

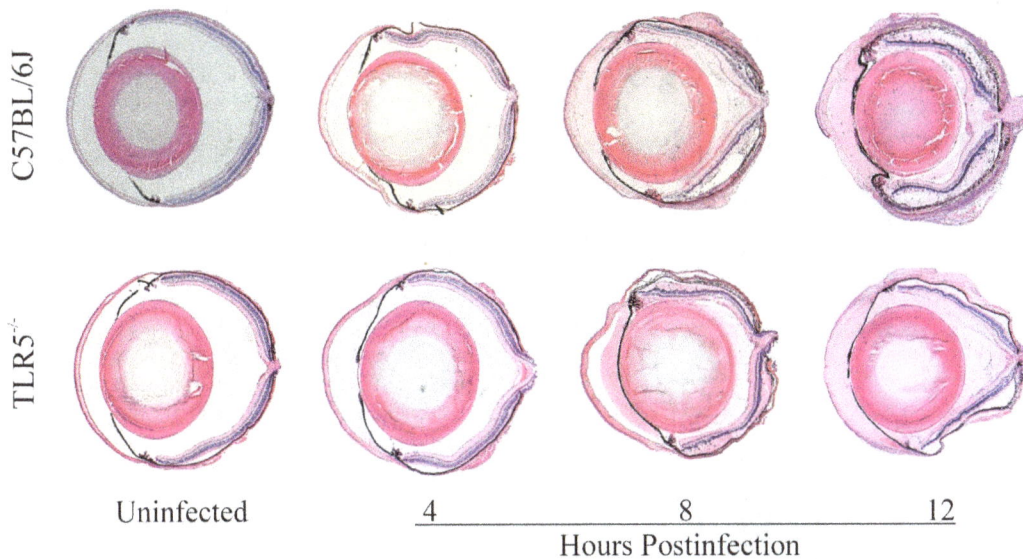

Figure 3. Whole eye histology of experimental *B. cereus* endophthalmitis. C57BL/6J and TLR5−/− mouse eyes were injected with 100 CFU *B. cereus*. Whole globes were harvested and processed for hematoxylin and eosin staining. Infected eyes of both groups had significant inflammation by 12 h postinfection, suggesting that the absence of TLR5 did not greatly impact intraocular inflammation. Sections are representative of 4 eyes per group. Magnification, 10X.

was present in the mouse eye at approximately 8 hours postinfection, a time when retinal function loss and inflammation were significant in infected eyes (Figures 2–4). At 8 hours following injection of 0.5 ng of purified flagellin, posterior segment inflammation was minimal and retinal function decreased slightly (but not significantly) from that at time 0 (Figure 6BC, P≥0.4). These results suggest that *B. cereus* flagellin alone may not have contributed significantly to retinal function loss or inflammation during an actual infection. Delayed intraocular inflammation was observed at 12 h following injection of 0.5 ng flagellin (Figure 6C); however, infected eyes at 12 h (Figure 3) demonstrated much greater pathology than that seen at 12 hours in eyes injected with

flagellin alone. No posterior segment inflammation was observed in eyes injected with 100-fold less purified flagellin (data not shown). Injection of 0.5 ng flagellin into TLR5−/− eyes resulted in slightly less but still significant inflammation and similar retained A-wave (P = 0.16) and B-wave (P = 0.76) amplitudes compared to that of wild type eyes at 12 h postinjection (Figure S1). These results suggest that TLR5 may not be essential to intraocular inflammation caused by flagellin, and that flagellin, when present, may induce inflammation through a different pathway when TLR5 is absent.

Because the degrees and timing of intraocular inflammation present in infected eyes versus those injected with purified flagellin differed so greatly, we analyzed whether *B. cereus* flagellin was an agonist of TLR5. TLR5 activation of purified *B. cereus* flagellin was compared with that of purified *Salmonella typhimurium* flagellin *in vitro*. Compared to *Salmonella* flagellin, *B. cereus* flagellin was a weak TLR5 agonist, resulting in significantly less NFκB activity at comparable flagellin concentrations (Figure 6D). Taken together, these results suggest that *B. cereus* flagellin/TLR5 interactions in the eye may not be significant enough to greatly impact the overall course of intraocular inflammation during experimental endophthalmitis.

Discussion

TLR5 is an important innate immune regulator of inflammation in infections, including those caused by *Salmonella* [47,48], *Legionella* [49,50], *Clostridium* [51], *Pseudomonas* [52,53], *E. coli* [54], and others. TLR5 is also an important mediator of gut homeostasis [55]. TLR5 has been detected in cells of the eye [27,56,57]. TLR5 has been reported to modulate corneal inflammation and the innate antimicrobial response *in vitro* [58–60] and is important in ocular inflammation during bacterial and fungal keratitis [52,61]. We therefore sought to determine the role of TLR5 in endophthalmitis caused by *B. cereus*, an organism which possesses the TLR5 ligand, flagella.

The majority of studies on flagellin/TLR5 interactions have been done with Gram-negative organisms. Few studies have

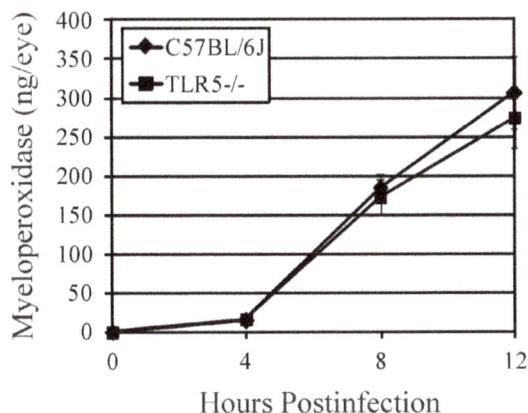

Figure 4. Influence of TLR5 on infiltration of PMN into mouse eyes during experimental *B. cereus* endophthalmitis. C57BL/6J and TLR5−/− mouse eyes were injected with 100 CFU *B. cereus*. PMN infiltration was estimated by quantifying MPO in whole eyes by sandwich ELISA. MPO levels were similar in these groups at all times points postinfection (P≥0.17), suggesting that the absence of TLR5 did not alter the PMN response during infection. Values represent the mean ±SD for N≥4 per group for at least 2 separate experiments. *P≤0.05.

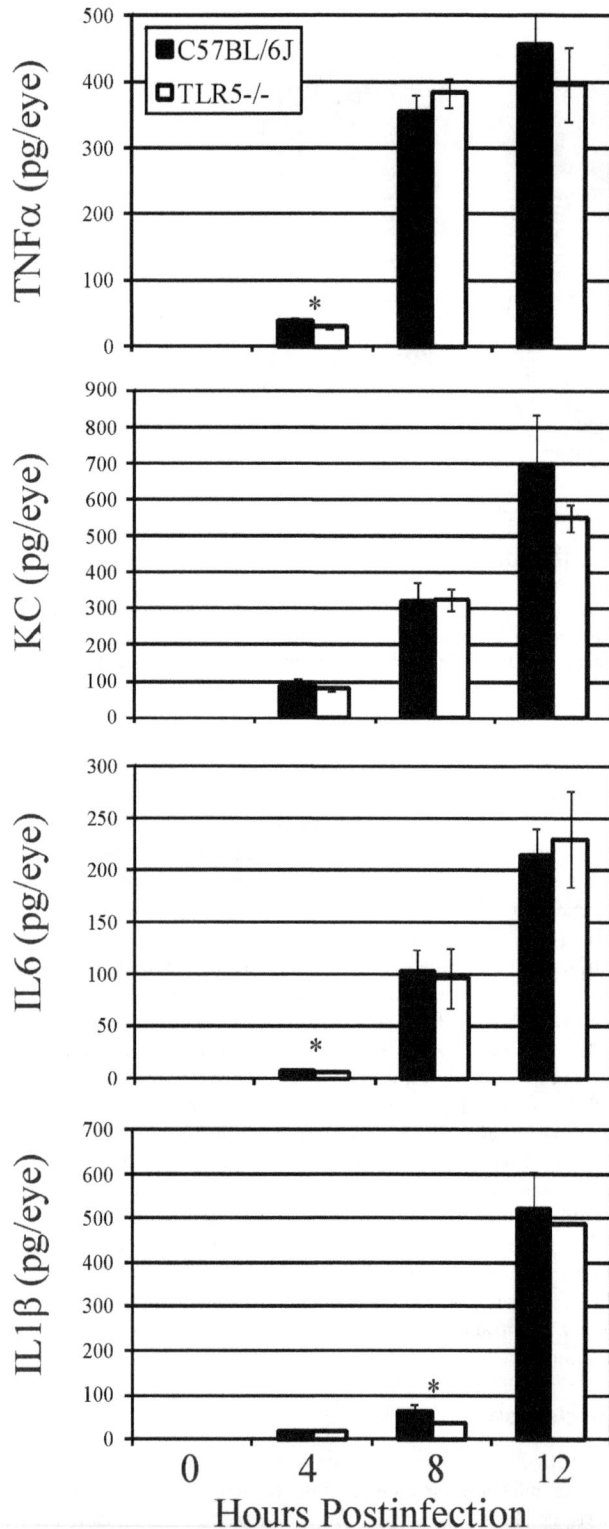

Figure 5. Influence of TLR5 on proinflammatory mediator expression during experimental *B. cereus* endophthalmitis. C57BL/6J and TLR5−/− mouse eyes were injected with 100 CFU *B. cereus*. Ocular proinflammatory cytokines and chemokines were analyzed by sandwich ELISA. Overall, similar levels of TNFα, KC, IL6, and IL1β were synthesized in infected eyes of C57BL/6J mice compared with that in infected eyes of TLR5−/− mice, suggesting that the absence of TLR5 did not alter the inflammatory mediator response during infection. Values represent the mean ±SD for N≥4 per group for at least 2 separate experiments. *P≤0.05.

analyzed interactions between Gram-positive flagella and TLR5. *Listeria* and *Clostridium* flagellin have been shown to be TLR5 agonists [35,62]. Although *B. subtilis* flagellin is commercially marketed as a TLR5 agonist, *in vitro* results disagree on its ability to activate TLR5 [56,63–65]. The *B. cereus sensu lato* group, including pathogens *B. cereus*, *B. anthracis*, and *B. thuringiensis*, have peritrichous flagella. Because the motility of *B. cereus*, and therefore its flagella, are important in the virulence of *B. cereus* during endophthalmitis [8,66], we hypothesized that flagellin/TLR5 interactions also contributed to the pathogenicity of infection.

Despite our finding that *B. cereus* flagellin was a weak TLR5 agonist *in vitro*, physiological concentrations of purified B. cereus flagellin monomers caused inflammation in wild type and TLR5−/− mouse eyes. However, this inflammation was dissimilar to the degree of inflammation caused by active infection when that concentration of flagellin would have been present in the eye. *B. cereus* migrates through all parts of the eye during endophthalmitis [31], so its flagella are likely polymerized. The disparity could therefore be explained by the fact that flagellin monomers of other organisms, but not polymerized flagellin, have been shown to activate TLR5 [42,43]. If flagellin monomers were not present in the eye during infection, this may explain why the absence of TLR5 did not significantly impact intraocular inflammation during infection. The results also suggest that unlike the environment in the inflamed gut where high levels of flagellin monomers exist [67,68], flagellin monomers are either not present or are present at non-inflammogenic concentrations in the eye during endophthalmitis.

The potential lack of a significant role for polymerized flagellin in intraocular inflammation also brings forth an interesting question about the physiological state of *B. cereus* in the eye during infection. We demonstrated that mutant *B. cereus* which cannot swarm do not migrate into the anterior segment and cause a less virulent infection than wild type *B. cereus* that can swarm [33]. *In vitro*, swarming cells are elongated and hyperflagellated on media [69], but the swarming state of *B. cereus* in the eye during endophthalmitis has not been analyzed. If *B. cereus* is in a physiological state of swarming in the eye, then our concentration of flagellin injected into the eye may have been too low, as 40-fold increases in flagellin have been reported for swarming *B. cereus* [70]. However, if flagellin monomers were not present in the eye during infection, the increased number of flagella present in swarming organisms would be irrelevant, and TLR5/flagellin interactions would still not be as important to the outcome of infection.

B. cereus and *B. subtilis* do not fall into the category of organisms whose flagellin is not recognized by TLR5 [71]. Therefore, a TLR5 evasion mechanism similar to that demonstrated by *Helicobacter* or *Campylobacter* [71] may not be occurring here. ClustalW alignments of the *B. cereus* ATCC 14579 and *S. typhimurium* H1-A flagellin sequences demonstrated significant similarity (81%) in a region shown to be important for IL8 activity in Caco2 cells (amino acids 30−52) (Figure 7) [72]. However, ClustalW comparisons of these flagellin sequences also demonstrated that *B. cereus* flagellin contains differences in the TLR5 recognition and binding sites. The *S. typhimurium* FliC-TLR5 stimulatory activity lies within amino acids 89-96 in the N-terminal D1 domain [43]. Important residues for TLR5 activation also exist in the C-terminal conserved domain (430–445) [43]. Additional residues located between the IL8 activity region and the N-terminal D1 domain (58, 59 of *S. typhimurium*) and within the C-terminal D1 domain (411 of *S. typhimurium*) are also required for TLR5 recognition, as these residues are in physical contact with the N-terminal TLR5 binding region [43]. *B. cereus* shares 62.5%

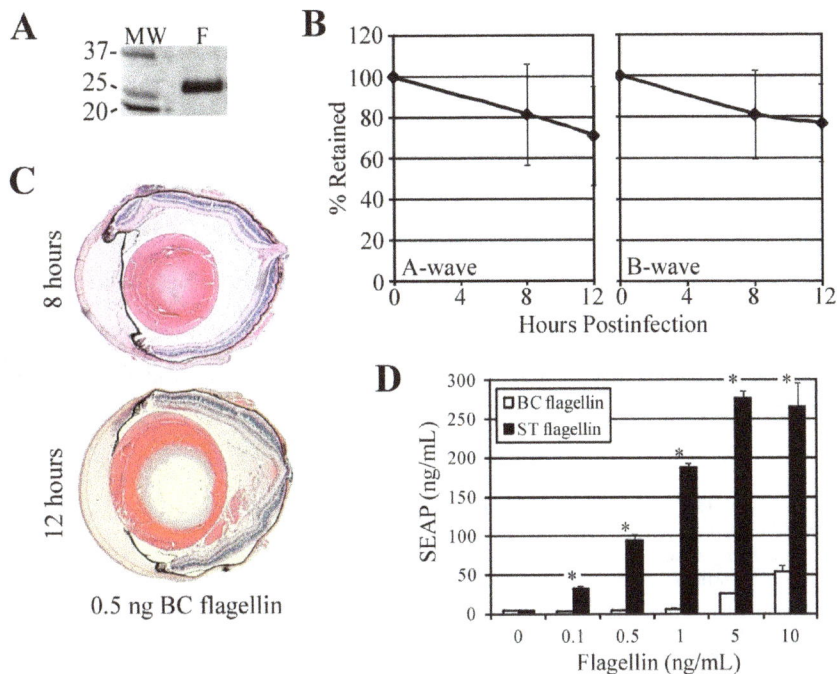

Figure 6. Role of flagellin in intraocular inflammation. A) Purified *B. cereus* flagellin was injected into C57BL/6J mouse eyes. *B. cereus* flagellin was purified as described in the Methods and analyzed for purity on a SDS-PAGE/Coomassie gel. A single band of 29 kD was recovered (lane F). **B and C)** Following injection of 0.5 ng/0.5 µL of purified flagellin, eyes were analyzed by ERG and whole globe histology at 8 and 12 hours. At 8 hours, posterior segment inflammation was minimal, retinas were intact, and retinal function decreased slightly (but not significantly) from that at time 0 (Figure 6B, P≥0.4). Compared with the significant inflammation and retinal function loss observed during infection, these results suggest that *B. cereus* flagellin alone may not have contributed significantly to the process. Infected eyes at 12 hours (Figure 3) demonstrated much greater pathology than that seen at 12 hours in eyes injected with flagellin alone. **D)** TLR5 activation by purified *B. cereus* (BC) flagellin was compared with that of purified *Salmonella typhimurium* (ST) flagellin. *B. cereus* flagellin was a weak TLR5 agonist, resulting in significantly less NFκB activity at comparable flagellin concentrations (mean ±SD for two repeated experiments, *P≤0.001).

identity with the 89-96 region of *S. typhimurium*, including identity with three amino acids deemed essential for TLR5 binding activity, protofilament assembly, and motility [43]. Only 25% of the residues in the C-terminal conserved domain are identical between these flagellins. Residues 58, 59, and 411 were not identical, suggesting that the three dimensional structure of TLR5 binding by *B. cereus* flagellin is different from that of *S. typhimurium*. This is not a surprise, as *B. cereus* flagellin is 221 residues shorter than *S. typhimurium* FliC. Whether or not these differences account for the reduced TLR5 agonism of *B. cereus* flagellin or whether this lack of agonism extends to other members of the *B. cereus sensu lato* group (Figure 7) is an open question. A recent report supports the idea of differential activation of TLR5 and NAIP5/NLRC4 inflammasome receptors by the flagellins of different organisms [73]. In evaluating the use of *Bacillus cereus sensu lato* group flagellins for vaccine development, species-specific differences in these domains are important to consider.

Our results demonstrated that *B. cereus* flagella/TLR5 interactions, if present, did not contribute significantly to endophthalmitis pathogenesis. Although *B. cereus* flagella may not have contributed to inflammation during infection, its role in migration throughout the eye during infection is clearly important. We previously demonstrated that non-motile and non-swarming flagellated mutants are significantly less virulent than their motile and swarming wild type parental strains [32,33]. Therefore, immobilization of the organism is paramount during the early stages of infection. Realistically, this would be achieved with appropriate administration of bactericidal antibiotics at the site of infection as early as possible during the infection course to sterilize the eye

[25]. However, antibiotics do not inactivate the multitude of toxins synthesized by *B. cereus* or other organisms in the eye during infection which contribute to intraocular virulence. Future efforts to improve the visual outcome of patients with endophthalmitis caused by *B. cereus* and other virulent pathogens should include anti-toxin strategies with sterilization and better anti-inflammatory drugs to prevent the inflammation and tissue damage which results in vision loss during this disease.

Supporting Information

Figure S1 Flagellin causes similar inflammation and retinal function changes in wild type and TLR5−/− mice. A) Purified *B. cereus* flagellin (0.5 ng) was injected into C57BL/6J mouse eyes as depicted in Figure 6. Injection of flagellin resulted in slightly less but still significant inflammation in TLR5−/− eyes compared to that of wild type eyes (representative of N = 3 TLR5−/− eyes at 12 h postinjection. **B)** Eyes underwent electroretinography as depicted in Figure 6. At 12 h postinjection, retained A-wave (P = 0.16) and B-wave (P = 0.76) amplitudes were similar between wild type and TLR5−/− eyes (mean ±SD, N≥ 2/group).

Acknowledgments

The authors thank Dr. Richard Flavell (Yale University, New Haven CT) for the kind gift of the original breeding pair of TLR5−/− mice. The authors thank Jennifer Thurman, Nanette Wheatley, Bo Novosad (University of Oklahoma Health Sciences Center, Oklahoma City OK),

```
              1                      2                  3
St 30 ERLSSGLRINSAKDDAAGQAIAN 89 QRVRELAVQ 430 NRFNSAITNLGNTVNN
Bc 28 NRLSSGKSINSAADDAAGLAIAT 87 LRMRDLATQ 209 NRLDHNLNNVTSQATN
      *****  **** ***** ***      * * ** *     **     *       *


Bc 28 NRLSSGKSINSAADDAAGLAIAT 87 LRMRDLATQ 209 NRLDHNLNNVTSQATN
Bt 08 DRLSSGKRINNASDDAAGLAIAT 68 LRMRDIANQ 283 NRLDFNVENLKSQSAS
Ba 28 NRLSSGKRINSAADDAAGLAIAT 87 TRMRDIAVQ 224 NRLDRNVENLNNQATN
      ****** ** * **********     **** * *    **** *  *     *
```

Figure 7. Multiple sequence alignments of *S. typhimurium*, *B. cereus*, *B. thuringiensis*, **and** *B. anthracis* **flagellins.** The amino acid sequences of *B. cereus* (Bc) and *S. typhimurium* (St) (top) and *Bc*, *B. thuringiensis* (Bt), and *B. anthracis* (Ba) (bottom) were aligned by ClustalW [71], focusing on regions important for IL8 activity (Region 1) and TLR5 stimulation and recognition (Regions 2 and 3). Asterisks and red letters identify amino acids conserved between Bc and St sequences or among Bc/Ba/Bt sequences. Blue amino acids have strongly similar properties, while green amino acids have weakly similar properties. Bc and St had 81% conserved residues in Region 1, 62.5% conserved residues in Region 2, and 25% conserved residues in Region 3. Bc, Ba, and Bt had 86% conserved residues in Region 1, 62.5% conserved residues in Region 2, and 44% conserved residues in Region 3.

Amanda Roehrkasse (Oklahoma Christian University, Edmond OK), Rachel Staats (Oklahoma State University, Stillwater OK) Mark Dittmar (Dean A. McGee Eye Institute Animal Research Facility), and Dr. Feng Li (OUHSC Live Animal Imaging Core, Dean A. McGee Eye Institute Animal Research Facility) for their invaluable technical assistance. We also thank Paula Pierce (Excalibur Pathology, Moore OK) for histology expertise and Dr. Phillip Coburn (University of Oklahoma Health Sciences Center) for his helpful comments.

Author Contributions

Conceived and designed the experiments: SMP MCC. Performed the experiments: SMP RA. Analyzed the data: SMP MCC. Contributed reagents/materials/analysis tools: MCC. Wrote the paper: SMP MCC.

References

1. Bottone EJ (2010) *Bacillus cereus*, a volatile human pathogen. Clin Microbiol Rev 23: 382–398.
2. Hsueh P-R, Teng L-J, Yang P-C, Pan H-L, Ho S-W, et al. (1999) Nosocomial pseudoepidemic caused by *Bacillus cereus* traced to contaminated ethyl alcohol from a liquor factory. J Clin Microbiol 37: 2280–2284.
3. Berger SA (1983) Pseudobacteremia due to contaminated alcohol swabs. J Clin Microbiol 18: 974–975.
4. Hernaiz C, Picardo A, Alos J, Gomez-Garces J (2003) Nosocomial bacteremia and catheter infection by *Bacillus cereus* in an immunocompetent patient. Clin Microbiol Infect 9: 973–975.
5. Bryce E, Smith J, Tweeddale M, Andruschak B, Maxwell M (1993) Dissemination of *Bacillus cereus* in an intensive care unit. Infect Control Hosp Epidemiol: 459–462.
6. Kalpoe J, Hogenbirk K, van Maarseveen N, Gesink-Van der Veer B, Kraakman M, et al. (2008) Dissemination of *Bacillus cereus* in a paediatric intensive care unit traced to insufficient disinfection of reusable ventilator air-flow sensors. J Hosp Infect 68: 341–347.
7. Dolan SA, Littlehorn C, Glodé MP, Dowell E, Xavier K, et al. (2012) Association of *Bacillus cereus* Infection with Contaminated Alcohol Prep Pads. Infect Control Hosp Epidemiol 33: 666–671.
8. Callegan MC, Gilmore MS, Gregory M, Ramadan RT, Wiskur BJ, et al. (2007) Bacterial endophthalmitis: therapeutic challenges and host–pathogen interactions. Prog Retin Eye Res 26: 189–203.
9. Sadaka A, Durand ML, Gilmore MS (2012) Bacterial endophthalmitis in the age of outpatient intravitreal therapies and cataract surgeries: Host–microbe interactions in intraocular infection. Prog Retin Eye Res 31: 316–331.
10. Callegan MC, Cochran DC, Kane ST, Gilmore MS, Gominet M, et al. (2002) Contribution of membrane-damaging toxins to Bacillus endophthalmitis pathogenesis. Infect Immun 70: 5381–5389.
11. Beecher DJ, Olsen TW, Somers EB, Wong ACL (2000) Evidence for contribution of tripartite hemolysin BL, phosphatidylcholine-preferring phospholipase C, and collagenase to virulence of *Bacillus cereus* endophthalmitis. Infect Immun 68: 5269–5276.
12. Callegan MC, Kane ST, Cochran DC, Novosad B, Gilmore MS, et al. (2005) Bacillus endophthalmitis: roles of bacterial toxins and motility during infection. Invest Ophthalmol Vis Sci 46: 3233–3238.
13. Callegan MC, Kane ST, Cochran DC, Gilmore MS, Gominet M, et al. (2003) Relationship of plcR-Regulated Factors to Bacillus Endophthalmitis Virulence. Infect Immun 71: 3116–3124.
14. Callegan MC, Jett BD, Hancock LE, Gilmore MS (1999) Role of hemolysin BL in the pathogenesis of extraintestinal *Bacillus cereus* infection assessed in an endophthalmitis model. Infect Immun 67: 3357–3366.
15. Booth MC, Cheung AL, Hatter KL, Jett BD, Callegan MC, et al. (1997) Staphylococcal accessory regulator (sar) in conjunction with agr contributes to *Staphylococcus aureus* virulence in endophthalmitis. Infect Immun 65: 1550–1556.
16. Booth MC, Atkuri RV, Nanda SK, Iandolo JJ, Gilmore MS (1995) Accessory gene regulator controls *Staphylococcus aureus* virulence in endophthalmitis. Invest Ophthalmol Vis Sci 36: 1828–1836.
17. Jett BD, Jensen HG, Atkuri RV, Gilmore MS (1995) Evaluation of therapeutic measures for treating endophthalmitis caused by isogenic toxin-producing and toxin-nonproducing *Enterococcus faecalis* strains. Invest Ophthalmol Vis Sci 36: 9–15.
18. Jett B, Jensen H, Nordquist R, Gilmore M (1992) Contribution of the pAD1-encoded cytolysin to the severity of experimental *Enterococcus faecalis* endophthalmitis. Infect Immun 60: 2445–2452.
19. Sanders ME, Norcross EW, Moore III QC, Onwubiko C, King LB, et al. (2008) A comparison of pneumolysin activity and concentration in vitro and in vivo in a rabbit endophthalmitis model. Clin Ophthalmol (Auckland, NZ) 2: 793–800.
20. Sanders ME, Norcross EW, Robertson ZM, Moore QC, Fratkin J, et al. (2011) The *Streptococcus pneumoniae* capsule is required for full virulence in pneumococcal endophthalmitis. Invest Ophthalmol Vis Sci 52: 865–872.
21. Das T, Jalali S, Gothwal VK, Sharma S, Naduvilath TJ (1999) Intravitreal dexamethasone in exogenous bacterial endophthalmitis: results of a prospective randomised study. Br J Ophthalmol 83: 1050–1055.
22. Shah GK, Stein JD, Sharma S, Sivalingam A, Benson WE, et al. (2000) Visual outcomes following the use of intravitreal steroids in the treatment of postoperative endophthalmitis. Ophthalmol 107: 486–489.
23. Meredith TA, Aguilar H, Drews C, Sawant A, Gardner S, et al. (1996) Intraocular dexamethasone produces a harmful effect on treatment of experimental *Staphylococcus aureus* endophthalmitis. Trans Am Ophthalmol Soc 94: 241–257.
24. Yoshizumi MO, Lee GC, Equi RA, Kim I-T, Pitchekian-Halabi H, et al. (1998) Timing of dexamethasone treatment in experimental *Staphylococcus aureus* endophthalmitis. Retina 18: 130–135.
25. Wiskur BJ, Robinson ML, Farrand AJ, Novosad BD, Callegan MC (2008) Toward improving therapeutic regimens for Bacillus endophthalmitis. Invest Ophthalmol Vis Sci 49: 1480–1487.
26. Kumar MV, Nagineni CN, Chin MS, Hooks JJ, Detrick B (2004) Innate immunity in the retina: Toll-like receptor (TLR) signaling in human retinal pigment epithelial cells. J Neuroimmunol 153: 7–15.
27. Chang J, McCluskey P, Wakefield D (2006) Toll-like receptors in ocular immunity and the immunopathogenesis of inflammatory eye disease. Br J Ophthalmol 90: 103–108.
28. Lambiase A, Micera A, Sacchetti M, Mantelli F, Bonini S (2011) Toll-like receptors in ocular surface diseases: overview and new findings. Clin Sci 120: 441–450.
29. Kumar A, Singh CN, Glybina IV, Mahmoud TH, Fu-Shin XY (2010) Toll-like Receptor 2 Ligand—Induced Protection against Bacterial Endophthalmitis. J Infect Dis 201: 255–263.
30. Novosad BD, Astley RA, Callegan MC (2011) Role of Toll-Like Receptor (TLR) 2 in Experimental *Bacillus cereus* Endophthalmitis. PloS one 6: e28619.
31. Callegan MC, Booth MC, Jett BD, Gilmore MS (1999) Pathogenesis of Gram-positive bacterial endophthalmitis. Infect Immun 67: 3348–3356.
32. Callegan MC, Kane ST, Cochran DC, Gilmore MS (2002) Molecular mechanisms of Bacillus endophthalmitis pathogenesis. DNA Cell Biol 21: 367–373.

33. Callegan MC, Novosad BD, Ramirez R, Ghelardi E, Senesi S (2006) Role of swarming migration in the pathogenesis of Bacillus endophthalmitis. Invest Ophthalmol Vis Sci 47: 4461–4467.

34. Senesi S, Celandroni F, Salvetti S, Beecher DJ, Wong AC, et al. (2002) Swarming motility in Bacillus cereus and characterization of a fliY mutant impaired in swarm cell differentiation. Microbiol 148: 1785–1794.

35. Hayashi F, Smith KD, Ozinsky A, Hawn TR, Yi EC, et al. (2001) The innate immune response to bacterial flagellin is mediated by Toll-like receptor 5. Nature 410: 1099–1103.

36. Moyer AL, Ramadan RT, Novosad BD, Astley R, Callegan MC (2009) Bacillus cereus–Induced Permeability of the Blood–Ocular Barrier during Experimental Endophthalmitis. Invest Ophthalmol Vis Sci 50: 3783–3793.

37. Ramadan RT, Moyer AL, Callegan MC (2008) A role for tumor necrosis factor-α in experimental Bacillus cereus endophthalmitis pathogenesis. Invest Ophthalmol Vis Sci 49: 4482–4489.

38. Ramadan RT, Ramirez R, Novosad BD, Callegan MC (2006) Acute inflammation and loss of retinal architecture and function during experimental Bacillus endophthalmitis. Curr Eye Res 31: 955–965.

39. DePamphilis M, Adler J (1971) Purification of intact flagella from Escherichia coli and Bacillus subtilis. J Bacteriol 105: 376–383.

40. Motzel SL, Riley LK (1991) Bacillus piliformis flagellar antigens for serodiagnosis of Tyzzer's disease. J Clin Microbiol 29: 2566–2570.

41. Thompson JD, Higgins DG, Gibson TJ (1994) CLUSTAL W: improving the sensitivity of progressive multiple sequence alignment through sequence weighting, position-specific gap penalties and weight matrix choice. Nucleic Acids Res 22: 4673–4680.

42. Campodónico VL, Llosa NJ, Grout M, Döring G, Maira-Litrán T, et al. (2010) Evaluation of flagella and flagellin of Pseudomonas aeruginosa as vaccines. Infect Immun 78: 746–755.

43. Smith KD, Andersen-Nissen E, Hayashi F, Strobe K, Bergman MA, et al. (2003) Toll-like receptor 5 recognizes a conserved site on flagellin required for protofilament formation and bacterial motility. Nat Immunol 4: 1247–1253.

44. Salvetti S, Ghelardi E, Celandroni F, Ceragioli M, Giannessi F, et al. (2007) FlhF, a signal recognition particle-like GTPase, is involved in the regulation of flagellar arrangement, motility behaviour and protein secretion in Bacillus cereus. Microbiol 153: 2541–2552.

45. Guttenplan SB, Shaw S, Kearns DB (2013) The cell biology of peritrichous flagella in Bacillus subtilis. Mol Microbiol 87: 211–229.

46. Ivanova N, Sorokin A, Anderson I, Galleron N, Candelon B, et al. (2003) Genome sequence of Bacillus cereus and comparative analysis with Bacillus anthracis. Nature 423: 87–91.

47. Fournier B, Williams IR, Gewirtz AT, Neish AS (2009) Toll-like receptor 5-dependent regulation of inflammation in systemic Salmonella enterica Serovar typhimurium infection. Infect Immun 77: 4121–4129.

48. Lai MA, Quarles EK, López-Yglesias AH, Zhao X, Hajjar AM, et al. (2013) Innate Immune Detection of Flagellin Positively and Negatively Regulates Salmonella Infection. PloS one 8: e72047.

49. Zhang F, Gao X-D, Wu W-W, Gao Y, Zhang Y-W, et al. (2013) Polymorphisms in toll-like receptors 2, 4 and 5 are associated with Legionella pneumophila infection. Infection: 1–8.

50. Hawn TR, Berrington WR, Smith IA, Uematsu S, Akira S, et al. (2007) Altered inflammatory responses in TLR5-deficient mice infected with Legionella pneumophila. J Immunol 179: 6981–6987.

51. Jarchum I, Liu M, Lipuma L, Pamer EG (2011) Toll-like receptor 5 stimulation protects mice from acute Clostridium difficile colitis. Infect Immun 79: 1498–1503.

52. Sun Y, Karmakar M, Roy S, Ramadan RT, Williams SR, et al. (2010) TLR4 and TLR5 on corneal macrophages regulate Pseudomonas aeruginosa keratitis by signaling through MyD88-dependent and-independent pathways. J Immunol 185: 4272–4283.

53. Morris AE, Liggitt HD, Hawn TR, Skerrett SJ (2009) Role of Toll-like receptor 5 in the innate immune response to acute P. aeruginosa pneumonia. Am J Physiol Lung Cell Mol Physiol 297: L1112–L1119.

54. Andersen-Nissen E, Hawn TR, Smith KD, Nachman A, Lampano AE, et al. (2007) Cutting edge: Tlr5−/− mice are more susceptible to Escherichia coli urinary tract infection. J Immunol 178: 4717–4720.

55. Cullender TC, Chassaing B, Janzon A, Kumar K, Muller CE, et al. (2013) Innate and Adaptive Immunity Interact to Quench Microbiome Flagellar Motility in the Gut. Cell Host Microbe 14: 571–581.

56. Kojima K, Ueta M, Hamuro J, Hozono Y, Kawasaki S, et al. (2008) Human conjunctival epithelial cells express functional Toll-like receptor 5. Br J Ophthalmol 92: 411–416.

57. Lin X, Fang D, Zhou H, Su SB (2012) The expression of Toll-like receptors in murine Müller cells, the glial cells in retina. Neurol Sci: 1–8.

58. Zhang J, Xu K, Ambati B, Fu-Shin XY (2003) Toll-like receptor 5-mediated corneal epithelial inflammatory responses to Pseudomonas aeruginosa flagellin. Invest Ophthalmol Vis Sci 44: 4247–4254.

59. Kumar A, Yin J, Zhang J, Fu-Shin XY (2007) Modulation of corneal epithelial innate immune response to pseudomonas infection by flagellin pretreatment. Invest Ophthalmol Vis Sci 48: 4664–4670.

60. Redfern RL, Reins RY, McDermott AM (2011) Toll-like receptor activation modulates antimicrobial peptide expression by ocular surface cells. Exp Eye Res 92: 209–220.

61. Gao N, Kumar A, Guo H, Wu X, Wheater M, et al. (2011) Topical flagellin-mediated innate defense against Candida albicans keratitis. Invest Ophthalmol Vis Sci 52: 3074–3082.

62. Yoshino Y, Kitazawa T, Ikeda M, Tatsuno K, Yanagimoto S, et al. (2012) Clostridium difficile flagellin stimulates toll-like receptor 5, and Toxin B promotes flagellin-induced chemokine production via TLR5. Life sci 92: 211–217.

63. Im J, Jeon JH, Cho MK, Woo SS, Kang S-S, et al. (2009) Induction of IL-8 expression by bacterial flagellin is mediated through lipid raft formation and intracellular TLR5 activation in A549 cells. Mol Immunol 47: 614–622.

64. de Vos AF, Pater JM, van den Pangaart PS, de Kruif MD, van't Veer C, et al. (2009) In vivo lipopolysaccharide exposure of human blood leukocytes induces cross-tolerance to multiple TLR ligands. J Immunol 183: 533–542.

65. Hozono Y, Ueta M, Hamuro J, Kojima K, Kawasaki S, et al. (2006) Human corneal epithelial cells respond to ocular-pathogenic, but not to nonpathogenic-flagellin. Biochem Biophys Res Commun 347: 238–247.

66. Callegan MC, Engelbert M, Parke DW, Jett BD, Gilmore MS (2002) Bacterial endophthalmitis: epidemiology, therapeutics, and bacterium-host interactions. Clin Microbiol Rev 15: 111–124.

67. Sanders D (2005) Mucosal integrity and barrier function in the pathogenesis of early lesions in Crohn's disease. J Clin Pathol 58: 568–572.

68. McCole DF, Barrett KE (2003) Epithelial transport and gut barrier function in colitis. Curr Opin Gastroenterol 19: 578–582.

69. Senesi S, Salvetti S, Celandroni F, Ghelardi E (2010) Features of Bacillus cereus swarm cells. Res Microbiol 161: 743–749.

70. Ghelardi E, Celandroni F, Salvetti S, Beecher DJ, Gominet M, et al. (2002) Requirement of flhA for swarming differentiation, flagellin export, and secretion of virulence-associated proteins in Bacillus thuringiensis. J Bacteriol 184: 6424–6433.

71. Andersen-Nissen E, Smith KD, Strobe KL, Barrett SLR, Cookson BT, et al. (2005) Evasion of Toll-like receptor 5 by flagellated bacteria. Proc Natl Acad Sci USA 102: 9247–9252.

72. Donnelly MA, Steiner TS (2002) Two nonadjacent regions in enteroaggregative Escherichia coli flagellin are required for activation of toll-like receptor 5. J Biol Chem 277: 40456–40461.

73. Yang J, Zhang E, Liu F, Zhang Y, Zhong M, et al. (2014) Flagellins of Salmonella typhi and nonpathogenic Escherichia coli are differentially recognized through the NLRC4 pathway in macrophages. J Innate Immun 6: 47–57.

2

Antagonistic Role of CotG and CotH on Spore Germination and Coat Formation in *Bacillus subtilis*

Anella Saggese[1], Veronica Scamardella[1], Teja Sirec[1], Giuseppina Cangiano[1], Rachele Isticato[1], Francesca Pane[2], Angela Amoresano[2], Ezio Ricca[1], Loredana Baccigalupi[1]*

1 Department of Biology, Federico II University of Naples, Naples, Italy, **2** Department of Chemistry, Federico II University of Naples, Naples, Italy

Abstract

Spore formers are bacteria able to survive harsh environmental conditions by differentiating a specialized, highly resistant spore. In *Bacillus subtilis*, the model system for spore formers, the recently discovered crust and the proteinaceous coat are the external layers that surround the spore and contribute to its survival. The coat is formed by about seventy different proteins assembled and organized into three layers by the action of a subset of regulatory proteins, referred to as morphogenetic factors. CotH is a morphogenetic factor needed for the development of spores able to germinate efficiently and involved in the assembly of nine outer coat proteins, including CotG. Here we report that CotG has negative effects on spore germination and on the assembly of at least three outer coat proteins. Such negative action is exerted only in mutants lacking CotH, thus suggesting an antagonistic effect of the two proteins, with CotH counteracting the negative role of CotG.

Editor: Riccardo Manganelli, University of Padova, Medical School, Italy

Funding: This work was supported by EU grants (contract number 613703 and 614088) to ER. The funders had no role in study design, data collection and analysis, decision to publish, or preparation of the manuscript.

Competing Interests: The authors have declared that no competing interests exist.

* Email: lorbacci@unina.it

Introduction

Spore formers are Gram-positive bacteria belonging to different genera and including more than 1,000 species [1]. The common feature of these organisms is the ability to differentiate a spore, a dormant cell type that can survive for long periods in the absence of water and nutrients and resisting to a vast range of stresses (high temperature, dehydration, absence of nutrients, presence of toxic chemicals) [2]. When the environmental conditions ameliorates the spore germinates originating a cell able to grow and eventually sporulate [3]. Spore resistance to lytic enzymes and toxic chemicals is in part due to the presence of the spore coat, a multilayered structure composed by more than 70 proteins that surrounds the spore [4,5]. Development of the mature spore is finely controlled through different mechanisms acting at various levels. The synthesis of coat proteins (Cot proteins) is regulated by a cascade of transcription factors controlling the timing of expression of their structural genes (*cot* genes) while coat assembly is controlled by a subset of Cot protein with a morfogenetic role [5]. Among the morphogenetic proteins, CotH plays a role in the assembly of at least 9 other coat components, including CotG, CotC/U and CotS, [6–9]. In addition, CotH contributes to the formation of spores able to germinate efficiently and to resist to lysozyme treatment [9]. CotH action is strictly connected with that of the major outer coat regulator CotE and mutant spores lacking both CotH and CotE germinate less efficiently and showed an increased sensitivity to lysozyme than single *cotE* null spores [9]. A recent report has shown that, when over-expressed, CotH bypasses the requirement for CotE, and suggests that CotE acts by localizing CotH on the spore coat and thus allowing its activity. In the presence of high CotH concentrations, due to the gene over-expression, CotH does not require CotE anymore and is able to drive the assembly of CotH-dependent proteins in a CotE-independent way [10].

The *cotH* structural gene is clustered with two other *cot* genes: *cotB*, transcribed in the same direction, and *cotG* divergently oriented with respect to *cotH*. A recent paper [11] has shown that the *cotH* promoter maps more than 800 bp upstream of its coding region, that this region is not translated and entirely contains the divergently transcribed *cotG* gene. A direct consequence of this peculiar chromosomal organization is that *cotG* insertion/deletion mutations so far analyzed [12], should also affect *cotH* expression leading to double *cotG cotH* mutants. If this is the case, then, the role of CotG has never been studied in an otherwise wild type strain and induces us to reconsider some previously reported results. Indeed, *cotG* spores have been previously reported as identical to isogenic wild type spores for both germination efficiency and lysozyme-resistance [12], while *cotH* spores have been shown to be about 35% less efficient than isogenic wild type spores upon induction of germination [8]. However, if an insertion-deletion within *cotG* impairs also the expression of *cotH* [11], those data imply that when both CotG and CotH are both lacking spores germinate normally but when only CotH is lacking spore germination is defective. In order to clarify the role CotG and its interaction with CotH, we first verified that CotH is not produced in a strain with an insertion/deletion mutation in *cotG* and then constructed for the first time a single *cotG* null mutant. The phenotypic analysis of the mutant spores is reported.

Results and Discussion

Construction of a *cotG* mutant

To verify whether a strain with an insertion/deletion mutation in *cotG* produced CotH, coat proteins extracted from a wild type strain (PY79) and of two isogenic mutants in *cotG* (ER203) or in *cotH* (ER220) were compared. As previously reported [8], both mutants have on SDS-PAGE a strongly altered pattern of coat proteins with several minor differences characteristic of the two strains [8] (Fig. 1A). A western blot analysis with anti-CotH antibody of the coat proteins of the three strains confirmed that CotH is not produced in a strain with an insertion/deletion mutation in *cotG* (Fig. 1B).

In order to obtain a *cotG* null mutation that does not affect *cotH* transcription, we introduced a single nucleotide in the *cotG* coding region by gene-soeing [13], thus causing the formation of a stop codon 21 bp downstream of the *cotG* translation start site (Fig. 2A). The entire *cotG_{stop}cotH* region was PCR amplified, cloned into an integrative vector and inserted at the *amyE* locus on the *B. subtilis* chromosome of strain AZ603 carrying a deletion of the entire *cotG cotH* locus, yielding strain AZ604. An identical strategy was followed to PCR amplify, clone, integrate at the *amyE* locus and transfer into strain AZ603 a wild type copy of the *cotG cotH* region (AZ608). To verify the production of CotG and CotH in AZ604 (*ΔcotG ΔcotH amyE::cotG_{stop}cotH*) and AZ608 (*ΔcotG ΔcotH amyE::cotGcotH*) western blots with anti-CotG or anti-

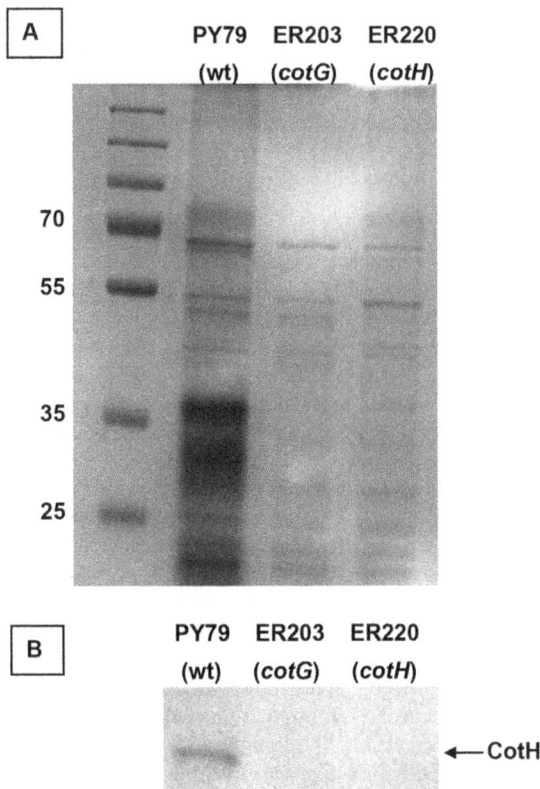

Figure 1. Production of CotH in a *cotG* null mutant. (A) SDS-PAGE fractionation of coat proteins from a wild type strain (PY79) and isogenic strains carrying null mutations in *cotG* (ER203) or in *cotH* (ER220). A molecular weight marker is also present and the size of relevant bands indicated. (B) Western blot with anti-CotH antibody of the same three strains analyzed in panel A. The arrow points to the CotH specific band.

CotH antibodies were performed. As shown in Fig. 2BC, the ectopic expression of a wild type copy of the *cotG cotH* region (lane 4 in both panels) in strain AZ603 complemented the deletion of the *cotG cotH* locus (lanes 2 in both panels). As expected, the ectopic expression of *cotG_{stop}cotH* in strain AZ603 did not affect CotH production (panel B, lane 3) and did not produce CotG (panel C, lane 3).

Role of CotG on spore germination and resistance to lysozyme

We used the single *cotG* null mutant strain (AZ604) to analyze the efficiency of germination and the resistance to lysozyme. Together with AZ604 we considered for our analysis spores of three other isogenic strains: a wild type (PY79) containing both CotG and CotH [8,12], *cotH* null (ER220) containing only CotG [7] and *cotH cotG* null (AZ603) lacking both proteins. As shown in Fig. 3A, AZ604 spores (*cotG*) showed an efficiency of germination identical to that of wild type spores (white and gray circles in the figure). As previously reported [8], spores of the *cotH* null strain were slightly less efficient in germination than wild type spores (white squares in Fig. 3A). With spores of strain AZ603 (*cotG cotH*) the germination efficiency was restored to wild type levels (black squares in Fig. 3A). These results indicate that the germination defect observed with spore lacking only CotH was rescued in spores lacking both CotH and CotG. As a consequence they suggest that the germination impairment is not directly due to the absence of CotH as previously believed [8] but instead to the presence of CotG in a *cotH* null background. This finding also suggest a protective role for CotH in counteracting the CotG negative effect. The same four strains were also used to analyze the spore resistance to lysozyme and were all identical to wild type spores (Fig. 3B).

Role of CotG on coat protein assembly

We then analyzed the assembly of various coat proteins in the presence and in the absence of CotG and/or CotH. For our analysis we compared by western blot a wild type strain (PY79) and isogenic strains with an insertion/deletion in *cotH* (ER220, *cotH::spc*), or deleted of the entire *cotH cotG* locus (AZ603) and expressing either a wild type (AZ608) or a *cotG* mutant (AZ604) copy of the *cotH cotG* locus. As shown in Fig. 4, our analysis confirmed that levels of CotA (a CotH-independent protein) is not affected by CotH and/or CotG and that CotB maturation is dependent on the presence of both CotG and CotH [14]. Indeed, in spores of strains lacking CotG or CotH or both, CotB is assembled within the coat in its immature 43 kDa form. Only when both CotG and CotH are present the mature protein of 66 kDa is formed (Fig. 4A).

CotC and CotU are two CotH-dependent proteins that are homologous and recognized by both anti-CotC and anti-CotU antibodies [15]. CotC is present within the spore coat as a monomer (12 kDa), homodimer (21 kDa) and as two additional forms of 12.5 and 30 kDa [16]. CotU is found as a 17 kDa monomer [15] and as a heterodimer with CotC of 23 kDa [17]. As expected, all the CotC/CotU forms are found when both CotG and CotH are present (Fig. 4B, lanes 1 and 3) and none of them is observed when CotH is not expressed (Fig. 4B, lane 5). However, when both CotH and CotG are lacking (Fig. 4B, lane 2) as well as when only CotG is lacking (Fig. 4B, lane 4) all CotC/CotU proteins are normally assembled on the spore. These data indicate that, as for the germination phenotype, CotG has a negative role on CotC/CotU assembly and that its role is counteracted by CotH. To confirm the negative effect of CotG in a *cotH* background, we inserted an ectopic copy of *cotG* allele at *amyE*

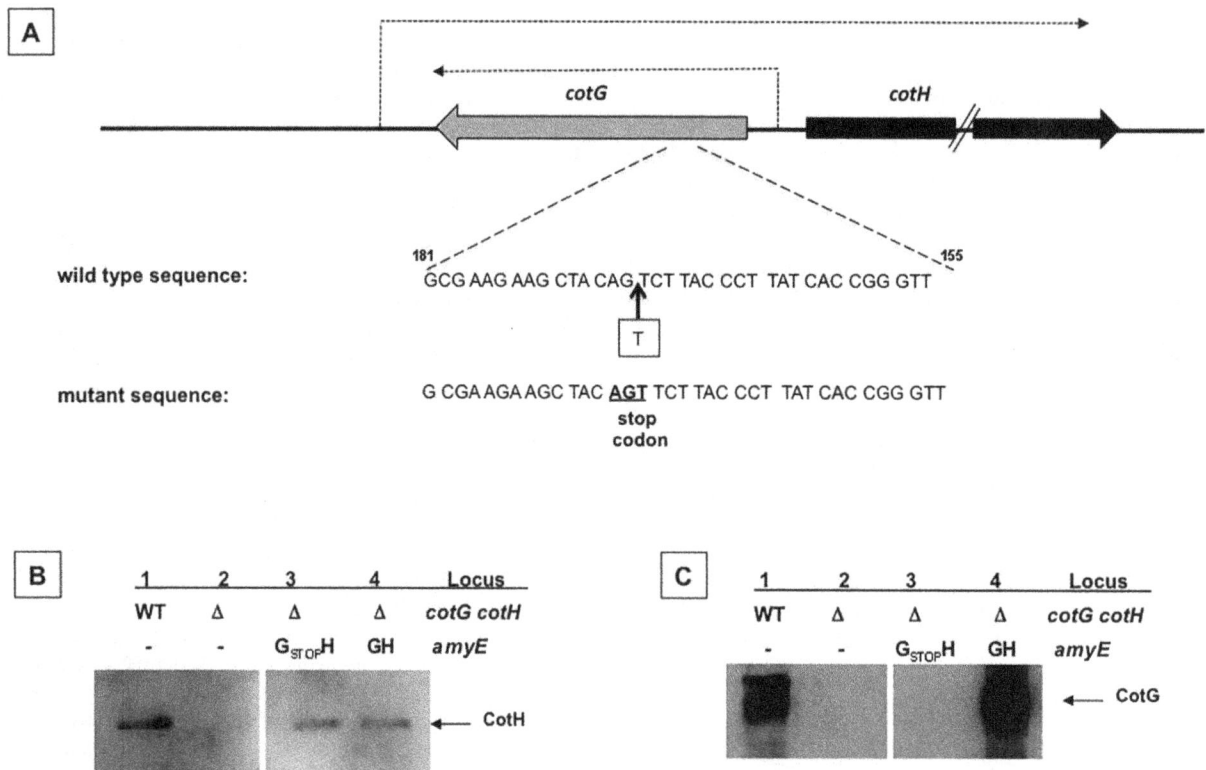

Figure 2. Construction of a single *cotG* mutant. (A) Thick gray and black arrows indicate the coding parts of *cotG* and *cotH*, respectively. Dashed arrow indicates the mRNA produced from the *cotG* and *cotH* promoters, as already reported. Site of insertion of the additional base in the *cotG* coding sequence (wild type sequence) that causes the formation of a premature stop codon (mutant sequence). Western blot analysis with anti-CotH (B) and anti-CotG (C) antibodies of proteins extracted by SDS treatment from wild type and isogenic mutant spores. The mutants genotype relative to the *cotG cotH* and *amyE* loci is indicated. Arrows point the CotH and CotG specific bands.

locus in the double *cotGcotH* mutant and also in this case all the CotC/CotU forms are no more assembled in the coat (Fig. 4B, lane 6).

CotS is 41 kDa, *cotH*-dependent spore coat protein [18], clearly identified by SDS-PAGE and western blot [19]. As shown in Fig. 5A, a protein absent in the *cotS* null mutant (AZ541, lane 2), is not present in the *cotH* mutant (ER220, lane 5) but is present in both the single *cotG* mutant (AZ604, lane 4) and in the double *cotH cotG* mutant (AZ603, lane 3). To confirm this SDS-PAGE analysis we constructed a *cotS::gfp* fusion and integrated it on the chromosome of a wild type strain (PY79). By chromosomal DNA-mediated transformation we then moved the fusion into strains AZ603 (*ΔcotG ΔcotH*), AZ604 (*ΔcotG ΔcotH amyE::cotG_{stop}cotH*) and ER220 (*cotH::spc*) and analyzed all resulting strains by fluorescence microscopy. A fluorescence signal was observed around mature and forming spores in a wild type strain and in isogenic strains lacking both CotH and CotG (AZ603) or lacking only CotG (AZ604) (Fig. 5B). However, when CotG is present and CotH is lacking (ER220) [7] a fluorescence signal was observed around forming spores but never around mature, free spores (Fig. 5B). This result is in agreement with the SDS-PAGE of Fig. 5A, performed with proteins extracted from mature spores, and indicates that, also for CotS assembly, CotG has a negative role antagonized by CotH.

On the nature of CotG-CotH interaction

The nature of the antagonistic action of CotH on CotG negative role, suggested by results of Fig. 3, 4 and 5, is not clear. However some hints come from a recent bioinformatic analysis that has identified CotH as a putative kinase [20]. In addition, another previous report has shown that a *B. anthracis* protein with some similarities with CotG of *B. subtilis* is highly phosphorylated [21]. These literature data induced us to hypothesize that CotH is a kinase and CotG one of its substrates. To partially support this hypothesis we performed a mass spectrometry analysis of CotG. Coat proteins extracted from wild type spores were fractionated on SDS polyacrylamide gel and a region of the gel containing CotG used to reduce, alkylate and digest the proteins *in situ* with trypsin (see Material and Methods). The peptide mixture was divided in two aliquots and submitted to MALDIMS and nanoLCMSMS analyses and then directly analyzed by nanoHPLC-chip MS/MS. Due to the low resolution of the SDS-PAGE, more than one protein was identified in the same region of the gel but CotG exhibited the highest MASCOT score (not shown). Several phosphorylation sites were identified within CotG, some detected in the MALDIMS runs and some by a manual interpretation of the MS/MS spectra (Table S2 in File S1). Fig. 6 reports a summary of the phosphorylation sites identified in CotG. The occurrence of phosphorylation sites at level of Ser15, Ser39 and Thr147 was unambiguous and suggests that a kinases belonging to Serine-threonine kinase family is involved in CotG modification. Other phosphorylation sites occurred in amino acid sequences repeated several times within the CotG central region (for example, the tripeptides SYK underlined or SYR double-underlined in Fig. 6), thus impairing the exact localization of the modifications. Although we cannot definitely conclude that all of

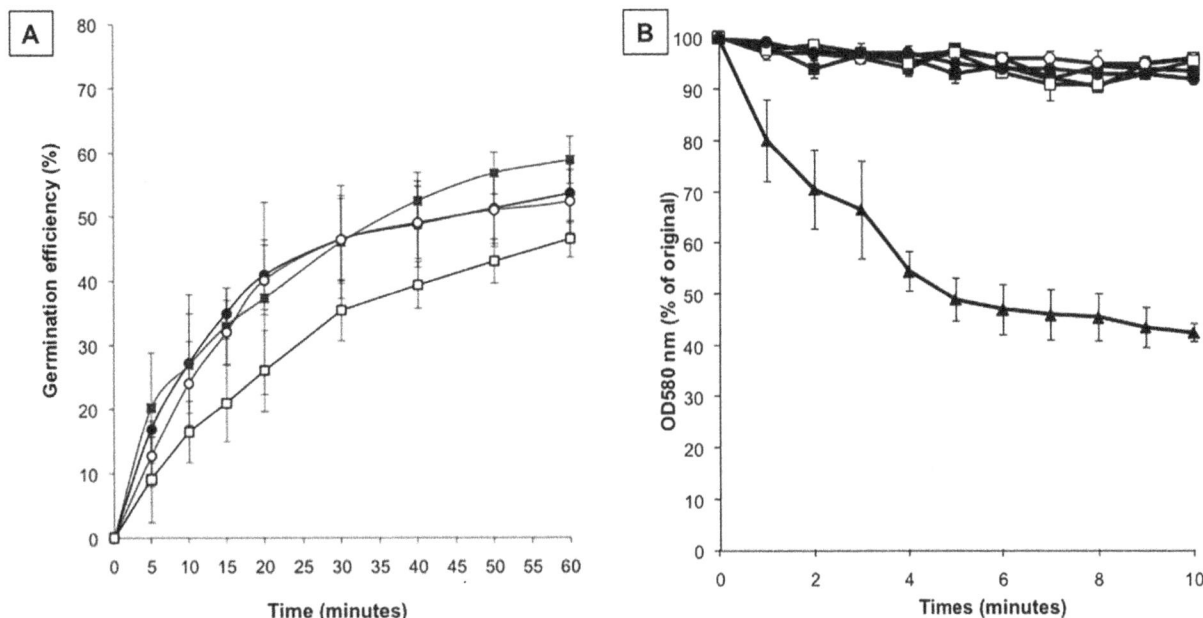

Figure 3. Germination efficiency and lysozyme-resistance assays. Spores derived from wild type (PY79, black circles), *cotG* null (AZ604, white circles), *cotH* null (ER220, white squares) and *cotGcotH* null (AZ603, black squares) were tested for germination efficiency (A) and for lysozime resistance (B). Germination was induced by Asn-GFK and measured as percentage of loss of optical density at 580 nm. Similar results were obtained by using L-Ala to induce germination. A *cotE* null strain (black triangles) known to be sensitive to lysozyme has been used as positive control during the lysozyme treatment. Error bars are based on the standard deviation of 4 independent experiments.

the underlined and double-underlined tripeptides are phosphorylated, the absence of the same tryptic fragments among the unmodified peptides strongly suggests that most, if not all of them are phosphorylated and that serine, always present in those tripeptides, is the most probable amino acid interested by the post-translational modification.

While Ser15 is in the N-terminal part of CotG, Ser39, Thr147 and all the other possible sites of phosphorylation are located in the repeated central region (Fig. 6 and Table S2 in File S1). This region is composed by random coiled repeats [11], each containing serine residues surrounded by positively charged amino acids (Fig. 6). In a bioinformatic analysis of known phosphorylated proteins [22] all these features have been indicated as typical of intrinsically disordered structures and have been identified as predictor of phosphorylation substrates.

In a *cotH* mutant CotG is not present around both the mature and the forming spore [8] but accumulates in the mother cell compartment of the sporulating cell [7]. However, its peculiar structure has so far impaired CotG isolation from the mother cell compartment of sporulating *B. subtilis* cells as well as from a heterologous host (*E. coli*), therefore not allowing further analysis. Although additional experiments, beyond the aims of this manuscript, will be needed to confirm that CotH is a kinase and CotG one of its substrates, we speculate that in a wild type strain CotG would be mainly present in a phosphorylated form and that, in this form, it plays its structural role as a coat component. In a *cotH* mutant, we predict that CotG would not be phosphorylated and have a negative effect on the assembly of some coat proteins and on spore germination.

Conclusions

Because of the peculiar chromosomal organization of the *cotG cotH* locus [11], in a *cotG* null mutant also the expression of the

cotH gene is impaired and, as a consequence, the presumed *cotG* mutant is a double mutant lacking both CotG and CotH. In this work we constructed for the first time a *cotG* null mutant in which CotH is produced. A phenotypic analysis of this mutant has shown that it does not differ significantly from the isogenic wild type strain but has also shown that phenotypes previously attributed to the lack of CotH are only observed when in the *cotH* strain is present CotG. When both CotH and CotG are absent the defects observed in the single *cotH* mutant are completely restored and the double mutant is indistinguishable from the isogenic wild type strain. This is the case of the germination defect of *cotH* spores that is rescued in a *cotG cotH* double mutant; is the case of CotC/U and CotS assembly within the coat. CotG has a peculiar primary structure: it has several repeats in its central part and has a high positive charge (pI 10.26). In a wild type strain CotG is highly phosphorylated and this post-translational modification is probably important to neutralize the positive charges and, consequently to guarantee protein stability and ability to interact with other coat components. The kinase responsible of this modification has not been identified yet. A recent bioinformatic data has indicated that CotH has some homology with eukaryotic Ser-Thr kinases [20] and our results functionally linking CotG to CotH, point to CotH as the kinase responsible of CotG phosphorylation. Future site-directed mutagenesis experiments will be needed to support this hypothesis.

Methods

Bacterial strains and transformation

B. subtilis strains are listed in Table 1. Plasmid amplification for nucleotide sequencing, subcloning experiments, and transformation of *E. coli* competent cells were performed with *Escherichia coli* strain DH5α [23]. Bacterial strains were transformed by previously described procedures: CaCl₂-mediated transformation

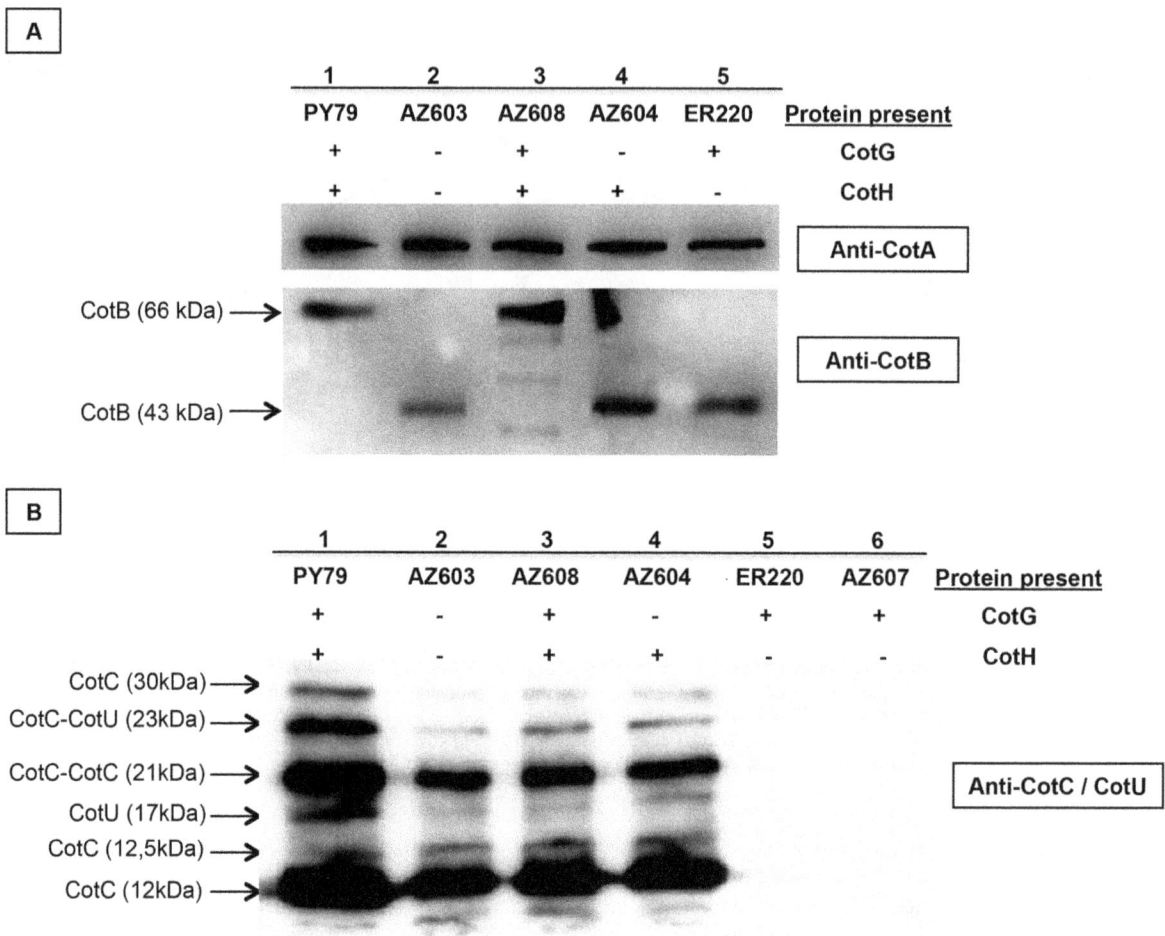

Figure 4. Western blot analysis. Western blot analysis of proteins extracted from mature spores of wild type (PY79, lane 1), *ΔcotGΔcotH* (AZ603, lane 2), *ΔcotGΔcotH amyE::cotGcotH* (AZ608, lane 3), *ΔcotGΔcotH amyE::cotG_{stop}cotH* (AZ604, lane 4), *cotH::spc* (ER220, lane 5) and *ΔcotGΔcotH amyE::cotG* (AZ607, lane 6 of panel B) strains. For CotA and CotB detection (panel A) the proteins have been extracted by SDS treatment while for CotC and CotU detection (panel B) the NaOH treatment has been used. Proteins (25 μg) were reacted with CotA, CotB and CotC specific rabbit antibodies and then with peroxidase-conjugated secondary antibodies and visualized by the Pierce method. The estimated size of CotB, CotC and CotU is indicated.

of *E. coli* competent cells [23] and two-step transformation of *B. subtilis* [24].

Genetic and molecular procedures

Isolation of plasmids, restriction digestion and ligation of DNA, were carried out by standard methods [23]. Chromosomal DNA from *B. subtilis* was isolated as described elsewhere [24].

Deletion of the *cotG cotH* locus

The *cotG cotH* locus was entirely deleted and substituted by a neomycin-resistance (*neo*) gene cassette. Chromosomal DNA of strain PY79 was used as a template and oligonucleotide pairs Del3-H18 and H29-B-anti (Table S1 in File S1) were used to prime the PCR amplification of two DNA fragments of 361 bp and 704 bp, respectively located upstream and downstream of the *cotH* gene. The two DNA fragments were separately cloned in the pBEST501 vector [25] at 5′ or 3′ ends of the *neo* gene. The resulting plasmid, pVS6, was then linearized by restriction digestion with *Sca*I and used to transform competent cells of the PY79 strain of *B. subtilis*. Replacement of the *cotH cotG* locus on the chromosome with the *neo* gene occurred by double cross-over

between homologous DNA sequences originating strain AZ603 (*ΔcotG ΔcotH*) and was verified by PCR.

Construction of a single *cotG* mutant

The entire *cotH cotG* locus was PCR amplified using oligonucleotides Del5 and H28 (Table S1 in File S1) to prime the reaction and PY79 chromosomal DNA as a template. The resulting DNA fragment was cloned into plasmid pDG364 [24], yielding plasmid pVS8. To insert a single nucleotide within the cotG coding part (at position +22, considering as +1 the first nucleotide of the first *cotG* codon) we used a *gene soeing* approach [13]. Two partially overlapping DNA fragments were PCR amplified priming the reaction with oligonucleotide pairs Gstop/Del5 (743 bp) and Gstop-anti/H (317 bp) (Table S1 in File S1) and using chromosomal DNA of PY79 as a template. The obtained PCR products were used as templates to prime a third linear PCR of 7 cycles using only the external primers Del5 and H (Table S1 in File S1). The single-strand products thus obtained were mixed and used to perform a standard PCR program of 20 cycles that led to their cohesion. The recombinant fragment was cloned in pGemT easy vector (Promega) and controlled by

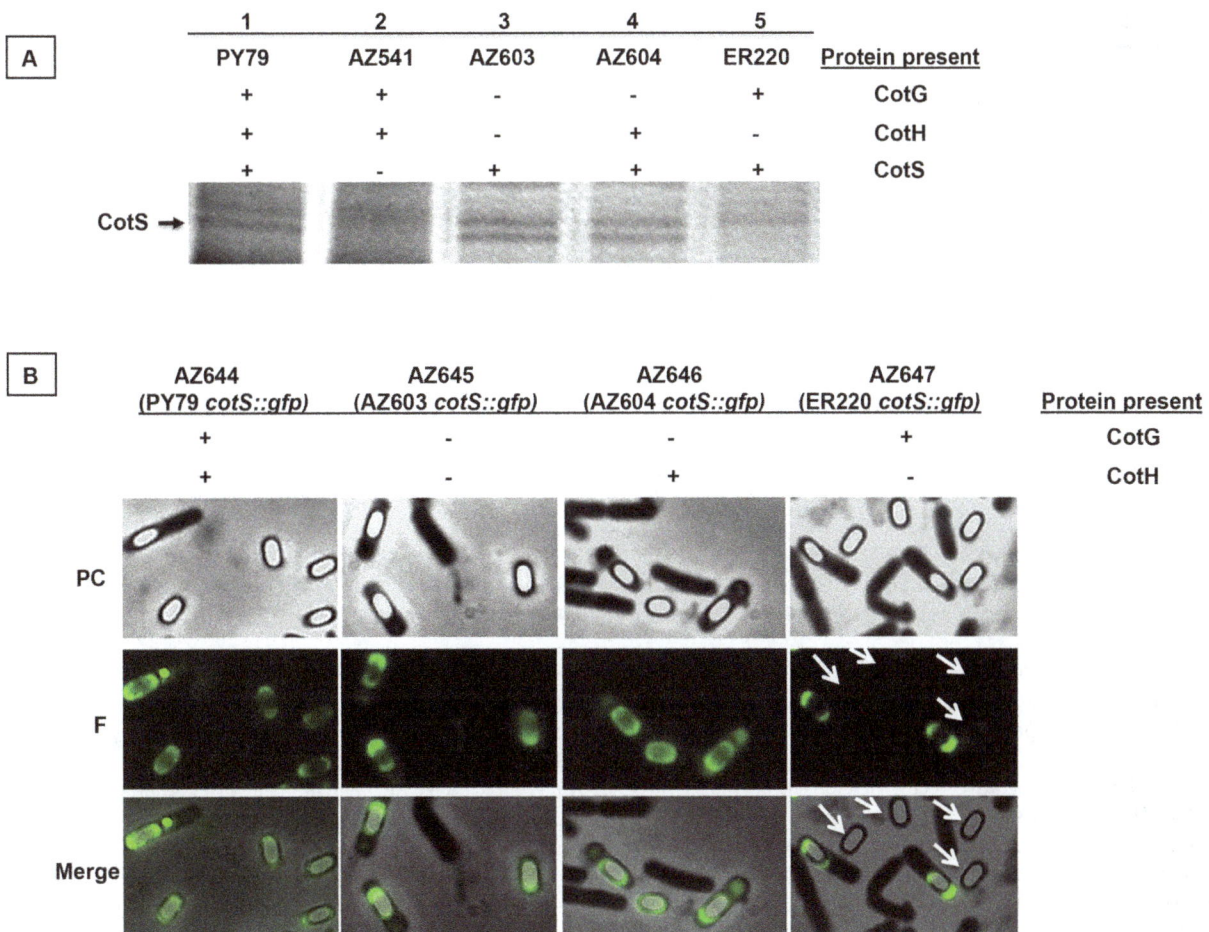

Figure 5. SDS-PAGE and Fluorescence analysis. (A) Proteins released after treatment with SDS of spores of the indicated strains were fractionated on a 12,5% polyacrilamide gel. The arrow indicates the 41 kDa band correspoding to CotS (18). The gel was stained with Coomassie brilliant blue. (B) Strains carrying the *cotS::gfp* fusion were analyzed by phase-contrast (PC) and fluorescence (F) microscopy. The bottom panel reports a merge of the two images. Exposure time was 588 ms in all cases.

sequencing to confirm the presence of the point mutation resulting in the substitution of the 8th *cotG* codon with a stop codon. The mutant *cotG* allele (here called *cotG*_{stop}) was digested with *Bam*HI-*Bgl*II and cloned into pVS8 to replace the wild type *cotG* allele, yielding plasmid pVS7. Both plasmids pVS7 (carrying the

*cotG*_{stop}*cotH* locus) and pVS8 (carrying the wild type *cotG cotH* locus) were separately used to transform competent cells of AZ603 (*ΔcotG ΔcotH*). The occurrence of a single reciprocal (Campbell-like) recombination event between homologous DNA on the

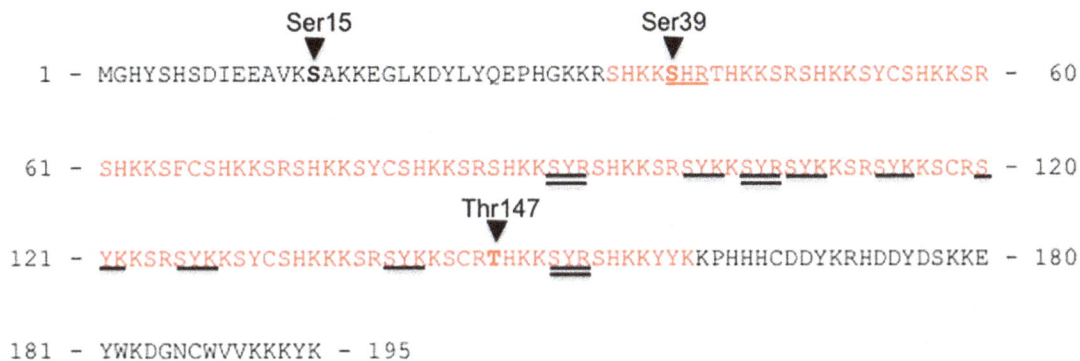

Figure 6. CotG and phosphorylation sites. Results of a mass spectrometry analysis of peptides derived from trypsine digestion of CotG are reported. Unambiguosly identified sites of phosphorylation are indicated. Tripeptides containing a phosphate moiety are underlined; the random coiled tandem repeats region is in red.

Table 1. *Bacillus subtilis* strains used in this study.

Strain	Relevant genotype	Reference
PY79	wild type	[32]
ER220	*cotH::spec*	[8]
AZ541	*cotS::cm*	[33]
AZ603	*ΔcotG ΔcotH::neo*	This work
AZ604	*ΔcotG ΔcotH::neo amyE::cotG$_{stop}$cotH*	This work
AZ608	*ΔcotG ΔcotH::neo amyE::cotGcotH*	This work
AZ607	*ΔcotG ΔcotH::neo amyE::cotG*	This work
AZ644	*cotS::gfp*	This work
AZ645	*ΔcotG ΔcotH::neo cotS::gfp*	This work
AZ646	*ΔcotG ΔcotH::neo amyE::cotG$_{stop}$cotH cotS::gfp*	This work
AZ647	*cotS::gfp cotH::spec*	This work

plasmids and on the chromosome (*amyE* locus) was verified by PCR.

Ectopic expression of cotG

The entire *cotG* gene was PCR amplified priming the reaction with oligonucleotide pairs G22 and H19 (774 bp), cloned in pGEM-T Easy vector (Promega), controlled by sequencing and transferred into the integrative vector pDG364 [24] using *Eco*RI and *Bam*HI restriction sites.

The plasmid was used to transform the double mutant AZ603 (*ΔcotG ΔcotH*). The occurrence of a single reciprocal (Campbell-like) recombination event between homologous DNA sequences present on the plasmid and on the chromosome (*amyE* locus) was verified by PCR and yielded strain AZ607 (*ΔcotG ΔcotH, amyE::cotG*).

Construction of cotS::gfp fusion

The *gfp mut3a* gene, encoding the green fluorescent protein (GFP) [26] was PCR amplified using plasmid pAD123 (Bacillus Genetic Stock Center, BGSC, www.bgsc.org) as a template and priming the reaction with oligonucleotides GFPfor and GFPrev (Table S1 in File S1). The *gfp mut3a* gene was cloned in pGEM-T Easy vector (Promega), controlled by sequencing and transferred into the integrative vector pER19 [27] using *Pst*I and *Bam*HI restriction sites. The region containing the entire *cotS* gene except the stop codon, was PCR amplified using chromosomal DNA of strain PY79 as a template and priming the reaction with oligonucleotides cotS-for and cotS-rev (table S1 in File S1), and cloned in frame with *gfp* using the *Sph*I restriction site located at 5′ end of *gfp*. The resulting plasmid pcotS-gfp was used to transform competent cells of strain PY79. The occurrence of a single reciprocal (Campbell-like) recombination event between homologous DNA sequences present on the plasmid and on the chromosome (*cotS* locus) yielded strain AZ644 (*cotS::gfp*) was verified by PCR. Chromosomal DNA of strain AZ644 was then used to transfer the *cotS-gfp* fusion into strains AZ603 (*ΔcotG ΔcotH*), AZ604 (*cotG$_{stop}$*) and ER220 (*cotH::spec*), yielding respectively AZ645 (*ΔcotG ΔccotH cotS::gfp*), AZ646 (*cotG$_{stop}$ cotS::gfp*), AZ647 (*cotH::spec cotS::gfp*). Fluorescence microscopy analysis was performed with an Olympus BX51 fluorescence microscope using a Fluorescein-Isothiocyanate (FITC) filter as previously reported [28]. Typical acquisition times were 588 ms

and the Images were captured using a Olympus DP70 digital camera and processed.

Spore purification, extraction of spore coat proteins and western blot analysis

Sporulation was induced by exhaustion by growing cells in DSM (Difco Sporulation Medium) as described elsewhere [24]. After a 30 hours of incubation at 37°C, spores were collected, washed four times, and purified as described by Nicholson and Setlow [29] using overnight incubation in H_2O at 4°C to lyse residual sporangial cells. Spore coat proteins were extracted from a suspension of spores by SDS-dithiothreitol (DTT) [24], or NaOH [29] treatment as previously described. The concentration of extracted proteins was determined by using Bio-Rad DC protein assay kit (Bio-Rad), and 20 μg of total spore coat proteins were fractionated on 12,5% SDS polyacrylamide gels and electro-transferred to nitrocellulose filters (Bio-Rad) for Western blot analysis following standard procedures. CotH-, CotA-, CotC-, CotB- and CotG-specific antibodies were used at a working dilutions of 1:150 for CotH detection and 1:7000 for CotA, CotC, CotB and CotG detection. Then an horseradish peroxidase (HRP)-conjugated anti-rabbit secondary antibody was used (Santa Cruz). Western blot filters were visualized by the SuperSignal West Pico chemiluminescence (Pierce) method as specified by the manufacturer.

Germination efficiency and lysozyme resistance

Purified spores were heat activated as previously described [24] and diluted in 10 mM Tris-HCl (pH 8.0) buffer containing 1 mM glucose, 1 mM fructose, and 10 mM KCl. After 15 min at 37°C, germination was induced by adding 10 mM L-alanine or 10 mM L-asparagine and the optical density at 580 nm was measured at 5-min intervals for 60 minutes [24].

Sensitivity to lysozyme was measured as described by Zheng *et al.* [30]. Spores were prepared as previously described [24], omitting the lysozyme step and eliminating vegetative cells by heat treatment (10 min at 80°C). Purified spores were then suspended in 10 mM Tris-HCl (pH 7.0) buffer containing lysozyme (50 mg/ml), and the decrease in optical density was monitored at 595 nm at 1-min intervals for 10 min. Spore viability was measured after 30 min as CFU on TY agar plates.

In situ digestion and mass spectral analyses

Protein bands corresponding to CotG were excised from the gel and destained by repetitive washes with 0.1 M NH_4-HCO_3 pH 7.5 and acetonitrile. Samples were then submitted to in situ trypsin digestion and analyzed by MALDI mass spectrometry and LCMSMS as previously described [31]. The acquired MS/MS spectra were transformed in *mzData* (.XML) format and used for protein identification with a licensed version of MASCOT software (www.matrixscience.com) version 2.4.0. Raw data from nanoLC-MS/MS analysis were used to query the NCBInr database NCBInr 20121120 (21,582,400 sequences; 7,401,135,489 residues). Mascot search parameters were: trypsin as enzyme; 3, as allowed number of missed cleavage; carboamidomethyl as fixed modification; oxidation of methionine; phosphorylation of serine/threonine/tyrosine; pyro-Glu N-term Q as variable modifications; 10 ppm MS tolerance and 0.6 Da MS/MS tolerance; peptide charge from +2 to +3. Peptide score threshold provided from MASCOT software to evaluate quality of matches for MS/MS data was 25.Spectra with MASCOT score of <25 having low quality were rejected.

Supporting Information

File S1   Table S1: list of oligonucleotides used in this study. Table S2: Mass spectral analyses of CotG trypsin digest.

Acknowledgments

We thank L. Di Iorio for technical support.

Author Contributions

Conceived and designed the experiments: LB ER AA. Performed the experiments: AS VS TS GC RI FP. Analyzed the data: LB ER AA RI. Contributed to the writing of the manuscript: ER AA LB.

References

1. Fritze D (2004) Taxonomy and systematics of the aerobic endospore forming bacteria: Bacillus and related genera. In Bacterial Spore Formers E. Ricca, A.O. Henriques, S.M. Cutting (Eds) 17–34 Norfolk, UK, Horizon Biosience.
2. Higgins D, Dworkin J (2012) Recent progress in *Bacillus subtilis* sporulation. FEMS Microbiol. Rev. 36: 131–148.
3. Dworkin J, Shah IM (2010) Exit from dormancy in microbial organisms. Nat. Rev. Microbiol. 8: 890–896.
4. Henriques AO, Moran Jr CP (2007) Structure, assembly and function of the spore surface layers. Ann. Rev. Microbiol. 61: 555–588.
5. McKenney PT, Driks A., Eichemberger P (2013) The *Bacillus subtilis* endospore: assembly and functions of the multilayered coat. Nat Rev Microbiol. 11: 33–44.
6. Baccigalupi L, Castaldo G, Cangiano G, Isticato R, Marasco R, et al (2004) GerE-independent expression of *cotH* leads to CotC accumulation in the mother cell compartment during *Bacillus subtilis* sporulation. Microbiol. 150: 3441–3449.
7. Kim H, Hahn M, Grabowski P, McPherson D, Otte MM, et al (2006) The *Bacillus subtilis* spore coat protein interaction network. Mol. Microbiol. 59: 487–502.
8. Naclerio G, Baccigalupi L, Zilhao R, De Felice M, Ricca E. (1996). *Bacillus subtilis* spore coat assembly requires cotH gene expression. J Bacteriol. 178: 4375–4380.
9. Zilhao R, Naclerio G, Baccigalupi L, Henriques A, Moran C et al (1999) Assembly requirements and role of CotH during spore coat formation in *Bacillus subtilis*. J. Bacteriol. 181: 2631–2633.
10. Isticato R, Sirec T, Giglio R, Baccigalupi L, Rusciano G et al (2013) Flexibility of the programme of spore coat formation in *Bacillus subtilis*: bypass of CotE requirement by over-production of CotH. PLoS ONE 8(9): e74949.
11. Giglio R, Fani R, Isticato R, De Felice M, Ricca E et al (2011) Organization and evolution of the *cotG* and *cotH* genes of *Bacillus subtilis*. J. Bacteriol. 193: 6664–6673.
12. Sacco M, Ricca E, Losick R, Cutting S (1995) An additional GerE-controlled gene encoding an abundant spore coat protein from *Bacillus subtilis*. J. Bacteriol. 177: 372–377.
13. Horton RM, Hunt HD, Ho SN, Pullen JK, Pease LR (1989) Engineering hybrid genes without the use of restriction enzymes: gene splicing by overlap extension. Gene 77: 61–68.
14. Zilhao R, Serrano M, Isticato R, Ricca E, Moran Jr CP et al (2004) Interactions among CotB, CotG, and CotH during assembly of the *Bacillus subtilis* spore coat. J. Bacteriol. 186: 1110–1119.
15. Isticato R, Pelosi A, Zilhao R, Baccigalupi L, Henriques AO et al (2008) CotC-CotU heterodimerization during assembly of the *Bacillus subtilis* spore coat. J. Bacteriol. 190: 1267–1275.
16. Isticato R, Esposito G, Zilhão R, Nolasco S, Cangiano G et al (2004) Assembly of Multiple CotC Forms into the *Bacillus subtilis* Spore Coat. J. Bacteriol. 186: 1129–1135.
17. Isticato R, Pelosi A, De Felice M, Ricca E (2010) CotE binds to CotC and CotU and mediates their interaction during spore coat formation in *Bacillus subtilis*. J. Bacteriol. 192: 949–954.
18. Takamatsu H, Chikahiro Y, Kodama T, Koide H, Kozuka S et al (1998) A spore coat protein, CotS, of *Bacillus subtilis* is synthesized under the regulation of sigmaK and GerE during development and is located in the inner coat layer of spores. J. Bacteriol. 180: 2968–2974.
19. Little S, Driks A (2001) Functional analysis of the *Bacillus subtilis* morphogenetic spore coat protein CotE. Mol. Microbiol. 42: 1107–1120.
20. Galperin MY, Mekhedov SL, Puigbo P, Smirnov S, Wolf YI, et al. (2012) Genomic determinants of sporulation in Bacilli and Clostridia: towards the minimal set of sporulation-specific genes. Environ. Microbiol. 14: 2870–2890.
21. McPherson SA1, Li M, Kearney JF, Turnbough CL Jr (2010) ExsB, an unusually highly phosphorylated protein required for the stable attachment of the exosporium of Bacillus anthracis. Mol. Microbiol. 76: 1527–38.
22. Iakoucheva LM, Radivojac P, Brown CJ, O'Connor TR, Sikes JG et al (2004) The importance of intrinsic disorder for protein phosphorylation. Nucl Acid Res. 32: 1037–1049.
23. Sambrook J, Fritsch EF, Maniatis T (1989) Molecular cloning. A laboratory manual. Second edition. Cold Spring Harbor Laboratory Press, Cold Spring Harbor, NY, US.
24. Cutting S, Vander Horn PB (1990) Genetic analysis In: C. Harwood and S. Cutting (Eds.), Molecular Biological Methods for *Bacillus*. John Wiley and Sons, Chichester, UK. 27–74.
25. Itaya M, Kondo K, Tanaka T (1989) A neomycin resistance gene cassette selectable in a single copy state in the *Bacillus subtilis* chromosome. Nucl Acids Res. 17: 4410.
26. Cormack BP, Valdivia RH, Falkow S (1996) FACS-optimized mutants of the green fluorescent protein. Gene 173: 33–38.
27. Ricca E, Cutting S, Losick R (1992) Characterization of *bofA*, a gene involved in inter-compartmental regulation of pro-s^K processing during sporulation in *Bacillus subtilis*. J. Bacteriol. 174: 3177–3184.
28. Manzo N, Di Luccia B, Isticato R, D'Apuzzo E. De Felice M et al (2013) Pigmentation and sporulation are alternative cell fates in *Bacillus pumilus* SF214. PLoS ONE 8(4): e62093.
29. Nicholson WL, Setlow P (1990) Sporulation, germination and outgrowth. In C. Harwood and S. Cutting (Eds.), Molecular Biological Methods for *Bacillus*. John Wiley and Sons, Chichester, UK. 391–450.
30. Zheng L, Donovan WP, Fitz-James PC, Losick R (1988) Gene encoding a morphogenic protein required in the assembly of the outer coat of the *Bacillus subtilis* endospore. Genes Develop. 2: 1047–1054.
31. Amoresano A, Di Costanzo A, Leo G, Di Cunto F, La Mantia G et al (2010) Identification of DeltaNp63alpha protein interactions by mass spectrometry. J Proteome Res. 9: 2042–2048.
32. Youngman P, Perkins JB, Losick R (1984) A novel method for the rapid cloning in *Escherichia coli* of *Bacillus subtilis* chromosomal DNA adjacent to Tn917 insertion. Mol. Gen. Genet. 195: 424–433.
33. Sirec T, Strazzulli A, Isticato R, De Felice M, Moracci M et al (2012) Adsorption of β-galactosidase of *Alicyclobacillus acidocaldarius* on wild type and mutants spores of *Bacillus subtilis*. Microbial Cell Factories. 11: 100.

3

The ESX System in *Bacillus subtilis* Mediates Protein Secretion

Laura A. Huppert[1,9], **Talia L. Ramsdell**[2,9], **Michael R. Chase**[2], **David A. Sarracino**[3], **Sarah M. Fortune**[2], **Briana M. Burton**[1]*

1 Department of Molecular and Cellular Biology, Harvard University, Cambridge, Massachusetts, United States of America, 2 Department of Immunology and Infectious Diseases, Harvard School of Public Health, Boston, Massachusetts, United States of America, 3 Thermo Fisher Scientific, BRIMS Unit, Cambridge, Massachusetts, United States of America

Abstract

Esat-6 protein secretion systems (ESX or Ess) are required for the virulence of several human pathogens, most notably *Mycobacterium tuberculosis* and *Staphylococcus aureus*. These secretion systems are defined by a conserved FtsK/SpoIIIE family ATPase and one or more WXG100 family secreted substrates. Gene clusters coding for ESX systems have been identified amongst many organisms including the highly tractable model system, *Bacillus subtilis*. In this study, we demonstrate that the *B. subtilis yuk/yue* locus codes for a nonessential ESX secretion system. We develop a functional secretion assay to demonstrate that each of the locus gene products is specifically required for secretion of the WXG100 virulence factor homolog, YukE. We then employ an unbiased approach to search for additional secreted substrates. By quantitative profiling of culture supernatants, we find that YukE may be the sole substrate that depends on the FtsK/SpoIIIE family ATPase for secretion. We discuss potential functional implications for secretion of a unique substrate.

Editor: Mickaël Desvaux, INRA Clermont-Ferrand Research Center, France

Funding: This work was supported by the Harvard University Milton Fund (BMB) (www.harvard.edu), an HHMI Physician Scientist Early Career Award (SMF) (www.hhmi.org), The Hood Foundation (SMF) (http://hria.org/tmfservices.html), and by NIH grant 1DP20D001378 (SMF) (www.nih.gov). The funders had no role in study design, data collection and analysis, decision to publish, or preparation of the manuscript.

Competing Interests: One of the authors of this study is employed by a commercial company (Thermo Fisher Scientific).

* E-mail: bburton@mcb.harvard.edu

9 These authors contributed equally to this work.

Introduction

Bacterial secretion systems play a critical role in the ability of bacterial cells to interface with their environment. In addition to the Sec (secretory) and Tat (twin-arginine translocation) systems that are involved in protein export (i.e. transport across the cytoplasmic membrane) [1–3], several outer membrane machineries have been described that complete protein secretion [4–7]. These secretion systems are less widely conserved and have more specific functions, such as horizontal gene transfer, nutrient uptake, and enabling virulence [8]. Recent studies identified a novel, dedicated export system called the Esat-6 secretion system (ESX or Ess), which is now known to be present in many bacteria including the archtypical Gram-positive bacterium *Bacillus subtilis* [9–12].

ESX protein secretion systems were initially identified in *Mycobacterium tuberculosis*, where it was demonstrated that the ESX-1 secretion system is responsible for the export of the small proteins ESAT-6 and CFP-10 (also named EsxA and EsxB respectively)[13,14]. EsxA is a 100-amino acid peptide that lacks an N-terminal signal sequence and has a helix-turn-helix structure with a WXG motif in the central turn, so it is also known as a WXG100 protein [11]. Bioinformatic studies using *in silico* methods to search for WXG100 family genes in other bacterial species have predicted the existence of ESX secretion systems in other Actinobacteria, some Firmicutes, and several Chloroflexi [11,12,15]. These predictions have been validated in several species, including *Staphylococcus aureus* [16–19], *Bacillus anthracis* [20], and *Streptomyces coelicolor* [21]. Intriguingly, genes homologous to some ESX components are sporadically distributed more broadly, including among the Proteobacteria [15]. ESX secretion systems are now defined by the presence of one or more WXG100 family substrates in addition to an FtsK/SpoIIIE family ATPase, often called EccC/EssC, that is required for substrate secretion [10].

The primary function of the proteins exported by ESX secretion systems remains unknown and therefore it is unclear whether the ESX systems share a conserved function(s). Numerous studies have demonstrated that the *M. tuberculosis* ESX-1 secretion system is essential for the virulence of this human pathogen; some studies suggest that the ESX-1 substrates compromise the integrity of the phagosomal membranes during macrophage infection [22–25], while other work suggests that the ESX secreted substrates are important for bacterial cell wall maintenance [23,26,27]. In addition, several of the recently identified ESX systems play a role in bacterial pathogenesis, including the ESX systems in *S. aureus* and *B. anthracis* [16–20,28]. However, there are also examples of ESX systems that do not play a role in virulence, such as the ESX system in the plant pathogen *Streptomyces scabies* that modulates sporulation and development [29]. Furthermore, ESX systems are predicted in non-pathogenic bacteria, and such systems have been validated in the soil bacterium *S. coelicolor* [11,21] and in *M. smegmatis* [30].

Bioinformatic analysis predicted that the *yuk* operon in the non-pathogenic bacterium *Bacillus subtilis* may encode an ESX protein secretion system [11]. Currently, there are five annotated genes in the *yuk* operon: *yukE, yukD, yukC, yukBA,* and *yueB* [31,32] (Figure 1A). The current annotation of the *yuk* operon suggests a terminator after *yueB*, but recent high throughput transcriptomics data implicates *yueC* and/or *yueD* as potential members of the *yuk/yue* locus as well [33]. By sequence analysis, the signature ESX/Ess proteins are represented in this system: YukE is homologous to the secreted virulence factor EsxA in *M. tuberculosis* and YukBA is predicted to be an FtsK/SpoIIIE family ATPase homologous to EccCa and EccCb in *M. tuberculosis* and EssC in *S. aureus* [11,16].

In this study, we demonstrate that the *yuk/yue* locus in *B. subtilis* encodes functional components of an ESX protein secretion system. We demonstrate that the small WXG100 protein, YukE, is secreted from cells. The secretion of YukE depends upon the other gene products encoded by the locus, including the other signature member of ESX secretion systems, the FtsK/SpoIIIE family ATPase YukBA. These results confirm a recent study of the *yuk/yue* locus components [34], and expand on that work by establishing the specificity of each of the locus components. Using an unbiased mass spectrometry approach, we find YukE to be the

only measurable YukBA-dependent substrate. Further, we demonstrate that the presence of the locus and the constitutive secretion of YukE provide neither a growth disadvantage nor a competitive advantage for the strain.

Results

The Bacillus subtilis yuk/yue locus encodes a secreted protein, YukE

All ESX protein secretion systems that have been studied to date have been shown to secrete at least one WXG100 family protein homologous to the prototypic ESX-1 substrate EsxA [13,16,20,21]. In *B. subtilis*, this protein is encoded by *yukE*. Therefore, our first experimental objective was to determine whether YukE is secreted from the *B. subtilis* cell. To address this question, we grew cultures of the wild-type domesticated strain of *B. subtilis* (PY79) in nutrient-rich LB medium to mid-exponential phase, harvested whole cell pellets, and filtered the culture supernatants. Proteins in the culture supernatant were concentrated by TCA precipitation and analyzed by SDS-PAGE. Presence of YukE was assessed using a primary antibody raised against recombinant full-length YukE. As a lysis control, we tested for the presence of the cytosolic protein RNA polymerase sigma factor SigmaA by immunoblotting with α-SigmaA antibodies [35]. In these experiments, we detected YukE in both the pellet and supernatant fractions (Figure 1B). These data confirm the

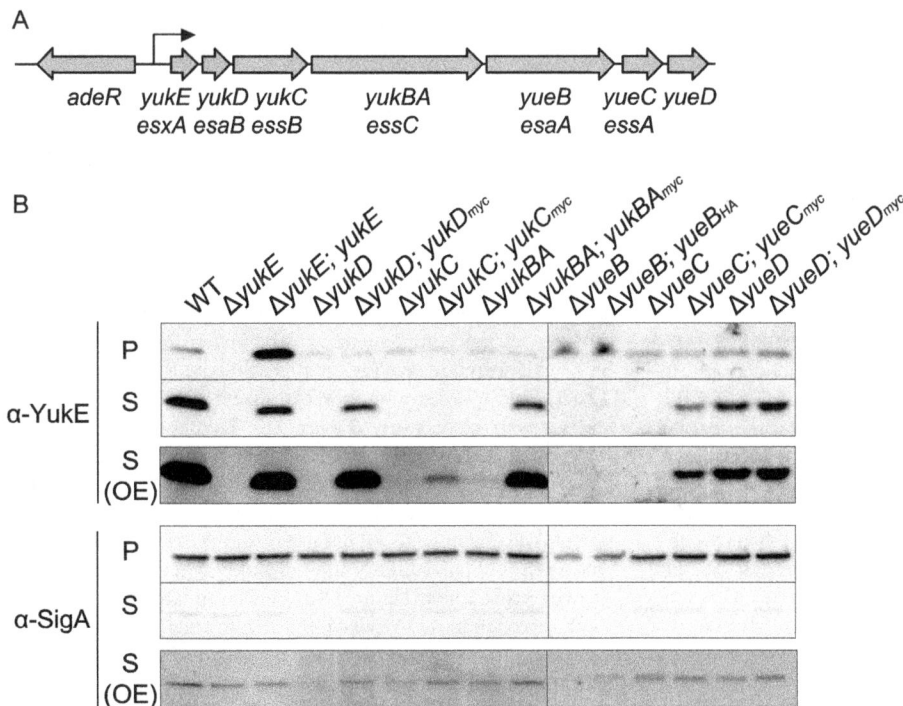

Figure 1. YukE is secreted, and secretion of YukE depends on other proteins encoded by the *yuk/yue* locus. A: Schematic depicting the *yuk/yue* locus and surrounding genes. Currently, there are five annotated genes in the *yuk* operon: *yukE, yukD, yukC, yukBA,* and *yueB* [31,32]. Recent high throughput transcriptomics data implicates *yueC* and/or *yueD* as potential members of the *yuk/yue* locus as well [33]. The predicted promoter (Pyuk) is indicated with an arrow. Homology to genes of other ESX/Ess systems is indicated below the corresponding *yuk/yue* gene name. B: Secretion assay for YukE. Cells were grown in LB medium to OD600nm of approximately 1.0–1.3. The cell pellet (P) was separated from the culture supernatant (S) by centrifugation. The pellet fractions were prepared into whole cell lysates and the supernatant fractions were filtered through a 0.2 micron filter and TCA precipitated. Samples were analyzed by SDS-PAGE under reducing conditions and immunoblot analysis with an α-YukE antibody and an α-SigmaA antibody as a loading/lysis control. The supernatants are shown in two exposures; the overexposed α-YukE blot (OE) allows visualization of faint bands. Data are representative of at least three biologically independent experiments. Pellet samples are equivalent to 0.1 OD and twenty-fold more was loaded for supernatant samples. Equivalent loading of precipitated supernatant samples was confirmed by densitometry of the Coomassie-stained gel.

Table 1. Quantification of secreted YukE.

STRAIN	% SigA in pellet	% SigA in supernatant	% YukE in pellet	% YukE in supernatant
Wildtype	99.97	0.03	81.06	18.94
ΔyukE	99.99	0.01	N/A	N/A
ΔyukE; yukE	100.00	0.00	97.19	2.81
ΔyukD	100.00	0.00	100.00	0.00
ΔyukD; yukD-myc	99.99	0.01	65.20	34.80
ΔyukC	99.99	0.01	100.00	0.00
ΔyukC; yukC-myc	99.98	0.02	99.65	0.35
ΔyukBA	99.98	0.02	100.00	0.00
ΔyukBA; yukBA-myc	99.98	0.02	78.44	21.56
ΔyueB	99.94	0.06	99.49	0.51
ΔyueB; yueB-HA	99.84	0.16	99.67	0.33
ΔyueC	99.74	0.26	100.00	0.00
ΔyueC; yueC-myc	99.77	0.23	88.41	11.59
ΔyueD	99.86	0.14	87.15	12.85
ΔyueD; yueD-myc	99.79	0.21	87.97	12.03

Densitometric analysis of the YukE and SigmaA proteins from the blots shown in Figure 1.

prediction and recent demonstration that YukE is secreted from the cell [34]. In contrast to the previous work, we were able to detect YukE secretion in a domesticated laboratory strain. We found that YukE was secreted in all conditions tested, ranging from growth in nutrient-rich media to the nutrient-limiting conditions that promote competence and biofilm formation (Figure S1).

YukE secretion depends upon other yuk/yue locus components

Next, we asked whether YukE secretion depends upon the other gene products in the *yuk/yue* locus. To address this question, we created a series of *yuk/yue* knockout strains. Each *yuk/yue* gene was individually replaced with an antibiotic resistance cassette and the *yuk* promoter (*Pyuk*) was reinserted after the resistance cassette to drive expression of the downstream operon genes. We used the intergenic region between *yukE* and *adeR* as the *yuk* promoter, and confirmed that *Pyuk* was transcriptionally active by inserting a *Pyuk-lacZ* construct at an ectopic integration site (*amyE::Pyuk-lacZ*) and assessing transcriptional activity. The β-galactosidase activity in this strain was approximately three-fold lower than the β-galactosidase activity in a strain with *lacZ* integrated at the endogenous *yuk* operon start site (*ΩPyuk-lacZ*) (Figure S2). This was ultimately useful, because genome-wide expression studies indicate that *yukE* expression is at least twice as high as the expression of other *yuk* operon genes [36]. Therefore, we reasoned that using our weaker *Pyuk* should result in approximately wild-type levels of transcription of the downstream genes. We confirmed that the reinserted *Pyuk* drove expression of downstream *yuk* genes, although resulting protein levels were approximately two-fold higher than native levels, as assessed by semi-quantitative immunoblotting (Figure S2).

To determine whether the genes of the *yuk/yue* locus are required for YukE secretion, we tested whether YukE is produced and secreted in each of the *yuk/yue* knockout strains. Currently, there are five annotated genes in the *yuk* operon: *yukE*, *yukD*, *yukC*, *yukBA*, and *yueB* [31,32]. Knocking out each gene in the annotated *yuk* operon (*yukE-yueB*) individually abolished YukE secretion in all

five of these strains (Figure 1B). Recently, transcriptomic profiling has implicated *yueC* and/or *yueD* as potential members of the *yuk/yue* operon as well [33]. Therefore, we also tested whether YukE is secreted in ΔyueC and ΔyueD strains. YukE was not secreted in the ΔyueC strain, demonstrating that YueC is required for YukE export, but it was secreted in the ΔyueD strain, suggesting that YueD is not required for YukE export (Figure 1B).

To demonstrate the specificity of these results, we constructed complementation strains by inserting the corresponding *yuk/yue* gene at an ectopic integration site under the control of an inducible promoter. We attached a C-terminal Myc or HA tag to each of the complementation constructs (except for the untagged YukE complementation construct), thereby allowing us to verify presence of the complementing protein by immunoblot (Figure S3). YukE secretion was restored to wild-type levels in the ΔyukD, ΔyukBA, and ΔyueC strains upon expression of *yukD-myc*, *yukBA-myc*, and *yueC-myc* respectively (Figure 1B). Densitometric analysis of secretion levels in each strain is presented in Table 1; values indicate the percentage of total YukE in each strain that is localized to the pellet versus culture supernatant. Complementation of ΔyukC with *yukC-myc* did not restore YukE secretion to wild-type levels, but partial restoration of YukE secretion can be seen in an overexposed blot (Figure 1B). We were unable to complement YukE secretion in the ΔyueB strain, despite attempts with untagged and several tagged versions of YueB. Nonetheless, YukE secretion appears dependent upon the *yueB* gene product and a recent study produced a complementing construct which confirms the specificity of a *yueB* deletion [34]. Thus we conclude that YukE secretion requires the full *yuk* operon as well as *yueC*, but not *yueD*.

The divergently transcribed gene *adeR* (formerly annotated as *yukF*) is a predicted transcription factor. Since regulatory proteins are often coded in the general vicinity of the genes they regulate, we also tested for YukE secretion in an *adeR* knockout strain, and found that YukE was still secreted in this background (Figure S4). This result is consistent with the idea that *yuk/yue* activity is perhaps principally regulated through stress response pathways including those governed by DegS/U and Spo0A [33,34,37–40], although inputs from other regulatory pathways may remain to be discovered.

YukE is the only protein detected to be dependent upon YukBA for secretion

To gain insight into possible function(s) of the *yuk/yue* system, we next sought to determine whether there are additional secreted proteins dependent upon the *yuk/yue* locus for secretion. Besides YukE, there is one other predicted WXG100 protein encoded in the *B. subtilis* genome, YfjA, and therefore this protein was a candidate *yuk/yue* substrate. [11]. In addition, secretion of LXG-motif proteins and non-WXG100 proteins has been reported in other ESX secretion systems, and these proteins are often encoded away from the primary ESX/Ess locus [20,41]. Therefore, we decided to use an unbiased, quantitative proteomics approach to analyze the full profile of *yuk/yue*-dependent proteins in the culture supernatant.

In addition to the virulence factor polypeptides, the FtsK/SpoIIIE family ATPases are a signature of ESX loci. Thus, using quantitative mass spectrometry, we compared the proteins in culture supernatants of the wild-type domesticated strain and the ATPase deletion strain Δ*yukBA* grown in defined media. Consistent with our immunoblot assay, we detected YukE in the supernatant of the wild-type strain in a manner that was dependent upon *yukBA* (Figure 2A, 2B). YukE secretion was restored in the YukBA complementation strain (Figure 2B). Ninety-five YukE-specific peptide spectra were detected in the supernatant from the wild-type strain, no peptides were detected in the Δ*yukBA* strain and 116 YukE-specific peptide spectra were detected in the Δ*yukBA*; *yukBA-myc* complementation strain. We detected high levels of YueB peptides in the culture supernatant of the Δ*yukBA* and complement strains (Figure 2A, 2B), which is an expected consequence of the strain design. Briefly, the *yuk* promoter was reinserted after the *yukBA* deletion to drive expression of the downstream genes, as otherwise this would be a polar mutation. Most surprisingly, we did not detect any other proteins with the same secretion profile as YukE in these conditions. Therefore, by this method and under these growth conditions, we found YukE to be the only protein that requires the ATPase YukBA for secretion.

The yuk/yue locus does not confer a growth or competition phenotype

The biological function of the *yuk/yue* locus remains unknown but it is highly unusual for a secretion system to have only a single substrate. Further, since all conditions we tested yielded secreted YukE, we speculated that the *yuk/yue* knockout strains might display a growth or competition phenotype. We first tested whether various *yuk/yue* knockout strains have a growth defect compared to the wild-type domesticated strain by conducting growth assays. The growth curves of the *yuk/yue* knockout strains were statistically indistinguishable from the growth curve of the wild-type domesticated strain, indicating that the *yuk/yue* knockout strains do not have a growth defect under standard, nutrient-rich laboratory conditions (Figure 3A). Next, we performed competition assays between the wild-type domesticated strain and *yuk/yue* knockout strains. We found that the *yuk/yue* knockout strains did not have a statistically significant competitive advantage or disadvantage compared to the wild-type domesticated strain in nutrient-rich or nutrient-limiting media (Figure 3B and Figure S5).

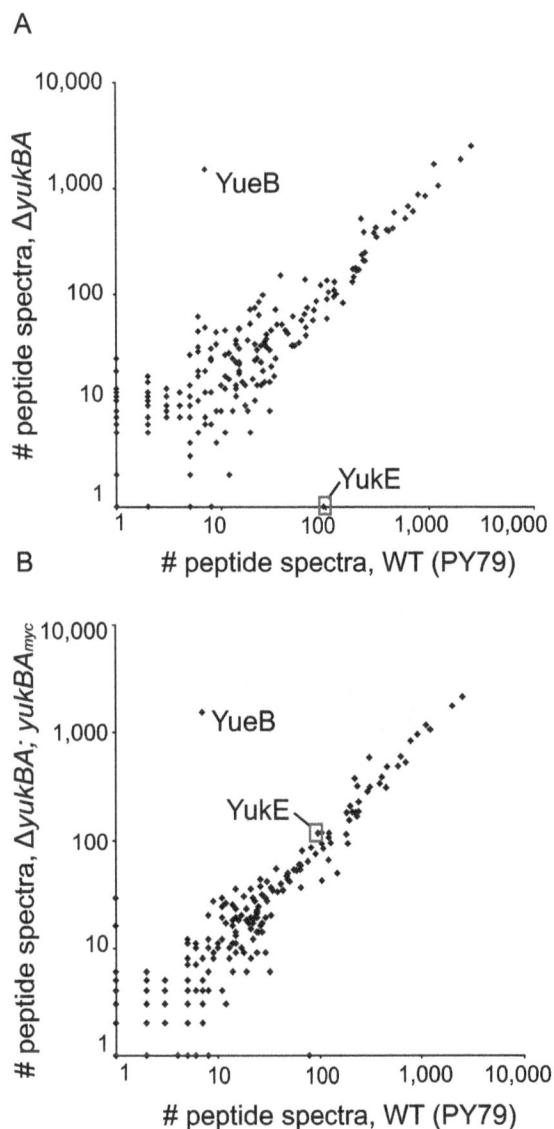

Figure 2. YukE is the only protein dependent upon YukBA for secretion. (A). and (B). The relative abundance of proteins detected in the culture supernatant of the wild-type strain (PY79) versus the Δ*yukBA* strain (A) or the complemented Δ*yukBA*; *yukBA-myc* strain (B). Cells were grown in nutrient-limiting 1XMC medium to mid-exponential phase, and the supernatant fractions were filtered through a 0.2 micron filter and TCA precipitated. The proteins in the culture supernatant were analyzed by mass spectrometry. Protein abundance was determined by spectral count analysis; spectral count data are combined totals from three biologically independent samples for each strain. Where no spectra were identified, an arbitrary value of 1 was assigned. The data point for YukE is circled in each graph. The point for YukE is at (95,1) in Figure 2A and at (95, 116) in Figure 2B. The complementation strain was constructed with the ectopically expressed *yukBA* gene disrupting the native *amyE* locus. Thus, as expected, AmyE peptides are underrepresented in the complementation strain as compared to both wild-type and Δ*yukBA* strains; the point located at (77, 1) in Figure 2B corresponds to the peptides assigned to AmyE. High levels of YueB peptides in the Δ*yukBA* and complement strains is a consequence of strain design; the *yuk* promoter was reinserted after the *yukBA* deletion to drive expression of the downstream genes.

A

B

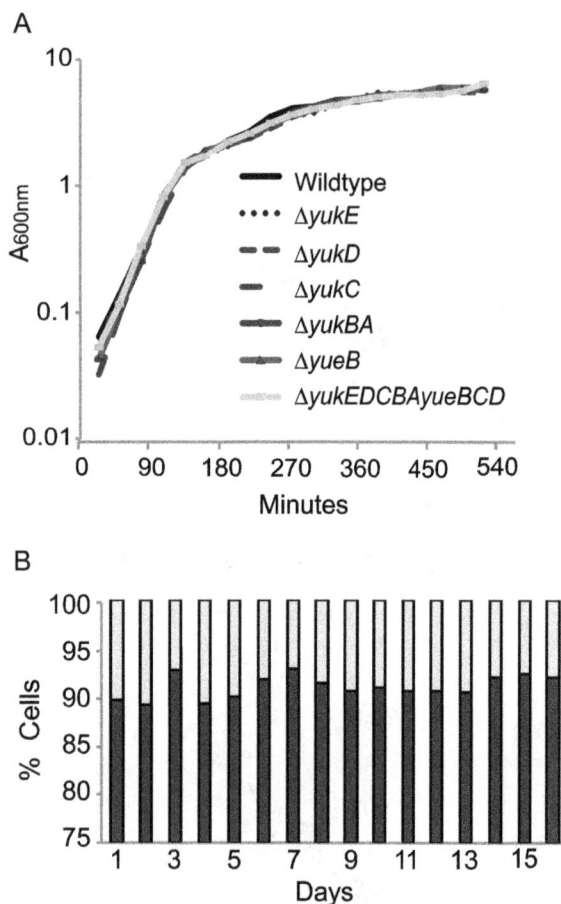

Figure 3. *yuk/yue* **knockout strains do not have a growth or competition defect compared to the wild-type strain.** A: Growth curve of the wild-type strain (PY79) and *yuk/yue* knockout strains grown in LB medium shaking at 37°C. The OD600nm was taken every 30 minutes for a total of 540 minutes. The following *yuk/yue* knockout strains were tested: ΔyukE, ΔyukD, ΔyukC, ΔyukBA, ΔyueB, and ΔyukEDCBAyueBCD. B: The results of a representative competition experiment between ΔyukEDCBA (light gray) versus the wild-type reporter strain (dark gray) in nutrient-rich LB medium. This competition had a starting ratio of 10% ΔyukEDCBA cells to 90% wild-type cells. The percentages were determined by counting the number of blue and white colonies on a single plate each day (typically 150–250 colonies per plate) and then calculating the percentage of colonies from each strain. Shown are the mean percentages averaged from triplicate platings for each day.

Discussion

Here, we have confirmed that the WXG100 protein, YukE, is a secreted protein, as predicted by its homology to the secreted virulence factor EsxA of *M. tuberculosis* and EsxA of *S. aureus*. YukE secretion is dependent upon each of the four other genes encoded within the annotated *yuk* operon as well as *yueC*, and we have confirmed the specificity of these dependencies by complementation. Most notably, secretion of YukE depends on the conserved FtsK/SpoIIIE family ATPase YukBA, the other signature member of ESX secretion systems. Furthermore, YukE secretion depends on YukD and YukC, which are homologous to proteins EsaB and EssB respectively in the Ess secretion system of *S. aureus*. Together with another recent study, these results suggest that the *yuk/yue* locus in *B. subtilis* encodes a *bona fide* ESX protein secretion system [34]. The predicted topologies and subcellular localizations of the

Yuk/Yue proteins suggest a membrane-bound secretion complex. Indeed, the envelope protein YueB has been implicated as a phage receptor (28), but this information has yet to provide additional clues as to the complete architecture of the system.

We have found YukE to be the only dedicated substrate of this secretion system thus far; we detected the other predicted WXG100 protein, YfjA, to be equally secreted in all strains tested, suggesting that it is not a YukBA-dependent substrate. Further profiling studies with different strain backgrounds or under different conditions may yet reveal additional substrates. For example, a recent study also detected YukE as a secreted product, although that report suggested that the strain background affects the conditions under which secreted YukE is detected [34].

ESX protein secretion systems are conserved throughout pathogenic and non-pathogenic species. It is currently unclear what the primary function of these systems is and whether ESX secretion systems share a conserved function(s). All ESX systems studied to date have been shown to be responsible for the secretion of a conserved EsxA-like protein substrate [13,16,20,21]; however, these proteins do not have an obvious effector function, and it is unclear how the secretion of a single conserved substrate could be beneficial to bacterial species representing such a wide range of lifestyles and environmental niches.

In *M. tuberculosis*, the ESX-1 system is required for pathogenesis [22-24] and several secreted substrates have been identified [13,14,41–45], but the specific functions of the secreted proteins are unknown. The prevailing hypothesis is that the secreted protein EsxA acts as a pore-forming toxin and induces damage to host cell membranes [22,25]. *B. subtilis* is not a human pathogen, but it likely encounters eukaryotes in its natural environment so it may similarly play a role in bacterial-eukaryotic interactions. For example, other *B. subtilis* systems have been demonstrated to have anti-nematodal and anti-fungal properties [46,47], so the Yuk/Yue proteins may have a similar function. Alternatively, components of the ESX systems have been implicated in DNA transfer in both mycobacterial species and in *B subtilis* [48,49] so the *yuk/yue* system may play a role in bacterial-environmental interactions by aiding with competence and DNA transfer.

An alternative hypothesis is that the ESX secreted proteins are required for a housekeeping function such as the maintenance of the bacterial cell wall [23,26,27]. In our study, we detect secretion of YukE under all tested conditions so it is possible that YukE is constitutively secreted to provide a function required for cell wall integrity or maintenance. It remains formally possible that YukE is in fact a component of the secretion apparatus itself. Further studies are needed to evaluate these hypotheses.

In this study, we find that YukE is the only identified substrate that is secreted under the conditions we tested. We also find that the *yuk/yue* system is not essential under these conditions. Therefore, it is possible that in response to some other stimulus, additional substrates will be identified and the *yuk/yue* system may be essential for bacterial growth or survival. This notion is further supported by a few lines of evidence that link regulation of the *yuk/yue* locus to the cell's stress response systems. A recent study implicated the two-component DegUS system in regulating YukE secretion, and numerous studies have pointed to the role of the master regulator Spo0A in upregulating *yuk/yue* genes [33,34,37–40]. Together these studies suggest that further work with undomesticated strains may ultimately yield vital clues to the biological role of the *B. subtilis* ESX machinery.

Materials and Methods

Strain construction

General methods for molecular cloning and strain construction were performed according to published protocols [50]. Chromosomal DNA isolated from the prototrophic domesticated strain PY79 was used as a template for all PCR amplification. Introduction of DNA into PY79 derivatives was conducted by transformation [51]. The bacterial strains used in this study are listed in Table 2. Complete strain construction information including oligonucleotide primers is included in Supporting Information.

Media and growth conditions

For general propagation, *B. subtilis* strains were grown at 37°C in LB (lysogeny broth) [52] (10 g tryptone per liter, 5 g yeast extract per liter, 5 g NaCl per liter) or on LB plates containing 1.5% Bacto agar. Where indicated, *B. subtilis* strains were grown in the nutrient-limiting medium *B. subtilis* Medium for Competence (1XMC) [53]. When appropriate, antibiotics were included in the growth medium as follows: 100 µg mL^{-1} spectinomycin, 5 µg mL^{-1} chloramphenicol, 5 µg mL^{-1} kanamycin, 10 µg mL^{-1} tetracycline, and 1 µg mL^{-1} erythromycin plus 25 µg mL^{-1} lincomycin (mls). When required, 100 µM IPTG (isopropyl-β-D-thiogalactopyranoside) was added to cultures or solid media to induce protein expression.

Bacillus lysates and TCA precipitation

Bacterial strains were grown in LB medium to an OD$_{600}$ of approximately 1.0–1.3. The cells were pelleted and the supernatant was collected. The pellet samples were processed to make whole cell lysates according to standard protocols [53]. Briefly, one milliliter of cells was harvested, lysed in the presence of lysozyme and then boiled for 15 minutes in 1× sample buffer (4% SDS, 250 mM Tris pH 6.8, 20% glycerol, 10 mM EDTA, 1% bromophenol blue, 10% β-mercaptoethanol (BME)). The culture supernatant samples were first filtered through a 0.2 micron filter and then incubated in 10% tricholoracetic acid (TCA) for 12–15 hours at 4°C. The following day, the samples were spun at 15,000xg for 20 minutes to pellet the precipitated proteins, the liquid was poured off, and the pellets were washed with ice-cold acetone. The pellets were resuspended in 100 µL of 1× sample buffer and the samples were boiled for 15 minutes. After processing the pellet and supernatant samples, the proteins were separated by SDS-polyacrylamide gel electrophoresis (SDS-PAGE) and analyzed by immunoblot analysis with appropriate antibodies. Pellet samples are equivalent to 0.1 OD units and twenty-fold more was loaded for supernatant samples. Precipitated supernatant samples were normalized based on Coomassie staining.

YukE polyclonal antibody generation

A hexahistidine-tagged version of YukE was utilized for antibody production. YukE was PCR-amplified with primers oLH067 and oLH068 using genomic DNA from the wild-type

Table 2. Strains used in this study.

Strain	Genotype	Source, Reference
PY79	Prototrophic domesticated laboratory strain	[56]
bLH015	*yukE::erm-Pyuk*	This work
bLH018	*yukEDCBA::erm-Pyuk*	This work
bLH019	*amyE::Pyuk-lacZ (spec)*	This work
bLH021	*ΩPyuk-lacZ (cat)*	This work
bLH027	*amyE::Phyperspank-lacZ (spec)*	RL2508 (Gift of Losick Lab)
bLH049	*amyE::kan*	pER82 (Gift of Rudner Lab)
bLH078	*adeR::erm; amyE::Pyuk-lacZ (spec)*	This work
bLH107	*yukEDCBAyueB::erm*	This work
bLH110	*yukBA::erm-Pyuk*	This work
bLH404	*yukBA::erm-Pyuk; amyE::Phyperspank-yukBA-myc (spec)*	This work
bLH421	*yukD::erm-Pyuk*	This work
bLH422	*yukC::erm-Pyuk*	This work
bLH458	*yukD::erm-Pyuk; amyE::Phyperspank-yukD-myc (spec)*	This work
bLH500	*yukC::erm-Pyuk; amyE::Phyperspank-yukC-myc (spec)*	This work
bLH533	*yukE::erm-Pyuk; amyE::Phyperspank-yukE (spec)*	This work
bLH579	*yueB::erm-Pyuk*	This work
bLH581	*yueC::erm-Pyuk*	This work
bLH585	*yueD::erm*	This work
bLH589	*yueB::erm-Pyuk; amyE::Phyperspank-yueB-HA (spec)*	This work
bLH590	*yueB::erm-Pyuk; amyE::Phyperspank-yueB (spec)*	This work
bLH591	*yueC::erm-Pyuk; amyE::Phyperspank-yueC-myc (spec)*	This work
bLH593	*yueD::erm; amyE::Phyperspank-yueD-myc (spec)*	This work

domesticated strain PY79 as a template. The sequence was inserted into an inducible *E. coli* expression vector to make pLH054, which was then transformed into *E. coli* BL21 cells. The cells were induced and YukE was purified from the *E. coli* extracts by nickel-affinity chromatography. Finally, a rabbit polyclonal serum was raised against this protein (Covance).

Immunoblot analysis

Proteins were separated by SDS-PAGE and transferred to nitrocellulose membrane. The membrane was probed with affinity-purified α-YukE (polyclonal), α-GFP (polyclonal), α-Myc (Novus Biologicals), and/or α-SigmaA (polyclonal) antibodies. Primary antibodies were diluted 1:1000 (α-YukE), 1:5,000 (α-GFP), 1:10,000 (α-Myc) or 1:1,000,000 (α-SigmaA) in 5% nonfat milk in TBS-0.05% Tween20. The primary antibody was detected using horseradish peroxidase-conjugated goat, α-rabbit immunoglobulin G (Bio-Rad or Jackson Laboratories). Supersignal West Femto chemiluminescent substrate (Thermo Scientific) was used to create a visible chemical reaction. The blots were imaged and densitometric quantitation of YukE secretion was performed using a FlourChem FC2 gel documentation system (Alpha Innotech) and provided software. The densitometry values in Table 1 indicate the proportion of total YukE in each strain that is localized to the pellet versus supernatant; values reflect normalization based on loading of an equivalent of 0.1 OD unit for pellet samples and twenty-fold more sample loaded for supernatant samples.

Mass spectrometry

Bacterial strains were grown in MC media to an OD_{600} of ~2.0. The cells were pelleted and the supernatant was collected and filtered through a 0.2 micron filter. Total proteins in the supernatant were obtained by TCA precipitating 30 mL of sample as described above. The samples were prepared for mass spectrometry analysis as described previously [27]. Briefly, samples were separated by molecular weight on a 10–20% Tricine gel (Invitrogen), each lane of the gel was sectioned into 10 roughly equal sized segments, followed by in-gel reduction, alkylation and trypsin digestion. Samples were run on a Thermo Fisher Scientific LTQ Veloz Mass Spectrometer (Thermo Fisher Scientific, Cambridge, MA). Samples were injected onto a Proxeon Easy nLC system configured with a 5 cm × 100 μm trap packed with 15–20 μm PS-DVB 300A media, and a 25 cm × 100 μm ID resolving column packed with 200A C18AQ media. Buffer A was 96% water, 4% methanol, and 0.2% formic acid. Buffer B was 10% water, 10% isopropanol, 80% acetonitrile, and 0.2% formic acid; loading buffer (sample loading/rinsing buffer) was 96% water, 4% methanol, and 0.2% formic acid. Samples were loaded at 5 μL min^{-1} for 9 min, and a gradient from 0–60% B at 375 nL min^{-1} was run over 70 min, for a total run time of 115 min (including regeneration and sample loading). Injection standards (Michrom Medium Molecule test mix, 5 angios, and the TP4 peptides) were injected at 61 fmoles per sample. Velos was run in a data dependent 15 configuration, with a full scan run in the in enhance scan mode (3^{e}4 target), with up to 15MS2 events. Rejection of +1 ions was used in precursor ion selection.

Resulting spectra were searched against a composite database which contained the predicted open reading frames annotated in the genome of *Bacillus subtilis* 168 supplemented with common contaminates using SEQUEST (Thermo Scientific, San Jose, CA). Peptides were filtered at a 1% FDR with PeptideProphet and grouped into proteins with ProteinProphet [54] with a cutoff of 0.95. Spectral counts across the gel slices for three biological replicates were pooled, and then levels of protein abundance

between strains were compared using an extended G-test [55]. Data was corrected for multiple testing (Benjamini and Hochberg) using a p value of ≤0.01; for a given protein, a criterion of having ≥5 peptides in at least one strain was set.

Supporting Information

Figure S1 YukE is secreted in LB, MC, and MSGG media. Secretion assays were performed to test YukE secretion from the domesticated PY79 laboratory strain under nutrient-rich growth conditions (LB medium) and nutrient-limiting growth conditions that promote competence (MC medium) or biofilm production (MSGG medium). Cells were grown in LB, MC, or MSGG medium to OD600nm of approximately 1.0–1.3. The cell pellet was separated from the culture supernatant (S) by centrifugation. Supernatant fractions were filtered through a 0.2 micron filter, TCA precipitated, and secretion was analyzed by SDS-PAGE under reducing conditions and immunoblot analysis with an α-YukE antibody and an α-SigmaA antibody as a loading/lysis control.

Figure S2 *yuk* knockout strain schematic and *Pyuk* promoter activity. A: Expression from the *yuk* promoter (*Pyuk*) was measured using *Pyuk-lacZ* transcriptional fusions. Two *Pyuk-lacZ* transcriptional fusion reporter strains were used: Ω*Pyuk-lacZ* and *amyE::Pyuk-lacZ*. Because the *yuk* promoter has not been previously characterized, we used the intergenic region between *yukE* and *adeR* as the *yuk* promoter for the latter construct. Strains were grown in LB medium to mid-exponential phase, and then transcriptional activity from P*yuk* was monitored by quantitative β-galactosidase assays. Shown are the mean ± SE of measurements from three independent experiments. B: Schematic showing the native *yuk* operon (top panel with white background) and the *yuk* knockout strains constructed by double crossover recombination (bottom panel with grey background). The *yuk* knockout strains used throughout this work include: Δ*yukE*, Δ*yukD*, Δ*yukC*, Δ*yukBA*, Δ*yueB*, and Δ*yueC*. The predicted *yuk* promoter (*Pyuk*) is indicated with a black arrow, the predicted terminator is indicated with a circle, and *erm* is an antibiotic resistance cassette. *Pyuk* is inserted after the antibiotic resistance cassette to drive expression of downstream genes in the Δ*yukE*, Δ*yukD*, Δ*yukC*, Δ*yukBA*, Δ*yueB* and Δ*yueC* strains. We confirmed that the re-inserted *Pyuk* drives expression of downstream *yuk* genes by inserting Ω*yueB-gfp* into each of these strains and assessing protein levels by semi-quantitative immunoblot with an α-GFP antibody. Compared to YueB-GFP levels detected in the wild-type background (+), YueB-GFP levels in the knockout strains were approximately two-fold higher than native levels (++).

Figure S3 Expression of epitope-tagged complementing constructs. Complementation strains were constructed by inserting each corresponding *yuk/yue* gene at an ectopic integration site (*amyE*) under the control of an inducible promoter. Immunoblot analysis with α–Myc (YukB-Myc, YukC-Myc, YukBA-Myc, YueC-Myc, YueD-Myc) or α-HA (YueB-HA) antibodies was used to verify the expression of each complementing protein. Astrisks indicate the protein-specific band for each full-length protein. Predicted molecular weight for each protein is as follows: *yukD*, 9 kDa; *yukC*, 52 kDa; *yukBA*, 171 kDa; *yueB*, 120 kDa; *yueC*, 16 kDa; *yueD*, 26 kDa.

Figure S4 YukE is secreted in an *adeR* knockout strain. Secretion assays were performed to test YukE secretion in a wildtype and *adeR* knockout background (bLH078). Cells were grown in LB medium to OD600nm of approximately 1.0–1.3. The cell pellet (P) was separated from the culture supernatant (S) by centrifugation. Supernatant fractions were filtered through a 0.2 micron filter, TCA precipitated, and secretion was analyzed by SDS-PAGE under reducing conditions and immunoblot analysis with an α-YukE antibody and an α-SigmaA antibody as a loading/lysis control. Deletion of *adeR* may have affected the *yuk* operon promoter, possibly causing reduced levels of intracellular YukE in the Δ*adeR* strain as compared to PY79.

Figure S5 The *yukBA* knockout strain does not have a competition defect compared to the wild-type strain in MC media. The results of a representative competition experiment between Δ*yukBA* (light gray) versus the wild-type reporter strain (dark gray) in Media for Competence (MC). This competition had a starting ratio of 90% wildtype cells to 10% Δ*yukBA* cells. The percentages were determined by counting the number of blue and white colonies on a single plate each day (typically 150–250 colonies per plate) and then calculating the percentage of colonies from each strain. Shown are the mean percentages averaged from triplicate platings for each day.

Table S1 Strains used in this study.

Table S2 Oligos used in this study.

Acknowledgments

The authors thank A. Garces for help with mass spectrometry, and members of the Burton, Fortune, Losick, and Rubin laboratories for discussion and comments.

Author Contributions

Conceived and designed the experiments: LAH TLR SMF BMB. Performed the experiments: LAH TLR DAS. Analyzed the data: LAH TLR MRC SMF BMB. Wrote the paper: LAH TLR SMF BMB.

References

1. Papanikou E, Karamanou S, Economou A (2007) Bacterial protein secretion through the translocase nanomachine. Nat Rev Microbiol 5: 839–851.
2. Driessen AJ, Nouwen N (2008) Protein translocation across the bacterial cytoplasmic membrane. Annu Rev Biochem 77: 643–667.
3. Robinson C, Matos CF, Beck D, Ren C, Lawrence J, et al. (2011) Transport and proofreading of proteins by the twin-arginine translocation (Tat) system in bacteria. Biochim Biophys Acta 1808: 876–884.
4. Chagnot C, Zorgani MA, Astruc T, Desvaux M (2013) Proteinaceous determinants of surface colonization in bacteria: bacterial adhesion and biofilm formation from a protein secretion perspective. Front Microbiol 4: 303.
5. Economou A, Christie PJ, Fernandez RC, Palmer T, Plano GV, et al. (2006) Secretion by numbers: Protein traffic in prokaryotes. Mol Microbiol 62: 308–319.
6. Desvaux M, Hebraud M, Talon R, Henderson IR (2009) Secretion and subcellular localizations of bacterial proteins: a semantic awareness issue. Trends Microbiol 17: 139–145.
7. Desvaux P, Corman A, Hamidi K, Pinton P (2004) [Management of erectile dysfunction in daily practice—PISTES study]. Prog Urol 14: 512–520.
8. Finlay BB, Falkow S (1997) Common themes in microbial pathogenicity revisited. Microbiol Mol Biol Rev 61: 136–169.
9. Brodin P, Majlessi L, Marsollier L, de Jonge MI, Bottai D, et al. (2006) Dissection of ESAT-6 system 1 of Mycobacterium tuberculosis and impact on immunogenicity and virulence. Infect Immun 74: 88–98.
10. Abdallah AM, Gey van Pittius NC, Champion PA, Cox J, Luirink J, et al. (2007) Type VII secretion—mycobacteria show the way. Nat Rev Microbiol 5: 883–891.
11. Pallen MJ (2002) The ESAT-6/WXG100 superfamily — and a new Gram-positive secretion system? Trends Microbiol 10: 209–212.
12. Gey Van Pittius NC, Gamieldien J, Hide W, Brown GD, Siezen RJ, et al. (2001) The ESAT-6 gene cluster of Mycobacterium tuberculosis and other high G+C Gram-positive bacteria. Genome Biol 2: RESEARCH0044.
13. Sorensen AL, Nagai S, Houen G, Andersen P, Andersen AB (1995) Purification and characterization of a low-molecular-mass T-cell antigen secreted by Mycobacterium tuberculosis. Infect Immun 63: 1710–1717.
14. Berthet FX, Rasmussen PB, Rosenkrands I, Andersen P, Gicquel B (1998) A Mycobacterium tuberculosis operon encoding ESAT-6 and a novel low-molecular-mass culture filtrate protein (CFP-10). Microbiology 144 (Pt 11): 3195–3203.
15. Sutcliffe I (2011) New insights into the distribution of WXG100 protein secretion systems. Antonie Van Leeuwenhoek 99: 127–131.
16. Burts ML, Williams WA, DeBord K, Missiakas DM (2005) EsxA and EsxB are secreted by an ESAT-6-like system that is required for the pathogenesis of Staphylococcus aureus infections. Proc Natl Acad Sci U S A 102: 1169–1174.
17. Burts ML, DeDent AC, Missiakas DM (2008) EsaC substrate for the ESAT-6 secretion pathway and its role in persistent infections of Staphylococcus aureus. Mol Microbiol 69: 736–746.
18. Anderson M, Chen YH, Butler EK, Missiakas DM (2011) EsaD, a secretion factor for the Ess pathway in Staphylococcus aureus. J Bacteriol 193: 1583–1589.
19. Chen YH, Anderson M, Hendrickx AP, Missiakas D (2012) Characterization of EssB, a protein required for secretion of ESAT-6 like proteins in Staphylococcus aureus. BMC Microbiol 12: 219.
20. Garufi G, Butler E, Missiakas D (2008) ESAT-6-like protein secretion in Bacillus anthracis. J Bacteriol 190: 7004–7011.
21. Akpe San Roman S, Facey PD, Fernandez-Martinez L, Rodriguez C, Vallin C, et al. (2010) A heterodimer of EsxA and EsxB is involved in sporulation and is secreted by a type VII secretion system in Streptomyces coelicolor. Microbiology 156: 1719–1729.
22. Hsu T, Hingley-Wilson SM, Chen B, Chen M, Dai AZ, et al. (2003) The primary mechanism of attenuation of bacillus Calmette-Guerin is a loss of secreted lytic function required for invasion of lung interstitial tissue. Proc Natl Acad Sci U S A 100: 12420–12425.
23. Pym AS, Brodin P, Brosch R, Huerre M, Cole ST (2002) Loss of RD1 contributed to the attenuation of the live tuberculosis vaccines Mycobacterium bovis BCG and Mycobacterium microti. Mol Microbiol 46: 709–717.
24. Stanley SA, Raghavan S, Hwang WW, Cox JS (2003) Acute infection and macrophage subversion by Mycobacterium tuberculosis require a specialized secretion system. Proc Natl Acad Sci U S A 100: 13001–13006.
25. de Jonge MI, Pehau-Arnaudet G, Fretz MM, Romain F, Bottai D, et al. (2007) ESAT-6 from Mycobacterium tuberculosis dissociates from its putative chaperone CFP-10 under acidic conditions and exhibits membrane-lysing activity. J Bacteriol 189: 6028–6034.
26. Sani M, Houben EN, Geurtsen J, Pierson J, de Punder K, et al. (2010) Direct visualization by cryo-EM of the mycobacterial capsular layer: a labile structure containing ESX-1-secreted proteins. PLoS Pathog 6: e1000794.
27. Garces A, Atmakuri K, Chase MR, Woodworth JS, Krastins B, et al. (2010) EspA acts as a critical mediator of ESX1-dependent virulence in Mycobacterium tuberculosis by affecting bacterial cell wall integrity. PLoS Pathog 6: e1000957.
28. Renshaw PS, Lightbody KL, Veverka V, Muskett FW, Kelly G, et al. (2005) Structure and function of the complex formed by the tuberculosis virulence factors CFP-10 and ESAT-6. EMBO J 24: 2491–2498.
29. Fyans JK, Bignell D, Loria R, Toth I, Palmer T (2012) The ESX/type VII secretion system modulates development, but not virulence, of the plant pathogen Streptomyces scabies. Mol Plant Pathol.
30. Converse SE, Cox JS (2005) A protein secretion pathway critical for Mycobacterium tuberculosis virulence is conserved and functional in Mycobacterium smegmatis. J Bacteriol 187: 1238–1245.
31. Barbe V, Cruveiller S, Kunst F, Lenoble P, Meurice G, et al. (2009) From a consortium sequence to a unified sequence: the Bacillus subtilis 168 reference genome a decade later. Microbiology 155: 1758–1775.
32. Sao-Jose C, Baptista C, Santos MA (2004) Bacillus subtilis operon encoding a membrane receptor for bacteriophage SPP1. J Bacteriol 186: 8337–8346.
33. Nicolas P, Mader U, Dervyn E, Rochat T, Leduc A, et al. (2012) Condition-dependent transcriptome reveals high-level regulatory architecture in Bacillus subtilis. Science 335: 1103–1106.
34. Baptista C, Barreto HC, Sao-Jose C (2013) High Levels of DegU-P Activate an Esat-6-Like Secretion System in Bacillus subtilis. PLoS One 8: e67840.
35. Fujita M (2000) Temporal and selective association of multiple sigma factors with RNA polymerase during sporulation in Bacillus subtilis. Genes Cells 5: 79–88.
36. Rasmussen S, Nielsen HB, Jarmer H (2009) The transcriptionally active regions in the genome of Bacillus subtilis. Mol Microbiol 73: 1043–1057.
37. Rosenberg A, Sinai L, Smith Y, Ben-Yehuda S (2012) Dynamic expression of the translational machinery during Bacillus subtilis life cycle at a single cell level. PLoS One 7: e41921.

38. Marchadier E, Carballido-Lopez R, Brinster S, Fabret C, Mervelet P, et al. (2011) An expanded protein-protein interaction network in Bacillus subtilis reveals a group of hubs: Exploration by an integrative approach. Proteomics 11: 2981–2991.

39. Garti-Levi S, Eswara A, Smith Y, Fujita M, Ben-Yehuda S (2013) Novel modulators controlling entry into sporulation in Bacillus subtilis. J Bacteriol 195: 1475–1483.

40. Kobayashi K (2007) Gradual activation of the response regulator DegU controls serial expression of genes for flagellum formation and biofilm formation in Bacillus subtilis. Mol Microbiol 66: 395–409.

41. Fortune SM, Jaeger A, Sarracino DA, Chase MR, Sassetti CM, et al. (2005) Mutually dependent secretion of proteins required for mycobacterial virulence. Proc Natl Acad Sci U S A 102: 10676–10681.

42. Xu J, Laine O, Masciocchi M, Manoranjan J, Smith J, et al. (2007) A unique Mycobacterium ESX-1 protein co-secretes with CFP-10/ESAT-6 and is necessary for inhibiting phagosome maturation. Mol Microbiol 66: 787–800.

43. McLaughlin B, Chon JS, MacGurn JA, Carlsson F, Cheng TL, et al. (2007) A mycobacterium ESX-1-secreted virulence factor with unique requirements for export. PLoS Pathog 3: e105.

44. Raghavan S, Manzanillo P, Chan K, Dovey C, Cox JS (2008) Secreted transcription factor controls Mycobacterium tuberculosis virulence. Nature 454: 717–721.

45. MacGurn JA, Raghavan S, Stanley SA, Cox JS (2005) A non-RD1 gene cluster is required for Snm secretion in Mycobacterium tuberculosis. Mol Microbiol 57: 1653–1663.

46. Xia Y, Xie S, Ma X, Wu H, Wang X, et al. (2011) The purL gene of Bacillus subtilis is associated with nematicidal activity. FEMS Microbiol Lett 322: 99–107.

47. Ruiz A, Neilson JB, Bulmer GS (1982) Control of Cryptococcus neoformans in nature by biotic factors. Sabouraudia 20: 21–29.

48. Coros A, Callahan B, Battaglioli E, Derbyshire KM (2008) The specialized secretory apparatus ESX-1 is essential for DNA transfer in Mycobacterium smegmatis. Mol Microbiol 69: 794–808.

49. Rosch TC, Golman W, Hucklesby L, Gonzalez-Pastor JE, Graumann PL (2014) The Presence of Conjugative Plasmid pLS20 Affects Global Transcription of Its Bacillus subtilis Host and Confers Beneficial Stress Resistance to Cells. Appl Environ Microbiol 80: 1349–1358.

50. Sambrook JR, DW (2006) The condensed protocols from molecular cloning: a laboratory manual. Cold Spring Harbor, NY: Cold Spring Harbor Laboratory Press.

51. Gryczan TJ, Contente S, Dubnau D (1978) Characterization of Staphylococcus aureus plasmids introduced by transformation into Bacillus subtilis. J Bacteriol 134: 318–329.

52. Bertani G (1951) Studies on lysogenesis. I. The mode of phage liberation by lysogenic Escherichia coli. J Bacteriol 62: 293–300.

53. Cutting SVH, PB (1990) Genetic analyses. In: Hardwood CR CS, editor. Molecular biological methods for Bacillus. New York: John Wiley & Sons. pp. 27–61.

54. Keller A, Eng J, Zhang N, Li XJ, Aebersold R (2005) A uniform proteomics MS/MS analysis platform utilizing open XML file formats. Mol Syst Biol 1: 2005 0017.

55. Zhang B, VerBerkmoes NC, Langston MA, Uberbacher E, Hettich RL, et al. (2006) Detecting differential and correlated protein expression in label-free shotgun proteomics. J Proteome Res 5: 2909–2918.

56. Youngman PJ, Perkins JB, Losick R (1983) Genetic transposition and insertional mutagenesis in Bacillus subtilis with Streptococcus faecalis transposon Tn917. Proc Natl Acad Sci U S A 80: 2305–2309.

57. Guerout-Fleury AM, Frandsen N, Stragier P (1996) Plasmids for ectopic integration in Bacillus subtilis. Gene 180: 57–61.

4

Effects of Slag-Based Silicon Fertilizer on Rice Growth and Brown-Spot Resistance

Dongfeng Ning[1], Alin Song[1], Fenliang Fan[1], Zhaojun Li[1], Yongchao Liang[1,2]*

1 Ministry of Agriculture Key Laboratory of Crop Nutrition and Fertilization, Institute of Agricultural Resources and Regional Planning, Chinese Academy of Agricultural Sciences, Beijing, China, **2** Ministry of Education Key Laboratory of Environment Remediation and Ecological Health, College of Environmental and Resource Sciences, Zhejiang University, Hangzhou, China

Abstract

It is well documented that slag-based silicon fertilizers have beneficial effects on the growth and disease resistance of rice. However, their effects vary greatly with sources of slag and are closely related to availability of silicon (Si) in these materials. To date, few researches have been done to compare the differences in plant performance and disease resistance between different slag-based silicon fertilizers applied at the same rate of plant-available Si. In the present study both steel and iron slags were chosen to investigate their effects on rice growth and disease resistance under greenhouse conditions. Both scanning electron microscopy (SEM) and transmission electron microscopy (TEM) were used to examine the effects of slags on ultrastructural changes in leaves of rice naturally infected by *Bipolaris oryaze*, the causal agent of brown spot. The results showed that both slag-based Si fertilizers tested significantly increased rice growth and yield, but decreased brown spot incidence, with steel slag showing a stronger effect than iron slag. The results of SEM analysis showed that application of slags led to more pronounced cell silicification in rice leaves, more silica cells, and more pronounced and larger papilla as well. The results of TEM analysis showed that mesophyll cells of slag-untreated rice leaf were disorganized, with colonization of the fungus (*Bipolaris oryzae*), including chloroplast degradation and cell wall alterations. The application of slag maintained mesophyll cells relatively intact and increased the thickness of silicon layer. It can be concluded that applying slag-based fertilizer to Si-deficient paddy soil is necessary for improving both rice productivity and brown spot resistance. The immobile silicon deposited in host cell walls and papillae sites is the first physical barrier for fungal penetration, while the soluble Si in the cytoplasm enhances physiological or induced resistance to fungal colonization.

Editor: Guoping Zhang, Zhejiang University, China

Funding: This work was jointly supported by The 12th Five-year Key Programs entitled "Techniques for Agricultural Use of Steel and Iron Slag: Research and Demonstration" and "Study of Key Technologies for Alleviating Obstacle Factors and Improving Productivity of Low-Yield Cropland" supported by Ministry of Science and Technology, China and The National Natural Science Foundation of China entitled "The quantitative study of influence of straw returning on silicon releasing in typical rice soil of Southern China (41301310)". The funders had no role in study design, data collection and analysis, decision to publish, or preparation of the manuscript.

Competing Interests: The authors have declared that no competing interests exist.

* Email: ycliang@zju.edu.cn

Introduction

Silicon (Si) is the second most abundant element in soils [1,2]. Although Si has not been proven to be an essential element for plant growth and development, its beneficial roles in stimulating plant growth, grain yield and resistance to abiotic (metal toxicity, salt and drought stress, nutrient imbalance, extreme temperature) and biotic stress (plant diseases and insect pests) have been well documented [3–7].

Rice (*Oryza sativa* L.) is the second most widely grown crop in the world, and the major staple food for more than half of the world's population [8,9]. Rice is also a typical Si hyper-accumulating plant species, containing Si up to 10% in shoots on a dry weight basis [2]. Rice roots take up Si in the form of silicic acid (H_4SiO_4) from the soil solution [10]. In tropical and subtropical areas, because of heavy desilication-aluminization arising from high temperature and rainfall, plant-available Si is low in these highly-weathered soils [11]. In addition, repeated mono-cropping with rice may greatly decrease plant-available Si in soil. It is estimated that producing a total rice grain yield of 5000 kg ha^{-1} will remove Si at 230–470 kg ha^{-1} from the soil [5],

and Si may then become a yield-limiting element for rice production [12–14]. Therefore, it maybe is necessary to provide exogenous Si fertilizer for an economic and sustainable rice production system [15–18].

Brown spot caused by the fungus (*Bipolaris oryzae*) is one of the most devastating and prevalent diseases of rice. Brown spot may cause significant yield losses [19,20]. The major method to control brown spot in agriculture is through application of fungicides [19]. However, there is a need to explore more eco-friendly management practices in consideration of the public's concerns with health and environmental issues. The physiological condition of rice plant, which is strongly influenced by soil conditions, particularly soil nutrient status (e.g. potassium, calcium, magnesium, manganese, iron, and silicon etc.), is one of the main factors governing brown spot severity [19,21]. Some authors suggest that application of Si fertilizer to rice fields is an alternative approach to control brown spot, especially in soils where plant-available Si is very low [22–24].

Steel slags or iron slags are byproducts of steel or iron industries, which account for 15–20% of total steel production. Large

Table 1. The main chemical characteristics of two slag-based silicon fertilizers tested in the present study (%).

Si fertilizer	CaO	SiO$_2$	MgO	Al$_2$O$_3$	Fe$_2$O$_3$	MnO	TiO$_2$
Q	43.6	26.9	8.1	10.9	3.1	0.9	1.2
H	50.9	21.0	7.7	6.0	5.0	1.5	0.6

amounts of slag are produced in China annually [25]. Slags are not merely metallurgical wastes, but they have been successfully used in agriculture in many developed counties [26,27]. In contrast, only 10% of the total slag is recycled in China [28]. Slags contain sufficient amounts of Si (10–28%); therefore, they may potentially be used as a Si fertilizer source. Application of such kind of Si fertilizer has been shown to improve degraded paddy soils, as well as rice growth and disease resistance [2,29–34]. So, slag applied to paddy rice fields as Si fertilizer is beneficial not only for rice health and growth, but also from economic and environmental perspectives. However, variation exists in the ore and coke, as well as in the cooling process; consequently, the composition and property of slags may vary widely [35]. Therefore, plant-available Si content and Si availability in slags vary widely too [36]. Previous studies have demonstrated positive effects of wollastonite or calcium silicate as Si resource on rice growth and disease resistance [30,32]. However, there are only a few reports to compare the agronomic benefits of different sources of slag used in rice.

The objective of this study was to assess the effects of steel slag and iron slag applied at the same rates of plant-available Si on rice growth and brown spot development in rice and to investigate the relationship between Si-mediated ultrastructural changes and brown spot disease infection in rice.

Materials and Methods

Soil and plant material preparation

The soil used was sampled from Qionghai, Hainan province of South China (N 19°09′16.2″, E 110°17′35.3″) (no specific permissions were required for soil sampling in this location and the field in this study did not involve endangered or protected species). It was a latosol derived from basalt with a plant-available Si concentration of 41.8 mg kg^{-1} (extracted by 0.025 M citric acid) and a pH value of 5.16. The soil was air-dried and sieved (2.0 mm). The rice variety tested is a hybrid (*Oryza sativa* L. cv. Fengyuanyou 299), characterized by its mid-late maturity. Seeds

were sterilized with 10% (v/v) H$_2$O$_2$ for 15 min, rinsed with distilled water, soaked in water for 24 hours, and then transferred into culture dishes for germination at 25°C in the dark. Two days later, the germinated seeds were placed on a float tray (10×15 cm) in a controlled environment with a day/night temperature of 25°C (12 h): 25°C (12 h).

Experimental design

A pot experiment factorially arranged in a 2×4 randomized, complete block design was conducted with three replicates per treatment, giving a total of 24 pots. The entire experiment was duplicated. Two different Si fertilizers were chosen for the pot experiment. One was derived from air-cooling steel slag, with HCl-soluble Si content of 7.61%, referred to as H, and the other was based on water-cooling iron slag, with HCl-soluble Si content of 9.35%, referred to as Q. The main chemical properties of the two slags are presented in Table 1. Four Si treatments with three replicates each were established. The rate of Si applied, equivalent to 0.5 M HCl-soluble Si, was 0 (Si$_0$), 187 (Si$_1$), 560 (Si$_2$) and 935 (Si$_3$) mg Si kg^{-1}. The Si fertilizer was thoroughly mixed with soil prior to potting. Basal fertilizers supplied were 0.2 g N kg^{-1} as urea, 52 mg P kg^{-1} as potassium dihydrogen phosphate, and 84 mg K kg^{-1} as potassium sulfate. Each plastic pot was filled with 5 kg of air-dried and sieved (2.0 mm) soil. Uniform seedlings with three leaves fully expanded were transplanted at two seedlings per pot. During the rice growing period, distilled water was applied to maintain a 2-cm water layer but no pesticides were applied.

Plant sampling

Rice plants were harvested at maturity, and separated into stem, leaf, and grain, and then washed thoroughly with distilled water. The dry weight of these tissues was recorded after being oven-dried at 75°C till a constant weight. These tissues were then ground to pass through a 0.5-mm sieve for Si analysis.

Table 2. Effects of different silicon treatments on dry weight of rice organs (%).

Fertilizer	Rate	Leaf	Stem	Grain
Control	Si$_0$	7.72±1.60 b	13.6±4.73 b	4.71±1.52 c
Q	Si$_1$	13.1±2.02 a	20.8±1.11 a	13.7±0.69 b
	Si$_2$	13.7±2.02 a	20.8±1.07 a	13.4±2.37 b
	Si$_3$	13.0±2.27 a	20.2±1.46 a	11.8±1.99 b
H	Si$_1$	12.8±2.71 a	19.4±1.30 b	15.1±2.33 ab
	Si$_2$	11.5±1.08 a	20.2±1.24 a	15.0±1.53 ab
	Si$_3$	12.5±1.15 a	23.7±2.40 a	16.9±2.08 a

Si$_0$: no Si fertilizer; Si$_1$: slag fertilizer applied at a rate of 187 mg plant-available Si per kg soil; Si$_2$: slag fertilizer applied at a rate of 560 mg plant-available Si per kg soil; Si$_3$: slag fertilizer applied at a rate of 935 mg plant-available Si per kg soil; H: slag fertilizer H, Q: slag fertilizer Q; Data are means ± SD of three replicates; mean values followed by different letters (a, b, c) are significantly different (P≤0.05).

Table 3. Analysis of variance of the effects slag-based silicon fertilizer (slag) and application rate of Si (Si-R) on dry weight of rice organs (%).

Sources of variation	Df	F values		
		Leaf	Stem	Grain
Slag	1	0.816 ns	0.142 ns	6.41*
Si-R	3	9.93*	11.61*	36.30**
Slag×Si-R	3	0.388 ns	1.03 ns	1.90 ns

Levels of probability: ns = non significant, significantly different *p≤0.05 and **p≤0.01.
Levels of probability: ns = non significant and *p≤0.05, **p≤0.01.

Disease index survey

Rice leaves were naturally infected by *Bipolaris oryzae*, the causal agent of brown spot at the joining stage. Disease severity, based on the percentage of infected leaf surface area and the percentage of infected leaves per pot, was determined two weeks after infection. In this study, disease severity (DS) was classified into nine grades based on the following: DS0 = healthy plants, DS1≤1%, DS3 = 2–5%, DS5 = 6–15%, DS7 = 16–25% and DS9≥25%. Disease index (%) = $[\sum(S*n_s)/(9*Ns)] *100$. Where S is the severity value, n_s is the number of infected leaves with a severity of S and Ns is the number of leaves evaluated [37].

Scanning electron microscopy

The deposition of Si in the leaf was observed using scanning electron microscopy (SEM). Since similar slag effects on plant growth and brown spot resistance were observed for both Si sources, only leaf samples of rice plants treated with slag H were collected for microscopic examination. At the anthesis stage, fresh specimens of the top-second leaf of rice plants grown without slag (control) or with slag (H) applied at a rate of 935 mg plant-available Si per kg soil were randomly sampled from two plants per pot. They were first fixed with 2.5% (v/v) glutaraldehyde in 0.1 M phosphate buffer solution (pH 7.4) under vacuum for 2–3 h at 20°C, and then post-fixed with 1% (w/v) osmium tetroxide in the phosphate buffer solution for 30 min [38]. Afterwards, they were dehydrated through a graded series of ethanol [50, 70, 80, 90 and 100% (v/v)], dried by a critical-point drying method with liquid CO_2 and coated with metal and then loaded onto the instrument [38]. The surface scan was performed using a scanning electron microscope (FEI QUANTA200, Japan).

Transmission electron microscopy

Squares were excised with scissors from the top-second leaf at the anthesis stage. The leaf samples of rice plants grown without slag (control) or with slag (H) applied at a rate of 935 mg plant-available Si per kg soil were collected and fixed immediately with 2% (v/v) glutaraldehyde and 2% (v/v) paraformaldehyde in 0.05 M sodium cacodylate buffer (pH 7.2) at room temperature overnight and then washed with the same buffer three times for 10 min each [38]. Afterwards, samples were postfixed with 1% (w/v) osmium tetroxide in the same buffer at room temperature for 2 h and washed twice with distilled water. The post-fixed samples were stained with 0.5% (w/v) uranyl acetate at 4°C overnight. They were then dehydrated in a graded series of ethanol [30, 50, 70, 80, 95, and 100% (v/v)] and three times in 100% ethanol for 10 min each [38]. Ultrathin sections (approximately 50 nm in thickness) were made with a diamond knife by an ultramicrotome (LKBVI). The sections were mounted on copper grids and stained for 7 min each with 2% (w/v) uranyl acetate and Reynolds' lead citrate [38]. The sections were examined by transmission electron microscopy (Phillips EM 400 ST, the Netherlands).

Chemical analysis

The main chemical components of slag fertilizers were measured by SEM. Scanning electron microscopy was performed in a JSM-6510 SEM at accelerating voltage of 20 kV attached with an X-ray energy-dispersive spectrometer, EDS (Genesis XM2). Before the scanning process, all samples were dried and coated with gold to enhance the electron conductivity.

The available Si content in slag was determined following extraction with 0.5 M HCl [slag/(HCl) ratio of 1:50, shaking at

Table 4. Effects of different silicon (Si) treatments on Si concentrations in rice organs (SiO_2%).

Fertilizer	Rate	Leaf	Stem	Grain
Control	Si_0	11.1±0.31 c	5.17±0.42 d	0.19±0.031 c
Q	Si_1	12.3±0.36 b	5.57±0.47 cd	0.22±0.04 bc
	Si_2	12.3±0.35 b	5.34±0.27 cd	0.23±0.04 bc
	Si_3	12.1±0.31 b	5.76±0.14 c	0.27±0.02 ab
H	Si_1	11.9±0.22 b	5.54±0.30 cd	0.21±0.03 c
	Si_2	12.6±0.13 ab	6.31±0.39 b	0.26±0.02 ab
	Si_3	12.9±0.27 a	6.87±0.37 a	0.28±0.04 a

Si_0: no Si fertilizer; Si_1: slag fertilizer applied at a rate of 187 mg plant-available Si per kg soil; Si_2: slag fertilizer applied at a rate of 560 mg plant-available Si per kg soil; Si_3: slag fertilizer applied at a rate of 935 mg plant-available Si per kg soil; H: slag fertilizer H, Q: slag fertilizer Q; Data are means ± SD of three replicates; mean values followed by different letters (a, b, c) are significantly different (P≤0.05).

Table 5. Analysis of variance of the effects of slag-based silicon fertilizer (slag) and application rate of Si (Si-R) on Si concentrations in rice organs (SiO$_2$%).

Sources of variation	Df	F values		
		Leaf	Stem	Grain
Slag	1	2.09 ns	21.05**	0.602 ns
Si-R	3	30.93**	18.51**	13.74*
Slag×Si-R	3	1.80 ns	7.56*	0.786 ns

Levels of probability: ns = non significant, significantly different *p≤0.05 and **p≤0.01

300 rpm for 1 h] and analyzed by the colorimetric silicon molybdenum blue method [39]. Slag pH and EC were measured at a water/soil ratio of 2.5.

Plant-available Si content in soil was extracted by 0.25 M citric acid [soil/(citric acid) ratio of 1:5] for 5 hrs, and analyzed by the colorimetric silicon molybdenum blue method [40]. The soil pH was measured at a water/soil ratio of 2.5.

The silicon content in rice plants was determined by the colorimetric silicon molybdenum blue method [41–42]. Briefly, 100 mg of plant tissue was mixed with 3 mL of 50% (w/v) NaOH in a polyethylene tube. These tubes were covered with loose-fitting plastic caps and autoclaved at 125°C for 1 h and analyzed by the colorimetric silicon molybdenum blue method.

Statistical analysis

All data in figures and tables are shown as means ± SD of three replicates. Two -way ANOVA was used for statistical analysis and Fisher's L.S.D. test was adopted to detect the significant difference ($p \leq 0.05$) between the means of different treatments. All statistical analyses were done using the Excel 2007 and SPSS (PASW Statistics 18.0).

Results

Dry weight and silicon concentration of different rice tissues

Table 2–3 show that application of both iron slag (Q) and steel slag (H) fertilizers significantly increased dry weight of leaf and stem, and grain yield compared with the control (Si$_0$) treatment. However, there was no significant difference among different application rates of silicon (except Si$_0$). Dry weight of leaf and stem showed no significant difference between the two Si fertilizers

tested, but Si fertilizer H produced significantly more grain weight than Si fertilizer Q.

Table 4–5 show that the Si concentration was significantly different among different organs, with the order of leaf > stem > grain. Application of both Si fertilizers significantly increased the Si concentration in leaf, stem and grain compared with the Si$_0$ treatment. The Si concentration in rice organs tended to increase with increasing application rate of Si, and there was a significant difference between the Si$_3$ treatment and Si$_1$ treatment. The Si concentration of stem was significantly higher in Si fertilizer H than in Si fertilizer Q. However, no significant difference in leaf or grain Si concentration was noted between the two Si fertilizers used.

Disease severity

Under greenhouse conditions, rice leaves were naturally infected with brown spot caused by *Bipolaris oryzae*. At the anthesis stage, disease severity showed visible differences among treatments. The data in Table 6–7 demonstrate that rice leaf lesion of the control treatment (Si$_0$) was most severe with an incidence of 39.6%, and a disease index of 56.0%. Application of both Si fertilizers significantly decreased brown spot development. Meanwhile, disease severity of fertilizer H treatments was lower than that of fertilizer Q treatments.

Transmission electron microscopic analysis of silica cell

Ultrathin sections of leaf samples were observed by transmission electron microscope (TEM). The ultrastructural details demonstrated that the numbers of fungal cells and fungal colonization in the leaf epidermis were different between Si-untreated and Si-treated rice plants. The leaf mesophyll cells of silicon-untreated

Table 6. Effects of different silicon (Si) treatments on rice brown spot development at the anthesis stage (%).

Fertilizer	Rate	Incidence of disease	Disease index
Control	Si$_0$	39.7±2.11a	56.1±2.60 a
Q	Si$_1$	4.67±0.60 b	22.0±3.50 b
	Si$_2$	1.33±1.53 b	8.64±2.71bc
	Si$_3$	0.33±0.58 b	2.22±3.85 c
H	Si$_1$	3.67±2.08 b	18.0±2.26 bc
	Si$_2$	1.33± 0.58 b	8.89±2.67bc
	Si$_3$	0.00±0.0 b	0.00±0.00 c

Si$_0$: no Si fertilizer; Si$_1$: slag fertilizer applied at a rate of 187 mg plant-available Si per kg soil; Si$_2$: slag fertilizer applied at a rate of 560 mg plant-available Si per kg soil; Si$_3$: slag fertilizer applied at a rate of 935 mg plant-available Si per kg soil; H: slag fertilizer H, Q: slag fertilizer Q; Data are means ± SD of three replicates; mean values followed by different letters (a, b, c) are significantly different (*P*≤0.05).

Table 7. Analysis of variance of the effects of slag-based silicon fertilizer (slag) and application rate of Si (Si-R) on rice brown spot development at the anthesis stage (%).

Sources of variation	Df	F values	
		Incidence of disease	Disease index
Slag	1	0.0145 ns	13.20*
Si-R	3	47.05**	3545**
Slag×Si-R	3	0.00727 ns	5.85 ns

Levels of probability: ns = non significant, significantly different *p≤0.05 and **p≤0.01.

plants were disorganized at the stage of the fungal (*Bipolaris oryzae*) colonization. The cytoplasm was disintegrated with a consequence of chloroplast degradation and cell-wall alterations. Abundant amorphous materials were noticed in a mesophyll cell colonized by the fungus (Figure 1). The Si layers were observed in Si-treated epidermal cell walls, and the thickness of the silicon layer was seen to be increased by Si application (Figure 2). The chloroplast thylakoid lamella of mesophyll cells of Si-untreated leaves became swollen, and stroma lamellae and grana lamellae of chloroplasts were distorted. In contrast, the chloroplast structure of mesophyll cells of Si-treated rice leaves was relatively intact, with thylakoid lamellae stacked in order, grana lamellae accumulated compactly and some starch grains visible (Figure 3).

Scanning electron microscopic analysis of silica cells

Morphology of silica cells on the surface of the top-second leaf at the anthesis stage differed among treatments (Figure 4, 5). There were many silica cells, wart-like protuberances (papillae) and stomata on the leaf surface. The silica cells had a dumbbell shape and were distributed in rows along the leaf veins. However, the morphology and number of these silica cells varied among treatments. Silicon application led to more pronounced cell silicification in rice leaves, more silica cells and larger papillae.

Discussion

In this study, the concentration of plant-available Si in the soil tested was 41.8 mg (Si) kg^{-1}. In China, the critical value for plant-

available Si concentration in acid paddy soil is 44.4–51.4 mg kg^{-1} (Si), below which positive rice responses to silicate fertilizer can be expected [29]. Our results show that silicon fertilizers from steel slag and iron slag both significantly promoted rice growth and rice yield. Silicon fertilizer H produced significantly higher grain weight than silicon fertilizer Q at the same plant-available Si application rate. Two factors may account for this observation. First, the composition and cooling process of slags influence Si-dissolution from slags. Slag H, which was cooled slowly, had higher Si-availability to plants compared with slag Q, which was more rapidly cooled in water. This result was consistent with a report by Takahashi (1981) [43], suggesting that Si availability of slag to plants cannot be precisely determined only by the extraction method using 0.5 M HCl. It is necessary to estimate the Si-releasing process from slags in paddy soils and to analyze the factors affecting the solubility of the slags in future studies. Second, other nutrients provided by slag might be also beneficial for rice growth, such as Ca, Mg, Fe and Mn etc. In this study, the plant-available Si concentration was lower in slag H than in slag Q, thus, at the same available-Si application rate, the real application rate of the slag H was higher than that of slag Q. In this case, the amount of other nutrients such as Ca, Fe and Mn provided by slag H might be higher than that by slag Q because not only the real application rate of slag H was higher than that of slag Q but also the content of Ca, Fe and Mn was higher in slag H than in slag Q (Table 1). It could be supposed that other nutrients provided by slags also contributed to the final rice performance,

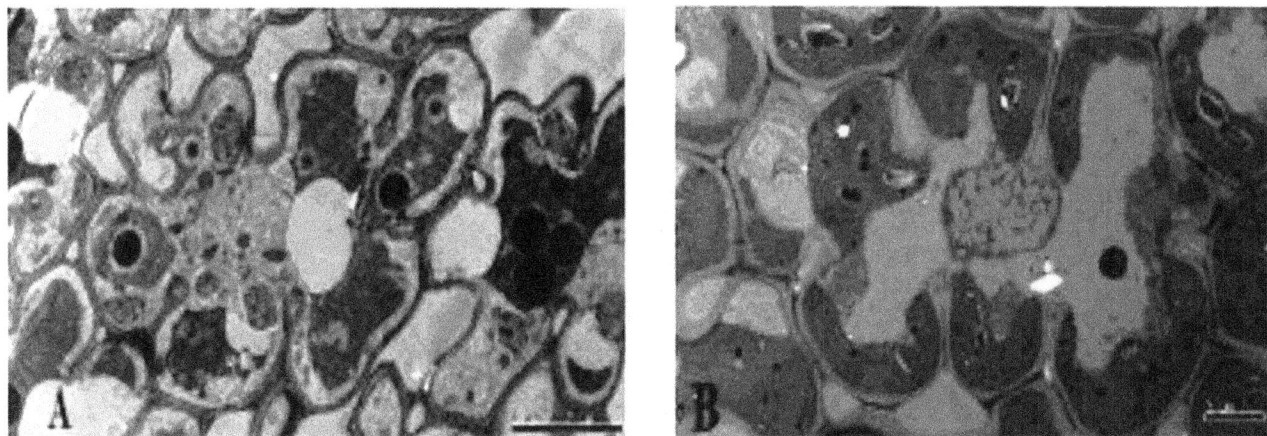

Figure 1. Transmission electron micrographs of mesophyll cells of rice leaves. Scale bars = 5 μm. **A**: Mesophyll cells of a control plant grown without silicon fertilizer at the anthesis stage; **B**: Mesophyll cells of a silicon-treated plant grown with slag (H) applied at a rate of 935 mg plant-available Si per kg soil at the anthesis stage.

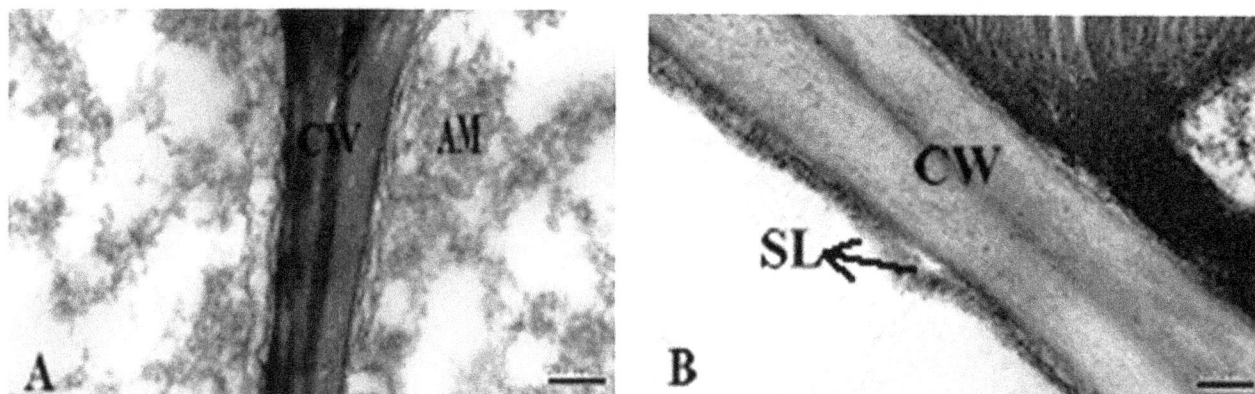

Figure 2. Transmission electron micrographs of cell wall from leaves of rice. CW, cell wall; AM, amorphous material; SL, silicon layer. A: Leaf epidermis of a control plant grown without silicon fertilizer at the anthesis stage; B: Leaf epidermis of a silicon treated plant with slag (H) applied at a rate of 935 mg plant-available Si per kg soil at the anthesis stage.

dry weight and rice yield, which, however, needs further validation.

Silicon fertilizer could be an environmentally-friendly alternative to control rice diseases [14,44–46]. In this study, rice leaves were naturally infected with brown spot disease caused by *Bipolaris oryzae* at the jointing stage. The leaves of rice plants that were not treated with slag showed disease symptoms 5 days earlier than those treated with slag. At anthesis, visible differences in disease severity appeared among treatments. We found that application of both steel slag and iron slag fertilizers showed significantly lower brown spot incidence and severity (Table 6–7). Lesion areas of leaves showed a decreasing, but non-significant trend with increasing Si application rates (Table 4–5). This result was consistent with rice yield (Table 2). The ultrastructural characteristics showed that the chloroplast thylakoid lamellae of mesophyll cells of untreated rice leaves became swollen, and stroma and grana lamellae of chloroplast were distorted at anthesis. However, the chloroplast structure of mesophyll cells of Si-treated leaves was relatively intact (Figure 3).

Brown spot severity has been reported to be negatively correlated with Si concentration in rice tissue [15,24,47]. An active Si uptake by lateral roots of rice plants plays a key role in

rice resistance to brown spot [24]. Application of the two Si fertilizers significantly increased the Si concentration in leaves (Table 2). There have been debates of the mechanisms involved in Si-mediated plant disease resistance. Some authors suggest that a mechanical or physical barrier provided by Si deposition in cell walls contributes to enhanced resistance [38,48–49], while more recent studies suggest that Si plays a biochemical role in mediating plant resistance to pathogens [33,46,50]. Our results show that Si application led to more pronounced cell silicification in rice leaves and more elaborate and larger papillae (Figure 4, 5). The elaborate papillae formed in Si-treated leaf epidermal surface might increase the resistance to fungal penetration [37,51]. The Si layers were observed in Si-treated epidermal cell walls, and their thickness was increased by Si treatment (Figure 2). The Si layers in epidermal cell walls supposedly confer enhanced host resistance to brown spot, which is in line with the previous reports that the cuticular Si double layer developed on rice leaf cells constituted a physical barrier to impede fungal penetration and colonization [38,48–49].

In this study, we also found apparent differences in the number of fungal cells and fungal colonization in the leaf epidermis between Si-untreated and Si-treated plants (Figure 1). We surmise that soluble Si may induce physiological resistance to restrain the

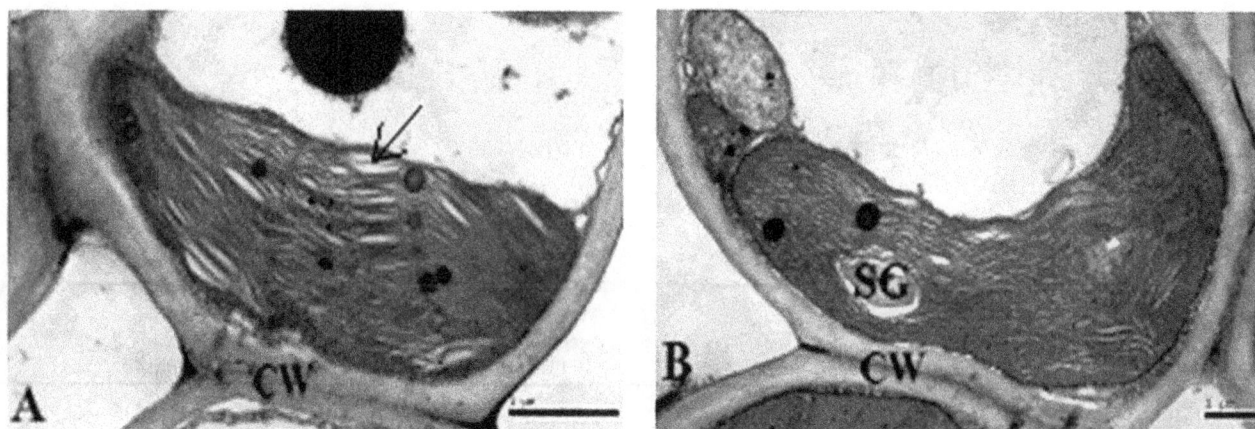

Figure 3. Transmission electron micrographs of chloroplasts from leaves of rice. Scale bars = 1 μm. CW, cell wall; SG, starch grain. **A:** Chloroplast of a control plant grown without silicon fertilizer at the anthesis stage; **B:** Chloroplast of a silicon treated plant grown with slag (H) applied at a rate of 935 mg plant-available Si per kg soil at the anthesis stage.

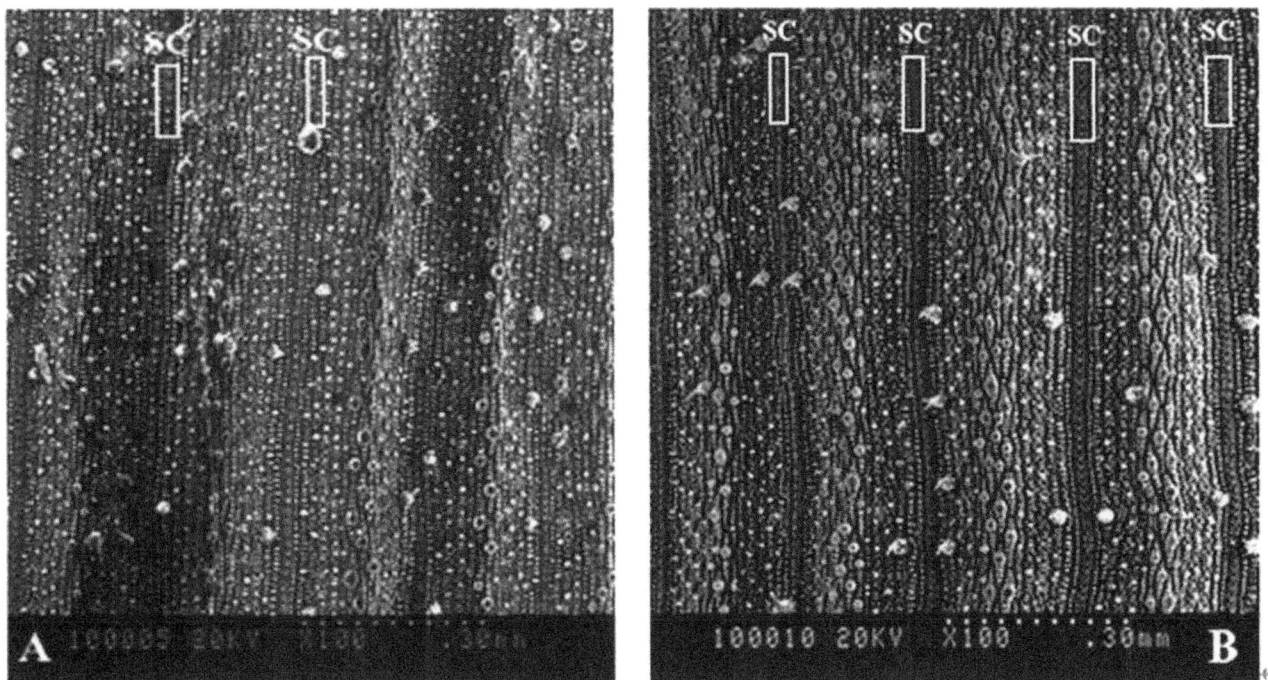

Figure 4. Scanning electron micrographs with 100 magnification of rice leaves. Scale bars = 30 mm. SC, silica cell. **A**: the top second rice leaf epidermis of a control plant without silicon fertilizer at the anthesis stage; **B**: the top second rice leaf epidermis of a silicon-treated rice plant grown with slag (H) applied at a rate of 935 mg plant-available Si per kg soil at the anthesis stage.

growth of *Bipolaris oryzae* and keep host cells relatively intact. Rodrigues et al. (2003a, 2005b) suggested that Si induced accumulation of phenolic compounds or phytoalexins, which played a primary role in rice defense against infection by

Magnaporthe grisea [31,52]. Dallagnol et al. (2011) found that the concentrations of soluble phenolics and lignin and activities of peroxidase and chitinase were higher in Si-treated rice leaves infected by *Bipolaris oryzae*, which contributed to rice resistance to

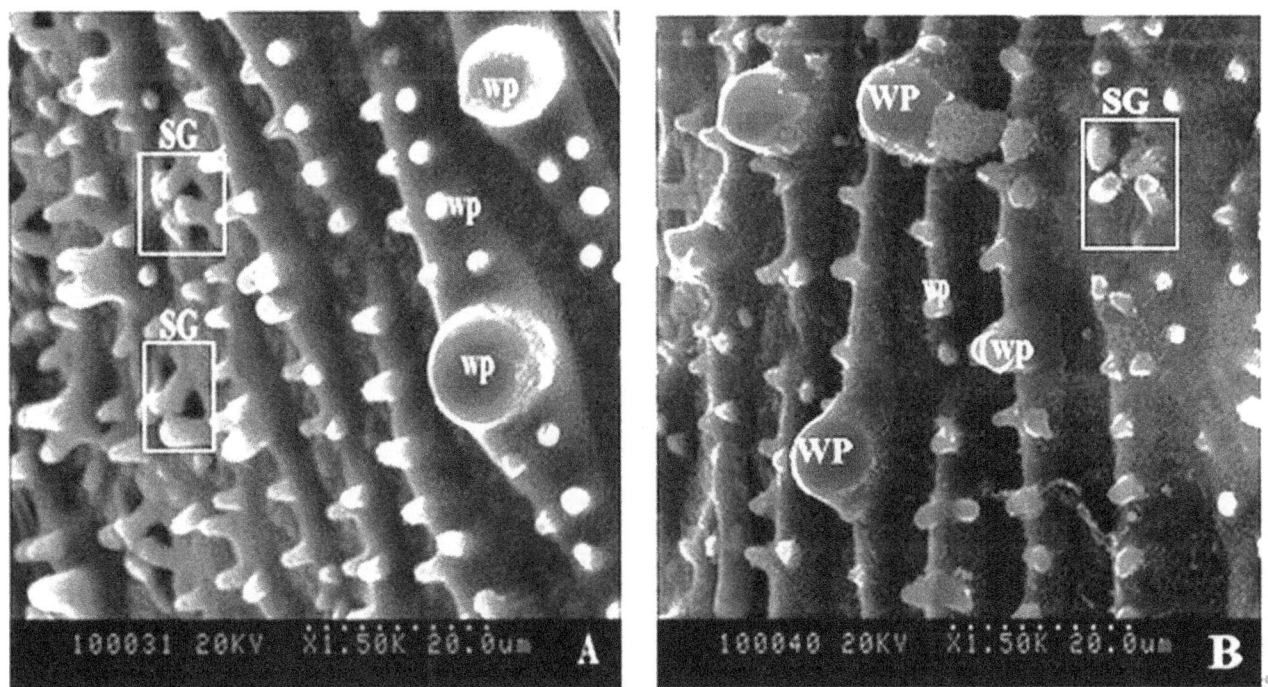

Figure 5. Scanning electron micrographs with 150 K magnification of rice leaves. Scale bars = 20 μm. SC, silica cell; WP, wart-like protuberance; SG, stomatal guard cell. **A**: the top second leaf epidermis of a control plant without silicon fertilizer at the anthesis stage; **B**: The top second leaf epidermis of a silicon-treated rice plant grown with slag (H) applied at a rate of 935 mg plant-available Si per kg soil at the anthesis stage.

brown spot [47]. Other reports suggest that after inoculation with *M. grisea*, Si-treated rice plants significantly increased the activities of pathogenesis-related proteins (PRs) in leaves, such as peroxidase (POD), polyphenol oxidase (PPO), phenylalanine ammonia lyase (PAL), and catalase (CAT) [37,46]. Therefore, we believe that Si-enhanced plant disease resistance plus the role of Si as physical barrier as suggested by Sun et al. (2010) in rice blast resistance [46] also contributed to the Si-enhanced resistance to rice brown spot observed in the present study.

Conclusions

Applying Si fertilizer to Si-deficient paddy soil is necessary for both high rice yield and brown spot resistance. Both steel slag and iron slag are effective in this regard. In this experiment, silicon fertilizer H produced significantly higher grain weight than silicon fertilizer Q at the same plant-available Si application rate. Composition and cooling process of slags influence Si-dissolution from slags. Si availability of slag to plants cannot be precisely determined only by the extraction method using 0.5 M HCl. The immobile silicon deposited in host cell walls and papillae sites is the first physical barrier for fungal (*Bipolaris oryzae*) penetration and soluble Si in the cytoplasm enhances physiological or induced resistance to restrain fungal colonization.

Author Contributions

Conceived and designed the experiments: DFN ALS FLF ZJL YCL. Performed the experiments: DFN. Analyzed the data: DFN YCL. Contributed reagents/materials/analysis tools: YCL. Wrote the paper: DFN YCL.

References

1. Epstein E (1994) The anomaly of silicon in plant biology. Proc Natl Acad Sci USA 91: 11–17.
2. Ma JF, Takahashi E (2002) Soil, Fertilizer, and Plant Silicon Research in Japan. Elsevier, Amsterdam, pp. 1–2.
3. Epstein E (1999) Silicon. Ann Rev Plant Physiol Plant Mol Biol 50: 641–664.
4. Ma JF (2004) Role of silicon in enhancing the resistance of plants to biotic and abiotic stresses. Soil Sci Plant Nutr 50: 11–18.
5. Rodrigues FÁ, Datnoff LE (2005a) Silicon and rice disease management. Fitopatol Bras 30: 457–469.
6. Liang YC, Sun WC, Zhu YG, Christie P (2007) Mechanisms of silicon-mediated alleviation of abiotic stresses in higher plants: A review. Environ Pollut 147: 422–428.
7. Catherine Keller FG, Meunier JD (2012) Benefits of plant silicon for crops: a review. Agron Sust Develop 32: 201–213.
8. Wailes EJ, Cramer GL, Chavez EC, Hansen JM (1997) Arkansas global rice model: international baseline projections for 1997–2010. Arkansas Agric Exp Stat, Arkansas. pp. 1–46.
9. van Nguyen N, Ferrero A (2006) Meeting the challenges of global rice production. Paddy Water Environ 4: 1–9.
10. Ma JF, Yamaji N (2006) Silicon uptake and accumulation in higher plants. Trends Plant Sci 11: 392–397.
11. Raven JA (2003) Cycling silicon–the role of accumulation in plants. New Phytol 158: 419–430.
12. Foy CD (1992) Soil chemical factors limiting plant root growth. Adv Soil Sci 19: 97–149.
13. Winslow MD, Okada K, Correa-Victoria F (1997) Silicon deficiency and the adaptation of tropical rice ecotypes. Plant Soil 188: 239–248.
14. Datnoff LE, Deren CW, Snyder GH (1997) Silicon fertilization for disease management of rice in Florida. Crop Prot 16: 525–531.
15. Deren CW, Datnoff LE, Snyder GH, Martin FG (1994) Silicon concentration, disease response, and yield components of rice genotypes grown on flooded organic histosols. Crop Sci 34: 733–737.
16. Savant NK, Snyder GH, Datnoff LE (1996) Silicon management and Sustainable rice production. Adv Agron 58: 151–199.
17. Alvarez J, Datnoff LE (2001) The economic potential of silicon for integrated management and sustainable rice production. Crop Prot 20: 43–48.
18. Bocharnikova EA, Loginov SV, Matychenkov VV, Storozhenko PA (2010) Silicon fertilizer efficiency. Russ Agric Sci 36: 446–448.
19. Ou SH (1985) Rice diseases, 2nd ed. Kew, Surrey, UK, Commonwealth Mycological Institute.
20. Motlagh MR, Kaviani B (2008) Characterization of new bipolaris spp.: the causal agent of rice brown spot disease in the North of Iran. Int J Agric Biol 10: 638–642.
21. Marchetti MA, Peterson HD (1984) The role of *Bipolaris oryzae* in floral abortion and kernel discoloration in rice. Plant Dis 68: 288–291.
22. Lee TS, Hsu LS, Wang CC, Jeng YH (1981) Amelioration of soil fertility for reducing brown spot incidence in the patty field of Taiwan. J Agric Res Chin 30: 35–49.
23. Datnoff LE, Snyder GH, Raid RN, Jones DB (1991) Effect of calcium silicate on blast and brown spot intensities and yields of rice. Plant Dis 75: 729–732.
24. Dallagnol LJ, Rodrigues FÁ, Mielli MVB, Ma JF, Datnoff LE (2009) Defective active silicon uptake affects some components of rice resistance to brown spot. Phytopathol 99: 116–121.
25. Wu SP, Xue YJ, Ye QS, Chen YC (2007) Utilization of steel slag as aggregates for stone mastic asphalt (SMA) mixtures. Build Environ 42: 2580–2585.
26. Motz H, Geiseler J (2001) Products of steel slags an opportunity to save natural resources. Waste Manage 21: 285–293.
27. Shen HT, Forssberg E (2003) An overview of recovery of metals from slags. Waste Manage 23: 933–949.
28. Zhu GL (2010) The current state and developing of comprehensive disposal of steel and iron slag. Iron Steel Scrap 1: 12–16 (in Chinese).
29. Wang HL, Li CH, Liang YC (2001) Agricultural utilization of silicon in China. In: Silicon in Agriculture. Datnoff LE, Snyder GH, Korndorfer GH. ed. Elsevier, Amsterdam, pp. 343–358.
30. Seebold KW, Kucharek TA, Datnoff LE, Correa Victoria FJ, Marchetti MA (2001) The influence of silicon on components of resistance to blast in susceptible, partially resistant, and resistant cultivars of rice. Phytopathol 91: 63–69.
31. Rodrigues FÁ, Benhamou N, Datnoff LE, Jones JB, Bélanger RR (2003a) Ultrastructural and cytochemical aspects of silicon-mediated rice blast resistance. Phytopathol 93: 535–546.
32. Rodrigues FÁ, Valeb FXR, Korndörfer GH, Prabhud AS, Datnoff LE, et al. (2003b) Influence of silicon on sheath blight of rice in Brazil. Crop Prot 22: 23–29.
33. Rodrigues FÁ, Vale FXR, Datnoff LE, Prabhu AS, Korndörfer GH (2003c) Effect of rice growth stages and silicon on sheath blight development. Phytopathol 93: 256–261.
34. Rodrigues FÁ, McNally DJ, Datnoff LE, Jones JB, Labbé C, et al. (2004) Silicon enhances the accumulation of diterpenoid phytoalexins in rice: A potential mechanism for blast resistance. Phytopathol 94: 177–183.
35. Cha W, Kim J, Choi H (2006) Evaluation of steel slag for organic and inorganic removals in soil aquifer treatment. Water Res 40: 1034–1042.
36. Naoto K, Naoto O (1997) Dissolution of Slag Fertilizers in a Paddy Soil and Si Uptake by Rice Plant. Soil Sci Plant Nutr43: 329–341.
37. Cai KZ, Gao D, Luo SM, Zeng RS, Yang JY, et al. (2008) Physiological and cytological mechanisms of silicon-induced resistance in rice against blast disease. Physiol Plant 134: 324–333.
38. Kim SG, Kim KW, Park EW, Choi D (2002) Silicon-induced cell wall fortification of rice leaves: A possible cellular mechanism of enhanced host resistance to blast. Phytopathol 92: 1095–1103.
39. Buck GB, Korndörfer GH, Datnoff LE (2011) Extractors for estimating plant available silicon from potential silicon fertilizer sources. J Plant Nutr 34: 272–282.
40. Lu RK (2000) Analytical methods for soil and agro-chemistry. Chinese Agricultural Technology, Beijing, pp. 201–203 (in Chinese).
41. Nanayakkara UN, Uddin W, Datnoff LE (2008) Effects of soil type, source of silicon, and rate of silicon source on development of gray leaf spot of perennial ryegrass turf. Plant Dis 92: 870–877.
42. Dai WM, Zhang KQ, Duan BW, Sun CX, Zheng KL, et al. (2005) Rapid determination of silicon content in rice (Oryza sativa). Chinese J Rice Sci 19: 460–462 (in Chinese).
43. Takahashi K (1981). Effects of slags on the growth and the silicon uptake by rice plants and the available silicates in paddy soils. Bulletin Shikoku Agric Exp Stat 38: 75–114.
44. Liang YC, Ma TS, Li FJ, Feng YJ (1994) Silicon availability and response of rice and wheat to silicon in calcareous soils. Commun Soil Sci Plant Anal 25(13&14): 2285–2297.
45. Seebold KW, Datnoff LE, Correa-Victoria FJ, Kucharek TA, Snyder GH (2004) Effects of silicon and fungicides on the control of leaf and neck blast in upland rice. Plant Dis 88: 253–258.
46. Sun WC, Zhang J, Fan QH, Xue GF, Li ZL, et al. (2010) Silicon-enhanced resistance to rice blast is attributed to silicon-mediated defence resistance and its role as physical barrier. Eur J Plant Pathol 128: 39–49.
47. Dallagnol LJ, Rodrigues FÁ, DaMatta FM, Mielli M B, Pereira SC (2011) Deficiency in silicon uptake affects cytological, physiological, and biochemical events in the rice–*Bipolaris oryzae* interaction. Phytopathol 101:92–104.
48. Yoshida S (1965) Chemical aspects of the role of silicon in physiology of the rice plant. Bull Shikoku Agr Exp Stat 15: 1–58.
49. Hayasaka T, Fujii H, Ishiguro K (2008) The role of silicon in preventing appressorial penetration by the rice blast fungus. Phytopathol 98: 1038–1044.

50. Liang YC, Sun WC, Si J, Römheld V (2005) Effect of foliar- and root-applied silicon on the enhancement of induced resistance in *Cucumis sativus* to powdery mildew. Plant Pathol 54: 678–685.

51. Zhang GL, Gen DQ, Zhang HC (2006) Silicon Application enhances resistance to sheath blight (*rhizoctoniasolani*) in rice. J Plant Physiol Mol 32: 600–606 (in Chinese).

52. Rodrigues FÁ, Jurick WM, Datnoff LE, Jones JB, Rollins JA (2005b) Silicon influences cytological and molecular events in compatible rice-*Magnaporthe grisea* interactions. Physiol Mol Plant Pathol 66: 144–159.

5

Extraction and Sensitive Detection of Toxins A and B from the Human Pathogen *Clostridium difficile* in 40 Seconds Using Microwave-Accelerated Metal-Enhanced Fluorescence

Lovleen Tina Joshi[1], Buddha L. Mali[2], Chris D. Geddes[2], Les Baillie[1]*

1 Cardiff School of Pharmacy and Pharmaceutical Sciences, Cardiff University, Cardiff, United Kingdom, 2 Institute of Fluorescence, University of Maryland, Baltimore County, Baltimore, Maryland, United States of America

Abstract

Clostridium difficile is the primary cause of antibiotic associated diarrhea in humans and is a significant cause of morbidity and mortality. Thus the rapid and accurate identification of this pathogen in clinical samples, such as feces, is a key step in reducing the devastating impact of this disease. The bacterium produces two toxins, A and B, which are thought to be responsible for the majority of the pathology associated with the disease, although the relative contribution of each is currently a subject of debate. For this reason we have developed a rapid detection assay based on microwave-accelerated metal-enhanced fluorescence which is capable of detecting the presence of 10 bacteria in unprocessed human feces within 40 seconds. These promising results suggest that this prototype biosensor has the potential to be developed into a rapid, point of care, real time diagnostic assay for *C. difficile*.

Editor: Markus M. Heimesaat, Charité, Campus Benjamin Franklin, Germany

Funding: This work was supported by: 1. Cardiff University Travel grant to fund travel and work of Lovleen Tina Joshi at the Institute of Fluorescence, Baltimore, USA; 2. Mid-Atlantic Regional Center of Excellence for Biodefence & Emerging Infectious diseases for funding; and 3. NIH/NIAII for funding: grant 2 US4AO57168-09. The funders had no role in study design, data collection and analysis, decision to publish or preparation of the manuscript.

Competing Interests: The authors have read the journal's policy and the authors of this manuscript have the following competing interests: The probes which the authors have defined within the document for toxins A and B of *Clostridium difficile* are now undergoing the process of intellectual property protection by Cardiff University (University College Cardiff Consultants Limited). The authors have filed two World Intellectual Property Organisation patents to protect these DNA oligonucleotides and these have been published as: WO2013167876 A1, WO2013167877 A1. These are publicly available and the Inventors are: Leslie William James Baillie (a co-author of this manuscript) and Lovleen Tina Joshi (First Author of this manuscript).

* Email: BaillieL@cardiff.ac.uk

Introduction

Clostridium difficile is a spore forming, toxin-producing bacterium which is currently the principal cause of healthcare associated diarrhea in the western world. In the USA in 2007 there were 284 875 infections while in the UK during the same period the pathogen was linked to the death of 8 324 individuals [1,2]. The bacterium currently presents a considerable challenge to healthcare professionals and has stimulated researchers to develop improved diagnostics and medical countermeasures.

In healthy individuals the spore form of the bacterium is carried in the gut with no ill effects. Subsequent disruption of the gut flora, usually by broad spectrum antibiotics, enables the bacterium to proliferate and release two toxins, a 308 kDa enterotoxin (toxin A) and a 270 kDa cytotoxin (toxin B) which are responsible for symptoms ranging from mild diarrhea to fatal colitis [3].

The genes encoding the production of these toxins, *tcdA* and *tcdB*, are located within a 19.6 kb pathogenicity locus (PaLoc) which is present in all toxin producing strains [4–6]. The relative contribution of each toxin to pathogenicity is currently a subject of

debate and thus the ability to detect the presence of both would greatly enhance diagnostic capability [7,8].

The current gold standard for *C. difficile* diagnosis is the cell cytotoxin neutralization assay (CCTA) which can only detect toxin B, has 99–100% specificity and 85–100% sensitivity, and takes between 48–72 hours to generate a result [3,9]. To reduce the time for detection, other methods have been developed such as enzyme immunoassays (EIAs) which detect toxins A and B, and glutamate dehydrogenase [9,10]. These assays are easy to perform and give rapid results (15–20 minutes); however, they currently suffer from low sensitivity [11–14].

DNA methods based on the Real Time Polymerase Chain Reaction (RT PCR) have been developed to detect the genes associated with toxin production and the current commercially available systems, the ProGastro assay (Cepheid Smart Cycler), GeneXpert (Cepheid), and the BD Gene Ohm assay, all target toxin B [15,16]. Although these assays are highly sensitive they are labor intensive and costly due to the need to include a purification step to remove inhibitory biological materials [11,16–18].

A new platform technology called microwave-accelerated metal-enhanced fluorescence (MAMEF) has recently been developed which eliminates the need for extensive sample purification. The approach has been successfully employed to detect a range of human pathogens including *Bacillus anthracis, Salmonella typhimurium* and *Chlamydia trachomatis* [19–22]. The approach combines two technologies, metal enhanced fluorescence (MEF) to optically amplify fluorescence signatures and low power microwave heating to accelerate the reaction kinetics of the assay. Combined together they yield a system with a level of sensitivity comparable to RT PCR, but with a much shorter turnaround time of ~60 seconds, and at a significantly lower cost.

The underlying principal of the technology is based on the selective heating of water, whereby the aqueous medium is heated via microwave power. As the metal is not heated, a temperature gradient between the cold metal surface and the warm aqueous solution is formed, which facilitates mass transport to the surface and allows DNA, or other biomarkers, to be recognized when the assay is complete and the fluorophore is in close proximity to the metal [22]. Non-radiative energy transfer occurs between fluorophores and plasmon electrons in a non-continuous film. The metal surface itself radiates the coupled fluorescence quanta [23]. As a consequence the assay does not require the prior removal of organic material such as blood and feces and is thus able to directly detect the presence of a bacterial target in clinical samples.

A system capable of detecting the presence of *C. difficile* spores in patient feces within 5 minutes would represent a considerable improvement over current capabilities. Such a system could potentially impact on the quality of clinical care as early treatment would limit disease progression and reduce cross transmission to other patients.

In this paper we describe our efforts to develop a MAMEF based system capable of detecting *C. difficile* spores in human feces. Using DNA probes specific to toxin A and B we were able to detect a few as 10 spores in 500 μl of unprocessed human feces within 40 seconds, suggesting that this approach has the potential to be developed into a rapid point of care, real time diagnostic assay.

Materials and Methods

Bacterial Strains, growth conditions and genomic DNA

C. difficile strain CD630 was obtained for culture from the NCTC, Public Health England, UK. BHI agar and broth (Oxoid; Remel Inc, KS, USA) supplemented with 0.1% of the bile salt sodium taurocholate (Sigma Aldrich, USA) was used for routine culture of *C. difficile* at 37°C in a 3.4 L anaerobic jar (Oxoid, UK) with an anaerobic gas generating kit (Oxoid, UK). Phosphate buffered saline (PBS; Sigma Aldrich, USA) constituting 0.01 M phosphate buffer, 0.0027 M potassium chloride and 0.137 M sodium chloride was made by dissolving 1 tablet in 200 ml diH$_2$O. Human feces was provided by a healthy volunteer. Genomic DNA (gDNA) from *C. difficile* strain CD630 (5 μg) was obtained from ATCC, USA as a control.

Genomic DNA Extraction from *C. difficile* using Chelex 100

Genomic DNA was extracted from *C. difficile* and other species (Table S1; Table S2) as described previously [24].

C. difficile toxin A and B Probe Design

Probes were designed to recognize nucleotide sequences within conserved regions of toxins A and B. As part of the design process we incorporated features which would enable us to utilize the

probes in a future MAMEF assay. The anchor probe for the *C. difficile* assay was designed to be 17 nucleotides in length and to be separated from the 22 nucleotide fluorescent detector probe by a stretch of 5 nucleotides. The anchor probe binds target DNA while the detector probe subsequently binds at a distance which positions the fluorophore optimally for biomolecular recognition to occur [22]. The use of anchor and detector probes also allows for two levels of sensitivity within the assay for each toxin (Table 1).

Designed probes were tested against a representative collection of 58 *C. difficile* isolates; including blood culture and variant strains (only produced toxin B; ribotypes 017, 047, 110), obtained from the National Anaerobic Reference Unit, Cardiff, Wales, courtesy of Dr. Val Hall.

Preparation of DIG-labeled DNA Hybridization Probes

PCR was performed using *Taq* DNA polymerase core kit (Qiagen, Crawley, United Kingdom). Hybridization probes were labeled with the DIG (Digoxigenin) PCR dNTP labeling mix (Roche Diagnostics, UK) in 20 μl reactions as described previously [25]. PCR conditions are as described previously (Figure S3; Table S2) [25]. PCR products were analyzed via gel electrophoresis on a 1% agarose gel at 85 V (Biorad Mini sub cell GT, UK).

Dot blot hybridization and macro-arraying gDNA onto a positively charged nylon membrane

For dot blot hybridization, gDNA was macroarrayed onto a positively charged nylon membrane using dot blot apparatus (Figure S3; Flexys robotic workstation, Genomic Solutions Ltd, UK). Dot blot hybridization was performed as described previously [26,27].

Dot blots of Human Metagenomic gut DNA & Ethics Statements

The human gut environment contains approximately 10^{12}/g bacteria [28]. Therefore the probes designed to detect *C. difficile* must be able to specifically detect the toxin genes amongst the numerous bacteria present. To further confirm the specificity of the designed probes, metagenomic DNA samples from ten human volunteers were obtained from Cardiff School of Biosciences, Wales, Cardiff, UK, courtesy of Dr. Julian Marchesi. The metagenome was extracted from feces from humans in Zanzibar, Cote d'Ivoire and the UK [29,30]. Original sample collection in the UK was approved by Cardiff University Biosciences Local Ethics Committee as healthy samples with informed consent. Original sample collection from Cote d'Ivoire was approved by the ethical review boards of the ETH Zurich, Switzerland (2006–23), the University of Basel (EKBB), Switzerland (224/06), and the Ministry of Health in Côte d'Ivoire (5782/MSHP/CAB/CNESVS/06) [29]. The original sample collection from Zanzibar was approved by the institutional research commission of the Swiss Tropical and Public Health Institute (Basel, Switzerland). Ethical clearance was obtained from the Ethics Committee of the Ministry of Health and Social Welfare (MoHSW) in Zanzibar (application number 16) [30].

Deposition of gold triangles onto glass substrates to microwave and lyse *C. difficile*

To focus the microwave power in the microwave, gold lysing triangles were used, provided by the Institute of Fluorescence (Figure 1). Deposition and preparation as described previously [22].

Table 1. Table of Oligonucleotides.

	Toxin A	Toxin B
Anchor Probe	Thiol-TTTTT-TTTAATACTAACACTGC	Thiol-TTTTTT-CAAGACTCTATTATAG
Capture Probe	Alexa-488-TGTTGCAGTTACTGGATGGCAA	Alexa-594-TAAGTGCAAATCAATATGAAG
Synthetic oligonucleotides	AAATTATGATTGTGACGTAATCCCAATACAACGTCAATGACCTACCGTT	AGTTCTGAGATAATATCTAATCCCAATATTCACGTTTAGTTATACTTG

Anchor and capture probes, and synthetic oligonucleotides target regions used in the MAMEF assay for toxins A and B are shown, along with the corresponding fluorophore used.

The release of *C. difficile* DNA from spores using microwave irradiation

C. difficile (1×10^5 cfu/ml) was suspended into PBS buffer and 500 µl added into the gold tie lysing chamber and exposed to a 15 s microwave pulse at 80% power in a GE microwave Model No. JE2160BF01, kW 1.65 (M/W). The microwaved solution was examined for the presence of viable organisms. Further experiments involved spiking human feces diluted with PBS buffer with varying concentrations of *C. difficile* and then irradiating. Each experiment was conducted in triplicate.

Bacterial quantification after focused microwave irradiation

Control samples (1.33×10^7 cfu/ml) were drop counted as described previously [31,32]. Before DNA release, samples were heated at 80°C for 10 min to ensure removal of any vegetative cells. After lysis the numbers of remaining viable spores (cfu/ml) were enumerated.

Gel electrophoresis of samples

Post microwave irradiation the sample was centrifuged at 3000 g for 15 min (Heraeus Primo R; Fisher Scientific, USA) and the supernatant resuspended in 1:2 ethanol. This supernatant was resuspended in 100 µl sterile deionized water. Samples were analyzed via gel electrophoresis on a 2% agarose gel at 70 V (Biorad Mini sub cell GT, UK).

Formation of Silver Island Films (SiFs) on glass substrates

Silver island Films (SiFs) were prepared at an OD_{450} of 0.4–0.5 on Silane-prep glass slides (Sigma Aldrich, USA; Figure 1; Figure 2) to enable silver adhesion, as described previously [33,34].

Anchor and fluorescent probes and target DNA

Probes specific for conserved regions of toxin A and toxin B of *C. difficile* were designed and validated (Table 1; Figures 3, 4). Probes were modified to enable incorporation into the MAMEF

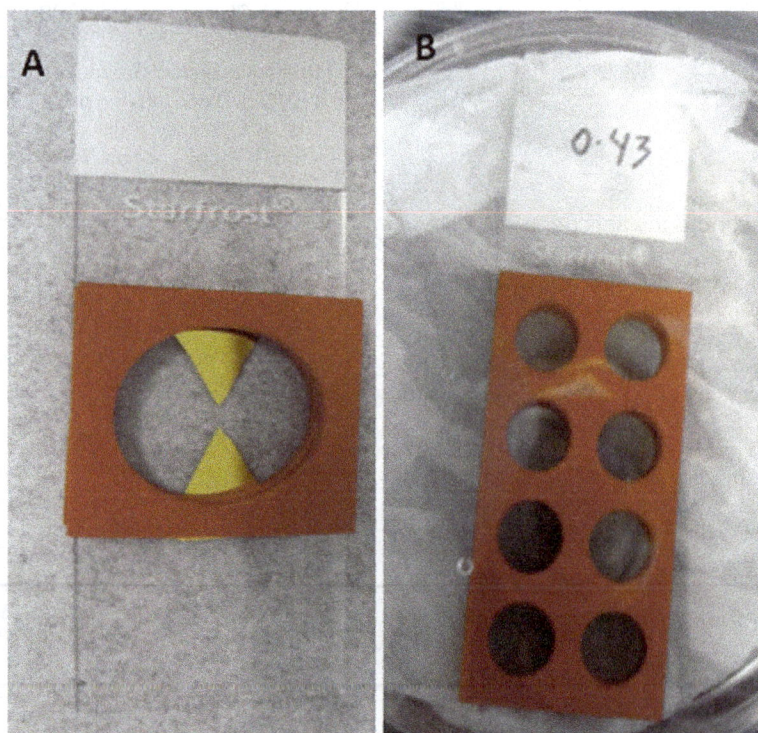

Figure 1. Gold lysing triangle on Starfrost slide and a typical Silver Island Film (SiFs). (A) Gold triangle with silicone isolators added to create lysing chamber. (B) The SiF here is at an OD_{450} of 0.43 and has multi-well silicone isolators added. Each well can hold an individual DNA assay reaction; hence multiple repeats can be conducted.

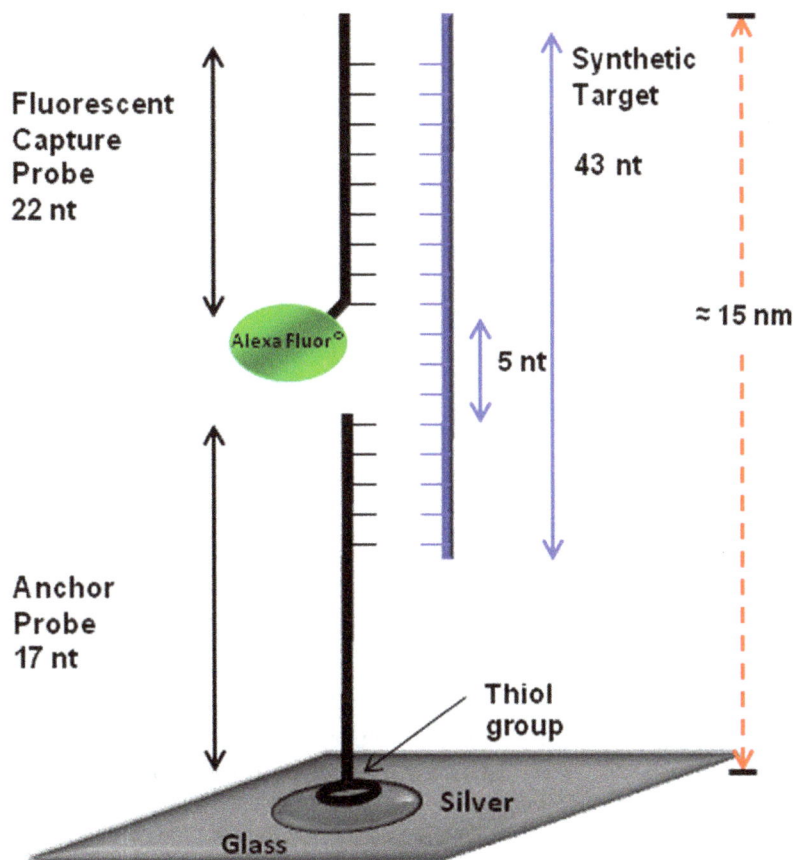

Figure 2. Schematic configuration of 3 piece DNA detection assay. This configuration was used for detection of both toxin A and toxin B probes. The anchor probe (17 oligonucleotides) was anchored to the SiFs by addition of the thiol group. The fluorescent probe (22 oligonucleotides) was attached to an Alexa at the 3′ end. Upon hybridization with target DNA, the 3 piece assay is formed, and the fluorophore labeled probe is plasmon enhanced.

detection platform (Figure 2); a thiol group was added to the 5′ region of the anchor probe to enable binding of the DNA to the surface of the SiF while the capture probe was labelled with an Alexa (Invitrogen, USA) fluorophore [20,21]. The Alexa fluorophore 488 (green) was used to label toxin A and the Alexa fluorophore 594 (red) was used to label toxin B. These fluorophores were chosen due to their wavelength separation and thus yield two different fluorescence emissions upon excitation. The common regions in the anchor probes were preceded by 5 consecutive thymine bases, included to increase the flexibility of the probe once bound to the silver surface, and enhanced by the inclusion of a C at the 3′ terminal to which a thiol group was subsequently added [20]. The negative strand of the target region was also synthesised to bind to the capture and anchor probes as a synthetic control sample (Figure S1; Figure S2; Invitrogen, USA).

Preparation of MAMEF assay platform for detection of C. difficile DNA

Glass slides with SiFs deposited were coated with self-adhesive silicone isolators as described previously [21]. The thiolated anchor probe was then covalently bound to the SiF in a self assembled monolayer chemistry [35]. This was achieved by diluting 40 μM anchor probe into 100 μl of 1 M Tris- EDTA buffer and adding 9 μl to 250 mM of 20 μl dithiothreitol. The mixture was incubated at room temperature for 60 min. The

anchor probe (1 μM) was diluted into 4 ml Tris- EDTA buffer and 100 μl of the anchor probe was added to and incubated in each well of the SiF for 75 min. After incubation the anchor was removed and 50 μl of 1 μM Alexa-Fluor probe was added to 50 μl of DNA from the target organism. The SiFs containing the bound probes and DNA were then incubated by heating for 25 s in a microwave cavity at 20% power, GE microwave Model No. JE2160BF01, 1.65 kiloWatts. Only in the presence of target DNA the three piece assay is complete and enhanced-fluorescence can be observed (Figure 2). The sample was removed from the well after MAMEF and the well washed with 100 μl TE buffer 3 times. Each experiment was conducted in triplicate.

Detection and fluorescence spectroscopy

The presence of target DNA was confirmed by the generation of a fluorescent signal following excitation with laser light. Fluorescence was emitted by the DNA MAMEF capture assay and measured using a Fiber Optic Spectrometer (HD2000) (Ocean Optics, Inc) by collecting the emission intensity (I).

Statistical Analysis

Statistical analysis was performed using GraphPad Prism version 5.04 for Windows, (GraphPad Software, La Jolla California USA, www.graphpad.com). Three replicates were performed for each detection experiment. Statistical significant

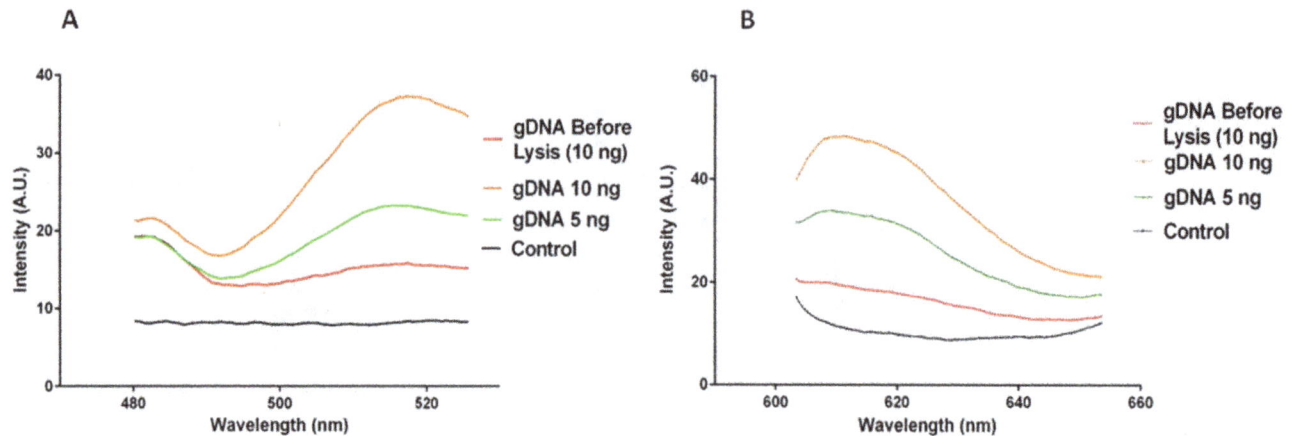

Figure 3. Detection of target DNA within various concentrations of gDNA by MAMEF. Various concentrations of *C. difficile* gDNA (strain CD630; ATCC, USA) were tested in the MAMEF platform. Concentrations of 10 ng and 5 ng were microwave irradiated for 8 seconds at 70% power. gDNA was subjected to disruption to increase the chances of the detection probes accessing the toxin specific sequences. The data presented is the result from a single reproduced assay. (A) The fluorescent signal intensities generated from toxin A detection within microwaved gDNA (excitation at 495 nm, emission at 519 nm). (B) The fluorescent signal intensities generated from toxin B detection within gDNA (excitation at 590 nm, emission at 617 nm). The control was PBS buffer.

differences were tested for using one way analysis of variance (ANOVA) at the 95% confidence interval in conjunction with a Kruskall-Wallis test. Dunnett's post test was employed to compare sample data to control data. A P value of <0.05 was considered significant [36].

Results

Specificity of DNA probes for Toxin A and Toxin B detection as determined via dot blot

DNA probes were designed to recognize sequences within conserved regions of the gene sequences of toxin A and B and were configured to take account of the requirements of the MAMEF

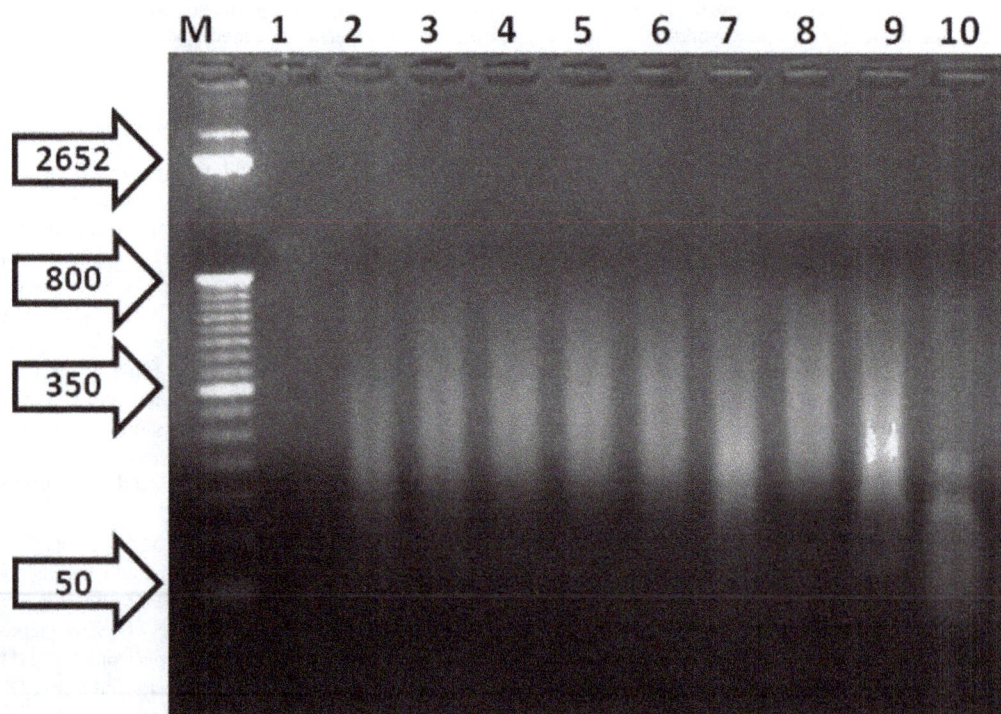

Figure 4. Electrophoresis gel of DNA released from *C. difficile* spores and vegetative cells following different periods of microwave treatment. *C. difficile* spores at a concentration of 1×10^5 cfu/ml, in combination with gold triangles, were exposed to 80% microwave power for periods ranging from 8 to 16 seconds (Lanes 2–10). Lane M shows the 1 kilobase molecular weight DNA marker ladder. Lane 1 shows control non-microwaved bacteria.

assay. Genomic DNA from 58 clinical isolates of *C. difficile*, representing a diverse range of ribotypes, was macroarrayed as shown in Figure S3. As expected each strain which contained a copy of the toxin A and B gene sequences gave a positive signal, the strength of which varied between isolates, likely due to an artifact of the experimental procedure. To confirm the specificity of the probes, variant isolates of *C. difficile* lacking toxin A (ribotypes 017; 047: $tcdA^-tcdB^+$) or toxin B (Toxinotype XIa; XIb; DS1684: $tcdA^-tcdB^-$) gene sequences were included in the panel. The DNA from each of these isolates was not recognized by the relevant probe.

To further confirm the specificity of the probes genomic DNA from other bacterial species, both close and distant relatives were subject to hybridization analysis (Figure S4; Table S2). The probes did not bind to bacterial species unrelated to *C. difficile*, further indicating that the probes were highly specific to toxins A and B of *C. difficile*. Species of the *Clostridium* genus, including species closely related to *C. difficile* (which possess toxins closely related to *tcdB*) did not show any probe hybridization which further validates probe specificity. We further confirm the specificity of these probes by demonstrating that they were unable to recognise metagenomic DNA from the human gut flora of ten healthy human volunteers (July 2011). Based on these results we concluded that the probes were highly specific and thus were suitable for inclusion in a MAMEF- based detection assay.

Detection of toxin A and B sequences in genomic DNA isolated from *C. difficile* using the MAMEF assay

After confirming the specificity of the toxin A and B probes, their ability to bind to toxin gene sequences in the context of the MAMEF assay was assessed. gDNA isolated from a known toxin producing strain of *C. difficile* (CD630) was suspended in PBS buffer and subjected to microwave fragmentation for 8 seconds at 70% total microwave power [GE microwave Model No. - JE2160BF01, kW 1.65 (M/W)]. DNA was subjected to disruption to increase the chances of the detection probes accessing the toxin specific sequences. Untreated gDNA in PBS was included as a control. Both sets of toxin specific probes were able to detect 10 ng of intact and disrupted gDNA (Figure 3A & B). The intensity of the signals generated against gDNA at emissions wavelengths of 519 nm (toxin A) and 617 nm (toxin B) were similar to those seen for synthetic oligonucleotide targets (Figures S1; S2) suggesting

that microwave treatment is an efficient means of generating probe target DNA.

Release of DNA from *C. difficile* spore preparations using focused microwave irradiation

Bacterial gDNA is normally sequestered within the body of the host organism, making it difficult for detection probes to gain access. The process is more complicated with bacteria such as *B. anthracis* and *C. difficile* which have the ability to form spores. Thus we employed a microwave based approach, previously developed to release DNA from *B. anthracis* spores, to break open the spore and vegetative form of *C. difficile* to release target DNA [20]. This approach employs gold triangles with a 1 mm gap to focus microwaves which rapidly heat the spores leading to their physical disruption. Thus 500 μl aliquots comprising 1×10^5 cfu/ml of *C. difficile* spores suspended in PBS were subjected to microwave radiation for periods between 2 and 16 seconds, and then analyzed for DNA release via gel electrophoresis. Exposure to microwaves for 15 seconds at 80% total microwave power resulted in optimal release of DNA, with the majority of DNA disrupted to produce fragments between 50–200 bp in size (Figure 4). In contrast, irradiation for 16 seconds resulted in further fragmentation of DNA and yielding fragments too small to be detected by the toxin probes.

Bacterial quantification following focused microwave irradiation

To confirm that microwave irradiation had resulted in disruption of bacterial structure and that the DNA detected did not simply represent free-floating extracellular DNA, the effect of microwave treatment on bacterial viability was determined. Exposure to microwaves for 15 seconds resulted in a 99% reduction in viable spore numbers from 1.33×10^7 cfu/ml to 3.67×10^3 cfu/ml.

Detection of target DNA in spores of *C. difficile* suspended in PBS

Spores of *C. difficile* CD630 were suspended in PBS and microwaved for 15 seconds at 80% power. Following treatment, the cell suspension was diluted to give a range of bacterial concentrations (Figure 5A & B). At the highest spore concentra-

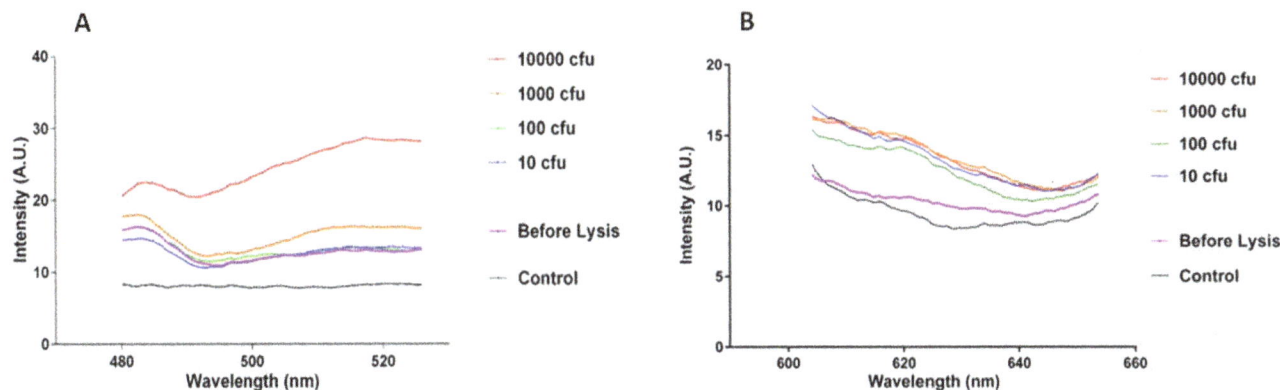

Figure 5. Detection of DNA from microwaved *C. difficile* spore preparation in PBS using MAMEF. Spores of *C. difficile* from strain CD630 at a range of concentrations 10, 100, 1000 and 10 000 cfu were suspended in PBS and subjected to microwave irradiation and screened for the presence of toxin A and toxin B using MAMEF. (A) Various fluorescent signal intensities generated from detection of toxin A within gDNA from irradiated spores (excitation at 495 nm, emission at 519 nm). (B) The various fluorescent signal intensities generated from detection of toxin B within gDNA from irradiated spores (excitation at 590 nm, emission at 617 nm).

A

B

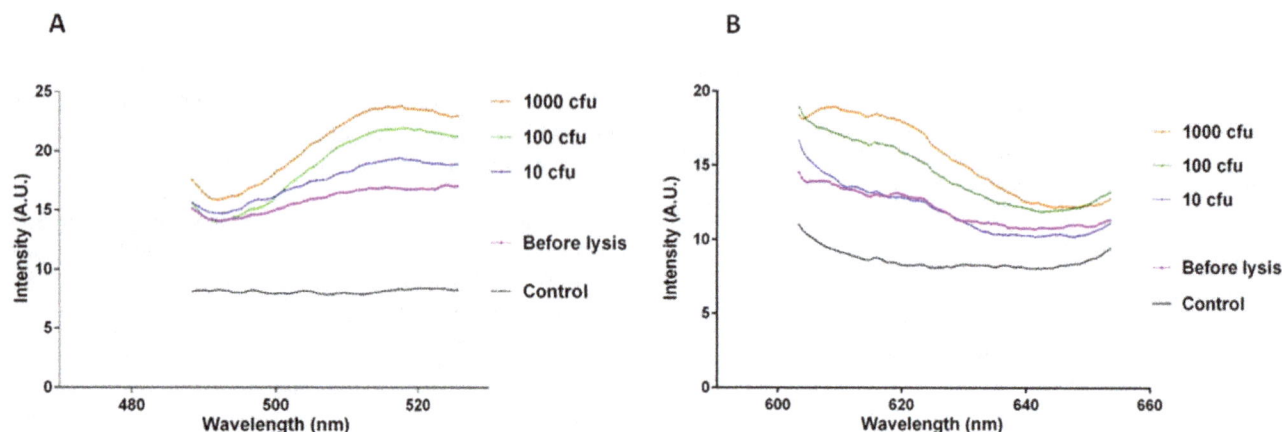

Figure 6. Detection of DNA from lysed *C. difficile* in feces using MAMEF. Spores of *C. difficile* from strain CD630 at a range of concentrations 10, 100, and 1000 cfu were suspended in human feces and subjected to microwave irradiation, then screened for the presence of toxin A and toxin B using MAMEF. (A) The various fluorescent signal intensities generated from detection of toxin A within gDNA from irradiated spores of *C. difficile* in human feces (excitation at 495 nm, emission at 519 nm). (B) Graph demonstrating the fluorescent signal intensities generated from detection of toxin B within gDNA from irradiated spores of *C. difficile* in human feces (excitation at 590 nm, emission at 617 nm).

tion tested, 10000 cfu, the toxin A and B probes gave signals of 15 AU (519 nm) and 15 AU (617 nm) respectively. The lowest concentration of spores tested, 10 cfu, also gave detectable signals with both probes- (toxin A 12 AU, toxin B 10 AU).

Detection of spores of *C. difficile* in human feces

Spores of *C. difficile* CD630 were suspended in human feces and microwaved for 15 seconds at 80% power. Prior to treatment, the spore suspension was diluted to give a range of bacterial concentrations (Figure 6A & B). At 1000 cfu, the toxin A and B probes gave signals of 23 AU (519 nm) and 19 AU (617 nm) respectively. The lowest concentration of spores tested, 10 cfu, also gave detectable signals with both probes- 15 AU (519 nm) and 13 AU (617 nm).

Discussion

In this communication we describe the development of a bench top assay based on MAMEF which is capable of detecting as few as 10 bacteria of *C. difficile* in a fecal sample within 40 seconds. To put this into context, during infection patients excrete between 10 000 to 10 000 000 spores per gram of feces [37].

Thus, in addition to detecting active infection, the assay could also be used to identify patients who carry small numbers of *C. difficile* in their gut but are asymptomatic. The *C. difficile* asymptomatic carriage rate in humans varies between 4–20% and treatment of these individuals with broad spectrum antibiotics could precipitate an active infection [38]. Therefore information concerning the presence or absence of the bacterium in a patient's gut would assist the clinician in determining the most appropriate treatment option for that individual.

The level of sensitivity and speed of this prototype assay compares favorably with all of the currently available *C. difficile* detection methods [39]. The sensitivity of assays within the *C. difficile* detection market range from 90–100% for the cell cytotoxin, EIAs and PCR assays; however, to increase the sensitivity and accuracy of diagnosis these assays are employed in algorithms [40].

The ability of the assay to directly detect the pathogen in the presence of fecal material is a major advantage over other DNA based methods. Current PCR based assays can take between 2–

8 hrs to generate a result; due in part to the need to remove enzyme inhibitors in feces which can prevent the amplification of target DNA [9]. In contrast, the MAMEF assay does not employ an enzymic amplification step and thus can operate in the presence of organic material such as blood and feces [22,41]. This is due to the low power microwave acceleration which increases the rate of DNA hybridization within the biological sample via metal enhanced fluorescence; whereby the SiF amplifies the fluorescence of the labeled reporter DNA [42].

A further advantage of this prototype assay is its ability to detect the genes encoding both toxins A and B. The contribution of both toxins to the pathology of the infection is currently a subject of debate [8,43]. Until relatively recently it was thought that only toxin B was essential for virulence. As a consequence commercial PCR based methods such as BD Gene Ohm and Xpert Cepheid are configured to detect only toxin B [44]. Given this uncertainty, the ability to detect both toxins would provide an obvious advantage until the relative importance of each toxin has been clarified.

To realize the potential of our promising technology, future studies will focus on optimizing the MAMEF platform. This will involve the development of an integrated microwave and laser platform miniaturization, reduced power signature and a simple readout would need to be incorporated into a hand held device capable of being used at the bedside. Studies are already in progress to multiplex the assay and thus detect the presence of both toxins A and B in a single sample well.

We believe our prototype *C. difficile* detection assay has the potential to be developed into a real-time assay capable of identifying infected and colonized individuals within 40 seconds. Such an assay would fulfill the pressing need for a rapid and sensitive detection system capable of detecting both toxin A and B. In the future we believe this assay could be used as a means of screening patients upon admission to hospital to enable appropriate treatment regimens and clinical management decisions to be made.

Supporting Information

Figure S1 Detection of various concentrations of synthetic oligonucleotide in TE buffer by MAMEF. The concentration of anchor probe attached to the silver island film surface for toxin A was 10 nM, whereas for toxin B the anchor concentration was 100 nM. Thus there is variation in the signal intensity generated from the toxin A and toxin B probes. A range of synthetic oligonucleotide concentrations from 1 nM to 100 000 nM were used to determine the ability of the probes to fluoresce in response to excitation via laser light. The fluorescent intensity data presented in the above graphs is the result from a single reproduced assay (A) Graph demonstrating the various fluorescent signal intensities of a range of toxin A synthetic oligonucleotides bound to the toxin A anchor and detector probes. (B) Real color photographs of the fluorescent signal produced at each concentration from the sample wells. The laser light used to excite the fluorescent toxin A probe was at a wavelength of 495 nm excitation which produced an emission at 519 nm. (C) Graph demonstrating the various fluorescent signal intensities of a range of toxin B synthetic oligonucleotides bound to the toxin B anchor and detector probes. (D) Real color photographs of the fluorescent signal produced at each concentration from the sample wells. The laser light used to excite the fluorescent toxin B probe was at a wavelength of 590 nm excitation which produced an emission at 617 nm.

Figure S2   Detection of various concentrations of synthetic oligonucleotides in feces by MAMEF. Human feces was diluted by 50% in PBS and mixed with the synthetic target concentrations and tested in the MAMEF platform. The concentration of anchor probe attached to the silver island film surface for toxin A and toxin B was 10 nM. A range of synthetic oligonucleotide concentrations from 1 nM to 1000 nM were used to determine the ability of the probes to fluoresce in response to excitation via laser light. The fluorescent intensity data presented in the above graphs is the result from a single assay which was repeated three times. (A) Graph demonstrating the various fluorescent signal intensities of a range of toxin A synthetic oligonucleotides bound to the toxin A anchor and detector probes in the presence of human feces. (B) Graph demonstrating the various fluorescent signal intensities of a range of toxin B synthetic oligonucleotides bound to the toxin B anchor and detector probes in the presence of human feces.

Figure S3   Dot blot of hybridization of *C. difficile* gDNA. Genomic DNA from the panel of 58 *C. difficile* isolates tested were macroarrayed as shown (3A+B) and tested against our DIG-labeled probes. To confirm probe specificity, variant isolates of *C. difficile* ($tcdA^- tcdB^+$) lacking either the toxin A (ribotypes 017; 047: $tcdA^- tcdB^+$) or toxin B (Toxinotype XIa; XIb; DS1684: $tcdA^- tcdB^-$) gene sequences were included. DNA from these isolates did not bind to the probes. (C) $tcdA76$ anchor probe (D) $tcdA76$ detector probe, (E) $tcdB$ anchor probe (F) $tcdB$ detector

probe vs. *C. difficile* isolates. Bromophenol blue+ lambda phage DNA was added to the first well orientate the membrane.

Figure S4   Dot blot of related and unrelated bacterial species by DNA hybridization. Species related and unrelated to *C. difficile* were tested against the probes for specificity. There was no hybridization of the probes to the gDNA on the membrane. Positive control for each probe was CD630 and a variant strain control R22680 ($tcdA^- tcdB^+$) was also included. (A) $tcdA50$ anchor probe, (B) $tcdA50$ detector probe, (C) $tcdA76$ anchor probe, (D) $tcdA76$ detector probe, (E) $tcdB$ anchor probe, (F) $tcdB$ detector probe.

Table S1   Table of isolates used in this study. The isolates of *C. difficile* are listed. Panel also includes 21 isolates as described previously [30]. Isolates from blood culture are listed as (B/C). Toxin production for each strain and its PCR ribotype are shown, with additional information including the source. The isolates and information above was provided courtesy of Dr. Jon Brazier and Dr. Val Hall at the Anaerobic Reference Unit, University Hospital Wales, Cardiff, UK, 2008.

Table S2   Additional bacterial species used in this study. The additional species and their strain designations used in this study are listed above. The species related to, and those not related to, *C. difficile* are shown. The species were obtained from the NTCC (National Type Culture Collection, HPA, London, UK), unless otherwise stated. Other isolates obtained from the anaerobic reference unit (ARU) at the University Hospital Wales (UHW) are listed as ARU, UHW. Those isolates and information was provided courtesy of Dr. Jon Brazier and Dr. Val Hall at the Anaerobic Reference Unit, University Hospital Wales, Cardiff, UK.

Table S3   PCR Thermocycle annealing temperatures per probe. The probes for each toxin were found to have the above optimal temperatures for PCR to occur. These are the temperature at which all further PCR reactions and dot blot reactions were conducted.

Acknowledgments

Authors would like to thank Dr. J. Marchesi, Cardiff University School of Biosciences, for supplying the metagenomic human gut DNA samples.

Author Contributions

Conceived and designed the experiments: LTJ CDG LB. Performed the experiments: LTJ. Analyzed the data: LTJ LB. Contributed reagents/materials/analysis tools: LTJ BLM CDG LB. Contributed to the writing of the manuscript: LTJ CDG LB.

References

1. Office of National Statistics Website. Available: http://www.statistics.gov.uk/hub/index.html. Accessed 2011 Aug 3.
2. Kuntz JL, Johnson ES, Raebel MA, Petrik AF, Yang X, et al. (2012) *Clostridium difficile* Infection, Colorado and the Northwestern United States, 2007. Emerging Infect Dis 18(6): 960.
3. Poutanen SM, Simor AE (2004) *Clostridium difficile*-associated diarrhea in adults. Can Med Assoc J 171: 51–58.
4. Cohen SH, Tang YJ, Silva JR (2000) Analysis of the pathogenicity locus in *Clostridium difficile* strains. J Infect Dis 181: 659–663.
5. Spigaglia P, Mastrantonio P (2002) Molecular analysis of the pathogenicity locus and polymorphism in the putative negative regulator of toxin production (TcdC) among *Clostridium difficile* clinical isolates. J Clin Microbiol 40(9): 3470–3475.
6. Rupnik M, Wilcox MH, Gerding DN (2009) *Clostridium difficile* infection: new developments in epidemiology and pathogenesis. Nat Rev Microbiol 7: 526–536.
7. Bongaerts GPA, Lyerly DM (1994) Role of toxins A and B in the pathogenesis of *Clostridium difficile* disease. Microb Pathog 17: 1–12.
8. Kuehne SA, Cartman ST, Heap JT, Kelly ML, Cockayne A, et al. (2010) The role of toxin A and toxin B in *Clostridium difficile* infection. Nature 467: 711–713.
9. Goldenberg SD, Cliff PR, Smith S, Milner M, French GL (2009) Two-step glutamate dehydrogenase antigen real-time polymerase chain reaction assay for detection of toxigenic *Clostridium difficile*. Journal of Hospital Infection 74(1): 48–5.

10. Arnold A, Pope C, Bray S, Riley P, Breathnach A, et al. (2010) Prospective assessment of two-stage testing for *Clostridium difficile*. Journal of Hospital Infection 76: 18–22.

11. Belanger SD, Boissinot M, Clairoux N, Picard FJ, Bergeron MG (2003) Rapid detection of *Clostridium difficile* in feces by Real-Time PCR. J Clin Microbiol 41: 730–734.

12. Planche T, Achaizu A, Holliman R, Riley P, Poloniecki J, et al. (2008) Diagnosis of *Clostridium difficile* infection by toxin detection kits: a systematic review. Lancet Infect Dis 8: 777–784.

13. Tenover FC, Novak-Weekley S, Woods CW, Peterson LR, Davis T, et al. (2010) Impact of strain type on detection of toxigenic *Clostridium difficile*: comparison of molecular diagnostic and enzyme immunoassay approaches. J Clin Microbiol 48: 3719–3724.

14. Carman RJ, Wickham KN, Chen L, Lawrence AM, Boone JH, et al. (2012) Glutamate dehydrogenase (GDH) is highly conserved among *Clostridium difficile* ribotypes. J Clin Microbiol 50(4): 1425–1426.

15. Doing KM, Hintz MS, Keefe C, Horne S, Levasseur S, et al. (2010) Reevaluation of the Premier *Clostridium difficile* toxin A and B immunoassay with comparison to glutamate dehydrogenase common antigen testing evaluating Bartels cytotoxin and Prodesse ProGastro Cd polymerase chain reaction as confirmatory procedures. Diagn Microbiol Infect Dis 66: 129–134.

16. Knetsch CW, Bakker D, De Boer RF, Sanders I, Hofs S, et al. (2011) Comparison of real-time PCR techniques to cytotoxigenic culture methods for diagnosing *Clostridium difficile* infection. J Clin Microbiol 49: 227

17. Rinttila T, Kassinen A, Malinen E, Krogius L, Palva A (2004) Development of an extensive set of 16S rDNA-targeted primers for quantification of pathogenic and indigenous bacteria in faecal samples by real-time PCR. J Appl Microbiol 97: 1166–1177.

18. Eastwood K, Else P, Charlett A, Wilcox M (2009) Comparison of nine commercially available *Clostridium difficile* toxin detection assays, a real-time PCR assay for *C. difficile tcd*B, and glutamate dehydrogenase detection assay to cytotoxin testing and cytotoxigenic culture methods. J Clin Microbiol 47: 3211.

19. Aslan K, Zhang Y, Hibbs S, Baillie L, Previte MJR, et al. (2007) Microwave-accelerated metal-enhanced fluorescence: application to detection of genomic and exosporium anthrax DNA in <30 seconds. The Analyst 132: 1130–1138.

20. Aslan K, Previte MJ, Zhang Y, Gallagher T, Baillie L, et al. (2008) Extraction and detection of DNA from *Bacillus anthracis* spores and the vegetative cells within 1 min. Anal Chem 80(11): 4125–4132.

21. Zhang Y, Agreda P, Kelley S, Gaydos C, Geddes CD (2011) Development of a microwave-accelerated metal-enhanced fluorescence 40 second, <100 cfu/mL point of care assay for the detection of *Chlamydia trachomatis*. Biomedical Engineering, IEEE Transactions 1-1.

22. Tennant SM, Zhang Y, Galen JE, Geddes CD, Levine MM (2011) Ultra-Fast and sensitive detection of non-typhoidal *Salmonella* using Microwave-Accelerated Metal-Enhanced Fluorescence (MAMEF). PLoS ONE 6: e18700.

23. Geddes CD, Lakowicz JR (2002) Editorial: Metal-enhanced fluorescence. J Fluoresc 12: 121–129.

24. Stubbs SLJ, Brazier JS, O'Neill GL, Duerden BI (1999) PCR targeted to the 16S-23S rRNA gene intergenic spacer region of *Clostridium difficile* and construction of a library consisting of 116 different PCR ribotypes. J Clin Microbiol 37:461

25. Mahenthiralingam E, Bischof J, Byrne SK, Radomski C, Davies JE, et al. (2000) DNA-based diagnostic approaches for identification of *Burkholderia cepacia* complex, *Burkholderia vietnamiensis*, *Burkholderia multivorans*, *Burkholderia stabilis*, and *Burkholderia cepacia* genomovars I and III. J Clin Microbiol 38(9): 3165–3173.

26. Mahenthiralingam E, Simpson DA, Speert DP (1997) Identification and characterization of a novel DNA marker associated with epidemic *Burkholderia cepacia* strains recovered from patients with cystic fibrosis. J Clin Microbiol 35(4): 808–816.

27. Drevinek P, Baldwin A, Lindenburg L, Joshi LT, Marchbank A, et al. (2010) Oxidative stress of *Burkholderia cenocepacia* induces insertion sequence-mediated genomic rearrangements that interfere with macrorestriction-based genotyping. J Clin Microbiol 48(1):34–40.

28. Shoemaker NB, Vlamakis H, Hayes K, Salyers AA (2001) Evidence for extensive resistance gene transfer among *Bacteroides* spp. and among *Bacteroides* and other genera in the human colon. Appl Environ Microbiol 67: 561.

29. Knopp S, Mohammed KA, Stothard JR, Khamis IS, Rollinson D, et al. (2010) Patterns and risk factors of helminthiasis and anemia in a rural and a peri-urban community in Zanzibar, in the context of helminth control programs. PLoS Negl Trop Dis 4(5): 681.

30. Rohner F, Zimmermann MB, Amon JR, Vounatsou P, Tschannen AB, et al. (2010) In a randomized controlled trial of iron fortification, anthelmintic treatment, and intermittent preventive treatment of malaria for anemia control in Ivorian children, only anthelmintic treatment shows modest benefit. The Journal of Nutrition 140, 3: 635–641.

31. Miles AA, Misra SS, Irwin JO (1938) The estimation of the bactericidal power of the blood. J Hyg 732–749.

32. Joshi LT, Phillips DS, Williams CF, Alyousef A, Baillie L (2012) Contribution of Spores to the Ability of *Clostridium difficile* To Adhere to Surfaces. Appl Environ Microbiol 78(21): 7671–7679.

33. Aslan K, Leonenko Z, Lakowicz JR, Geddes CD (2005) Annealed silver-island films for applications in metal-enhanced fluorescence: Interpretation in terms of radiating plasmons. J Fluoresc 15: 643–654.

34. Aslan K, Gryczynski I, Malicka J, Matveeva E, Lakowicz JR, et al. (2005) Metal-enhanced fluorescence: an emerging tool in biotechnology. Curr Opin Biotechnol 16: 55–62.

35. Michota A, Kudelski A, Bukowska J (2001) Influence of electrolytes on the structure of cysteamine monolayer on silver studied by surface-enhanced Raman scattering. J Raman Spectrosc 32: 345–350.

36. Bowker DW, Randerson PF (2007) Practical Data Analysis Workbook for Biosciences (3rdEdn). Essex. Pearson Education Ltd. 299 p

37. Mulligan ME, Rolfe RD, Finegold SM, George WL (1979) Contamination of a hospital environment by *Clostridium difficile*. Curr Microbiol 3: 173–175.

38. Riggs MM, Sethi AK, Zabarsky TF, Eckstein EC, Jump RLP, et al. (2007) Asymptomatic carriers are a potential source for transmission of epidemic and nonepidemic *Clostridium difficile* strains among long-term care facility residents. Clin Infect Dis 45: 992–998

39. Planche T, Wilcox M (2011) Reference assays for *Clostridium difficile* infection: one or two gold standards? J Clin Pathol 64: 1–5.

40. Novak-Weekely SM, Marlowe EM, Miller JM, Cumpio J, Nomura JH, et al. (2010) *Clostridium difficile* testing in the clinical laboratory by use of multiple testing algorithms. J Clin Microbiol 48: 889.

41. Aslan K, Baillie L, Geddes CD (2010) Ultra fast and sensitive detection of biological threat agents using microwaves, nanoparticles and luminescence. Journal of Medical Chemical, Biological and Radiological Defense 8: 1–21

42. Melendez JH, Huppert JS, Jett-Goheen M, Hesse EA, Quinn N, et al. (2013) Blind Evaluation of the Microwave-Accelerated Metal-Enhanced Fluorescence Ultrarapid and Sensitive Chlamydia trachomatis Test by Use of Clinical Samples. J Clin Microbiol 51(9): 2913–2920.

43. Lyras D, O'Connor JR, Howarth PM, Sambol SP, Carter GP, et al. (2009) Toxin B is essential for virulence of *Clostridium difficile*. Nature 458: 1176–1179.

44. Deshpande A, Pasupuleti V, Rolston DD, Jain A, Deshpande N, et al. (2011) Diagnostic accuracy of real-time polymerase chain reaction in detection of *Clostridium difficile* in the stool samples of patients with suspected *Clostridium difficile* infection: a meta-analysis. Clin Infect Dis 53(7): 81–90.

6

Effects of Low-Level Deuterium Enrichment on Bacterial Growth

Xueshu Xie, Roman A. Zubarev*

Division of Physiological Chemistry I, Department of Medical Biochemistry and Biophysics, Karolinska Institutet, Stockholm, Sweden

Abstract

Using very precise ($\pm 0.05\%$) measurements of the growth parameters for bacteria *E. coli* grown on minimal media, we aimed to determine the lowest deuterium concentration at which the adverse effects that are prominent at higher enrichments start to become noticeable. Such a threshold was found at 0.5% D, a surprisingly high value, while the ultralow deuterium concentrations ($\leq 0.25\%$ D) showed signs of the opposite trend. Bacterial adaptation for 400 generations in isotopically different environment confirmed preference for ultralow ($\leq 0.25\%$ D) enrichment. This effect appears to be similar to those described in sporadic but multiple earlier reports. Possible explanations include hormesis and isotopic resonance phenomena, with the latter explanation being favored.

Editor: Vipul Bansal, RMIT University, Australia

Funding: This work was supported by the Swedish Research Council (Grant # 2011-3726). The funder had no role in study design, data collection and analysis, decision to publish, or preparation of the manuscript.

Competing Interests: The authors have declared that no competing interests exist.

* Email: Roman.Zubarev@ki.se

Introduction

Since the discovery of D_2O (heavy water) in 1932 by Urey, Brickwede and Murphy [1], its biological effects have attracted a great deal of researchers' interest. Already early experiments have revealed that deuterium has profound effect on living organisms. Between 1934 and the beginning of the second World war in 1939, a total of 216 publications appeared dealing with biological effects of deuterium [2]. Excess of deuterium in water was found to cause reduction in synthesis of proteins and nucleic acids, disturbance in cell division mechanism, changes in enzymatic kinetic rates and cellular morphological changes [3,4]. Yet it is possible to grow microorganisms (e.g. some variants of *Escherichia coli*) in a highly substituted medium, and achieve almost complete deuterium substitution [5,6].

The biological effects of stable isotopes are usually observed shortly after the microorganism is placed in an isotopically different medium [7]. At first, prokaryotic cells experiences an "isotopic shock" manifested through the growth arrest and morphology changes. After a period of adaptation ("lag phase"), growth resumes, but the rate is usually slower than in normal isotopic environment [8]. The changes in the growth rate can be explained by the impact of isotopic substitutions on the kinetics of enzymes [9], pattern of hydrogen bonds and similar relatively subtle but cumulatively potentially important effects.

The Katz group who have studied multiple heavy-isotope (^{13}C, ^{15}N and ^{18}O) substitutions in *Chlorella vulgaris* grown in heavy water found that all additional isotopic substitutions resulted into abnormal effects in cell size, appearance, growth rate and division [10]. The effects were progressively stronger as the isotopic composition deviated from the normal. The authors concluded that "*organisms of different isotopic compositions are actually different organisms, to the degree that their isotopic compositions are removed from naturally occurring compositions*" [10].

The question is whether this conclusion made for heavy enrichments levels (>50%) remains true for low levels of enrichment (<10%). Since isotopic compositions of microorganisms grown in a slightly enriched environment showed deficit of heavy isotopes [11], one can reasonably assume that even low levels of deuterium enrichment may cause biological effects. Indeed, the earliest works on biological effect of deuterium reported in the 1930s were performed using low levels of deuterium, as high enrichment was still not available. Barnes *et al.* studied the physiological effect of low deuterium concentration on the growth of *Spirogyra*, flatworms and *Euglena*. They found an increased growth of these three organisms at 0.06% of D in water compared to ordinary water that contains 156 ppm D (0.0156%). More specifically, they observed increased cell division for *Euglena*, longer longevity for flatworms, and both less cell disjunction and greater longevity for *Spirogyra* [12–15]. Lockemann and Leunig's study on the effect of heavy water with less than 0.54% D upon *E. coli* and *Pseudomonasa* revealed that concentrations as low as 0.04% D favored growth [16]. Curry *et al.* observed ca. 10% faster growth for *Aspergillus* at 0.05% D, but the result was not statically significant due to large errors of measurements [17]. After the first half of 1930s, the research focus has shifted to the effects of highly enriched deuterium, which were more pronounced and largely negative. The interested to low deuterium enrichment has returned in the 1970s and 1980s, when Lobyshev *et al.* studied the Na, K-ATPase activity at different concentration of deuterium and found it to increase at low deuterium concentrations, with a maximum reached at 0.04–0.05% [18,19]. Lobyshev *et al.* then performed experiments with regeneration of hydrioid pohyps *Obelia geniculata* in a wide range of deuterium added to sea water and found faster regeneration at and below 0.1% D [20]. Somlyai *et al.* have shown that 0.06% D in tissue culture activated the growth of L_{929} fibroblast cell lines [21]. Nikitin *et al.* have studied growth of bacteria with different membrane lipid composition in liquid media in a range of deuterium concentration in water varied from 0.01% to 90% and observed pronounced activation in growth for *Methylobacterium organopholium* and *Hyphomonas jannaschiane* at around 0.01% enrichment [22]. Recent studies on the impact of D_2O on the life span of *Drosophila melanogaster* revealed the biggest positive effect of deuterium at the lowest D concentration tested in that work, 7.5% of enrichment [23].

In the current study, we probed enrichment levels starting from 0.03%, i.e. double the normal deuterium abundance of 156 ppm. Disregarding the earliest reports from 1930s on the biological effects of 0.06% deuterium in water which suffered from the lack of statistical analysis and proper controls, we expected the size of the biological effects to be commensurable with the enrichment levels, i.e. to be exquisitely small. To address the expected minute size of the phenomena, we developed a very sensitive method for detecting the biological effects of unspecific deuterium incorporation in the model organism. *E. coli* was chosen as such because of the ease of handling, robustness and speed of growth as well as the ability to thrive on minimal media with easily changeable isotopic composition.

The applied method utilizes robotized sample preparation, automated and massively parallel data acquisition (hundreds of individual experiments per concentration point) and measures three independent parameters per growth curve: maximum growth rate, the lag phase duration and maximum density (Figure 1). Massive parallel measurement approach is achieved using the BioScreen C automated fermentor that affords up to 200 experiments run at the same time. Every second experimental well in a 100-well plate was kept at standard isotopic conditions (Figure S1), and the data from each "test" well was normalized by that of the neighboring standard well, providing the accuracy of relative measurements close to those achieved with internal standard. Using massive statistics, we achieved the precision of relative measurements close to 0.05% (standard error).

Materials and Methods

Chemicals and Materials

Glycerol stock of *E. coli* BL 21 strain (stored at $-80°C$) was obtained from the microbiology lab of our department. The M9 minimal media [24] was prepared using D-glucose ($C_6H_{12}O_6$), disodium hydrogen phosphate ($Na_2HPO_4.2H_2O$), monopotassium phosphate (KH_2PO_4), sodium chloride (NaCl), magnesium sulfate ($MgSO_4$), calcium chloride ($CaCl_2$), ammonia chloride (NH_4Cl), heavy water (99.9% of ^2H), all purchased from Sigma-Aldrich (Schnelldorf, Germany), and distilled water prepared with a Milli-Q device from Millipore (Billerica, MA, USA). Heavy water (99.9% of ^2H) was also purchased from Sigma-Aldrich (Schnelldorf, Germany).

Vacuum filtration system with 0.2 μm polyethersulfone (PES) membrane for bacteria media sterilization was purchased from VWR (Stockholm, Sweden). Petri dishes (90×14 mm) and inoculating sterile loops were purchased from Sigma-Aldrich. Sterile plastic conical tubes (50 mL and 15 mL) for sample preparation were purchased from Sarstedt (Nümbrecht, Germany). The BioScreen C automatic fermentor was obtained from Oy Growth Curves AB Ltd (Helsinki, Finland).

M9 minimal media preparation

5-time concentrated M9 minimal salts stock solution was prepared by dissolving 42.5 g $Na_2HPO_4.2H_2O$, 15 g KH_2PO_4 and 2.5 g NaCl in Milli-Q water to a final volume of 1000 mL. The solution was then sterilized by autoclaving and stored at 4°C for further use. To prepare M9 minimal media, the salts stock solution was diluted five times in Milli-Q water.

M9 minimal media were prepared by mixing the following components: 800 mL Milli-Q water, 200 mL M9 concentrated salts stock solution, 2 mL of 1 M $MgSO_4$ solution, 0.1 mL of 1 M $CaCl_2$ solution, 5 g D-glucose ($C_6H_{12}O_6$) and 1 g NH_4Cl.

Preparation of streak agar plates

M9 minimal media agar plates were prepared by dissolving 3 g of agar powder in 200 mL M9 minimal media. The obtained mixture was sterilized by autoclaving, then cooled down to ca. 60°C and finally poured into Petri dishes (ca. 15 mL agar solution per plate). The agar plates were allowed to solidify at room temperature for ca. 10 min, sealed with parafilm and stored at 4°C till further use.

E. coli streak agar plates were prepared by streaking [25] *E. coli* from $-80°C$ glycerol stock onto M9 minimal media agar plate followed by 40-hour incubation at 37°C to form visible isolated colonies. Streak agar plates were stored for experiments for maximum one week at 4°C.

Measurement of *E. coli* growth

From the *E. coli* agar plate, one isolated colony was picked with a sterile loop into 5 mL M9 minimal media and incubated at 37°C while shaking with 200 r.p.m for 5–6 hours until it reached its early exponential phase with optical density (OD_{590}) around 0.2, measured with Colorimeter WPA CO75 (York, UK).

Sample preparation workflow is shown in Figure 2. In each experiment, four stock solutions were used. Stock A for preparing sample S_A, and stock B for preparing sample S_B were obtained by mixing M9 minimal media with sterilized heavy water at a certain ratio (Table 1). For the preparation of stock solutions of standard A and standard B, M9 minimal media were mixed with sterile Milli-Q water at the same ratio as stock A and stock B. The final solutions were dispensed into the honeycomb well plates using

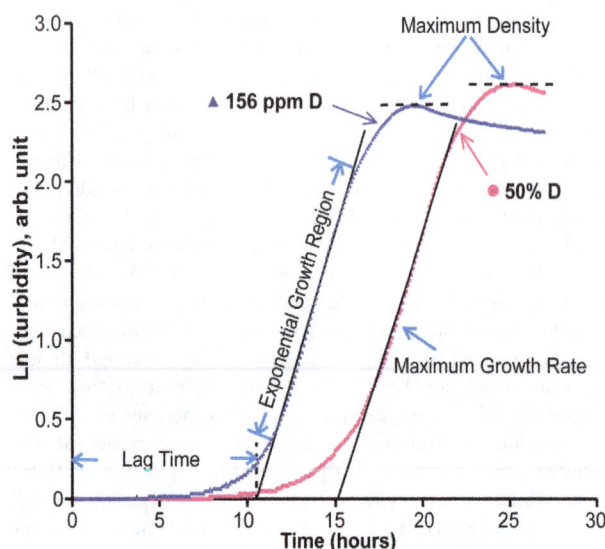

Figure 1. Typical growth curve and the three growth parameters derived from the curve.

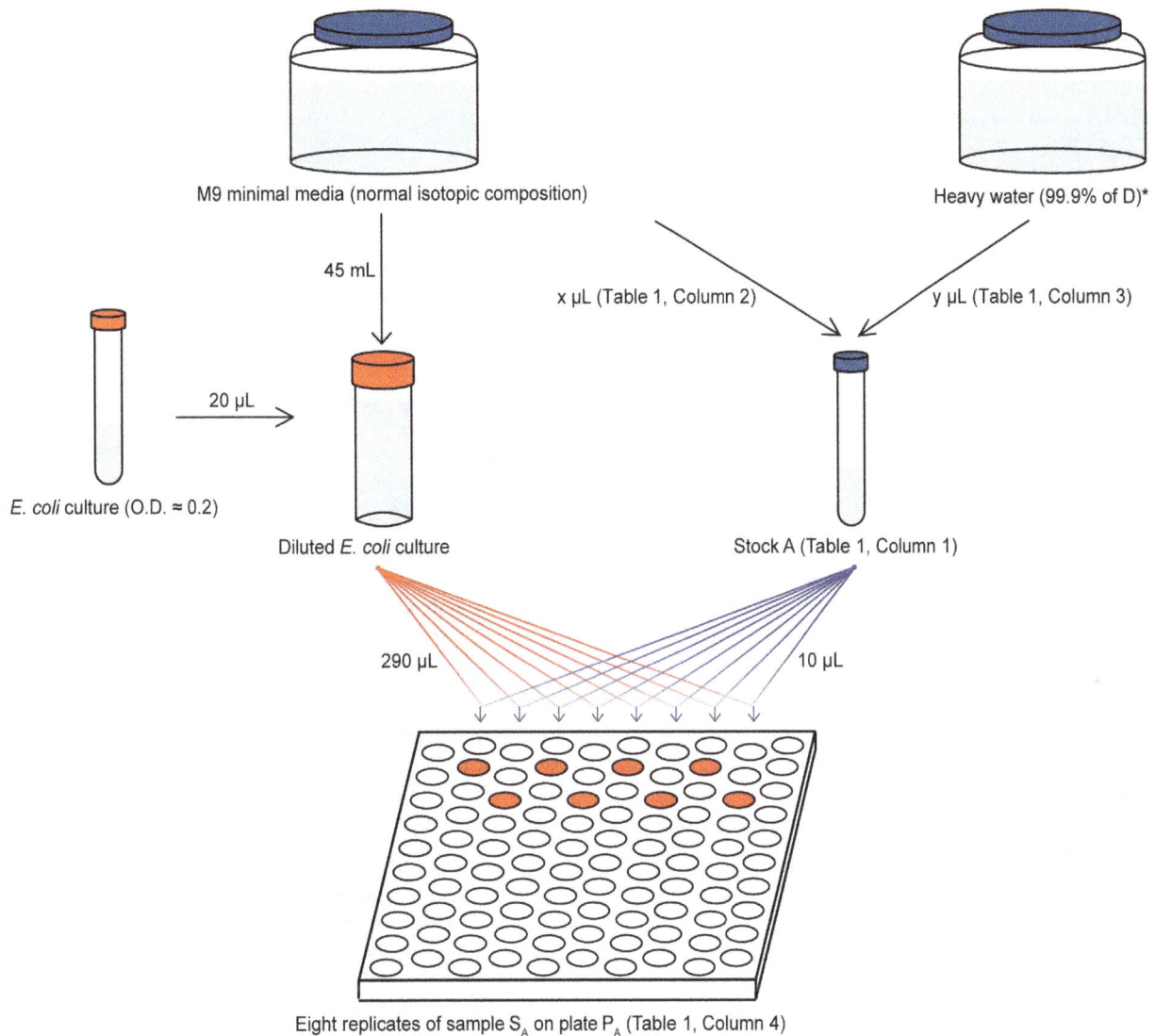

Figure 2. Experimental workflow. For each plate, 32 samples (S_A) and 32 standards were prepared. Stock A was used for preparing 32 samples on plate P_A. * Milli-Q water was used instead of heavy water to prepare stock solution for the preparation of standards.

programmed robotic system (Tecan, Genesis RSP 150, Männedorf, Switzerland).

A 20 µL aliquot of the incubated *E. coli* culture (O.D. ≈ 0.2) was diluted in 45 mL M9 minimal media for robotic sample preparation. First, 300 µL M9 minimal media without bacteria were introduced into each of the border wells ("edge cells") on both plates P_A and P_B (72 wells in total) to serve as blank samples (no color code in Figure S1). Second, 30 µL (10 µL for samples with <4% enrichment of deuterium) aliquot of stock A was dispensed into each "sample" well (marked with purple on plate P_A in Figure S1) to prepare 32 replicates of sample S_A. Third, 30 µL (10 µL for samples with <4% enrichment of deuterium) aliquot of stock solution of standard A was added into each "standard" well (marked with yellow on plate P_A in Figure S1) to prepare 32 reference standards on plate. In the same way, wells were filled on plate P_B. Finally, 270 µL (290 µL for <4% D) of the diluted *E. coli* culture was dispensed into each well except blank wells. In total, 32 replicate pairs of "sample" and "standard" wells were prepared on each plate. Sample configuration is shown in Figure S1.

Bioscreen C monitors *E. coli* concentration in each well by measuring turbidity (with wide band filter 420–580 nm) in it at 39°C with continuous bacterial agitation by shaking. In our experiments, turbidity was sampled every six minutes and was monitored for ca. 24 hours.

E. coli adaptation to different deuterium content

One *E. coli* colony was inoculated into 5 mL M9 minimal media (normal isotopic condition) from a fresh M9 minimal media agar plate, and cultured at 37°C until O.D. 1.0. For every deuterium concentration point (156 ppm, 0.03%, 0.25%, and 1%), 2 µL *E. coli* culture was diluted 2500 times into 5 mL M9 minimal media with the corresponding D content and cultured overnight at 37°C. After the culture reached its maximum density (O.D. ≈ 1.4), for each *E. coli* culture, 10 µL was then diluted 500 times into 5 mL media with the same D content, and culturing continued at 37°C. In this way, each *E. coli* culture was diluted 500 times twice a day, with nine generations grown between dilutions. The process continued until the 400th generation was obtained, after which the difference in the rate

Table 1. Stock solutions and their corresponding D content in the final samples.

D content in stock solution	M9 minimal media (µL)	Heavy water (µL)	D content in the final sample (honey comb well plate)
0.0156%	5948	28 (Milli-Q water)	0.0156%
0.48%	5948	28	0.03%
1.42%	5892	84	0.06%
3.29%	5957	202	0.12%
7.04%	5728	433	0.25%
14.52%	5620	955	0.5%
29.50%	4403	1844	1%
59.46%	2437	3581	2%
39.80%	3667	2427	4%
79.73%	1222	4830	8%

Stock solution (column one) was prepared by mixing M9 minimal media and heavy water at a certain ratio that resulted in the final D content in the sample (column four). For each sample, stock solution for its corresponding standards (below it is called stock solution of standard X) was prepared in the same way but using Milli-Q water instead of heavy water.

of growth of the adapted bacteria on the corresponding content of deuterium versus normal D content was measured using Bioscreen C.

Data Analysis

Using Microsoft Excel, the logarithm of turbidity was plotted against time (Figure 1). The slope for every 8-h interval was calculated, and the maximum value was determined. The extrapolation of the line with maximum slope to the background level of turbidity gave the lag time. The maximum turbidity for each replicate was taken as the maximum density. The obtained three values for each growth curve were treated in the same way as below.

For each "sample" A, the obtained value was normalized by that of the "standard" B. To minimize the influence of nonstatistical outliers that could arise due to gross errors in sample preparation and handling (e.g. robot or BioScreen C malfunction), the 32 replicates were divided into 4 groups according to their positions on the honey comb well plate (group 1: columns 1 and 2; …, group 4: columns 7 and 8). In each group, the median of the eight values was calculated and then the four medians were averaged to obtain the value for a given plate and its standard deviation.

Altogether, nine independent 32-replicate experiments were performed for each D content point and six experiments per each D point for adapted bacteria. To compensate for small systematic errors of measurement (e.g., differences in the geometry of the honeycomb wells, position-dependent sensitivity of the BioScreen C detector, etc.), the obtained average values of the maximum growth rate, lag time as well as the maximum density within each experiment were normalized to that of the sample with terrestrial D composition (156 ppm). The final result was obtained when the average of $9 \times 4 = 36$ ($6 \times 4 = 24$ for adapted bacteria) median values, and the corresponding standard error, were calculated. Since after normalization, the values for terrestrial composition were all unities, the p-values for non-terrestrial compositions were calculated using two-tailed, paired Student's t-test comparing the relative value R and its inverse $1/R$.

Results and Discussion

High enrichment (50% D)

To learn the behavior of the growth parameters at high enrichment, bacteria were grown first at 50% D. The growth curve showed significant deviation from the normal (Figure 1), with the growth rate reduced by ca. 5%, maximum density increased by ca. 15% and the lag phase extended by almost 40% (Figure S2). While reduction in the growth rate and lag time extension were expected based on abundant previous research, the increase in maximum density was surprising: at adverse growth conditions, the maximum density usually follows the declining growth rate trend. The increase in maximum density at 50% D, or at least the magnitude of the increase, could be overestimated by the optical density measurements if the optical properties of the deuterated bacteria are different than that of bacteria with normal D content. Indeed, literature suggests that the size and shape of the bacteria may be quite distorted when grown in a highly deuterated media, especially in the beginning of growth [3,4,7]. If this is the case, and the maximum density at 50% D is overestimated, then the growth rate reduction at 50% D may be underestimated in our measurements. However, both these possible effects should be relatively small at <10% D, and especially at <1% D. Besides, our prime goal was to identify the lowest concentration of deuterium at which biological effects become detectable. Thus we did not make any corrections for these hypothetical effects in low-enrichment measurements.

Low enrichment (<10% D)

Growth behavior of *E. coli* in minimal media under ten different content of deuterium (including 156 ppm, 0.03%, 0.06%, 0.1%, 0.25%, 0.5%, 1%, 2%, 4%, 8%) were investigated, starting with 156 ppm of deuterium (terrestrial condition) and increasing the content of deuterium by a factor of 2 until ca. 10% D is reached. In each run of the BioScreen C instrument, *E. coli* was grown in minimal media on two honey comb well plates each containing wells with two different contents of deuterium, one always being the normal D content (156 ppm), giving 32 replicates for each plate (*vide infra*). This way, five runs were required to cover the entire range of deuterium content, which constituted one experiment.

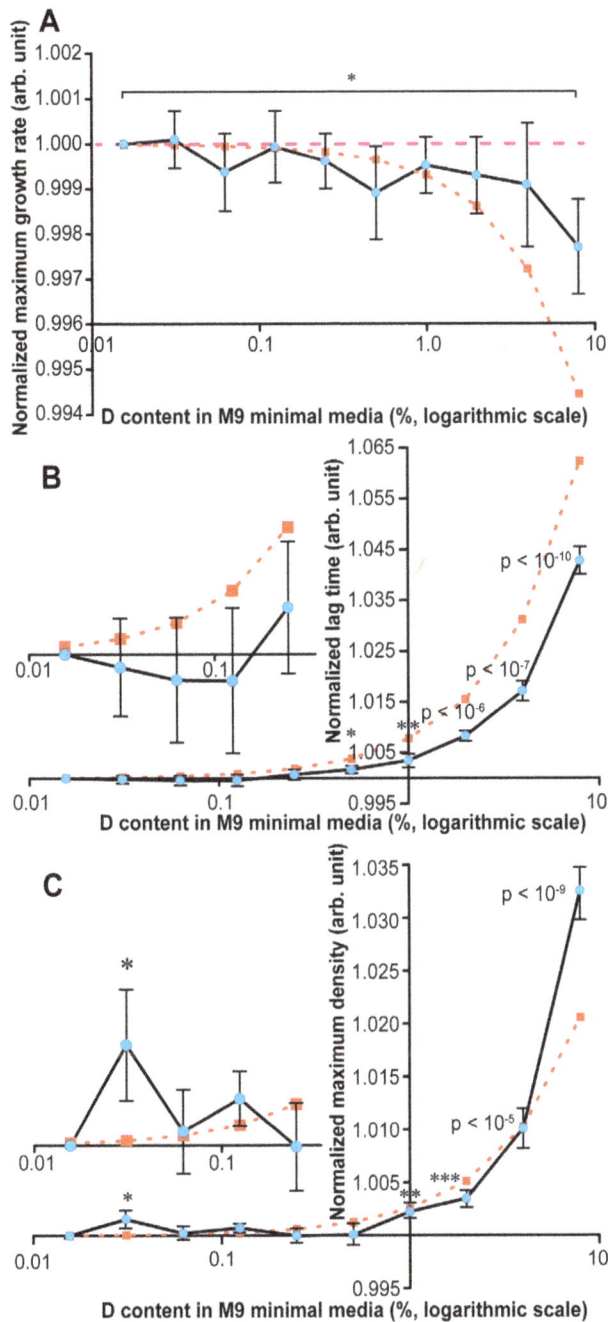

Figure 3. Maximum growth rate, lag time, maximum density of E. coli grown in minimal media. (**A**) Blue circles: maximum growth rate of E. coli grown in M9 minimal media normalized by that at normal deuterium content of 156 ppm. * denotes p<0.05, ** - p<0.005, etc. Brown squares: predicted maximum growth rate calculated according to the maximum growth rate of E. coli grown in 50% of deuterium. (**B**) Blue circles: lag time of E. coli grown in M9 minimal media normalized by that at terrestrial content of deuterium from 156 ppm (terrestrial value) to 8%. Inset shows a zoom-in of the ultralow enrichment region. Brown squares: predicted lag time calculated according to the lag time of E. coli grown in 50% of deuterium. (**C**) Blue circles: maximum density of E. coli grown in M9 minimal media normalized by that at terrestrial deuterium content from 156 ppm (terrestrial value) to 8%. Inset shows a zoom-in of the ultralow enrichment region. Brown squares: predicted maximum density calculated according to maximum density of E. coli grown in 50% of deuterium.

Figure 3A shows the normalized maximum growth rate. The curve is trending downwards, but only at the highest enrichment the statistical significance at p<0.05 is reached. The observed trend significantly (by a factor of 4) lags behind the linear extrapolation of the 50% D results to low-enrichment levels (i.e., a 5% slower growth at 50% D means growth rate ≈1–0.001*%D).

In contrast, the lag phase curve (Figure 3B) fluctuates at ultralow enrichment levels slightly below the normal level. But after 0.1% D, the lag time increases dramatically, reaching statistical significance at 0.5% D. The curve lags behind the linear prediction by a factor of two. Even if linear extrapolation is made based on 8% D results instead of 50% D, the actual trend remains below the predicted one.

The maximum density curve (Figure 3C) shows surprising behavior: at the lowest enrichment level of 0.03% D amounting to only twice the normal D content, there is a small but significant (p<0.05) peak corresponding to a ≈0.16% higher density. The increase in the maximum density continues at higher D content, and vanishes at 0.25% D, to increase again above the significance level at 1% D. For the interval 0.2%–4% D, the maximum density curves lags behind the linear extrapolation, but at 8% D it actually exceeds it, increasing by 3.2% compared to the predicted 2.4%.

Summarizing the above observations, the effects typical for high D enrichment (slower growth, longer lag and higher maximum density), start to be noticeable at 0.5% D. This is a surprisingly high threshold, given that the 0.05% precision of the measurements would allow us to detect with statistical significance as small changes in growth parameters as 0.1%. According to the 50% D results, if the effect of D content on E. coli growth was linear, the expected detection threshold for biological effects would be around 0.1–0.2% D. That no significant change was detected at 0.25% D indicates the presence of a previously unreported threshold for isotopic phenomena. Moreover, the results hint on the existence of an "inverted region" at ultralow enrichment levels with statistically significant fluctuations in the direction of higher comfort, rather than adverse reaction as at ≥0.5% D. The existence of such a region would be in agreement with previous studies, even though many of the actually found much bigger effects than were observed in our experiments [12–22].

Validity of ultralow enrichment effect

The biological effect at <0.25% D in our study does not result from one or two aberrantly high measurements: in six out of nine constituent experiments, the maximum density was enhanced compared to normal % D. Another way to test the validity is to investigate the variability of measurements: usually, stronger biological effect results in larger variability. Indeed, the standard error of the maximum density determination at normal D content was 0.05%, while for 8% D it was five times larger, 0.25%. At 0.03% D, the corresponding standard error was 0.08%, which is the highest value for all measurements below 1% D. Also for the other two growth parameters, the maximum growth rate and the lag phase, the standard error at 0.03% D was higher than at normal conditions. Curiously, of all the data points, only 0.03% D shows an increase (however slight and statistically insignificant) in the maximum growth rate on Figure 3A. Another supporting evidence in favor of the opposite trend at ultralow enrichment is the fact that, at such enrichment, all three growth parameters lag behind the trend extrapolated from 50% D measurements, and two out of three (lag phase and maximum density) – of the trend extrapolated from 8% D results. These results speak in favor of the validity of the biological effect at 0.03% D.

The effect at the ultralow enrichment levels, assuming it is real, in unlike that at high enrichments. While the growth rate does not change, the lag phase shortens (although insignificantly in statistical sense), and the maximum density increases above the expected trend. Overall, it's a signature of growth in a more comfortable environment. In contrast, at high enrichments the signature is definitely that of a more hostile environment, with a very long lag phase and slower growth. What is common for both regimes, though, is the increased maximum density, but the origins of this effect may differ (*vide supra*).

How to explain the presumed effect of ultralow deuterium enrichment on bacterial growth? The previous research has not produced any convincing explanation [12–22], which probably was one of the factors why this research has largely became forgotten. One of possible explanations is the effect of hormesis. The concept of hormesis suggests that, at low dosages, the effects of adverse agents (e.g. chemical toxins or radiation) can be stimulating for growth. Hormetic effect on *E. coli* growth has previously been detected at low concentrations of chemical toxins [26]. However, the size of the effect in our study (≈0.16% in maximum density) is disproportionally large compared to the enrichment degree (0.032% D, or +0.016% D compared to normal composition, Figure S3), which speaks against the role of hormesis in the studied situation.

Adaptation to ultralow enrichment conditions

Hormesis is usually explained by activation of protecting mechanisms that, in effect, boost the growth at low concentrations of adverse agents. But gradual adaptation of the organism to slightly adverse environment should increase the threshold for activation and thus reduce this effect. To test this hypothesis, we grew bacteria for ca. 400 generations at normal isotopic conditions, 0.03% D, 0.25% D and 1% D, changing the media every 12 hours (Figure S4). After that, the growth parameters in the same media were measured, using as control the same adapted bacteria growing at normal isotopic conditions. Adaptation has dramatically reduced the lag time from 10–11 h to 6–7 h, and increased the maximum growth rate as well as the maximum density by about 15–20% (Figure S5). The directions of the changes in growth parameters are consistent with bacteria feeling much more comfortable in the minimal media after adaptation.

In the isotopically different environment, the growth rate has increased by 0.10–0.15% at ultralow enrichment (p<0.05 at 0.25%, Figure 4A), while the lag time has increased insignificantly by the same amount (Figure 4B). The maximum density has shown most significant changes of ca. 0.2% (p<0.05, Figure 4C) at 0.25% D, but was elevated also at 0.03% D. Overall, two out of three parameters showed statistically significant changes characteristic for better adaptation in isotopically different environment compared to normal media. Note that at 1% D this effect disappeared.

Strictly speaking, deuterium is not toxic for *E. coli*, as this organism can grow at very high enrichment levels. Even at 50% D, the maximum growth rate is only slightly (5%) lower than at normal conditions. Therefore, deuterium enriched medium is more correctly characterized as an unusual, rather than toxic, environment. Thus the applicability of the hormetic explanation to ultralow deuterium enrichment is questionable.

An alternative to hermetic explanation could be provided by the isotopic resonance hypothesis [27]. Briefly, the hypothesis suggests that, at certain isotopic compositions of the elements C, H, N and O that together compose ≥96% of the living matter of prokaryotes and many eukaryotes, the rate of all chemical and biochemical reactions, including the bacterial growth, should increase. Normal terrestrial isotopic compositions are very close to one such resonance. The existing small deviation from the perfect

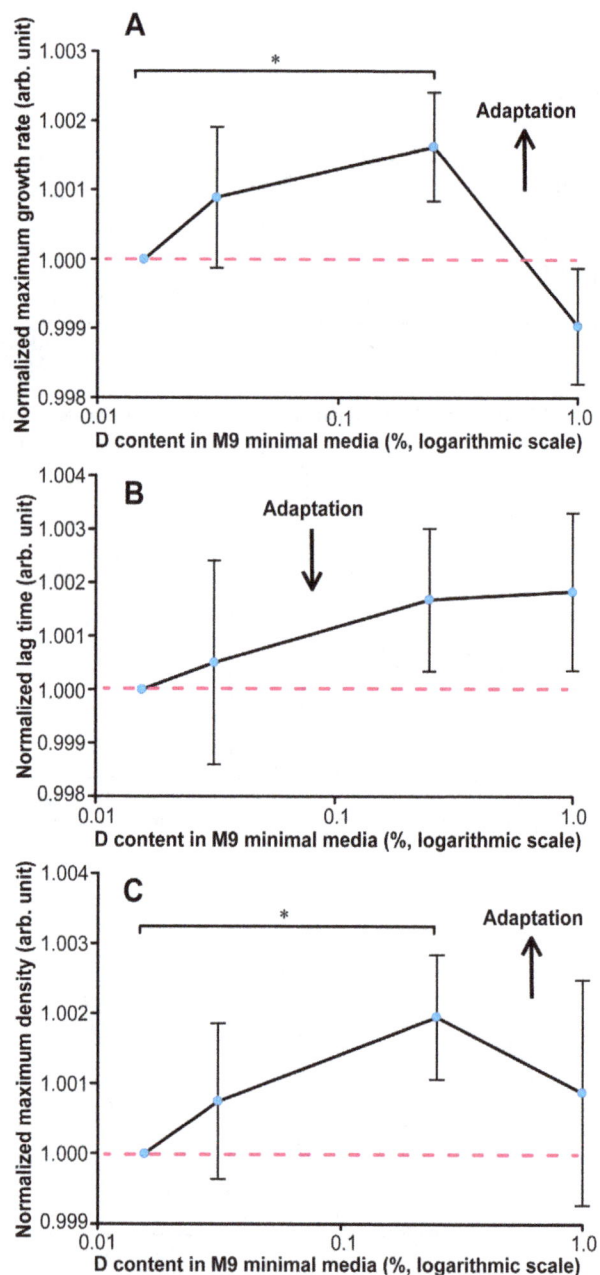

Figure 4. Maximum growth rate, lag time and maximum density of aged *E. coli*. (**A**) Maximum growth rate of aged *E. coli* grown in M9 minimal media with ultralow composition of deuterium. * denotes p<0.05. (**B**) Lag time of aged *E. coli* grown in M9 minimal media with ultralow composition of deuterium. (**C**) Maximum density of aged *E. coli* grown in M9 minimal media with ultralow composition of deuterium. * denotes p<0.05.

resonance, suggests the hypothesis, can be corrected by "tuning" the isotopic composition of any of the elements C, H, N and O. After such tuning the growth rate of all organisms should slightly increase. The size of the effect is not predicted, but the required value of the D content for the perfect "terrestrial" isotopic resonance is close to 0.03% [28]. Since the isotopic composition of water inside the *E. coli* bacteria has much lower deuterium enrichment compared to the growth media [29], the resonance conditions should occur in the region 0.03–0.06% of D. Such an agreement between the predictions of the hypothesis and the experiment could be spurious, but it receives strong support from earlier studies [12–22]. Taken together, these results suggest that the growth behavior of *E. coli* and other organisms at ultralow enrichment levels well worth exploring further, using even more precise measurements.

Conclusions

Here we studied the growth behavior of *E. coli* at low enrichment levels of deuterium, and found that the adverse effects characteristic for high enrichment levels become noticeable at 0.5% D, which is several times higher level than expected given the 0.05% precision of our measurements. This discrepancy highlights the "buffering capacity" of living organisms that can partially compensate the adverse nature of the isotopically altered growth medium by reduced incorporation levels of heavy isotopes [28]. On the other hand, there seems to be a small but significant fingerprint of more comfortable environment at an ultralow enrichment level of 0.03% D. This behavior, which begs for additional verification by more precise measurements, could be explained by both hormesis as well as the isotopic resonance hypothesis. The latter theory predicts the exact location of the "resonance" at $\approx 0.03\%$–0.06% D, which is in a broad agreement with sporadic but multiple earlier reports [12–22].

Supporting Information

Figure S1 Sample configuration on the honey comb well plates.

Figure S2 Maximum growth rate, maximum density and lag time of *E. coli* grown in M9 minimal media with deuterium content of 50% normalized by that at normal deuterium content of 156 ppm. ** is equivalent to $p < 0.005$.

Figure S3 Change of maximum density per percentage of deuterium for *E. coli* grown in M9 minimal media with content of deuterium from 156 ppm (terrestrial value) to 8%. * is equivalent to $p < 0.05$, ** is equivalent to $p < 0.005$, etc.

Figure S4 Workflow of adapting the bacteria to growth media.

Figure S5 Growth curves of *E. coli* grown in minimal media with 156 ppm of D before and after adaptation.

Acknowledgments

The authors thank the Swedish Research Council for funding this work.

Author Contributions

Conceived and designed the experiments: RAZ XX. Performed the experiments: XX. Analyzed the data: XX. Contributed reagents/materials/analysis tools: XX. Wrote the paper: XX RAZ.

References

1. Urey HC, Brickwedde FG, Murphy GM (1932) A hydrogen isotope of mass 2. Phys Rev 39: 164–166.
2. Koletzko B, Sauerwald T, Demmelmair H (1997) Safety of stable isotope use. Eur J Pediatr 156: S12–S17.
3. Katz JJ, Crespi HL, Hasterlik RJ, Thomson JF, Finkel AJ (1957) Some observations on biological effects of deuterium, with special reference to effects on neoplastic processes. J Natl Cancer Inst 18: 641–658.
4. Katz JJ, Crespi HL (1971) Isotope effects in biological systems. In: Collins, C J, Bowman NS, editor. Isotope effects in chemical reactions. New York: Van Nostrand Reinhold Co. 286–363.
5. Rokop S, Gajda L, Parmerter S, Crespi HL, Katz JJ (1969) Purification and characterization of fully deuterated enzymes. Biochim Biophys Acta 191: 707–715.
6. Paliy O, Bloor D, Brockwell D, Gilbert P, Barber J (2003) Improved methods of cultivation and production of deuteriated proteins from E. coli strains grown on fully deuterated minimal medium. J Appl Microbiol 94: 580–586.
7. Katz JJ, Crespi HL (1966) Deuterated organisms: Cultivation and uses. Science 151: 1187–1194.
8. Ernest B, Rittenberg D (1960) Anomalous growth of microorganisms produced by changes in isotopes in their environment. Proc Natl Acad Sci 46: 777–782.
9. Fowler EB, Adams WH, Christenson CW, Kollman VH, Buchholz JR (1972) Kinetic studies of C. pyrenoidosa using 94% 13C CO2. Biotechnol Bioeng 14: 819–829.
10. Uphaus RA, Flaumenhaft E, Katz JJ (1967) A living organism of unusual isotopic composition sequential and cumulative replacement of stable isotopes in Chlorella vulgaris. Biochim Biophys Acta 141: 625–632.
11. Kreuzer-Martin HW, Jarman KH (2007) Stable isotope ratios and forensic analysis of microorganisms. Appl Environ Microbiol 73: 3896–3908.
12. Barnes TC (1933) A possible physiological effect of the heavy isotope of H in water. J Am Chem Soc 55: 4332–4333.
13. Barnes TC (1933) Further experiments on the physiological effect of heavy water and of ice water. J Am Chem Soc 55: 5059–5060.
14. Barnes TC, Larson EJ (1935) The influence of heavy water of low concentration on Spirogyra, Planaria and enzyme action. Protoplasma 22: 431–443.
15. Barnes TC (1934) The effect of heavy water of low concentration on Euglena. Science 79: 370.
16. Lockemann G, Leunig H (1934) Über den Einfluß des "schweren Wassers" auf die biologischen Vorgänge bei Bakterien. Berichte der Dtsch Chem Gesellschaft (A B Series) 67: 1299–1302.
17. Curry J, Partt R, Trelease SF (1935) Does dilute heavy water influence biological processes? Science 81: 275–277.
18. Lobyshev VI, Tverdislov VA, Vogel J, Iakovenko LV (1978) Activation of Na, K-ATPase by small concentrations of D2O, inhibition by high concentrations. Biofizika 23: 390–391.
19. Lobyshev VI, Fogel' Iu, Iakovenko LV, Rezaeva MN, Tverdislov VA (1982) D2O as a modifier of ionic specificity of Na, K-ATPase. Biofizika 27: 595–603.
20. Lobyshev VI (1983) Activating influence of heavy water of small concentration on the regeneration of hydroid polyp obelia geniculata. Biofizika 28: 666–668.
21. Somlyai G, Jancsó G, Jákli G, Vass K, Barna B, et al. (1993) Naturally occurring deuterium is essential for the normal growth rate of cells. FEBS Lett 317: 1–4.
22. Nikitin DI, Oranskaya MN, Lobyshev VI (2003) Specificity of bacterial response to variation of isotopic composition of water. Biofizika 48: 678–682.
23. Hammel SC, East K, Shaka AJ, Rose MR, Shahrestani P (2013) Brief early-life non-specific incorporation of deuterium extends mean life span in drosophila melanogaster without affecting fecundity. Rejuvenation Res 16: 98–104.
24. Sambrook J, Russell DW (2001) Molecular cloning: A laboratory manual. 3rd ed. New York: Cold Spring Harbor Laboratory Press.
25. Madigan M, Martinko J, Stahl D, Clark D (2012) Brock biology of microorganisms. 13th ed. San Francisco: Pearson.
26. Calabrese EJ, Hoffmann GR, Stanek EJ, Nascarella MA (2010) Hormesis in high-throughput screening of antibacterial compounds in E coli. Hum Exp Toxicol 29: 667–677.
27. Zubarev RA, Artemenko KA, Zubarev AR, Mayrhofer C, Yang H, et al. (2010) Early life relict feature in peptide mass distribution. Cent Eur J Biol 5: 190–196.
28. Zubarev RA (2011) Role of stable isotopes in life-testing isotopic resonance hypothesis. Genomics Proteomics Bioinformatics 9: 15–20.
29. Kreuzer-Martin HW, Lott MJ, Ehleringer JR, Hegg EL (2006) Metabolic processes account for the majority of the intracellular water in log-phase Escherichia coli cells as revealed by hydrogen isotopes. Biochemistry 45: 13622–13630.

7

Surface Physicochemistry and Ionic Strength Affects eDNA's Role in Bacterial Adhesion to Abiotic Surfaces

Viduthalai R. Regina[1], Arcot R. Lokanathan[1¤a], Jakub J. Modrzyński[1¤b], Duncan S. Sutherland[1], Rikke L. Meyer[1,2]*

1 Interdisciplinary Nanoscience Center (iNANO), Aarhus C, Denmark, 2 Department of Bioscience, Aarhus University, Aarhus C, Denmark

Abstract

Extracellular DNA (eDNA) is an important structural component of biofilms formed by many bacteria, but few reports have focused on its role in initial cell adhesion. The aim of this study was to investigate the role of eDNA in bacterial adhesion to abiotic surfaces, and determine to which extent eDNA-mediated adhesion depends on the physicochemical properties of the surface and surrounding liquid. We investigated eDNA alteration of cell surface hydrophobicity and zeta potential, and subsequently quantified the effect of eDNA on the adhesion of *Staphylococcus xylosus* to glass surfaces functionalised with different chemistries resulting in variable hydrophobicity and charge. Cell adhesion experiments were carried out at three different ionic strengths. Removal of eDNA from *S. xylosus* cells by DNase treatment did not alter the zeta potential, but rendered the cells more hydrophilic. DNase treatment impaired adhesion of cells to glass surfaces, but the adhesive properties of *S. xylosus* were regained within 30 minutes if DNase was not continuously present, implying a continuous release of eDNA in the culture. Removal of eDNA lowered the adhesion of *S. xylosus* to all surfaces chemistries tested, but not at all ionic strengths. No effect was seen on glass surfaces and carboxyl-functionalised surfaces at high ionic strength, and a reverse effect occurred on amine-functionalised surfaces at low ionic strength. However, eDNA promoted adhesion of cells to hydrophobic surfaces irrespective of the ionic strength. The adhesive properties of eDNA in mediating initial adhesion of *S. xylosus* is thus highly versatile, but also dependent on the physicochemical properties of the surface and ionic strength of the surrounding medium.

Editor: Mickaël Desvaux, INRA Clermont-Ferrand Research Center, France

Funding: The authors gratefully acknowledge the Danish Strategic Research Council (Grant No. 2106-07-0013), Alfa Laval, the Lundbeck Foundation, and the Carlsberg Foundation for funding for this work. The funders had no role in study design, data collection and analysis, decision to publish, or preparation of the manuscript.

Competing Interests: The authors have the following interests. This study was partly funded by Alfa Laval. There are no patents, products in development or marketed products to declare.

* Email: rikke.meyer@inano.au.dk

¤a Current address: Department of Forest Products Technology, Aalto University, Aalto, Finland
¤b Current address: Department of Plant and Environmental Sciences, University of Copenhagen, Frederiksberg C, Denmark

Introduction

Bacteria adhere to almost all kinds of surfaces, enabling biofilm formation [1]. Biofilms can cause serious health problems as well as economic losses in many industries, such as the oil and gas industry where biofilms cause corrosion, and the food industry where spoilage, and contamination leading to food-borne illnesses are the main concern [2–10]. While antimicrobial surfaces releasing toxins are the traditional approach to antifouling solutions, development of non-toxic antifouling surfaces that intercept cell adhesion and biofilm formation rather than cell viability are gaining more interest [11]. However, development of such antifouling strategies relies on detailed understanding of the mechanisms behind bacterial adhesion and the subsequent establishment of a biofilm. Bacterial adhesion is a complex process involving long range Lifshitz van der Waals and electrostatic forces, as wells as short range acid-base interactions [12,13]. The physico-chemistry of the substrate surface and the bacterial surface decides the extents of these interactions and hence the adhesion of bacteria. However, bacterial adhesion cannot always be predicted by the average physicochemical properties of the cell surface. Specific extracellular components that might evade the interaction force barrier formed between the approaching surfaces are often the primary facilitator for adhesion, even if they do not contribute substantially to the average physicochemical surface properties of the cell [14–16]. Such extracellular components include carbohydrate polymers, single and fibrillar proteins, and extracellular DNA (eDNA) [17].

The active role of eDNA in biofilm formation was discovered by Whitchurch *et al.* [18], who showed that removal of eDNA by DNase treatment could dissolve young *Pseudomonas aeruginosa* biofilms and prevent biofilm formation on abiotic surfaces. Since then, the involvement of eDNA in biofilm formation has been studied in many different bacteria across several phyla, and an image of eDNA as a universal adhesin is emerging. But what makes eDNA adhesive? Only few studies have attempted to address how eDNA promotes bacterial adhesion to abiotic surfaces. Das *et al.* [19] studied adhesion of *Streptococcus mutans*, and used AFM force spectroscopy to investigate how eDNA increased adhesion forces of bacterial cells to hydrophobic and hydrophilic surfaces at varying ionic strength (I). They found that presence of eDNA on the cell surface increased the adhesion strength to both hydrophilic and hydrophobic surfaces. The overall adhesion strength in the presence of eDNA was higher to hydrophobic surfaces than to hydrophilic surfaces, and the effect was more pronounced at high ionic strength. Furthermore, AFM retraction-force distance curves revealed that eDNA-mediated acid-base interactions were more pronounced in the interaction with hydrophilic surfaces, and that these interactions were high in low ionic strength. Ionic strength can affect the electrostatic properties of surfaces and also influence the conformation of biopolymers such as DNA. The ability of eDNA to function as an adhesive thus seems to occur within certain boundaries, defined by the physico-chemical properties of the substrate and the surrounding environment.

To get a better understanding of eDNA's versatility as a universal bacterial adhesin, we investigated how surface chemistry and ionic strength affects eDNA's role in bacterial adhesion. We used *Staphylococcus xylosus* as a model organism. *S. xylosus* are Gram positive, biofilm forming, bacteria that have frequently been isolated from food and food related environments [20,21]. These bacteria can produce aroma [22] and enterotoxins that qualify them as potential contaminants in food processing industries [23]. Furthermore, some strains of *S. xylosus* have been reported to carry antibiotic resistance genes [24] which is a major concern in the emerging threat of antibiotic resistance among bacteria. These characteristics makes *S. xylosus* an attractive candidate to study the factors influencing biofilm formation *in vitro*. In this study we quantified initial bacterial adhesion to a range of chemically modified surfaces, representing surfaces with varying charge and hydrophobicity, and repeated experiments at three different ionic strengths. eDNA is critical for the adhesion of bacteria to both hydrophilic and hydrophobic surfaces. Electrostatic interactions, modulated by surface chemistry, depended on the ionic strength in facilitating eDNA mediated adhesion to hydrophilic surfaces. The amount of eDNA released by the cells, their ability to use eDNA and/or a combination of other adhesins for adhesin will be different for different types of bacteria. Therefore, the observed effects on eDNA mediated adhesion of *S. xylosus* may not be generalized to all biofilm forming bacteria. However, the findings presented in this study can be used to determine the range of physico-chemical characteristics wherein eDNA can influence adhesion of bacteria to abiotic surfaces.

Results and Discussion

eDNA is critical for adhesion to glass

The presence of eDNA was critical for adhesion of *S. xylosus* to glass surfaces, as removal of eDNA by DNase treatment almost fully impaired adhesion of cells (Figure 1). Interestingly, the cells quickly regained their adhesive properties if the DNase was removed from the cell suspension before the onset of the adhesion experiment. In this case, adhesion of *S. xylosus* to glass was only delayed by approximately 30 min, after which the adhesion rate was not significantly different from the untreated control (Figure 1) (Two-way ANOVA; $F = 0.03$, $p = 0.87$). Only if DNase was continuously present, adhesion was impaired for the entire duration of the experiment (Figure 1). Autolysis caused by the activity of the major autolysin AltE is responsible for the eDNA in biofilms formed by other *Staphylococcus* species [25–27], and the same mechanism may be present here. The short time needed for the cells to regain their adhesive properties after removing the DNase from the cell suspension suggests that eDNA was continuously released in the culture, and that the amount of

eDNA needed to facilitate cell adhesion is very low. We could not quantify the amount of eDNA adsorbed to the cells, but the eDNA concentration in the supernatant of the cell suspension was 110.2 ± 11.5 ng/ml for the untreated cells, whereas DNase treated cells contained 15.5 ± 0.2 ng/ml for bacterial suspensions of 1.3×10^8 cells.

eDNA increases cell surface hydrophobicity but does not affect surface charge

To better understand the adhesive properties of cells with eDNA, we investigated the average physicochemical properties of untreated and DNase-treated cells by measuring the cell surface hydrophobicity and charge as approximated by the water contact angle and zeta potential, which is a measure of potential at the diffuse layer of ions formed around particles in aqueous medium. To account for the effect of ionic strength on zeta potential, measurements were done in PBS and 10 times diluted PBS. The zeta potential of the cells was not affected by DNase treatment (Table 1). Cells suspended in 10 times diluted PBS (low ionic strength compared to PBS) had more negative zeta potential, but remained unaffected by DNase treatment (Table 1). Hence, the adsorption of eDNA to the cell surface does not appear to alter the average surface charge of *S. xylosus*.

Das *et al.* [28] did find a more negative zeta potential after adding DNA to *S. aureus*, *S. epidermidis* and *P. aureginosa* cultures, but only at relatively high concentrations (>4–6×10^{-9} µg DNA per bacterium). As we have alluded before, the concentration of naturally occurring eDNA in our cell suspensions was very low (approximately 110 ng/ml, which corresponds to 8.5×10^{-7} ng per cell), and the amount of eDNA present may not have been sufficient to cause a detectable difference in cell surface charge. Another explanation could be that nucleotides or short strands of eDNA present after DNase treatment remain associated with the cell surface, and thus could contribute to the zeta potential. If eDNA in this form is unable to promote cell adhesion, DNase treatment could affect the adhesive properties of the cells without affecting their surface charge.

The water contact angle was significantly lower after DNase treatment, indicating that cells with eDNA are more hydrophobic (Table 1). The same phenomenon was observed by Das *et al.* for other staphylococci [28,29] who showed that *S. epidermidis* cells became less hydrophobic when lacking eDNA due to either DNase treatment or deletion of the *altE* gene that facilitates eDNA release through autolysis. They also showed that adding increasing amounts of DNA to the culture resulted in increasing cell surface hydrophobicity. DNA in itself is not hydrophobic, and it is not yet understood why removal of eDNA makes the cell surface less hydrophobic. However, it has been suggested that DNA is associated with other components on the cell exterior [30–36], and these might contribute to the hydrophobic cell surface properties. One study simply showed that the adhesive properties of eDNA-free *Nisseria meningiditis* could be restored by addition of genomic DNA from crude extracts, but not from purified DNA [33], indicating a role of a non-DNA component in eDNA-mediated adhesion. More specific knowledge was obtained from a similar experiment on *Listeria monocytogenes* showing that N-acetyl glucosamine as well as DNA was needed to restore the adhesive properties of eDNA-free cells [32]. Several studies have sought to determine if proteins are involved in eDNA-mediated adhesion. For example, DNA-binding proteins, such as the pilin protein of Type IV pili [37], play a part in eDNA's role in biofilm formation by e.g. *Acidovorax temperans* [38] and *P. aeruginosa* [39]. Furthermore, Das *et al.* [35] showed that pyocyanin, a phenazine molecule produced by *P. aeruginosa*, affects cell surface

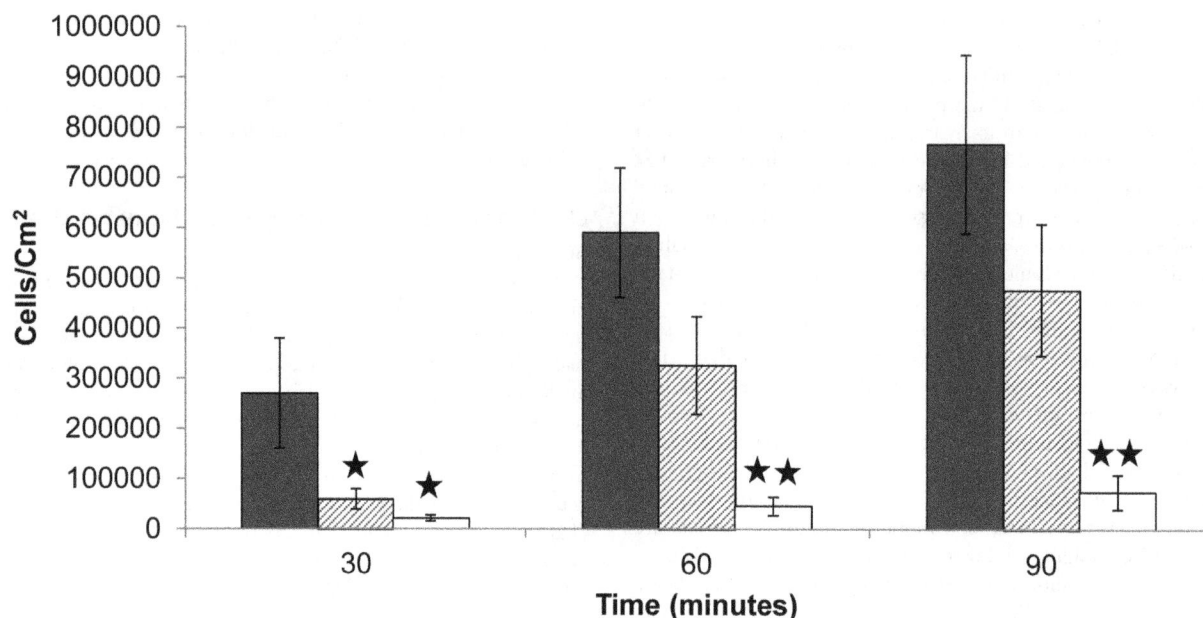

Figure 1. Effect of DNase on the adhesion of *S. xylosus* to glass surface in flow cell. Black bars indicate untreated cells. Crossed bars indicate cells treated with DNase (50 µg/ml), washed and resuspended in PBS. White bars indicate cells resuspended in PBS containing DNase (50 µg/ml). Asterisk indicates statistically significant differences between samples with and without eDNA (t-test, *p<0.05, **p<0.01).

properties and aggregation by facilitating the binding of eDNA binding to the cell surface [35]. Whether these are isolated examples or represent a general picture of eDNA being a partner in a multi-adhesin system is yet to be revealed, and the attention should be aimed at small molecules as well as macromolecules in the search for other extracellular components partnering with eDNA in mediating biofilm formation.

eDNA's effect on adhesion depends on surface chemistry and ionic strength

Bacterial adhesion is dictated by long and short range forces between the cell and the surface [13,40]. Liefshitz van der Waals and electric double layer forces are long range forces (several nanometers), where the former is attractive and the latter can be both attractive and repulsive. In contrast, Lewis acid-base interactions operate at short range [12]. The physico-chemistry defines the extent of these forces and thereby decides the interaction between approaching surfaces, but the interactions are also influenced by the properties of the surrounding liquid. Ionic strength of the liquid affects the thickness of the electric double layer, and thereby the electrostatic interactions [41]. The thickness of the electric double layer (the Debye length (δ)) is estimated by equation (1) where ε is the permittivity (determined

from $\varepsilon_r\varepsilon_0$, where ε_r is the relative permittivity at temperature, $T = 300$ K and ε_0 is the vacuum permittivity), k is the Boltzmann constant, e is the electron charge, c is the concentration, and z is the charge of the electrolyte.

$$\delta = \frac{1}{k} = \left[\frac{\mathcal{E}kT}{e^2 \sum_i c_i z_i^2} \right]^{\frac{1}{2}} \tag{1}$$

The Debye length at low, intermediate and high ionic strengths were 1.75 nm, 0.49 nm and 0.25 nm respectively.

The conformation of biopolymers, such as eDNA, can also be influenced by ionic strength and ionic composition of the surrounding medium [42], and we therefore hypothesized that the ionic strength of the surrounding media as well as the physico-chemical properties of the abiotic surface will affect eDNAs role in bacterial adhesion. In order to test this hypothesis, we quantified the adhesion of bacteria to glass surfaces with different surface chemistries in the presence or absence of DNase. We modified glass cover slips to obtain an array of surfaces with highly different chemistries, representing different hydrophobicity and charge, and

Table 1. Cell surface properties with and without eDNA.

S. xylosus cells	Water Contact angle (°)	Zeta potential (mV)	
		10 mM PBS	1 mM PBS
with eDNA	46.7±3.7	−15.1±0.9	−34.0±2.4
without eDNA	*33.5±2.6	−15.1±0.9	−34.4±1.3

Water contact angle and zeta potential measurements of *S. xylosus* cells with and without eDNA.
*indicates statistically significant difference (t-test, p<0.05).

submerged the cover slips in bacterial suspensions in 12 well plates under continuous shaking before rinsing and quantifying the adhered bacteria.

XPS analyses confirmed the presence of the functional groups on each surface (Figure S1). The Piranha treated and carboxyl functionalised surfaces were negatively charged with zeta potentials of -55.7 ± 2.5 mV and -70.4 ± 5.9 mV respectively, and highly hydrophilic with water contact angles of only $4.6 \pm 0.8°$ and $5.5 \pm 2.3°$, respectively. Amine functionalised surfaces were positively charged, $(84.6 \pm 4.9$ mV) and slightly less hydrophilic with a water contact angle of $30.3 \pm 4.8°$, and the fluoro silane functionalised surfaces were slightly negatively charged $(-13.7 \pm 0.9$ mV) and were the only hydrophobic surfaces tested, having a water contact angle of $73 \pm 3.1°$.

eDNA stimulated adhesion to all surfaces, but not to the same extent, and not at all ionic strengths. Very little adhesion occurred to the hydrophilic and negatively charged Piranha treated and carboxyl functionalised glass surfaces in the absence of eDNA (Figure 2). However, the stimulating effect of eDNA was only evident at low and intermediate I. Das et al. [28] showed that eDNA mediates bacterial adhesion through short range Lewis acid-base interactions and the same authors argued that these forces are more pronounced in eDNA-mediated adhesion at low I and toward hydrophilic as opposed to hydrophobic surfaces [19]. We speculate that loops of eDNA must extend from the cell surface and protrude the electric double layer to facilitate adhesion. Indeed Das et al. showed that eDNA extended approximately 400 nm from the cell surface of Streptococcus mutans cells [28], and the thickness of the electric double layer determine here was only a few nm. Furthermore, DLS measurements have confirmed that eDNA only increases the hydrodynamic radius of bacteria at low I [19], supporting our hypothesis that eDNA must extend from the bacterial cell surface to mediate adhesion to hydrophilic surfaces.

At the intermediate ionic strength, eDNA stimulated adhesion to Piranha treated but not carboxyl functionalised glass surfaces.

The negatively charged hydrophilic glass surfaces are known to contain localized positive charges [43], which are important for adhesion mediated by long range electrostatic interactions [15]. While these are masked by the thicker electric double layer and the overall negative charge at low electrolyte concentrations, they may become visible at high electrolyte concentrations. Such positive charges may not be accessible for the eDNA on carboxyl functionalized surfaces as they were modified by self-assembled monolayers of (Triethoxysilyl) propylsuccinic anhydride, but the availability of these charges on the Piranha treated glass may have contributed to eDNA-mediated adhesion despite the increase in I.

eDNA promoted adhesion of cells to positively charged aminated surfaces in medium and high I (Figure 2), but surprisingly, it had the opposite effect at low I. Electrostatic interactions are more predominant at low I, and we had therefore expected that the negatively charged eDNA would promote adhesion to amine-functionalised under these conditions. We did not investigate the underlying mechanism is behind this result further, but it is likely caused by the role of Mg^{2+} ions in the absence of other salts. Divalent cations like Mg^{2+} interact with the phosphate backbone of DNA and provide charge neutralization [44], which allows DNA to adsorb to negatively charged surfaces [45]. Mg^{2+} is therefore commonly used to facilitate DNA adsorption to negatively charged mica surfaces, e.g. for atomic force microscopy [46], and conversely, EDTA is used to chelate Mg^{2+} in order to keep DNA in suspension for molecular biology research. The charge neutralization of Mg^{2+} is approximately 100 fold more effective compared to monovalent ions [45], and we therefore expect that charge neutralization of eDNA on the bacterial cell surface is most significant in the low I incubation, where Mg^{2+} does not compete with Na^+ and K^+ for the interaction with the phosphate groups of the DNA backbone. Indeed, Nguyen et al. (2007) showed that the presence of Mg^{2+} enhanced DNA adsorption to natural organic matter, but to a lesser extent in the presence of 7 mM NaCl [47]. The NaCl concentration in 10 mM PBS is 149 mM, hence the effect of Mg^{2+} on eDNA-mediated

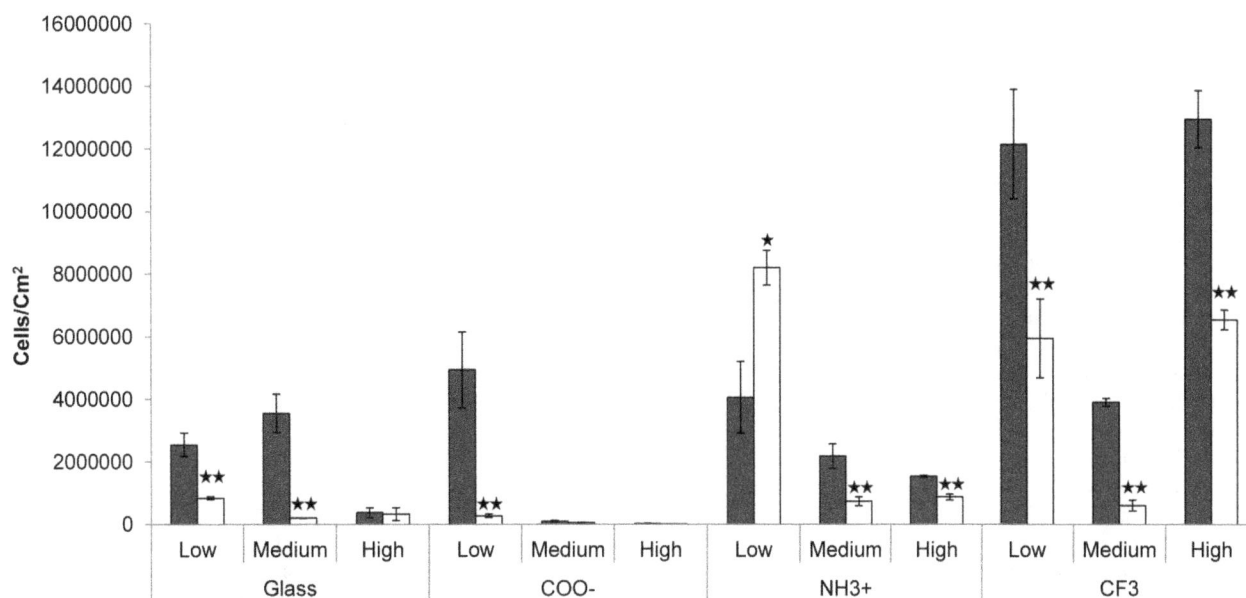

Figure 2. Effect of ionic strength and surface chemistries on eDNA mediated adhesion of S. xylosus. Black bars indicate untreated cells. White bars indicate cells treated with DNase. Experiments were carried out at low (I = 0.015 M), medium (I = 0.19 M) and high (I = 0.70 M) ionic strength. Values are average of 3 replicates (error bars = S.D.) Asterisk indicates statistically significant differences between samples with and without eDNA (t-test, *p<0.05, **p<0.01).

bacterial adhesion will be much less in PBS. The charge neutralization of DNA by Mg^{2+} in the absence of other ions can explain that eDNA lowered bacterial adhesion to the positively charged amine-functionalised surfaces and promoted adhesion to the negatively charged carboxyl-functionalised surfaces at low I. The presence of Mg^{2+} was required for the activity of DNase, and we could therefore not perform control experiments in the absence of Mg^{2+} to test the hypothesis.

In contrast to the hydrophilic surfaces, eDNA promoted adhesion of *S. xylosus* to the neutrally charged hydrophobic surfaces irrespective of the I of the surrounding liquid (Figure 2). Hydrophobic surfaces in general cannot contribute to acid-base interactions due to the lack of polar groups. *S. xylosus* cells were more hydrophobic in the presence of eDNA (Table 1), and when hydrophobic surfaces approach each other, Lifshitz-van der Waals forces increase due to the removal of interfacial water [48]. This increase in Lifshitz-van der Waals interaction facilitates faster approach of cells towards the surface, hence promoting adhesion. Lifshitz-van der Waals forces can be overcome by electrostatic forces at low ionic strength [49], however the hydrophobic surface, functionalized with fluorine end groups (-CF_3), cannot contribute to such forces. Das *et al.* [19] also found that eDNA creates a favorable conditions for bacterial adhesion to hydrophobic surfaces, and demonstrated this both theoretically through thermodynamic calculations, and experimentally by AFM adhesion force spectroscopy measurements [19,29]. DNA is an amphipathic molecule having a hydrophilic backbone and hydrophobic bases in the core. These nitrogenous bases might act as structures that participate in the hydrophobic interactions. This reinforces the idea that eDNA on the cell surface may be partially or completely single stranded, allowing the hydrophobic bases to get exposed. However, further investigation is needed to strengthen this hypothesis.

Collectively, our data and previous studies suggest that eDNA can participate both in short-range acid-base interactions and long-range electrostatic and Lifshitz-van der Waals interactions, and eDNA is thus able to stimulate bacterial adhesion to a wide range of surface chemistries. With eDNA being used in biofilm formation by a wide range of bacteria from several different phyla, an image of eDNA as a universal adhesin is emerging. In nature, eDNA-mediated biofilm formation is believed to occur at low ionic strength in aquatic environments, and as most surfaces in nature are negatively charged [50–52], our data support this hypothesis. The effect of ionic strength on the conformation of eDNA might be the key feature preventing its role in mediating the adhesion at high ionic strength. However, the ability of eDNA to aid adhesion to hydrophobic surfaces under any conditions raises new questions about the possible role of eDNA in e.g. cell aggregation or attachment to debris in marine environments. The versatility of eDNA as an adhesive may yet reveal new situations where eDNA-mediated adhesion increases survival of bacteria. Furthermore, the emerging image of eDNA as an important adhesin for many bacteria across several phyla points to eDNA as a potential global target in biofilm prevention. Understanding the mechanism by which eDNA mediates bacterial adhesion is therefore an fundamental basis for development of a new generation of antifouling surfaces or cleaning regimes that enzymatically degrade DNA or interfere with eDNA's adhesive properties.

Materials and Methods

Preparation of bacteria

A *Staphylococcus xylosus* (DSM 20266, DSMZ, Braunschweig, Germany) starter culture was inoculated from agar plates and grown in 3 ml of 1% TSB medium in 50 ml conical bottom tube by incubating overnight with shaking at 30°C. One ml from this culture was then used as inoculum for 100 ml of 1% TSB medium, which was incubated overnight, harvested by centrifugation (5 min at 3000×g), washed twice and resuspended in phosphate buffered saline (PBS, pH 7.4) containing 5 mM $MgCl_2$, with or without 50 μg/ml DNase I (Sigma), and incubated at 37°C for 1 h. The $MgCl_2$ is required for DNase activity. After incubation, cells were again washed three times in PBS and resuspended at OD_{600} of 0.05 in the buffer used for the subsequent experiment (see below).

Bacterial adhesion assay

To study bacterial adhesion under flow, glass cover slips, cleaned by dipping them sequentially in acetone, distilled water, ethanol, and distilled water for 1 minute each, were mounted on 3 channel flowcells, each channel measures (length×width×height) 40×4×4 mm (Denmark Technical University, Copenhagen, Denmark). Adhesion of untreated *S. xylosus* was compared to adhesion of *S. xylosus* that was continuously treated with DNase (50 μg/ml), or had temporarily been treated with DNase (50 μg/ml) for 1 h, washed and resuspended in PBS with 5 mM $MgCl_2$. All the three suspensions were adjusted to OD_{600} of 0.05 and passed through flowcells (flow rate ~1 ml per minute), and cells attached to the glass surface were enumerated by phase contrast microscopy after 30, 60 and 90 min.

Continuously shaking batch bacterial adhesion assays were performed in the presence or absence of DNase during the incubation of glass cover slips in bacterial suspensions for 100 min at room temperature with shaking at 120 rpm. The assay was performed on three replicate surfaces incubated individually in bacterial suspensions prepared from the same overnight culture and separated into aliquots before resuspending buffers with or without DNase. Buffers were prepared with three different ionic strengths: a) PBS with 5 mM $MgCl_2$ and 30 g l^{-1} NaCl ($I = 0.70$ M) (pH 6.9), b) PBS with 5 mM $MgCl_2$ ($I = 0.19$ M) (pH 7.0), and c) 5 mM $MgCl_2$ in deionized water ($I = 0.015$M) (pH 6.1). $MgCl_2$ was required for DNase activity. Glass cover slips were recovered and non-adhered bacteria were gently removed by dipping the slides in sterile PBS three times. The remaining bacteria were stained with 10 μl 20x SYBR Green II RNA stain (Sigma-Aldrich), placed on a glass slide and sealed with nail polish to avoid evaporation. Slides were kept in the dark at 4°C until quantification of adhered bacteria by epifluorescence microscopy (Zeiss Axiovert 200M, Carl Zeiss GmgH, Jena, Germany) using Zeiss filterset 10 and 40x or 63x oil immersion objectives. Cells were counted in 190 μm² or 120 μm² grids (depending on the magnification used) at random positions on the slide until a minimum of 1000 cells had been counted. Loosely adhered cells will be removed from the surfaces in the static assay, as large sheer forces are applied when passing the samples through an air/water interfaces. The assay therefore only enumerates the strongly adhered cells, which some studies define as "bacterial retention". For ease of language, we will refer to both assays as bacterial adhesion in the present study.

Surface preparation and modification

Glass surfaces with different chemistries were prepared by surface assembly of silanes on square glass cover slips (12×12 mm). Prior to all surface chemical modifications, the cover slips were cleaned by submersing them in Piranha reagent (Ammonium hydroxide (25%), Hydrogen peroxide (35%) and water at 1:1:4 volume/volume) for 5 minutes, rinsed with water and dried using jet of nitrogen [53]. Hydrophobic fluoro silane (-CF_3) functionalised surfaces were prepared by incubation in

100 mM Trimethoxy trifluoropropyl silane in toluene for 12 h and subsequently rinsed by sonication in toluene using a bath sonicator for 1 h, and dried using a jet of nitrogen. Negatively charged carboxyl (-COO$^-$) functionalised surfaces were prepared by incubation in 10% volume/volume 3-(Triethoxysilyl) propylsuccinic anhydride in toluene for 15 h. Positively charged amine (-NH$_3^+$) functionalised surfaces were prepared by incubating carboxyl terminated surfaces in 3% volume/volume polyethylene imine (PEI) in water for 12 h. The reaction was catalyzed by NHS-EDC by incubating the carboxyl terminated surfaces with 3 mg of NHS and 30 mg of EDC in 30 ml water prior to incubation with PEI solution. Both the surfaces were then rinsed by sonication in water using a bath sonicator for 1 h and dried using a jet of nitrogen. Piranha treated glass surfaces were used as reference in all the experiments. All surfaces were freshly prepared and used within 24 h.

Characterization of surfaces

The surface chemical composition of modified surfaces and the unmodified glass was characterized by X-ray photoelectron spectroscopy (XPS). XPS spectra were recorded using a Kratos Axis Ultra^{DLD} instrument (Kratos Ltd, Telford, UK) equipped with a monochromated aluminum anode (Al kα 1486 eV) operating at 150 W (15 kV and 10 mA) with pass energies of 20 eV and 160 eV for high resolution and survey spectra respectively. The charge neutralizer was used to neutralize any positive charge developed during the measurements on electrically non-conducting surfaces. A hybrid lens mode was employed during analysis (electrostatic and magnetic). XPS spectra of the surfaces were recorded at three different spots on each sample. Relative atomic percentages were calculated from the averages of three survey spectra recorded for each sample on three different spot. The take-off angle with respect to normal to the surface was 0° for all measurements. The measured binding energy positions were charge corrected with reference to 285.0 eV, corresponding to the C-C/C-H species. Quantification was conducted using CasaXPS software. A linear background was used to analyze all spectra.

Surface zeta potential measurements were carried out using the surface zeta potential cell in zetasizer nano ZS (Malvern Instruments limited), as described previously [54]. Polystyrene nanoparticles (~100 nm -PS100 sulphate modified Invitrogen DK) were used as tracer particles. Measurements were made at 125, 250, 375, and 500 μm from the sample surface. An additional measurement was made at 1000 μm to determine the zeta potential of the tracer particles. Surface zetapotentials were calculated from three replicate measurements and expressed as millivolts (mV). Surface zetapotential uncertainties calculated by the software from three measurements were considered as standard deviation.

Characterisation of bacterial cell surface

Static contact angles were measured on the different surfaces and bacterial lawns using water. Bacterial lawns were prepared as described previously [55]. Briefly, 10^9 cells per ml of planktonic *S. xylosus* cells from an overnight culture were harvested and treated with DNase or PBS (control) as described above. After treatment, the cells were washed 3 times and resuspended in distilled water, before depositing the suspension on to a cellulose acetate filter membrane containing 0.45 μm diameter pores under negative pressure. All images of liquid drops on surfaces were recorded using a KRUSS DSA100 (KRUSS GmbH, Hamburg, Germany) drop shape analysis system, followed by drop shape analysis using ImageJ software. Water contact angle measurements were done on at least 3 different places on triplicate samples of each surface and averaged.

The zeta potential of bacterial cells was also measured to investigate changes in cell surface charge after removal of eDNA. *S. xylosus* (+/– DNase treatment) were washed and resuspended to OD$_{600}$ of 1.0 in PBS. The conformation of the eDNA on cell surface, and the Zeta potential can be affected by the ionic strength of the buffer. PBS has a relatively high ionic strength. In order to account for the effect of ionic strength on zeta potential, we made the measurements in PBS with two different ionic strengths, 1 mM (I = ~0.019) and 10 mM (I = 0.19). Triplicate values of the zeta potential were measured for each sample in a zetasizer (Zetasizer nano, Malvern Instruments, UK) at 25°C. Each replicate value was an average of 10 measurements.

Statistical analysis

Bacterial adhesion was studied on three replicate surfaces incubated individually in bacterial suspensions that were prepared from the same overnight culture, and then separated into aliquots that were either suspended in PBS+MgCl$_2$ with and without DNase before submerging samples (one sampler per aliquot). Statistical analyses were done by student t-test and two-way analysis of variance (ANOVA) using R, the free software for statistical computing.

Supporting Information

Figure S1 XPS analysis of surfaces with different chemistries. A: Wide scan spectrum of Piranha-treated glass; B: Carboxyl-functionalised glass. The high resolution C 1s spectrum shows the carboxyl peak at B.E. 289.1 eV; C: Amine-functionalised glass. The wide scan spectrum shows the nitrogen (N 1s) peak at B.E. 397 eV; D: Fluoro-functionalised glass. The wide scan spectrum shows fluorine (F 1s) peak at B.E. 686 eV.

Acknowledgments

The authors thank Morten Ebbesen (Interdisciplinary Nanoscience Center, Aarhus University, Denmark) for his help in the zeta potential measurements.

Author Contributions

Conceived and designed the experiments: VRR ARL RLM. Performed the experiments: VRR ARL JJM. Analyzed the data: VRR ARL JJM DSS RLM. Contributed reagents/materials/analysis tools: RLM. Wrote the paper: VRR ARL JJM DSS RLM.

References

1. Costerton JW, Lewandowski Z, Caldwell DE, Korber DR, Lappin-Scott HM (1995) Microbial biofilms. Annual Review of Microbiology 49: 711–745.

2. Reid G (1999) Biofilms in infectious disease and on medical devices. International Journal of Antimicrobial Agents 11: 223–226.

3. Roberts AP, Mullany P (2010) Oral biofilms: A reservoir of transferable, bacterial, antimicrobial resistance. Expert Review of Anti-Infective Therapy 8: 1441–1450.

4. Stoodley P, Ehrlich GD, Sedghizadeh PP, Hall-Stoodley L, Baratz ME, et al. (2011) Orthopaedic biofilm infections. Current Orthopaedic Practice 22: 558–563.

5. Liu Y, Zhang W, Sileika T, Warta R, Cianciotto NP, et al. (2009) Role of bacterial adhesion in the microbial ecology of biofilms in cooling tower systems. Biofouling 25: 241–253.

6. Callow ME, Callow JA (2002) Marine biofouling: A sticky problem. Biologist 49: 10–14.

7. Lata S, Sharma C, Singh AK (2011) Microbial influenced corrosion by thermophilic bacteria. Journal of Corrosion Science and Engineering 14.

8. Granhall U, Welsh A, Throbäck IN, Hjort K, Hansson M, et al. (2010) Bacterial community diversity in paper mills processing recycled paper. Journal of Industrial Microbiology and Biotechnology 37: 1061–1069.

9. Van Houdt R, Michiels CW (2010) Biofilm formation and the food industry, a focus on the bacterial outer surface. Journal of Applied Microbiology 109: 1117–1131.

10. Cebula TA, Fricke WF, Ravel J (2011) Food-Borne Outbreaks: What's New, What's Not, and Where Do We Go from Here? pp. 29–41.

11. Banerjee I, Pangule RC, Kane RS (2011) Antifouling Coatings: Recent Developments in the Design of Surfaces That Prevent Fouling by Proteins, Bacteria, and Marine Organisms. Advanced Materials 23: 690–718.

12. Busscher HJ, Norde W, Van Der Mei HC (2008) Specific molecular recognition and nonspecific contributions to bacterial interaction forces. Applied and Environmental Microbiology 74: 2559–2564.

13. Chen Y, Busscher HJ, van der Mei HC, Norde W (2011) Statistical analysis of long- and short-range forces involved in bacterial adhesion to substratum surfaces as measured using atomic force microscopy. Applied and Environmental Microbiology 77: 5065–5070.

14. Fletcher M, Lessmann JM, Loeb GI (1991) Bacterial surface adhesives and biofilm matrix polymers of marine and freshwater bacteria†. Biofouling 4: 129–140.

15. Hermansson M (1999) The DLVO theory in microbial adhesion. Colloids and Surfaces B: Biointerfaces 14: 105–119.

16. Zita A, Hermansson M (1997) Effects of bacterial cell surface structures and hydrophobicity on attachment to activated sludge flocs. Applied and Environmental Microbiology 63: 1168–1170.

17. Marchand S, De Block J, De Jonghe V, Coorevits A, Heynrickx M, et al. (2012) Biofilm Formation in Milk Production and Processing Environments; Influence on Milk Quality and Safety. Comprehensive Reviews in Food Science and Food Safety 11: 133–147.

18. Whitchurch CB, Tolker-Nielsen T, Ragas PC, Mattick JS (2002) Extracellular DNA required for bacterial biofilm formation. Science 295: 1487.

19. Das T, Sharma PK, Krom BP, Van Der Mei HC, Busscher HJ (2011) Role of eDNA on the adhesion forces between streptococcus mutans and substratum surfaces: Influence of ionic strength and substratum hydrophobicity. Langmuir 27: 10113–10118.

20. Marino M, Frigo F, Bartolomeoli I, Maifreni M (2011) Safety-related properties of staphylococci isolated from food and food environments. Journal of Applied Microbiology 110: 550–561.

21. Moschetti G, Mauriello G, Villani F (1997) Differentiation of Staphylococcus xylosus strains from Italian sausages by antibiotyping and low frequency restriction fragment analysis of genomic DNA. Systematic and Applied Microbiology 20: 432–438.

22. Sondergaard AK, Stahnke LH (2002) Growth and aroma production by Staphylococcus xylosus, S. carnosus and S. equorum - A comparative study in model systems. International Journal of Food Microbiology 75: 99–109.

23. Vernozy-Rozand C, Mazuy C, Prevost G, Lapeyre C, Bes M, et al. (1996) Enterotoxin production by coagulase-negative staphylococci isolated from goats' milk and cheese. International Journal of Food Microbiology 30: 271–280.

24. Perreten V, Giampà N, Schuler-Schmid U, Teuber M (1998) Antibiotic resistance genes in coagulase-negative staphylococci isolated from food. Systematic and Applied Microbiology 21: 113–120.

25. Rice KC, Mann EE, Endres JL, Weiss EC, Cassat JE, et al. (2007) The cidA murein hydrolase regulator contributes to DNA release and biofilm development in Staphylococcus aureus. Proceedings of the National Academy of Sciences of the United States of America 104: 8113–8118.

26. Mann EE, Rice KC, Boles BR, Endres JL, Ranjit D, et al. (2009) Modulation of eDNA release and degradation affects Staphylococcus aureus biofilm maturation. PLoS ONE 4.

27. Qin Z, Ou Y, Yang L, Zhu Y, Tolker-Nielsen T, et al. (2007) Role of autolysin-mediated DNA release in biofilm formation of Staphylococcus epidermidis. Microbiology 153: 2083–2092.

28. Das T, Krom BP, Van Der Mei HC, Busscher HJ, Sharma PK (2011) DNA-mediated bacterial aggregation is dictated by acid-base interactions. Soft Matter 7: 2927–2935.

29. Das T, Sharma PK, Busscher HJ, Van Der Mei HC, Krom BP (2010) Role of extracellular DNA in initial bacterial adhesion and surface aggregation. Applied and Environmental Microbiology 76: 3405–3408.

30. Mackey-Lawrence N, Potter D, Cerca N, Jefferson K (2009) Staphylococcus aureus immunodominant surface antigen B is a cell-surface associated nucleic acid binding protein. BMC Microbiology 9: 61.

31. Fredheim EGA, Klingenberg C, Rohde H, Frankenberger S, Gaustad P, et al. (2009) Biofilm Formation by Staphylococcus haemolyticus. Journal of Clinical Microbiology 47: 1172–1180.

32. Harmsen M, Lappann M, Knøchel S, Molin S (2010) Role of Extracellular DNA during Biofilm Formation by Listeria monocytogenes. Applied and Environmental Microbiology 76: 2271–2279.

33. Lappann M, Claus H, Van Alen T, Harmsen M, Elias J, et al. (2010) A dual role of extracellular DNA during biofilm formation of Neisseria meningitidis. Molecular Microbiology 75: 1355–1371.

34. Goodman SD, Obergfell KP, Jurcisek JA, Novotny LA, Downey JS, et al. (2011) Biofilms can be dispersed by focusing the immune system on a common family of bacterial nucleoid-associated proteins. Mucosal Immunol 4: 625–637.

35. Das T, Kutty SK, Kumar N, Manefield M (2013) Pyocyanin Facilitates Extracellular DNA Binding to Pseudomonas aeruginosa Influencing Cell Surface Properties and Aggregation. PLoS ONE 8.

36. Nur A, Hirota K, Yumoto H, Hirao K, Liu D, et al. (2013) Effects of extracellular DNA and DNA-binding protein on the development of a Streptococcus intermedius biofilm. Journal of Applied Microbiology: n/a-n/a.

37. Jurcisek JA, Bakaletz LO (2007) Biofilms formed by nontypeable Haemophilus influenzae in vivo contain both double-stranded DNA and type IV pilin protein. Journal of Bacteriology 189: 3868–3875.

38. Heijstra BD, Pichler FB, Liang Q, Blaza RG, Turner SJ (2009) Extracellular DNA and Type IV pili mediate surface attachment by Acidovorax temperans. Antonie van Leeuwenhoek, International Journal of General and Molecular Microbiology 95: 343–349.

39. Barken KB, Pamp SJ, Yang L, Gjermansen M, Bertrand JJ, et al. (2008) Roles of type IV pili, flagellum-mediated motility and extracellular DNA in the formation of mature multicellular structures in Pseudomonas aeruginosa biofilms. Environmental Microbiology 10: 2331–2343.

40. Boks NP, Norde W, van der Mei HC, Busscher HJ (2008) Forces involved in bacterial adhesion to hydrophilic and hydrophobic surfaces. Microbiology 154: 3122–3133.

41. Loosdrecht MM, Lyklema J, Norde W, Zehnder AB (1989) Bacterial adhesion: A physicochemical approach. Microbial Ecology 17: 1–15.

42. van Loosdrecht MCM, Norde W, Lyklema J, Zehnder AJB (1990) Hydrophobic and electrostatic parameters in bacterial adhesion - Dedicated to Werner Stumm for his 65th birthday. Aquatic Sciences 52: 103–114.

43. Weiss L, Harlos JP (1972) Short-term interactions between cell surfaces. Progress in Surface Science 1: 355–405.

44. Robinson H, Gao YG, Sanishvili R, Joachimiak A, Wang AHJ (2000) Hexahydrated magnesium ions bind in the deep major groove and at the outer mouth of A-form nucleic acid duplexes. Nucleic Acids Research 28: 1760–1766.

45. Romanowski G, Lorenz MG, Wackernagel W (1991) Adsorption of plasmid DNA to mineral surfaces and protection against DNase I. Applied and Environmental Microbiology 57: 1057–1061.

46. Lyubchenko YL (2011) Preparation of DNA and nucleoprotein samples for AFM imaging. Micron 42: 196–206.

47. Nguyen TH, Chen KL (2007) Role of divalent cations in plasmid DNA adsorption to natural organic matter-coated silica surface. Environmental Science and Technology 41: 5370–5375.

48. Van Oss CJ (1995) Hydrophobicity of biosurfaces - Origin, quantitative determination and interaction energies. Colloids and Surfaces B: Biointerfaces 5: 91–110.

49. Huub HMR, Willem N, Edward JB, Johannes L, Alexander JBZ (1995) Reversibility and mechanism of bacterial adhesion. Colloids and Surfaces B: Biointerfaces 4.

50. Beckett R, Le NP (1990) The role or organic matter and ionic composition in determining the surface charge of suspended particles in natural waters. Colloids and Surfaces 44: 35–49.

51. Hunter KA, Liss PS (1979) The surface charge of suspended particles in estuarine and coastal waters [9]. Nature 282: 823–825.

52. Hunter KA, Liss PS (1982) Organic Matter and the Surface Charge of Suspended Particles in Estuarine Waters. Limnology and Oceanography 27: 322–335.

53. Seu KJ, Pandey AP, Haque F, Proctor EA, Ribbe AE, et al. (2007) Effect of surface treatment on diffusion and domain formation in supported lipid bilayers. Biophysical Journal 92: 2445–2450.

54. Corbett JCW, McNeil-Watson F, Jack RO, Howarth M (2012) Measuring surface zeta potential using phase analysis light scattering in a simple dip cell arrangement. Colloids and Surfaces A: Physicochemical and Engineering Aspects 396: 169–176.

55. Busscher HJ, Weerkamp AH, van Der Mei HC, van Pelt AW, de Jong HP, et al. (1984) Measurement of the surface free energy of bacterial cell surfaces and its relevance for adhesion. Applied and Environmental Microbiology 48: 980–983.

8

A Proteomic Approach to Understand the Role of the Outer Membrane Porins in the Organic Solvent-Tolerance of *Pseudomonas aeruginosa* PseA

R. Hemamalini, Sunil Khare*

Enzyme and Microbial Biochemistry Lab, Department of Chemistry, Indian Institute of Technology, Delhi, New Delhi, India

Abstract

Solvent-tolerant microbes have the unique ability to thrive in presence of organic solvents. The present study describes the effect of increasing hydrophobicity (log P_{ow} values) of organic solvents on the outer membrane proteome of the solvent-tolerant *Pseudomonas aeruginosa* PseA cells. The cells were grown in a medium containing 33% (v/v) alkanes of increasing log P_{ow} values. The outer membrane proteins were extracted by alkaline extraction from the late log phase cells and changes in the protein expression were studied by 2-D gel electrophoresis. Seven protein spots showed significant differential expression in the solvent exposed cells. The tryptic digest of the differentially regulated proteins were identified by LC-ESI MS/MS. The identity of these proteins matched with porins OprD, OprE, OprF, OprH, Opr86, LPS assembly protein and A-type flagellin. The reported pI values of these proteins were in the range of 4.94–8.67 and the molecular weights were in the range of 19.5–104.5 kDa. The results suggest significant down-regulation of the A-type flagellin, OprF and OprD and up-regulation of OprE, OprH, Opr86 and LPS assembly protein in presence of organic solvents. OprF and OprD are implicated in antibiotic uptake and outer membrane stability, whereas A-type flagellin confers motility and chemotaxis. Up-regulated OprE is an anaerobically-induced porin while Opr86 is responsible for transport of small molecules and assembly of the outer membrane proteins. Differential regulation of the above porins clearly indicates their role in adaptation to solvent exposure.

Editor: John R. Battista, Louisiana State University and A & M College, United States of America

Funding: RH gratefully acknowledges the Indian Council of Medical Research, New Delhi, India for providing research fellowship. ICMR Sanction Order No. No. 3/1/3/JRF-2010/MPD-47(36165). The funders had no role in study design, data collection and analysis, decision to publish, or preparation of the manuscript.

Competing Interests: The authors have declared that no competing interests exist.

* Email: skkhare@chemistry.iitd.ac.in

Introduction

Organic solvents have been in use as disinfectants and microbicidal agents for a long time. However, the discovery of solvent-tolerant *Pseudomonas putida* by Inoue and Horikoshi [1] initiated understanding of a class of microorganisms described as 'solvent-tolerant microorganisms' capable of growing at high concentrations of organic solvents [1,2]. For instance, strains of *Pseudomonas putida* tolerant to toluene, *Clostridium acetobutylicum* to butanol, *Escherichia coli* to ethanol are known among the Gram-negative bacteria [3]. On the other hand, *Arthrobacter* and *Flavobacterium* resistant to benzene, *Bacillus* resistant to chloroform, n-butanol and benzene and benzene-tolerant *Rhodococcus* strains are known from Gram-positive bacteria [4,5].

There is an increasing interest in such organisms due to their effectiveness in solvent bioremediation and biotransformations in non-aqueous media [2]. The unique properties of these microbes also kindled interest in understanding and elucidating their adaptation mechanism. In general, microorganisms adapt to organic solvents by (i) partitioning the solvent in the lipid layer, (ii) isomerisation of *cis*- unsaturated fatty acids to *trans*- unsaturated fatty acids leading to denser membranes, (iii) changing the saturated-to-unsaturated fatty acid ratio (iv) changes in length of

the acyl-chains and (v) changes in the phospholipid head groups [6]. Since the outer membrane proteins (OMPs) are the primary targets of solvents surrounding the cells, the outer membrane sub-proteome would be a sensitive response indicator for studying the effect of solvents [7].

So far, proteomics studies on the bacteria exposed to solvents have been focused on highly toxic solvents such as aliphatic and aromatic alcohols, and alkyl benzenes. For instance, the effect of phenol on the membrane proteome of *Pseudomonas*, have shown an increase in the expression levels of the solvent efflux pump systems (e.g. TtgA, TtgC, Ttg2A, Ttg2C, PP_1516-7) and a decreased content of porins OprB, OprF, OprG and OprQ [8]. An imp/ostA encoded 87 kDa, outer membrane protein has been implicated in n-hexane tolerance in *Escherichi coli* [9]. TolC, the outer membrane component of the AcrAB-TolC solvent efflux pump found in *Escherichia coli* is suggested to play a crucial role in imparting solvent tolerance [10]. While a substantial understanding of the mechanisms underlying the adaptation of solvent-tolerant bacteria towards aromatic compounds has emerged, the effect of alkanes has not been investigated systematically.

Alkanes are common environmental effluents as they constitute 20–50% of crude oil [11]. They occur as mixtures in crude oil and are also among the commonly encountered industrial solvents, yet systematic studies on their uptake by microbial cells, membrane adaptations, change in protein expression and proteomic profiles have not been much worked out. Hence, it seems quite worthwhile to investigate the interaction of alkanes with cells and understand the cellular response.

The present work aims at evaluating the effect of alkanes of increasing hydrophobicity (log P_{ow}) such as n-hexane, cyclohexane, n-heptane, n-octane, n-decane, n-dodecane and n-tetradecane on the proteome profile of a solvent-tolerant *Pseudomonas* strain, especially to see which proteins get differentially expressed in response to solvent exposure and thus play a role in the cellular adaptation. Since the outer membrane is the first to interface the outer environment, the focus of the present study is the outer membrane proteome. A well characterized solvent-tolerant *Pseudomonas aeruginosa* strain PseA (MTCC 10634), previously isolated by us has been used as a model system [12–14]. The cells grown in the medium containing high amount of alkanes were studied for the changes in the outer membrane proteome using two-dimensional gel electrophoresis followed by identification of the differentially expressed proteins by ESI-Mass. The results indicate that the outer membrane porins show significant differential expression in presence of alkanes. The differential expression of type A flagellin, in response to various alkanes is observed first time, although its involvement in toluene tolerance has been shown previously [15].

Methods

Chemicals

Yeast extract and peptone were purchased from HiMedia Laboratories (Mumbai, India). Acrylamide, bisacrylamide, iodoacetamide IPG strips, IPG buffer and DryStrip cover fluid were obtained from GE Healtcare (Uppsala, Sweden), DTT and DeStreak Reagent were from PlusOne (Germany). Formic acid was procured from Merck (Darmstadt, Germany). Molecular weight markers were from Bangalore Genei (Bangalore, India). In-gel tryptic digestion kit was from Sigma-Aldrich (St. Louis, USA). All other chemicals used were of analytical grade.

Bacterial strain and growth conditions

Pseudomonas aeruginosa strain PseA, (MTCC 10634), an organic solvent-tolerant microorganism was maintained and subcultured as described previously [12]. Inoculum was prepared by transferring a loopful of this stock culture to the growth medium prepared in nutrient broth (NB) containing (g/L) yeast extract 3.0, peptone 5.0 and NaCl 0.5 The pH was adjusted to 7.2 and the culture was grown at 30°C with shaking at 120 rpm until the absorbance at 660 nm (A_{660}) reached 1.0. To check the effect of organic solvents on the proteome of *P. aeruginosa*, 2% (v/v) of the inoculum was used to seed eight identical 500 ml Erlenmeyer flasks containing 100 ml of growth medium which was then overlaid with 50 ml of the solvent. The flasks were sealed with butyl rubber stoppers to prevent the evaporation of solvents and

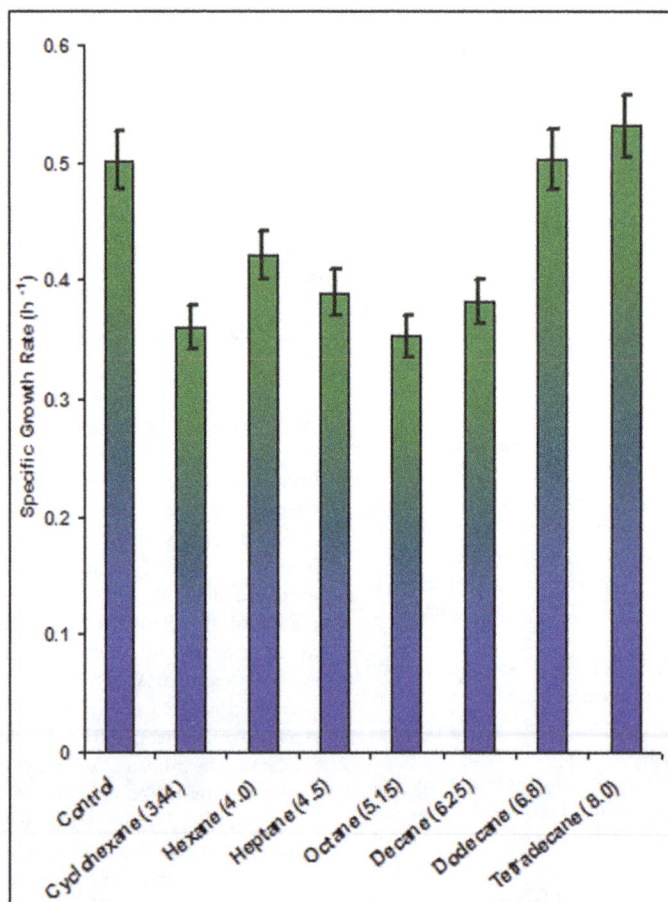

Figure 1. Effect of solvent polarity on the specific growth rate of *Pseudomonas aeruginosa* PseA. The log P_{ow} values have been indicated in parentheses.

(a)
(b)
(c)
(d)
(e)
(f)
(g)
(h)

Figure 2. Outer membrane proteome of *Pseudomonas aeruginosa* **PseA grown in presence of various alkanes.** The *Pseudomonas aeruginosa* cells were grown in NB medium containing 33% (v/v) solvents, pH 7.2, at 30°C and 120 rpm. The cells were harvested in late log phase and their outer membrane proteins extracted following alkaline extraction and ultracentrifugation. The proteins so obtained were then subjected to 2D electrophoresis. The results are of cells grown in (a) control, only NB medium. (b) NB medium + n-hexane (c) NB medium + cyclohexane (d) NB medium + n-heptane (e) NB medium + n-octane (f) NB medium + n-decane (g) NB medium + n-dodecane (h) NB medium + n-tetradecane.

incubated at 30°C and 140 rpm until late log phase was reached. *P. aeruginosa* was also grown in NB, in absence of solvent as a control under similar conditions. Growth was monitored by recording absorbance at 660 nm. The aqueous phase, containing the cells, was carefully pipetted out from beneath the organic layer and the cells were harvested by centrifugation for 15 minutes at $10,000 \times g$ at 4°C. The cells were washed with ice cold Tris-HCl buffer (0.02 M, pH 8.5) and the pellet was stored at −80°C.

Extraction of bacterial membrane proteins

The cell pellets were processed for membrane proteome sample preparation following a modified form of the method described previously [8]. Briefly, 3 stored cell pellets each from control and solvent treated cultures were resuspended in chilled lysis buffer (0.020 M Tris HCl, pH 8.5) containing 25 mM NaCl, 2 mM EDTA, 10% (v/v) glycerol, 0.5% (v/v) Triton X-100, 10 mM 2-mercaptoethanol, 1 mM DTT, 1 mM PMSF, 10 mM NaF. The suspension was sonicated twice on ice using a cycle of 40%, for 5 min, 120 watts at a frequency of 20 kHz on a Biologics Ultrasonic Homogenizer (Virginia, USA). The sonicated suspension was centrifuged for 30 min at $10,000 \times g$ at 4°C. The pellets were discarded. The supernatant was diluted with ice cold sodium carbonate solution (0.1 M, pH 11.0) and stirred slowly on ice for 1 h [14]. The carbonate treated membrane fraction was ultracentrifuged in a Beckman 70 Ti rotor for 1 h at $99,000 \times g$ at 4°C. The pellet was suspended in 16 ml of 50 mM Tris HCl, pH 7.2 and the ultracentrifugation step repeated. The pellets obtained at the end consisted of soluble membrane proteins. They were dissolved in DeStreak reagent (Plus One, GE Healthcare) and processed further.

2-D Electrophoresis

Proteins present in the total membrane fraction were quantified by Bradford assay [15]. Overnight passive rehydration of 13 cm, 3–10 NL IPG strips (GE Healthcare) were carried out in presence of 250 μl of sample containing 300 μg of protein and 2% (v/v) IPG buffer containing pharmalytes for the 3–10 NL IPG strips. Isoelectric focusing was carried out on an Ettan IPGphor 3 instrument (GE Healthcare, Buckinghamshire, UK) for 15,000 Vh at a maximum voltage of 8000 V. After the IEF run, the IPG strips were incubated in the SDS equilibration buffer (6 M Urea, 75 mM Tris-HCl pH 8.8, 29.3% Glycerol, 2% SDS, 0.002% bromophenol blue) containing DTT 1% (w/v) for 15 min followed by incubation in the same buffer containing iodoacetamide 2.5% (w/v) for 15 min. Second dimension electrophoresis was carried out on SE 600 Ruby (GE Healthcare) by placing the equilibrated IPG strips horizontally over 1.5 mm thick, 12.5% SDS-polyacrylamide gels and sealed in place with molten agarose sealing solution (25 mM Tris base, 192 mM glycine, 0.1% SDS, 0.5% agarose, 0.002% bromophenol blue). Electrophoresis was carried out in two steps: first step at 15 mA/gel for 15 min followed by the second step at 50 mA/gel till the dye front was 1 mm from the bottom of the gel. Gels were stained with Coomassie Brilliant Blue R-250 and then destained in a solution of 30% (v/v) methanol and 10% (v/v) acetic acid. All the gels were scanned on a BioRad Gel doc system, Universal Hood II and the images were saved for further spot analysis.

Analysis of protein expression levels

Gel images were analyzed using Image Master Platinum 7.0 software (GE Healthcare). Protein spots were identified by using the automatic spot detection algorithm. Individual spot volumes were normalized against total spot volumes for a given gel. Statistical analysis was also performed by the Image Master Platinum 7.0 software package. Averages of protein abundance for membrane fraction of each growth condition were also compared by their normalized volume using one-way ANOVA between test groups. This test returned a p-value that takes into account the mean difference and the variance of a matched spot between sample conditions and also the sample size. Only statistically significant spots ($p < 0.05$) were selected for analysis. Differential expression between the different solvents and control was quantified on the basis of volume percent. Spots that showed at least 1.5 fold change in their expression levels and clearly visible on the 2D gel were considered for further analysis. Protein spots that were absent from control or in all the solvent-treated samples, were also considered for analysis. Triplicate gels were run for control and each of the seven solvent treated samples.

Identification of proteins by liquid chromatography-electrospray ionization-tandem mass spectrometry (LC-ESI-MS/MS)

Protein spots that showed significant differential expression were excised and subjected to tryptic in-gel digestion using the Sigma-Aldrich in-gel tryptic digestion kit (Sigma-Aldrich, St. Louis, USA) as per manufacturer's instruction. The digests were kept at −20°C prior to analysis and then eluted on a 2.2 μm, 120°A (2.1×150 mm) Dionex column (Acclaim TM RSLC C18) using LC (Dionex Ultimate 3000) (Thermo Scientific, Geel, Belgium). Mobile phase A was composed of 0.1% (v/v) formic acid in water. Mobile phase B was 0.1% (v/v) formic acid in ACN. 40 μl of sample was injected and the percentage of mobile phase B was increased linearly from 10 to 75% (v/v) in 25 min and from 75 to 90% (v/v) in 10 min, maintained at 90% (v/v) for 10 more min before finally reducing to 10% (v/v) in 10 min. The column effluent was connected to an ESI nano-sprayer on microTOF-QII (Bruker, Bremen, Germany). In the survey scan, MS spectra were acquired for 1 min in the m/z range between 400 and 1400. The mass data was analyzed on the Compass Data Analysis software and a mascot generic file format was generated which was then used for automated protein identification through a library search using the Mascot 2.4.01 search engine (Matrix Science, London, UK) through the BioTools software version 3.0 (Bruker Daltonics, Billerica, USA). In the MS/MS Ion Search dialog box, the taxonomy was set to a 'All entries', NCBInr or SwissProt databases were used for the different searches, enzyme trypsin was selected from the list, partial (cleavages per peptide) was set to 1, global modification was set to carbamidomethyl (C) and the variable modification was set to Oxidation (M). The mono-isotopic peptide mass tolerance was set to 25 to 75 ppm depending on the sample while its fragment ion mass tolerance was set to 0.2 Da. Mascot scores above the zone of significance were taken to be valid for the protein identity.

Table 1. Differential expression of outer membrane proteins in *Pseudomonas aeruginosa* grown in presence of alkanes.

Protein spot no.	Effect on expression levels in the solvent treated cells*						
	n-Hexane	Cyclohexane	n-Heptane	n-Octane	n-Decane	n-Dodecane	n-Tetradecane
1		↓				↓	
3	↓				↓		
5		↓			↓	↓	↓
6							↓
7		↑	↑		↑	↑	
8		↑					
9	↑	↑	↑				
10	↓	↓	↓	↓	↓	↓	
18				↓	↓		
19		↑					
20		↑	↑			↑	
21		↑	↑			↑	

*Protein spots that are seen in control as well as solvent-treated cells but showing different level of expression.

↓ Solvent-treated cells showing significant reduction in expression levels.

↑ Solvent-treated cells showing significant increase in expression levels.

Table 2. Characteristics and function of outer membrane proteins of *Pseudomonas aeruginosa* PseA grown in alkanes.

Protein spot no.	Gene	Protein name	NCBI Accession number	Molecular weight kDa	pI	Mascot score	Number of matched peptides	Sequence coverage (%)	GRAVY	Function
1	oprD	Outer membrane porin OprD family porin	gi|386066727	46.9	5.45	791	12	29	−0.43	Allow entry of aminoacids, peptides and carbapenems into cell [25]
3	oprF	Outer membrane porin F	gi|15596974	37.8	4.98	202	4	19	−0.44	Involved in cell structure maintenance, outer membrane permeability, environment and host immune system sensing, and virulence [33–36]
5	oprF	Outer membrane porin F	gi|15596974	37.8	4.98	202	4	19	−0.44	Involved in cell structure maintenance, outer membrane permeability, environment and host immune system sensing, and virulence [33–36]
6	oprH	Chain A, Solution Structure Of Outer Membrane Protein H (OprH) In Dhpc Micelles	gi|344189438	19.5	8.11	1145	18	73	−0.63	Prevent uptake of antimicrobials [37]
7	oprE	Anaerobically-induced outer membrane porin OprE precursor	gi|15595488	49.6	8.67	264	4	9	−0.44	Involved in transport of small molecules and is anaerobically induced [38]
8	oprE	Anaerobically-induced outer membrane porin OprE precursor	gi|15595488	49.6	8.67	264	4	9	−0.44	Involved in transport of small molecules and is anaerobically induced [38]
9	oprE	Anaerobically-induced outer membrane porin OprE precursor	gi|15595488	49.6	8.67	264	4	9	−0.44	Involved in transport of small molecules and is anaerobically induced [38]
10	fliC	A-type flagellin	gi|126165550	40.0	4.94	904	12	37	−0.13	Structural units of flagella [39]
20	opr86	Outer membrane porin Opr86	gi|15598844	88.3	5.05	1197	15	30	−0.37	Responsible for outer membrane protein assembly, cell viability and biofilm formation [27]
21	lptD	LPS assembly protein	gi|25008883	104.6	5.37	314	4	5	−0.6	Role in lipopolysaccharide assembly in the outer membrane [28]
51	oprH	Chain A, Solution Structure Of Outer Membrane Protein H (OprH) In Dhpc Micelles	gi|344189438	19.5	8.11	1145	18	73	−0.63	Prevent uptake of antimicrobials [37]
53	oprH	Chain A, Solution Structure Of Outer Membrane Protein H (OprH) In Dhpc Micelles	gi|344189438	19.5	8.11	1145	18	73	−0.63	Prevent uptake of antimicrobials [37]
71	oprH	Chain A, Solution Structure Of Outer Membrane Protein H (OprH) In Dhpc Micelles	gi|344189438	19.5	8.11	1145	18	73	−0.63	Prevent uptake of antimicrobials [37]
75	oprH	Chain A, Solution Structure Of Outer Membrane Protein H (OprH) In Dhpc Micelles	gi|344189438	19.5	8.11	1145	18	73	−0.63	Prevent uptake of antimicrobials [37]
100	oprH	Chain A, Solution Structure Of Outer Membrane Protein H (OprH) In Dhpc Micelles	gi|344189438	19.5	8.11	1145	18	73	−0.63	Prevent uptake of antimicrobials [37]

Determination of protein hydrophobicity

The hydrophobicity of a protein is an important property of membrane proteins. In the present study, the theoretical hydrophobicity of the identified proteins was determined by an *in-silico* approach. The grand average hydropathy (GRAVY) value is a measure of the hydrophobicity of a protein [16]. For each of the identified proteins, the aminoacid sequence in the FASTA format was submitted to the online GRAVY calculator (www.gravy-calculator.de) in order to determine the GRAVY score. This value was then used to correlate the hydrophobic nature of the protein and its expression pattern in response to the hydrophobic solvent stress.

Results and Discussion

Effect of solvents on the growth

Solvent-tolerant Gram-negative bacteria adapt by making certain changes in outer membrane composition and structure in response to solvent exposure [3]. In the present study, the effect of alkanes was analyzed on a known solvent-tolerant *Pseudomonas aeruginosa* strain. The strain was grown in NB media containing 33% (v/v) alkanes of varying hydrophobicity and its outer membrane proteome was analyzed to elucidate the up-regulation and down-regulation of specific proteins involved in membrane-level response to the increasing hydrophobicity of solvent. The control consisted of cells grown in NB media without addition of any organic solvent.

The toxicity of a solvent and its effect on growth is generally monitored with respect to the hydrophobicity of that solvent, which is represented as the hydrophobicity index called log P_{ow} value (defined as the partition coefficient of a solvent in an equimolar mixture of octanol and water). Solvents with lower log P_{ow} values have been reported to be more toxic to living cells [17]. The effect of solvents of varying log P_{ow} on the growth of the solvent-tolerant *Pseudomonas aeruginosa* PseA is shown in Figure 1. The specific growth rate (μ) was reduced by 16–30% in presence of solvents cyclohexane, n-hexane, n-heptane, n-octane and n-decane (log P_{ow} 3.44–6.25), cyclohexane causing maximum reduction. The growth remained unaffected in presence of solvents of higher hydrophobicity index, n-dodecane (log P_{ow} 6.80) and n-tetradecane (log P_{ow} 8.00). Evidently, solvents with a lower log P_{ow} value affect growth more adversely.

The *Pseudomonas* strain used herein was previously isolated by cyclohexane enrichment from soil samples in the vicinity of a solvent extraction unit in New Delhi, India [12]. It has been well characterized for its solvent-tolerant traits [2,12,13]. A previous study on the cellular mechanisms for solvent tolerance of this strain revealed that cyclohexane (log P_{ow}) altered the cell growth, morphology, size, membrane integrity, permeability and surface hydrophobicity, whereas n-tetradecane (log P_{ow}) did not affect any of these parameters [18].

Loss of membrane integrity is reported to cause the leakage of ATP, ions, nucleic acids, phospholipids and proteins [3]. Similar leakage of dihydrolipoamide dehydrogenase, GroEL and other cytoplasmic proteins also occurred into the secretome of *Pseudomonas aeruginosa* PseA cells treated with solvents, n-hexane, cyclohexane, n-heptane and n-octane (data not shown). This observation further supports damaged/leaky membrane in presence of solvents of low log P_{ow} values. The reduction in the specific growth rate of the bacterial strain in presence of cyclohexane to n-decane can therefore be explained in the light of the more toxic effect of these solvents on the membrane, as compared with higher log P_{ow} solvents.

Effects of solvents on the outer membrane of *Pseudomonas aeruginosa*

Exposure to phenol has been shown earlier to lead to a coordinate increase in solvent efflux pump proteins such as TtgA, TtgC, Ttg2A, Ttg2C, and PP_1516-7 and a decreased content of porins OprB, OprF, OprG and OprQ [8]. Considering the fact that outer membrane is the first to be affected, the control and solvent-grown cells were harvested in the late log phase and processed for isolating the outer membrane proteins. The outer membrane proteins from each sample were subjected to 2-D gel electrophoresis in triplicates. The gels were scanned using BioRad Gel doc system. The outer membrane proteome map for control and each of the solvent-treated samples are shown in Figure 2 (a–h). The proteome spots were analyzed by Image Master Platinum 7.0 software. A total of 117 statistically significant spots (threshold p-value<0.05) were detected in the control and 7 gels. Number of proteins detected on the 2-D gels varied across the different solvent-treated samples. All these spots were seen to have a pI between 3–8 and molecular mass in the range of 14–40 kDa, as suggested by their position in the 2-D gel.

Further analysis by Image Master revealed that twelve spots were significantly, differentially expressed. The results are summarized in Table 1. Among these proteins spot numbers 1, 3, 5, 10 and 18 were down-regulated while 7, 8, 9, 19, 20 and 21 were up-regulated. Most of the spots show significant changes in case of cyclohexane-treated cells.

Protein spot 1 was down-regulated to the maximum extent in case of cyclohexane and n-dodecane exposed cells. The spot 10 showed a considerable variation in expression, mostly down-regulated in all the solvents, except in case of n-tetradecane. Protein spot 6 of the control was seen only in n-tetradecane with a significant down-regulation. This spot disappeared in all other solvents.

Interestingly, some spots which were not seen in control appeared as altogether new spots in response to solvents. These were 51 for dodecane, 53 for decane, 71 for octane, 75 for heptane and 100 for n-hexane. These spots were all significantly up-regulated and were analysed for determining their identity.

Characterization and functional correlation of differentially expressed proteins

Protein spots showing significant differential expression were excised and subjected to in-gel tryptic digestion and subjected to LC-ESI-MS/MS analysis of tryptic digests. The results were matched with databases and identified using the Mascot search engine. Only matches in the significant region and those with the top most score were considered. The identity, characteristics and possible functions of the differentially expressed proteins are summarized in Table 2. The theoretical pI values of these proteins, as per the database, were in the range of 4.94–8.67 and the molecular weights ranged between 19.5–104.5 kDa. All these proteins were localized to the outer membrane.

Protein spot 1 matched with OprD, 3 and 5 with OprF, 7,8 and 9 with OprE, 10 with A-type flagellin, 20 with Opr86 and 21 with LPS assembly protein LptD. OprH was the protein match found for protein spot 6 seen in control and tetradecane-treated cells. The protein spots 51, 53, 71, 75 and 100 that were absent from control, but were seen in n-dodecane, n-decane, n-octane, n-heptane and n-hexane also matched with OprH on the database search.

Outer membrane proteome of *Pseudomonas aeruginosa*, has been reported to have 104 Open Reading Frames (ORFs). Functionally, these proteins belong to three major categories namely, porins, receptors and proteins with unknown function. The biggest group is that of porins consisting of 17 proteins [19]. In the present study, eight differentially expressed proteins of *Pseudomonas aeruginosa* outer membrane belong to the porin classes, OprD, E, F, H, and Opr86. Porins are pore-forming proteins which are considered important for transport, both into and out of cells. General diffusion pores formed by porins allow the diffusion of hydrophilic molecules (<600 Da) without much substrate specificity [20]. Alkanes, despite their hydrophobic character also seem to have an effect on outer membrane porins which are usually known to form channels that enable the passage of hydrophilic molecules such as phenol [8,21], chlorophenol and antibiotics [22]. The entry of such solvents across the outer membrane into the cell, results in the leakage of protons across the outer membrane and a drop in the proton motive force across the inner bacterial membrane [24]. The change in the porin levels seen in presence of alkanes is likely to be an attempt by the cell to restore this proton imbalance.

Porins OprE, OprH, Opr86 and a lipopolysaccharide assembly protein, LptD were up-regulated in presence of solvents. OprE is an anaerobically induced outer membrane porin [38] predicted to take up either arginine or proline [39]. Its up-regulation in the present study can be explained by the distinct anaerobic environment created by the high concentration, 33% (v/v), of organic solvents used for growing the bacterial cells. This indicates that cells grown in solvent, face low oxygen and anaerobic like environment. OprH is up-regulated in all except n-tetradecane-treated cells, where it is infact down-regulated. OprH expression has been shown to result in the blocking of self-promoted uptake of antimicrobials [25]. The up-regulation of OprH in the present case indicates enhanced uptake of solvents into cell. Opr86 is responsible for transport of small molecules, cell viability and outer membrane protein assembly. When Opr86 is depleted, visible changes are evident in the morphology of the cell [26,27]. The up-regulation of Opr86 in the present case may be due to the damage to the outer membrane caused by the solvents and an imminent need thereof, for its repair. In our previous studies related to the solvent effect on the cells of *P. aeruginosa*, we have shown the damage caused to the membrane in presence of toxic solvents and also the intracellular accumulation of solvents [13]. Hence, the up-regulation of OprH in presence of organic solvents of lower log P_{ow} value and of Opr86 in all the solvent-treated samples, supports our previous observation. LptD along with LptE is an essential outer membrane protein involved with lipopolysaccharide assembly in the outer membrane as part of the outer membrane LPS assembly [28,29]. The up-regulation of this protein in presence of alkanes is likely to be another measure in the direction of restoration of normal structure of the cell.

The OprD, OprF, and A-type flagellin are down-regulated in response to all the alkanes. Their respective functions are that (i) OprD, as a component of a multi-drug resistance efflux pump, has been implicated in carbapenem resistance. It also acts as a specific channel for basic aminoacids and small peptides [23,30] and possibly alkanes in the present case (ii) OprF is the major non specific porin of *Pseudomonas aeruginosa* that is known to form large outer membrane channels that permit the passage of various solutes, albeit slowly [31]. Thus, the down-regulation of these porins appropriately explains the adaption of cells to block the solvent passage.

Besides porins, A-type flagellin, shows a significant decrease in expression across all the solvent treated cells. We hypothesize that this may be due to a need for the bacterial cells to reduce their permeable surface area. This is a new finding that has not been reported so far, in bacteria. However, studies on the algal flagellate, *Chlamydomonas* have shown that high concentrations of chemical agents in the cellular environment of the algal cells, caused the flagella to be shed. This enables the cells to reduce their permeable surface area [32]. This forms the highlight of this work.

Protein hydrophobicity

The subcellular localization of the proteins visualized on the 2-D gels was further confirmed by determination of their GRAVY values using the online GRAVY calculator (www.gravy-calculator. de). All the detected proteins had a negative GRAVY ranging from -0.13 to -0.63 indicating their hydrophilic nature and localization to the outer membrane protein.

The present study describes the changes in outer membrane proteome of a known solvent-tolerant *P. aeruginosa* cells as a result of solvent exposure. Briefly, the results show that seventeen proteins were differentially expressed when grown in presence of alkanes. These differentially expressed proteins were identified by LC-ESI MS/MS. Their identity matched with porins OprD, OprE, OprF, OprH, Opr86, LPS assembly protein and A-type flagellin. The results showed significant down-regulation of the flagellin A protein, OprF, OprD; up-regulation and modification of OprH, OprE, Opr86 and Lpt D in presence of organic solvents. These changes reflect upon the adaptation of the bacterial cells for survival under solvent-rich conditions in which the porins play an important role.

Acknowledgments

Authors are thankful to Dr. Sailesh Bajpai, GE Healthcare Lifesciences, Mr. Rajesh Vashisht, Bruker-Daltonics for their technical assistance, Dr. Manidipa Banerjee, Assistant Professor, Kusuma School of Biological Sciences, Indian Institute of Technology, Delhi (IITD), for permitting the use of the ultracentrifuge facility and Ms. Saumya Verma for her help.

Author Contributions

Conceived and designed the experiments: RH SK. Performed the experiments: RH. Analyzed the data: RH SK. Contributed reagents/materials/analysis tools: SK. Contributed to the writing of the manuscript: RH SK.

References

1. Inoue A, Horikoshi K (1989) A *Pseudomonas* thrives in high concentrations of toluene. Nature 338: 264–266.
2. Gupta A, Khare SK (2009) Enzymes from solvent tolerant microbes. Crit Rev Biotechnol 29: 44–54.
3. Segura A, Molina L, Fillet S, Krell T, Bernal P, et al. (2012) Solvent tolerance in Gram-negative bacteria. Curr Opin Biotechnol 23: 415–421.
4. Sardessai YN, Bhosle S (2004) Industrial potential of organic solvent tolerant bacteria. Biotechnol Prog 20: 655–660.
5. Na K-S, Kuroda A, Takiguchi N, Ikeda T, Ohtake H, et al. (2005) Isolation and characterization of benzene-tolerant *Rhodococcus opacus* strains. J Biosci Bioeng 99: 378–382.
6. Ramos JL, Duque E, Gallegos M-T, Godoy P, Ramos-Gonzalez MI, et al. (2002) Mechanisms of solvent tolerance in Gram-negative bacteria. Annu Rev Microbiol 56: 743–768.
7. Schliep M, Ryall B, Ferenci T (2012) The identification of global patterns and unique signatures of proteins across 14 environments using outer membrane proteomics of bacteria. Mol Biosyst 8: 3017–3027.
8. Roma-Rodrigues C, Santos PM, Benndorf D, Rapp E, Sá-correia I (2010) Response of *Pseudomonas putida* KT2440 to phenol at the level of membrane proteome. J Proteomics 73: 1461–1478.
9. Abe S, Okutsu T, Nakajima H, Kakuda N, Ohtsu I, et al. (2003) n-Hexane sensitivity of *Escherichia coli* due to low expression of imp/ostA encoding an

87 kDa minor protein associated with the outer membrane. Microbiology 149: 1265–1273.

10. Watanabe R, Doukyu N (2012) Contributions of mutations in acrR and marR genes to organic solvent tolerance in *Escherichia coli*. AMB Express 2: 58.

11. Van Beilen JB, Li Z, Duetz WA, Smilts THM, Witholt B (2003) Diversity of alkane hydroxylase systems in the environment. Oil Gas Sci Technol - RevIFP 58: 427–440.

12. Gupta A, Khare SK (2006) A protease stable in organic solvents from solvent tolerant strain of *Pseudomonas aeruginosa*. Bioresour Technol 97: 1788–1793.

13. Gaur R, Khare SK (2009) Cellular response mechanisms in *Pseudomonas aeruginosa* PseA during growth in organic solvents. Lett Appl Microbiol 49: 372–377.

14. Joshi C, Khare SK (2013) Purification and characterization of *Pseudomonas aeruginosa* lipase produced by SSF of deoiled Jatropha seed cake. Biocatal Agric Biotechnol 2: 32–37.

15. Segura A, Duque E, Hurtado ARJ (2001) Mutations in genes involved in the flagellar export apparatus of the solvent- tolerant *Pseudomonas putida* DOT-T1E strain impair motility and lead to hypersensitivity to toluene shocks. J Bacteriol 187: 4127–4133.

16. Molloy MP, Herbert BR, Slade MB, Rabilloud T, Nouwens AS, et al. (2000) Proteomic analysis of the *Escherichia coli* outer membrane. Eur J Biochem 267: 2871–2881.

17. Bradford MM (1976) A rapid and sensitive method for the quantitation of microgram quantities of protein utilizing the principle of protein-dye binding. Anal Biochem 72: 248–254.

18. Kyte J, Doolittle RF, Diego S, Jolla L (1982) A simple method for displaying the hydropathic character of a protein. J Mol Biol 157: 105–132.

19. Vermuë M, Sikkema J, Verheul A, Bakker R, Tramper J (1993) Toxicity of homologous series of organic solvents for the gram-positive bacteria *Arthrobacter* and *Nocardia* sp. and the gram-negative bacteria *Acinetobacter* and *Pseudomonas* sp. Biotechnol Bioeng 42: 747–758.

20. Nouwens AS, Walsh BJ, Cordwell S (2003) Applications of proteomics to Pseudomonas aeruginosa. In: Hecker M, Müllner S, editors. Proteomics of microorganisms Fundamental aspects and applications. Berlin Heidelberg: Springer Verlag. pp. 128–131.

21. Koebnik R, Locher KP, Van Gelder P (2000) Structure and function of bacterial outer membrane proteins: barrels in a nutshell. Mol Microbiol 37: 239–253.

22. Ceylan S, Akbulut BS, Denizci AA, Kazan D (2011) Proteomic insight into phenolic adaptation of a moderately halophilic *Halomonas* sp. strain AAD12. Can J Microbiol 57: 295–302.

23. Li H, Luo Y-F, Williams BJ, Blackwell TS, Xie C-M (2012) Structure and function of OprD protein in *Pseudomonas aeruginosa*: from antibiotic resistance to novel therapies. Int J Med Microbiol 302: 63–68.

24. Sikkema JAN, Jan AM, Poolman B (1995) Mechanisms of membrane toxicity of hydrocarbons. Microbiology 59: 201–222.

25. Chantal W Nde, Hyeung-Jin Jang FT and WEB (2009) Global transcriptomic response of *Pseudomonas aeruginosa* to chlorhexidine diacetate. Environ Sci Technol 43: 8406–8415.

26. Pseudomonas Genome Database. Available: http://www.pseudomonas.com. Accessed 2014 Jan 1.

27. Tashiro Y, Nomura N, Nakao R, Senpuku H, Kariyama R, et al. (2008) Opr86 is essential for viability and is a potential candidate for a protective antigen against biofilm formation by *Pseudomonas aeruginosa*. J Bacteriol 190: 3969–3978.

28. NCBI website. Available: http://www.ncbi.nlm.nih.gov/gene/945011. Accessed 2014 January 1.

29. Freinkman E, Chng S, Kahne D (2011) The complex that inserts lipopolysaccharide into the bacterial outer membrane forms a two-protein plug-and-barrel. PNAS 108: 2486–2491.

30. Trias J, Nikaido H (1990) Protein D2 Channel of the *Pseudomonas aeruginosa* outer membrane has a binding site for basic amino acids and peptides. J Biol Chem 265: 15680–156684.

31. Sugawara E, Nestorovich EM, Bezrukov SM, Nikaido H (2006) *Pseudomonas aeruginosa* porin OprF exists in two different conformations. J Biol Chem 281: 16220–16229.

32. Mastrobuoni G, Irgang S, Pietzke M, Assmus HE, Wenzel M, et al. (2012) Proteome dynamics and early salt stress response of the photosynthetic organism *Chlamydomonas reinhardtii*. BMC Genomics 13: 215.

33. Woodruff WA, Hancock RE (1989) *Pseudomonas aeruginosa* outer membrane protein F: structural role and relationship to the *Escherichia coli* OmpA protein. J Bacteriol 171: 3304–3309.

34. Rawling EG, Brinkman FS, Hancock REW (1989) Roles of the carboxy-terminal half of *Pseudomonas aeruginosa* major outer membrane protein OprF in cell shape, growth in low-osmolarity medium, and peptidoglycan association. J Bacteriol 180: 3556–3662.

35. Wu L, Estrada O, Zaborina O, Bains M, Shen L KJ (2005) Recognition of host immune activation by *Pseudomonas aeruginosa*. Science 309: 774–777.

36. Fito-Boncompte L, Chapalain A, Bouffartigues E, Chaker H, Lesouhaitier O, et al. (2011) Full virulence of *Pseudomonas aeruginosa* requires OprF. Infect Immun 79: 1176–1186.

37. Hancock REW, Siehnel R, Martin N (1990) Outer membrane proteins of *Pseudomonas*. Mol Microbiol 4: 1069–1075.

38. Siqueira R, Gonc SL, Alexander D, Domont GB, Maria L, et al. (2010) Effects of carbon and nitrogen sources on the proteome of *Pseudomonas aeruginosa* PA1 during rhamnolipid production. Process Biochem 45: 1504–1510.

39. Spangenberg C, Heuer T, Bürger C, Tijmmler B (1996) Genetic diversity of flagellins of *Pseudomonas aeruginosa*. Cloning 396: 213–217.

9

Cationic Synthetic Peptides: Assessment of Their Antimicrobial Potency in Liquid Preserved Boar Semen

Stephanie Speck[1][*][¤]**, Alexandre Courtiol**[1]**, Christof Junkes**[2]**, Margitta Dathe**[2]**, Karin Müller**[1]**, Martin Schulze**[3]

1 Leibniz Institute for Zoo and Wildlife Research, Berlin, Germany, 2 Leibniz Institute of Molecular Pharmacology, Berlin, Germany, 3 Institute for Reproduction of Farm Animals Schoenow e. V., Bernau, Germany

Abstract

Various semen extender formulas are in use to maintain sperm longevity and quality whilst acting against bacterial contamination in liquid sperm preservation. Aminoglycosides are commonly supplemented to aid in the control of bacteria. As bacterial resistance is increasing worldwide, antimicrobial peptides (AMPs) received lively interest as alternatives to overcome multi-drug resistant bacteria. We investigated, whether synthetic cationic AMPs might be a suitable alternative for conventional antibiotics in liquid boar sperm preservation. The antibacterial activity of two cyclic AMPs (c-WWW, c-WFW) and a helical magainin II amide analog (MK5E) was studied *in vitro* against two Gram-positive and eleven Gram-negative bacteria. Isolates included ATCC reference strains, multi-resistant *E. coli* and bacteria cultured from boar semen. Using broth microdilution, minimum inhibitory concentrations were determined for all AMPs. All AMPs revealed activity towards the majority of bacteria but not against *Proteus* spp. (all AMPs) and *Staphylococcus aureus* ATCC 29213 (MK5E). We could also demonstrate that c-WWW and c-WFW were effective against bacterial growth in liquid preserved boar semen *in situ*, especially when combined with a small amount of gentamicin. Our results suggest that albeit not offering a complete alternative to traditional antibiotics, the use of AMPs offers a promising solution to decrease the use of conventional antibiotics and thereby limit the selection of multi-resistant strains.

Editor: Axel Cloeckaert, Institut National de la Recherche Agronomique, France

Funding: The study was funded by AiF e.V. (www.aif.de) PRO INNO II Grant KF0376101MD6 and the Leibniz Institute of Zoo and Wildlife Research (www.izw-berlin.de). The funders had no role in study design, data collection and analysis, decision to publish, or preparation of the manuscript.

Competing Interests: The authors have declared that no competing interests exist.

* Email: stephanie.speck@vetmed.uni-leipzig.de

¤ Current address: Institute of Animal Hygiene and Veterinary Public Health, University of Leipzig, Leipzig, Germany

Introduction

Artificial insemination (AI) is the most commonly used assisted reproductive technology in swine industry [1]. For AI, short- or long-term semen extenders are used to process and store semen while maintaining sperm viability over days at 15 to 17°C. Bacteria are frequently found in freshly retrieved boar ejaculates but are detrimental to sperm quality and longevity particularly in liquid-preserved semen [2–4]. Up to 10^9 colony forming units/mL ejaculate have been reported [5–7]. The most prevalent bacteria were Gram-negative with the majority belonging to the family *Enterobacteriaceae* [7,8]. Bacterial contamination seems to have little effect on fecundity under natural mating conditions. However, processing and storage of extended semen for AI might facilitate bacterial growth and concentration-dependent spermicidal effects [9]. Besides a proper sanitation and hygiene management, antimicrobial substances, such as Aminoglycosides, are commonly supplemented to aid in the control of bacteria [7,9,10].

Bacteria are highly effective in adapting to changing environments [11] and due to an increasing spread of resistance to classic antibiotics there is a need for new antimicrobial alternatives [3,12].

In recent studies, antimicrobial peptides (AMPs) have received considerable attention as candidates to overcome bacterial resistance [13]. AMPs are naturally occurring molecules with a broad spectrum of antimicrobial activity that rapidly kill their target cells [14]. Well-known AMPs are mammalian defensins, amphibian magainins, and insect cecropins but even bacteria and fungi produce cationic AMPs (lantibiotics, bacteriocins) [14]. Roughly 5,500 AMPs have been discovered, predicted or synthesized so far [15]. Fortunately, most cationic peptides do not induce resistant mutant strains *in vivo* [14]. Among the large variety of AMPs, short arginine (R)- and tryptophan (W)-rich cyclic peptides demonstrated high antimicrobial activity and low toxic effects against eukaryotic cells [16]. Furthermore, the interaction of these R- and W-rich cyclic hexapeptides with *E. coli* rapidly permeabilised the outer membrane of *E. coli* [16,17].

The aim of our study was to evaluate whether selected synthetic AMPs are useful as substitutes for conventional antibiotics used in liquid boar sperm preservation. We describe the antimicrobial activity of two cationic cyclic peptides (c-WWW, c-WFW) [16] and a cationic helical magainin II amide analog (MK5E) [18] *in vitro* and in liquid preserved boar semen.

Materials and Methods

Synthetic cationic antimicrobial peptides

A helical magainin II amide derivative (MK5E) and two cyclic hexapeptides (c-WWW, c-WFW) were used in this study. The antimicrobial activity of these peptides against *E. coli* DH5α and *Bacillus subtilis* subsp. *spizizenii* DSM 347 (further referred to as *B. subtilis*) and their interaction with eukaryotic cells have been described in detail previously [16–18]. Peptides (Table 1) were obtained lyophilized from Biosyntan, Berlin, Germany. Stock solutions (400 μM) prepared in sterile distilled water were stored at –80°C until further use. The peptide synthesis was previously described in detail [16].

Antimicrobial susceptibility testing

***In vitro* antimicrobial activity of c-WFW, c-WWW, and MK5E.** For the determination of *in vitro* Minimum Inhibitory Concentrations (MICs), broth microdilution was performed according to the Clinical and Laboratory Standards Institute (CLSI) standard M31-A3 [19] using cation-adjusted Mueller-Hinton-II-Bouillon (MHIIB; Merck, Darmstadt, Germany). All antimicrobial substances were tested in 96-well plates in triplicate. These experiments were independently repeated twice. Selected Gram-negative bacteria isolated from native boar semen in preceding studies (unpublished data) were used: *Enterobacter cloacae*, hemolytic *E. coli* (further referred to as *E. coli* HE), *Klebsiella* (*K.*) *pneumoniae*, *Proteus* (*P.*) *myxofaciens*, *P. vulgaris*. In addition, AMPs were tested on *E. coli* DH5α, *B. subtilis* DSM 347, and four gentamicin-resistant *E. coli* (kindly provided by Stefan Schwarz, FLI, Mariensee, Germany). All strains were grown on Columbia sheep-blood (5%) agar (CSBA; Oxoid, Wesel, Germany). Briefly, MHIIB containing 5×10^5 CFU/mL was prepared for subsequent inoculation into 96-well plates containing the different peptide dilutions. The final peptide concentrations ranged from 100 μM − 0.05 μM (1:2 serial dilutions) as previously described [16]. Plates were sealed and incubated at 37°C for 18 to 24 h. The MIC of each tested AMP was defined as the lowest concentration exhibiting no visible growth compared to drug-free control wells. Turbidity was monitored with unaided eyes and a microplate reader at 600 nm. Gentamicin MICs were also determined (final concentration 0.113 μg/mL–116 μg/mL). As a quality control (QC) for broth microdilution, *E. coli* ATCC 25922 and *Staphylococcus* (*S.*) *aureus* ATCC 29213 were used as reference strains as recommended by CLSI [19]. Results were compared to the MIC QC ranges for broth microdilution (μg/mL) given by CLSI [19]. The test results were considered valid only when MICs for reference strains were within the QC ranges accepted by CLSI [19].

Evaluation of potency-enhancing effects: application of c-WWW and MK5E combined to gentamicin. The combination of AMPs and classical antibiotics has the potential to enhance the potency and target selectivity of AMPs [20]. We therefore combined c-WWW (2 μM) and MK5E (1 μM) but not c-WFW (as the latter was most promising for a stand-alone application) to gentamicin. AMP-concentrations were chosen according to sperm toxicity data as c-WWW and MK5E even at their lowest MIC (see results) would be harmful to boar spermatozoa (unpublished data). Gentamicin concentrations (i.e. 0.025 μg/mL–1 μg/mL) were selected according to MIC values defined in the first experiments and combined with c-WWW and MK5E. Determination of bacterial *in vitro* susceptibility was performed according to CLSI [19] and as outlined before. In addition, MICs were determined for gentamicin as a QC. The four multi-resistant *E. coli* were not included in these experiments.

Detection of bacteria in preserved semen

Ejaculates were collected from mature Pietrain boars housed at an EU-approved commercial insemination center during routine semen production and not as an animal experiment. The approval number according to Directive 90/429/EEC is KBS 085-EWG. Samples originated from a total of 39 boars and were retrieved by the gloved-hand technique. The gelatinous ejaculate fraction was removed using gauze. Boar ejaculates were diluted in Beltsville Thawing Solution (BTS) without additives (Minitüb, Tiefenbach, Germany), split, adjusted to 2×10^9 spermatozoa/portion (90 mL), and slowly cooled to 16°C over a 5 h-period.

The standard extender BTS containing 250 μg/mL gentamicin (BTS+G) was used as the control for all experiments. Ejaculates of ten individuals were comparatively investigated using BTS + c-WWW (2 μM) and BTS + c-WFW (4 μM). Samples of nine other individuals were prepared using BTS + MK5E (1 μM). In addition, a preparation using BTS without antimicrobial additive (BTS only) was available from three of these nine individuals. For the combined application of gentamicin (G) and AMPs, ejaculates from another 20 boars were prepared. BTS+G (16 μg/ml) was combined with c-WWW (2 μM), c-WFW (4 μM), and MK5E (1 μM), respectively. BTS+G (16 μg/mL) served as additional control. The latter concentration corresponded to the two-fold MIC breakpoint for gentamicin-resistant *Enterobacteriaceae* [19].

Each preparation was stored for 96 h at 16°C. Counting of bacteria and determination of bacterial species was performed after 12 h, 48 h, and 96 h of storage, respectively. To identify the different bacteria, a 50 μL-aliquot of the respective sample was each plated onto CSBA, Gassner medium (Oxoid), and McConkey agar (Oxoid). Plates were incubated for 48 h at 37°C. Bacterial species identification was carried out based on growth characteristics, Gram-staining, catalase- and oxidase-reaction, and conventional as well as commercially available (API® test system, bioMérieux, Nürtingen, Germany) biochemical tests. In addition, a serial dilution (10^{-1} to 10^{-5}) was prepared from each preparation after the respective storage time. 100 μL of each dilution were plated onto two nutrient agar plates (Oxoid), respectively. Plates were inspected after 24 h and 48 h of

Table 1. Cationic synthetic peptides used in this study.

Abbreviation	Peptide sequence	MW (g/mol)
c-WFW	Cyclic (RRWFWR)	989.5
c-WWW	Cyclic (RRWWWR)	1027.2
MK5E	Ac-GIGKF IHAVK KWGKT FIGEI AKS-NH2	2515.1

alanine (A), arginine (R), glutamic acid (E), glycine (G), histidine (H), isoleucine (I), lysine (K), phenylalanine (F), serine (S), threonine (T), tryptophan (W), valine (V), MW – molecular weight. The linear peptide, MK5E is N-terminally acetylated (Ac) and C-terminally amidated (NH2).

Table 2. Minimum inhibitory concentrations (MICs) determined for synthetic cationic peptides.

Bacteria	MICs (µM) determined for		
	c-WFW	c-WWW	MK5E
Escherichia coli ATCC 25922	6.3–12.5	50	25–50
Escherichia coli DH5α	6.3	12.5–25	25–50
Escherichia coli (hemolytic)	6.3–12.5	50	25–50
Escherichia coli 26	12.5	25–50	50
Escherichia coli 629	6.3	25	25
Escherichia coli 2078	12.5	25	50
Escherichia coli 2715	12.5	25	25
Enterobacter cloacae	25	25	25
Klebsiella pneumoniae	12.5–25	25–50	50
Proteus myxofaciens	>100	>100	>100
Proteus vulgaris	>100	>100	>100
Bacillus subtilis DSM 347	6.3	6.3	6.3–12.5
Staphylococcus aureus ATCC 29213	25	50	>100

incubation at 37°C. Colony forming units (CFU)/mL were calculated after 48 h of incubation.

Statistical analysis

To study the influence of AMPs on bacterial growth in preserved semen, we used the non-parametric test for longitudinal data in factorial experiments by Brunner *et al.* (2002) [21]. This test has specifically been designed to analyze time-dependent outcomes of an experiment performed on a small number of subjects. Analyses were implemented using the package nparLD version 2.1 [22] for the free statistical software R version 3.0.2 [23]. Following authors' terminology, our experiment setting corresponded originally to a F0-LD-F2 design. This means that for each semen sample, that we consider as subjects, we have no between-subject covariate and two within-subject covariates: time and treatment. The response variable was the number of CFU/mL.

In order to compare the effect of BTS only, gentamicin and the three AMPs on bacterial growth, we pooled the ten ejaculates treated with BTS+G (250 µg/mL), BTS + c-WWW (2 µM) and BTS + c-WFW (4 µM) and the nine ejaculates treated with BTS+G (250 µg/mL), BTS + MK5E (1 µM) and BTS only (for three of those nine ejaculates) in a first analysis. As preserved semen from each animal was not treated by all five treatments, we could not run the analysis as a F0-LD-F2 design. Instead, we randomly selected one treatment for each animal, making sure that the random sampling always included at least one sample for each treatment, and considered treatment as a between-subject covariate (F1-LD-F1 design). When testing of the effect of treatments on bacterial growth, the outcome is subject to variation due to the random sampling procedure. Therefore, we replicated the analysis 1000 times and report the median of all 1000 p-values obtained (hereafter reported *simulated p-value*). Importantly, making a separate analysis for each experiment and respecting the initial F0-LD-F2 study design led to same qualitative conclusions but precludes one to compare all treatments together (analysis performed without the treatment BTS only as this latter was not applied on all ejaculates, data not shown).

We also reran this analysis excluding the preparation BTS+G (250 µg/mL) to study differences between AMPs. Then, we performed a second analysis for the 20 ejaculates treated with BTS+G (250 µg/mL), BTS+G (16 µg/mL), BTS+G (16 µg/mL) + c-WWW (2 µM), BTS+G (16 µg/mL) + c-WFW (4 µM) and BTS+G (16 µg/mL) + MK5E (1 µM) to study the effect of a combined application of gentamicin and AMPs. For this latter analysis, directly fitting a F0-LD-F2 model was possible because each subject received all treatments.

Results

Antimicrobial susceptibility testing using c-WWW, c-WFW, and MK5E

MICs (µg/mL) defined for gentamicin using *S. aureus* ATCC 29213 (i.e. 0.225–0.7 µg/mL) and *E. coli* ATCC 25922 (i.e. 0.45–0.9 µg/mL) were within QC range recommended by CLSI (*S. aureus* ATCC 29213 0.12-1 µg/mL, *E. coli* ATCC 25922 0.25-1 µg/mL) [19]. Test results were reproducible in all experiments. Hence, systematic errors could be excluded. MICs determined for AMPs are given in Table 2. For most bacteria, the lowest MICs were defined for c-WFW followed by c-WWW and the linear magainin derivative MK5E. Using *Proteus*, MIC values for all peptides exceeded 100 µM and were not further specified. *Enterobacter cloacae* revealed identical values for all three AMPs. MIC values determined for a certain bacteria/peptide combination did not differ within one experiment but small variation was observed between experiments. This has been expected as approved QC MIC values for standard antibiotics also span over a range of concentrations in broth microdilution [19].

Combination of c-WWW, MK5E and gentamicin

Addition of 2 µM c-WWW or 1 µM MK5E to varying concentrations of gentamicin resulted in MIC values that did not considerably differ from those obtained solely for gentamicin (Table 3). Compared to the latter, a slight increase of MICs was noticed with the exception of *B. subtilis* DSM 347 as test organism.

Table 3. Minimum inhibitory concentrations (MICs) determined for gentamicin when combined with c-WWW or MK5E.

Bacteria	MIC (µg/mL) determined for gentamicin	MIC (µg/mL) determined for gentamicin when combined with	
		c-WWW (2 µM)	MK5E (1 µM)
Escherichia coli ATCC 25922	0.45–0.9*	0.6–0.7	0.6–0.8
Escherichia coli DH5α	0.113	0.3–0.5	0.2–0.5
Escherichia coli (hemolytic)	0.45	0.8–0.9	0.9
Enterobacter cloacae	0.113–0.225	0.2–0.4	0.3–0.4
Klebsiella pneumoniae	0.225–0.45	0.4–0.7	0.5–0.6
Proteus myxofaciens	0.45–0.9	0.7–0.9	0.7–0.9
Proteus vulgaris	0.45	0.6–0.8	0.5–0.8
Bacillus subtilis DSM 347	0.113	0.05–0.1	0.1
Staphylococcus aureus ATCC 29213	0.225–0.7*	0.6–0.7	0.5–0.6

*QC ranges as recommended by CLSI [19]: *S. aureus* (ATCC 29213) 0.12–1 µg/mL and *E. coli* (ATCC 25922) 0.25–1 µg/mL.

Effect of synthetic antimicrobial peptides on bacterial contamination in liquid preserved boar semen

Ejaculates of ten boars prepared with BTS+G (250 µg/mL), BTS + c-WWW (2 µM) and BTS + c-WFW (4 µM) and of nine boars prepared with BTS+G (250 µg/mL) and BTS+MK5E (1 µM) were investigated. In addition, BTS only-preserved samples from three boars were studied. As shown in Figure 1, treatments with AMPs or gentamicin presented fewer bacteria than the BTS only control. The number of CFU/mL did not significantly change with time for any preparations but BTS only and MK5E (simulated p-value for Anova Type Statistic [ATS] of the effect of time: for BTS only p = 0.021; for MK5E p<0.001; for all other treatments: p>0.38). Meanwhile, there was significantly less CFU/mL observed when using BTS+G (i.e. the standard semen extender) compared to when using any of the three AMP preparations (Figure 1). The comparison of AMPs showed that all three preparations did not differ significantly when the entire length of the experiment is considered (p = 0.11), but as bacteria grew with time for the MK5E treatment, once the bacteria count at 12 h is discarded the difference between treatments becomes significant (simulated p-value for modified ATS of the effect of preparation: p = 0.015). At 48 h and 96 h, MK5E was a less effective treatment against bacteria than c-WWW and c-WFW (p<0.001) and lost the initial improvement it had over the BTS only control observed at 12 h. During the entire experiment, c-WWW and c-WFW did not differ between each other in CFU/mL observed (p = 0.8).

Ejaculates of 20 boars were prepared to evaluate the effects of BTS+G (16 µg/mL) + c-WWW (2 µM), BTS + G + c-WFW (4 µM), and BTS + G + MK5E (1 µM) compared to the standard BTS+G (250 µg/mL) and BTS+G (16 µg/mL). Figure 2 shows that the amount of CFU/mL did not seem to change with time for any of the combined AMP/gentamicin-preparations (ATS for main effect of time: 0.57, df = 1.58, p = 0.52; ATS for time interacting with treatment: 0.65, df = 4.22, p = 0.63). In contrast, the number of CFU/mL was influenced by the preparation (ATS: 9.51, df = 3.33, p<0.0001) with BTS+G (16 µg/mL) being the less effective treatment, followed by BTS+G (16 µg/mL)+MK5E. Best results were obtained from preparations containing BTS+G (16 µg/mL)+c-WFW, BTS+G (16 µg/mL)+c-WWW, and BTS+ G (250 µg/mL). There was no significant difference in CFU/ml when using the latter three preparations (ATS: 1.63, df = 1.86, p = 0.20).

The amount of bacteria determined in different sperm preparations over time clearly varied between individuals. The CFU/mL counted for all preparations can be found in Table S1.

In total, 151 samples were investigated for bacterial growth. In the majority of samples (n = 125) more than one bacteria species was found. Scant growth of non-specific bacteria including mainly Gram-positive skin flora and Gram-negative bacteria commonly known as contaminants of distilled and stored water was found in 34% of all samples. Besides the non-specific bacteria, *Stenotrophomonas* (*S.*) *maltophilia* was predominant in samples treated solely by BTS+G (250 µg/mL, 16 µg/mL). Between three and five different Gram-negative and Gram-positive bacteria were isolated from the three BTS only preparations and identified as *S. maltophilia*, *Acinetobacter* sp., *Proteus* (*P.*) *vulgaris*, *Proteus* sp., *Serratia marcescens*, *Providencia rettgeri*, and *Staphylococcus* species. Preparations made of AMPs revealed ten different Gram-negative bacteria including *P. mirabilis*, *P. penneri*, *P. vulgaris*, *P. myxofaciens*, *Providencia alcalifaciens*, *Providencia rettgeri* or the non-fermentative bacteria *S. maltophilia*, *Ralstonia pickettii*, *Burkholderia cepacia*, and *Delftia acidovorans*. Of 29 samples treated solely with the single use of c-WWW, c-WFW, and MK5E, 21 (72%) revealed *Proteus* spp. and six (21%) were positive for *S. maltophilia*. In contrast, among the 60 samples obtained after the combined AMP/gentamicin treatment, we obtained eight (13%) *Proteus* spp.-positive specimens and 21 (35%) *S. maltophilia*-positive samples. Therefore, combining gentamicin (16 µg/mL) with an AMP significantly decreased the prevalence of *Proteus* spp. (proportion test: $X^2 = 28.4$, df = 1, p<0.0001), but did not significantly influence *S. maltophilia* counts ($X^2 = 1.28$, df = 1, p = 0.26).

Discussion

Alternatives to conventional antibiotics are in urgent need to combat multidrug-resistant bacteria. Because of their effectiveness, antimicrobial peptides have been suggested for antimicrobial therapy [24]. The aim of our study was to investigate whether cationic AMPs are effective against bacteria often found in boar semen and therefore might be a suitable alternative to antibiotics currently used in liquid sperm preservation.

MICs could be determined *in vitro* for c-WFW, c-WWW, and MK5E using eleven bacterial strains with the exception of *Proteus* spp. (all AMPs) and *S. aureus* ATCC 29213 (MK5E). These latter bacteria are known to produce proteases that cleave naturally

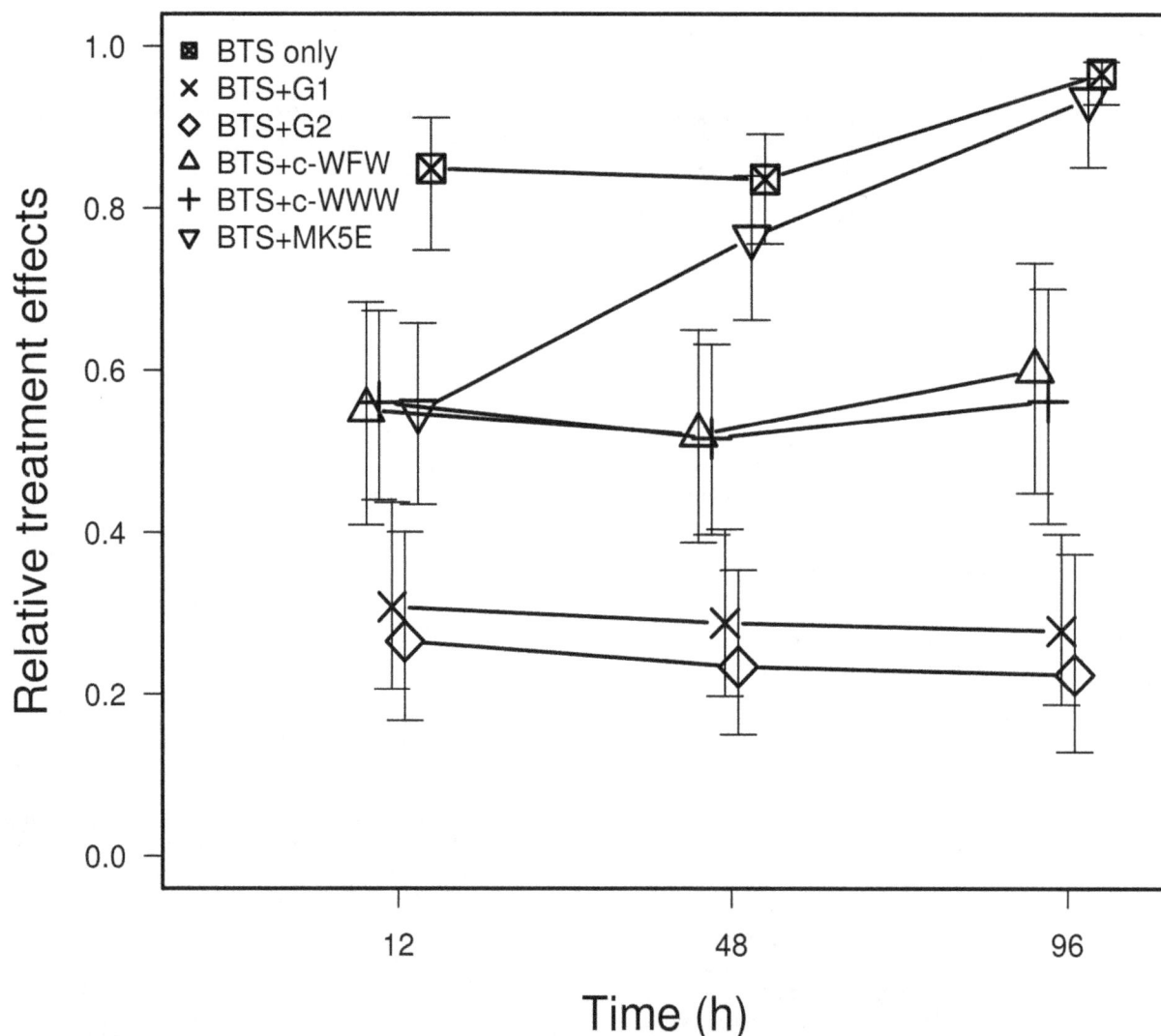

Figure 1. Relative effect of gentamicin or AMPs on the amount of bacteria in sperm preparation over time. Sperm preparations were made of BTS+G (250 µg/mL), BTS + c-WWW (2 µM) and BTS + c-WFW (4 µM) for ejaculates from ten individuals, and of BTS+G (250 µg/mL) and BTS + MK5E (1 µM) for ejaculates from nine other boars. Controls involving only BTS were also prepared from three of these nine individuals. The treatment BTS+G (250 µg/mL) is labeled BTS+G1 and BTS+G2 for the first and second experiment, accordingly. BTS+G1 and BTS + G2 were not distinguished in the analyses. The y-axis is the conventional graphical representation of the nonparametric method we used (see methods). It represents the relative marginal effect of the different treatments across time, i.e. the probability that the value being considered presents more CFU/mL than a random observation. The higher is the value on the y-axis, the higher is the corresponding value of CFU/mL, and the less effective is the treatment. Intervals represent 95% confidence intervals of the relative marginal effects and can here be used to compare treatments as the sample size is relatively similar for each point.

occurring linear cationic AMPs [24] and this mechanism might contribute to the results obtained in our experiments. Of the three peptides investigated, c-WFW resulted in lowest MIC values followed by c-WWW and MK5E. In former studies, hemolytic activity as well as toxicity against human cells at peptide concentrations up to 200 µM was negligible (c-WWW, c-WFW) [16] to non-existent (MK5E) [18]. However, our MIC-results revealed that only c-WFW might be applicable in liquid sperm preservation as negative effects on boar spermatozoa appeared at peptide concentrations higher than the MIC determined in this study (unpublished data). In contrast, even the lowest c-WWW and MK5E MIC determined for Gram-negative and -positive bacteria would be harmful to boar spermatozoa (unpublished data). We further investigated, whether a combined application of gentamicin and AMPs would result in enhanced antimicrobial

effectiveness. For these experiments sperm-compatible concentrations of c-WWW and MK5E but not c-WFW (as the latter was most promising for a stand-alone application) were used. Results of the combined application revealed bactericidal activity when c-WWW (2 µM) and MK5E (1 µM) were combined with gentamicin at a concentration of <1 µg/mL. However, MIC values defined for gentamicin in the combined application were slightly higher than those obtained solely for gentamicin. Hence, we cannot deduce an enhancing effect from the results of these experiments *in vitro*. In fact, the increase of gentamicin MICs in the presence of AMPs would rather indicate an antagonistic effect. Cell membrane interaction is the first and most crucial step for the antimicrobial activity of AMPs [17]. Cationic charge and amphipathicity of AMPs constitute the structural prerequisite for an initial electrostatic interaction with negatively charged lipid

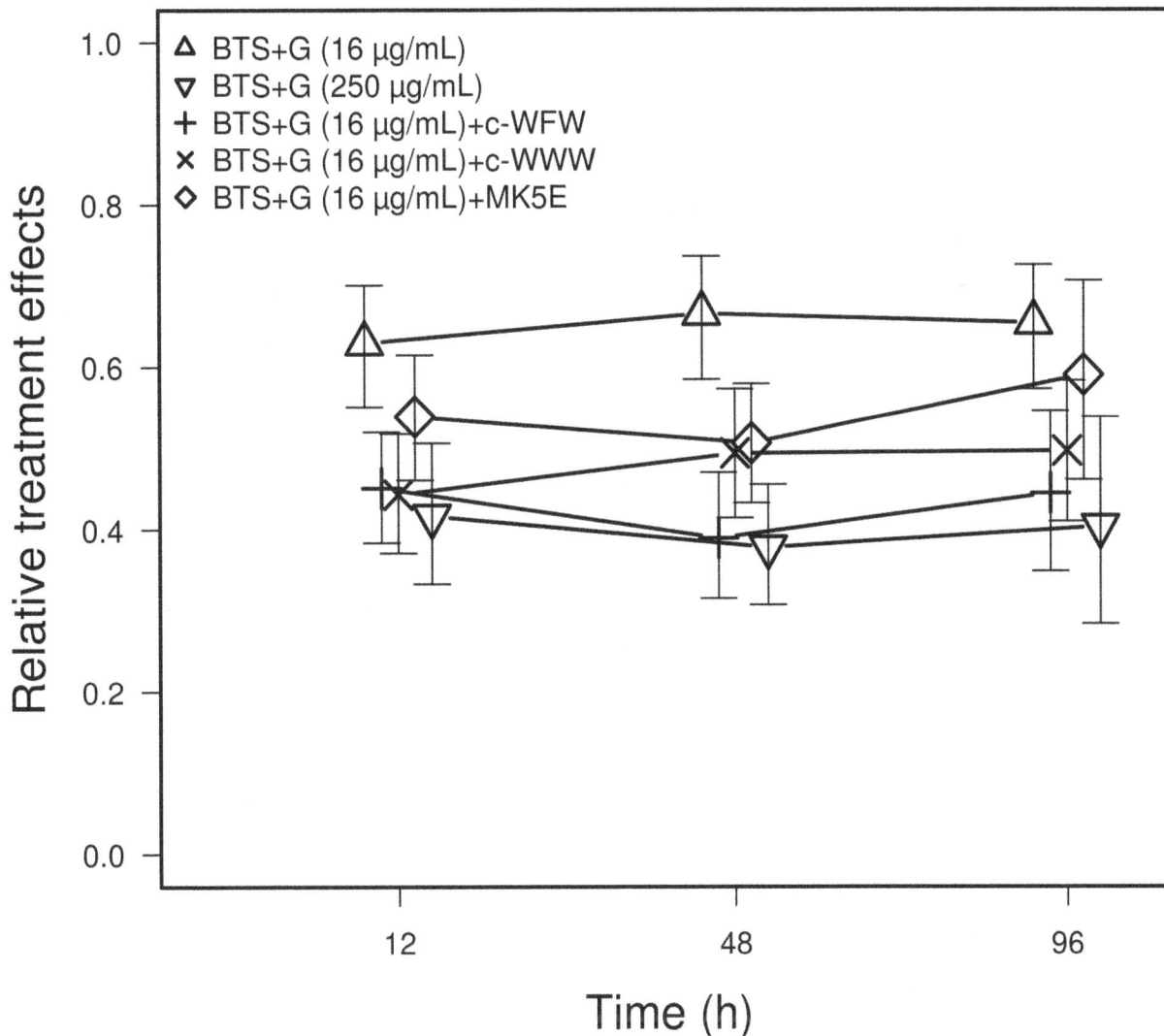

Figure 2. Relative effect of gentamicin or gentamicin combined with AMPs on the amount of bacteria in sperm preparation over time. Sperm preparations were made of BTS+G (250 µg/mL), BTS+G (16 µg/mL), BTS+G (16 µg/mL)+c-WWW (2 µM), BTS+G (16 µg/mL)+c-WFW (4 µM), and BTS+G (16 µg/mL)+MK5E (1 µM). See Figure 1 for legend details.

systems [25]. Electrostatic interactions are also the first step in aminoglycoside (e.g. gentamicin) action [26], hence a competing effect between both molecules might be assumed resulting in apparently higher MICs *in vitro*.

Based on the fact that, with the exception of c-WFW, MIC values determined for c-WWW and MK5E would be detrimental to spermatozoa, we decided to use sperm-compatible AMP concentrations to investigate whether their use in liquid sperm preservation would have any effect on bacterial contamination *in situ*. Treatments with AMPs or gentamicin presented fewer bacteria than the BTS only control. Interestingly, although used at concentrations below MIC determined *in vitro*, the different AMPs influenced the number of CFU/mL in liquid-preserved semen *in situ*. CFU/mL in preparations made of standard extender BTS containing 250 µg/mL gentamicin did not seem to change over time as was also the case for c-WFW and c-WWW that presented both the same antibacterial power (Figure 1). In contrast, preparations containing MK5E (1 µM) were less efficient and no longer prevented bacteria growth after 12 h. Enhancement of AMP-potency and target selectivity when combined to

conventional antibiotics has been described [20] and might be affirmed by our data regarding AMP/gentamicin-preserved sperm *in situ* although this is not supported by our *in vitro* data. Figure 2 clearly demonstrates that the combination of gentamicin and c-WFW as well as c-WWW enhanced the antimicrobial effectiveness *in situ*. In fact, the standard BTS+G (250 µg/mL) was as effective as preparations made of gentamicin (16 µg/mL) + c-WFW as well as gentamicin (16 µg/mL) + c-WWW. This effect cannot be attributed to gentamicin alone because BTS containing 16 µg/mL gentamicin without AMPs was significantly less effective than all other preparations in this study. Therefore, our results suggest that albeit not offering a complete alternative to traditional antibiotics, the use of adequate AMPs may allow for a substantial reduction in concentration of antibiotics used for semen preservation.

The bacteria isolated from liquid extended boar semen confirmed findings reported by others [7,9,27]. In their studies, also *Enterobacteriaceae*, *Xanthomonadaceae*, *Alcaligenaceae*, and *Burkholderiaceae* accounted for most of the Gram-negative contaminants. Most of the bacteria we isolated originate from the boar or occur ubiquitously and are often associated with water.

Many of them have an inherent ability to form biofilms and possess intrinsic or acquired resistance mechanisms. Overall, approximately one third of all samples contained *Enterobacteriaceae or S. maltophilia*. Althouse *et al.* [9] stated that ejaculates contaminated by bacteria only have little effect on fecundity under natural mating conditions. However, the presence of *S. maltophilia* was directly correlated to sperm agglutination and decreased gross motility [9]. Other Gram-negative bacteria may also act spermicidal thus negatively affecting litter size, when sows are inseminated with contaminated semen [27].

The usage of AMPs in liquid semen preservation was hindered by their sperm-toxicity at higher concentrations (unpublished data). Unexpectedly, we found AMPs effective *in situ* at concentrations that deemed to be ineffective during screening *in vitro*. We chose performance standards for antimicrobial dilution susceptibility tests according to CLSI [19] for quality assurance. Cation-adjusted MHIIB is recommended when using gentamicin as a reference [19] but may affect AMP properties. Cation-adjusted MHIIB contains 20 to 25 mg/L Ca^{2+} and 10 to 12.5 mg/L Mg^{2+} who might influence AMP-target structure-interactions. With regard to the magainin II amide analog MK5E this is supported by results of Matsuzaki *et al.* (1999) [28] who reported that Mg^{2+} tightens the lipopolysaccharide (LPS) packing by crosslinking adjacent phosphate groups. Their studies showed that 10 mM Mg^{2+} blocked the bactericidal action of magainin 2 on membrane models *in vitro* [28]. The *in situ* effect seen in our study might be explained by the finding that the antimicrobial activity of AMPs depends on an ionic milieu comparable to that in mammalian body fluids [29]. This was demonstrated on a structurally diverse panel of AMPs [29]. The presence of $NaHCO_3$ (27 mM) significantly enhanced antimicrobial activity against Gram-positive and -negative bacteria [29]. It has also been suggested that carbonate enhances AMP activity due to alterations in bacterial susceptibility [29]. Besides other components to preserve sperm metabolic activity (e.g. 3.7 mM EDTA), the standard extender BTS we used contained 15 mM $NaHCO_3$ thus possibly enhancing microbial susceptibility to AMPs in liquid-preserved semen.

Conclusions

Our results demonstrate activity of synthetic cationic antimicrobial peptides against different Gram-negative and Gram-positive bacteria *in vitro*. Furthermore, c-WWW and c-WFW suppressed bacterial growth in semen preparations *in situ*, especially when combined with a small concentration of gentamicin. As we also examined that AMPs did not impede the quality of sperm (unpublished data), they offer a promising solution to decrease the use of conventional antibiotics and thereby limit the selection of multi-resistant strains. In order to achieve comparable data for *in vitro* susceptibility testing and *in situ* studies, the implementation of a valid standardized method is in need. With regard to the application of AMPs in liquid boar sperm preservation further investigations should include the reduction of sperm toxicity, detection of possible enhancing effects using other conventional antibiotics, and analyses of peptide-stability in different standard semen extenders.

Supporting Information

Table S1   Bacterial counts given in CFU/mL in different sperm preparations determined after 12 h, 48 h, and 96 h of storage at 16°C.

Acknowledgments

The authors are deeply grateful to Lars Konkel and Nadine Jahn for their excellent technical assistance and laboratory work.

Author Contributions

Conceived and designed the experiments: SS MS KM MD CJ. Performed the experiments: SS MS. Analyzed the data: SS AC. Contributed reagents/materials/analysis tools: SS AC MS KM. Contributed to the writing of the manuscript: SS AC MS KM CJ MD. Peptide design: MD CJ.

References

1. Althouse GC, Rossow K (2011) The potential risk of infectious disease dissemination via artificial insemination in swine. Reprod Domest Anim 46: Suppl 2: 64–67.
2. Sone M, Ohmura K, Bamba K (1982) Effects of various antibiotics on the control of bacteria in boar semen. Vet Rec 111: 11–14.
3. Althouse GC, Lu KG (2005) Bacteriospermia in extended boar semen. Theriogenology 63: 573–584.
4. Maes D, Nauwynck H, Rijsselaere T, Mateusen B, Vyt P, et al. (2008) Diseases in swine transmitted by artificial insemination. Theriogenology 70: 1337–1345.
5. Dagnall GJR (1986) An investigation of the bacterial flora of the preputial diverticulum and of the semen of boars. London: Royal Veterinary College. M. Ph. Thesis.
6. Danowski KM (1989) Qualitative und quantitative Untersuchung zum Keimgehalt von Ebersperma und zur Antibiotikaempfindlichkeit des überwiegenden Keimspektrums (unter dem Aspekt der Samenkonservierung). Hannover: Tierärztliche Hochschule, Diss. med. vet. 87 p.
7. Althouse GC, Pierdon MS, Lu KG (2008). Thermotemporal dynamics of contaminant bacteria and antimicrobials in extended porcine semen. Theriogenology 70: 1317–1323.
8. Okazaki T, Mihara T, Fujita Y, Yoshida S, Teshima H, et al. (2010) Polymyxin B neutralizes bacteria-released endotoxin and improves the quality of boar sperm during liquid storage and cryopreservation. Theriogenology 74: 1691–1700.
9. Althouse GC, Kuster CE, Clark SG, Weisiger RM (2000) Field investigations of bacterial contaminants and their effects on extended porcine semen. Theriogenology 53: 1167–1176.
10. Althouse GC (2008). Sanitary procedures for the production of extended semen. Reprod Domest Anim 43 (Suppl. 2): 374–378.
11. Theuretzbacher U (2011) Resistance drives antibacterial drug development. Curr Opin Pharmacol 11: 433–438.
12. Wolska KI, Grzes K, Kurek A (2012) Synergy between novel antimicrobials and conventional antibiotics or bacteriocins. Pol J Microbiol 2: 95–104.
13. Arouri A, Dathe M, Blume A (2009) Peptide induced demixing in PG/PE lipid mixtures: a mechanism for the specificity of antimicrobial peptides towards bacterial membranes? Biomed Biochim Acta 1788: 650–659.
14. Hancock REW (1997) Peptide antibiotics. Lancet 349: 418–422.
15. Zhao X, Wu H, Lu H, Li G, Huang Q (2013) A database linking antimicrobial peptides. PLoS One 8(6): e66557.
16. Junkes C, Harvey RD, Bruce KD, Dölling R, Bagheri M, et al. (2011) Cyclic antimicrobial R-, W-rich peptides: the role of peptide structure and *E. coli* outer and inner membranes in activity and the mode of action. Eur Biophys J 40: 515–528.
17. Junkes C, Wessolowski A, Farnaud S, Evans RW, Good L, et al. (2008) The interaction of arginine- and tryptophan-rich cyclic hexapeptides with *Escherichia coli* membranes. J Pept Sci 14: 535–543.
18. Dathe M, Nikolenko H, Meyr J, Beyermann M, Bienert M (2001) Optimization of the antimicrobial activity of magainin peptides by modification of charge. FEBS Lett 501: 146–150.
19. Clinical and Laboratory Standards Institute (CLSI) (2008) Performance standards for antimicrobial disk and dilution susceptibility tests for bacteria isolated from animals; approved standard. 3rd ed. M31-A3. Vol. 28 No. 8. Wayne Pennsylvania: CLSI. 99 p.
20. Anantharaman A, Rizvi MS, Sahal D (2010) Synergy with rifampin and kanamycin enhances potency, kill kinetics, and selectivity of de novo-designed antimicrobial peptides. Antimicrob Agents Chemother 54(5): 1693–1699.
21. Brunner E, Domhof S, Langer F (2002) Nonparametric Analysis of Longitudinal Data in Factorial Experiments. New York: Wiley. 288 p.
22. Noguchi K, Gel YR, Brunner E, Konietschke F (2012) nparLD: An R Software Package for the Nonparametric Analysis of Longitudinal Data in Factorial Experiments. J Stat Softw 50(12): 1–23. Available: http://www.jstatsoft.org/v50/i12/.

23. R Core Team (2013) R: A language and environment for statistical computing. Vienna, Austria: R Foundation for Statistical Computing. URL http://www.R-project.org/.

24. Kraus D, Peschel A (2006) Molecular mechanisms of bacterial resistance to antimicrobial peptides. Curr Top Microbiol Immunol 306: 231–250.

25. Dathe M, Wieprecht T, Nikolenko H, Handel L, Maloy WL, et al. (1997) Hydrophobicity, hydrophobic moment and angle subtended by charged residues modulate antibacterial and haemolytic activity of amphipathic helical peptides. FEBS Lett 403: 208–212.

26. Magnet S, Blanchard JS (2005) Molecular insights into aminoglycoside action and resistance. Chem Rev 105: 477–497.

27. Maroto Martin LO, Cruz Munoz E, De Cupere F, Van Driessche E, Echemendia-Blanco D, et al. (2010) Bacterial contamination of boar semen affects the litter size. Anim Reprod Sci 120: 95–104.

28. Matsuzaki K, Sugishita K, Miyajima K (1999) Interactions of an antimicrobial peptide, magainin 2, with lipopolysaccharide-containing liposomes as a model for outer membranes of Gram-negative bacteria. FEBS letters 449: 221–224.

29. Dorschner RA, Lopez-Garcia B, Peschel A, Kraus D, Morikawa K, et al. (2006) The mammalian ionic environment dictates microbial susceptibility to antimicrobial defense peptides. FASEB J 20: 35–42.

10

Synthesis of RpoS Is Dependent on a Putative Enhancer Binding Protein Rrp2 in *Borrelia burgdorferi*

Zhiming Ouyang, Jianli Zhou, Michael V. Norgard*

Department of Microbiology, University of Texas Southwestern Medical Center, Dallas, Texas, United States of America

Abstract

The RpoN-RpoS regulatory pathway plays a central role in governing adaptive changes by *B. burgdorferi* when the pathogen shuttles between its tick vector and mammalian hosts. In general, transcriptional activation of bacterial RpoN (σ^{54})-dependent genes requires an enhancer binding protein. *B. burgdorferi* encodes the putative enhancer binding protein Rrp2. Previous studies have revealed that the expression of σ^{54}-dependent *rpoS* was abolished in an *rrp2* point mutant. However, direct evidence linking the production of Rrp2 in *B. burgdorferi* and the expression of *rpoS* has been lacking, primarily due to the inability to inactivate *rrp2* via deletion or insertion mutagenesis. Herein we introduced a regulatable (IPTG-inducible) *rrp2* expression shuttle plasmid into *B. burgdorferi*, and found that the controlled up-regulation of Rrp2 resulted in the induction of σ^{54}-dependent *rpoS* expression. Moreover, we created an *rrp2* conditional lethal mutant in virulent *B. burgdorferi*. By exploiting this conditional mutant, we were able to experimentally manipulate the temporal level of Rrp2 expression in *B. burgdorferi*, and examine its direct impact on activation of the RpoN-RpoS regulatory pathway. Our data revealed that the synthesis of RpoS was coincident with the IPTG-induced Rrp2 levels in *B. burgdorferi*. In addition, the synthesis of OspC, a lipoprotein required by *B. burgdorferi* to establish mammalian infection, was rescued in the *rrp2* point mutant when RpoS production was restored, suggesting that Rrp2 influences *ospC* expression indirectly via its control over RpoS. These data demonstrate that Rrp2 is required for the synthesis of RpoS, presumably via its action as an enhancer binding protein for the activation of RpoN and subsequent transcription of *rpoS* in *B. burgdorferi*.

Editor: Sven Bergström, Umeå University, Sweden

Funding: This work was funded by grant NIAID-NIH AI-059062. The funders had no role in study design, data collection and analysis, decision to publish, or preparation of the manuscript.

Competing Interests: The authors have declared that no competing interests exist.

* E-mail: michael.norgard@utsouthwestern.edu

Introduction

Borrelia burgdorferi, the causative agent of Lyme disease, is sustained in nature via a complex life cycle involving an arthropod tick vector (*Ixodes scapularis*) and mammals [1,2]. During its transit between these two markedly different host and tick milieus, *B. burgdorferi* undergoes significant adaptive changes. In *B. burgdorferi*, host adaptation is achieved by dramatic changes in gene expression in response to various tick or host stimuli [3–9]. Among a number of potential regulators that have been postulated to be present in *B. burgdorferi* [10–45], a novel genetic regulatory pathway, the RpoN-RpoS pathway (or the σ^{54}-σ^{S} cascade) [18], plays a central role in modulating *B. burgdorferi* host adaptive responses and virulence expression. In this pathway, one alternative sigma factor (σ^{54}, RpoN) controls the expression of another alternative sigma factor (σ^{S}, RpoS) through binding to a canonical −24/−12 promoter sequence [12,26,40]. Once RpoS is produced, it functions as a master regulator to modulate the expression of a number of virulence-associated outer membrane lipoproteins such as outer surface lipoproteins (Osp) C and A, and decorin binding proteins (Dbp) B and A [4,8,12–14,18,29,32,36,40,46–57].

Transcriptional activation of σ^{54}-dependent genes in bacteria requires a bacterial enhancer binding protein (bEBP), which is an AAA+ activator ATPase [58–63]. Sequence analyses have indicated that Rrp2 (BB0763) is composed of three structural domains, including an N-terminal regulatory domain (R), a central AAA+ ATPase core domain (C), and a C-terminal DNA binding domain (D), suggesting that Rrp2 may function as a bEBP to activate σ^{54}-dependent *rpoS* transcription in *B. burgdorferi* [15,45]. Previously, by exploiting a variant carrying a point mutation G239C in the C domain of Rrp2, we [10,29,45] reported that the *rrp2* point mutant was incapable of expressing *rpoS* and virulence-associated factors such as OspC and DbpA, suggesting that, as expected, Rrp2 is essential for activation of the RpoN-RpoS regulatory pathway. Despite this important finding, there remain many unanswered questions concerning the roles of Rrp2 in *B. burgdorferi* gene regulation. In particular, the finding that expression of *rpoS* was abolished in the *rrp2* point mutant has been hitherto the only evidence to support the role of Rrp2 in the activation of the RpoN-RpoS pathway. A direct link between Rrp2 protein levels produced in *B. burgdorferi* and the expression of *rpoS* has been lacking, primarily due to the inability to inactivate *rrp2* via deletion or insertion mutagenesis. The G239C point mutation in *rrp2* presumably abolishes the putative ATPase activity required for σ^{54}-dependent *rpoS* transcriptional activation. However, it also remains possible that the G239C point mutation causes a change in Rrp2's overall conformation, thereby preventing *rpoS* transcription in the *rrp2* point mutant. In addition, although the expression of *ospC* was found to be lost in the *rrp2* point mutant, how Rrp2 ultimately controls the expression of this key virulence-associated lipoprotein remains unknown. Rrp2 may indirectly modulate *ospC* expression via its control over RpoS. Alternatively, given that *ospC* was found to be constitutively expressed in *B. burgdorferi* when *ospAB* and *rpoS* were inactivated [64], Rrp2 may control *ospC* expression through another RpoS-independent factor(s). It is also possible that Rrp2 may directly modulate *ospC* expression by binding to its promoter. To address these questions, we employed an artificial gene expression system [65,66] to experimentally control the protein levels of Rrp2 synthesized in *B. burgdorferi*, and examined its impact on *rpoS* expression. Such a strategy affects only the levels of Rrp2 produced in *B. burgdorferi*, and does not alter the overall structure of the protein. Our data show that the expression level of Rrp2 correlates closely with the expression of *rpoS*, indicating that Rrp2 activates the expression of σ^{54}-dependent *rpoS* which, in turn, modulates *ospC* expression in *B. burgdorferi*.

Materials and Methods

Bacterial Strains and Culture Conditions

All strains and plasmids used in this study are described in Table 1. Low-passage infectious wild-type *B. burgdorferi* strain 297 [67], and the *rrp2* point mutant OY01 [29], were routinely cultured at 37°C and 5% CO_2 in either BSK-II medium [68] or BSK-H medium (Sigma) supplemented with 6% rabbit serum (Pel-Freeze). When appropriate, supplements were added to media at the following concentrations: erythromycin, 60 ng/ml; kanamycin, 150 μg/ml; streptomycin, 100 μg/ml. Spirochetes were enumerated by dark-field microscopy. *E. coli* strains were cultured in Lysogeny Broth (LB) supplemented with appropriate antibiotics at the following concentrations: ampicillin, 100 μg/ml; kanamycin, 50 μg/ml; spectinomycin, 100 μg/ml.

Generation of the IPTG-inducible *rrp2* Expression Construct

To experimentally control *rrp2* expression in *B. burgdorferi*, one *rrp2* expression construct pRrp2 was generated by using the *lac*-based gene inducible expression system [65,66]. Briefly, *rrp2* was amplified from *B. burgdorferi* using primers ZM303F and 303R (Table S1), and then cloned into pJSB275 [69] at the NdeI site. In pRrp2, *rrp2* transcription is directly controlled by the IPTG-inducible T5 promoter (a.k.a. the promoter PpQE30) from plasmid pQE30 (Qiagen).

Generation of the *rrp2* Conditional Lethal Mutant OY179

To create an *rrp2* conditional mutant, we first generated one shuttle plasmid pRrp2-FLAG in which *rrp2* expression was placed under the control of the IPTG-inducible promoter PpQE30. For creating pRrp2-FLAG, a DNA fragment encoding Rrp2-FLAG was amplified by using ZM303F and ZM263R (Table S1), and cloned into pJSB275 [69] at the NdeI site. This strategy adds a FLAG-tag (DYKDDDDK) to the C-terminus of Rrp2, thereby facilitating the detection of Rrp2 in *B. burgdorferi*. The plasmid pRrp2-FLAG was then electroporated into the *rrp2* point mutant OY01, yielding the streptomycin-resistant strain OY173. *B. burgdorferi* transformation was performed as previously described [29,70].

To inactivate *rrp2* in *B. burgdorferi* through homologous recombination, a suicide plasmid pOY202 was created. Briefly, the left arm for creating pOY202 was PCR-amplified using primers ZM86F and ZM86R, whereas the right arm was amplified using ZM87F and ZM87R (Table S1). After digestion with AscI, these two fragments were fused together. By using this ligated DNA as the template, PCR was employed to amplify a fragment comprising the upstream and downstream regions of *rrp2* by using primers ZM86F and ZM87R. The obtained fragment was cloned

Table 1. Strains and plasmids used in this study.

Strains or plasmids	Description	Source
B. burgdorferi		
297	Infectious *B. burgdorferi*, human spinal fluid isolate	[67]
OY01	297, *rrp2*(G239C) point mutant	[29]
AH206	297, *rpoS::ermC*	[18]
OY159	297 transformed with pRrp2	This study
OY160	OY01 transformed with pRrp2	This study
OY173	OY01 transformed with pRrp2-FLAG	This study
OY179	*rrp2* conditional lethal mutant, OY173 transformed with pOY202	This study
E. coli Top10F'	F'[*lacI*q Tn*10*(Tet^r)] *mcrA* Δ(*mrr-hsdRMS-mcrBC*) φ80*lacZ*ΔM15 ΔlacX74 recA1 araΔ*139* Δ (ara-leu)*7697 galU galK rpsL* (Str^r) *endA1 nupG*	Invitrogen
Plasmids		
pGEM-Teasy	TA cloning vector; Amp^r	Promega
pJSB275	Shuttle vector, Spec/Strep^r	[69]
pRpoS	IPTG-inducible *rpoS* expression construct	[30]
pOY100	PCR product of 86F/87R cloned into pGEM-Teasy, amp^r	This study
pJD55	Shuttle vector, Spec/Strep^r	[46]
pOY202	PflgB-kan cloned into pOY100, Amp/Kan^r	This study
pRrp2	PCR product of 303F/303R cloned into pJSB275, Spec/Strep^r	This study
pRrp2-FLAG	PCR product of 303F/263R cloned into pJSB275, Spec/Strep^r	This study

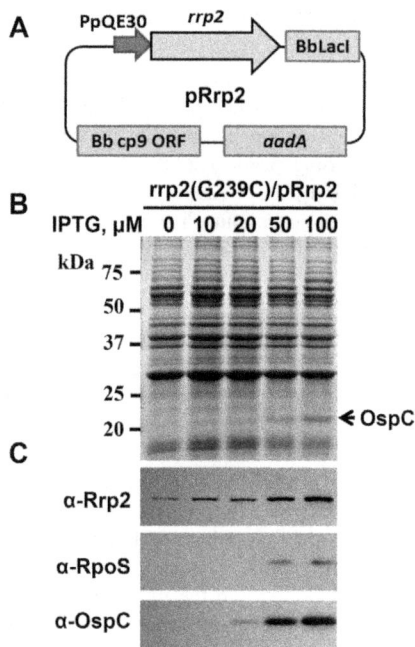

Figure 1. *Trans*-complementation of an *rrp2* point mutant using a shuttle plasmid. (A) Construction of an IPTG-inducible *rrp2* expression shuttle plasmid. To create pRrp2, *rrp2* was amplified from *B. burgdorferi* and cloned into pJSB275. The plasmid pRrp2 was then introduced into strain OY01 (*rrp2*[G239C]), yielding OY160. SDS-PAGE (B) and immunoblot (C) were performed to analyze gene expression in OY160. Bacteria were grown at 37°C in BSK-II medium with various concentrations of IPTG. When bacterial growth reached ~10^8 cells per ml, spirochetes were harvested. Approximately 4×10^7 spirochetes were loaded onto each lane of a 12.5% SDS-PAGE gel. In (B), approximate molecular masses are indicated at the left in kDa; concentrations of IPTG are indicated above the image; and the arrow indicates OspC. Specific antibodies, denoted as α- used in the immunoblot (C), are indicated on the left.

into pGEM-Teasy vector (Promega), yielding pOY100. The P*flgB*-Kan cassette, excised from pJD55 [46] using AscI, was cloned into pOY100 at the AscI site. In the resulting construct pOY202, the P*flgB*-Kan cassette was inserted in *rrp2* in the opposite direction as transcription of *csrA*. All constructs were confirmed using PCR amplification, restriction digestion, and sequence analysis. The plasmid pOY202 was then transformed into *B. burgdorferi* strain OY173. Transformants were isolated in the presence of kanamycin and streptomycin, along with 0.05 mM of IPTG to allow the production of Rrp2-FLAG during selection.

RNA Isolation and qRT-PCR

RNA isolation and qRT-PCR were performed as previously described [29–31,71]. Briefly, when total RNA was isolated from *B. burgdorferi* by using Trizol (Invitrogen), RNase-free DNase I (GenHunter Corporation) was used to digest genomic DNA. After RNA was further purified using RNeasy Mini Kit (Qiagen), 1 μg of RNA was used to synthesize cDNA using the SuperScript III Platinum Two-step qRT-PCR kit according to the manufacturer's protocol (Invitrogen). qRT-PCR was employed to examine gene expression, using the relative quantification method ($\Delta\Delta CT$) as described [29–31,71]. Gene expression fold change was presented as mean ± SE values from three independent experiments. Statistical analyses of the data were performed using the Student's *t* test.

SDS-PAGE and Semi-quantitative Immunoblot Analyses

A volume of whole cell lysate equivalent to 4×10^7 spirochetes was loaded per lane onto a 12.5% acrylamide gel. Resolved proteins were either stained with Coomassie brilliant blue or transferred to nitrocellulose membrane for immunoblot analysis as previously described [29–31,71]. Rrp2, FLAG, RpoS, OspC, and DbpA were detected using anti-Rrp2 monoclonal antibody 5B8-100-A1, anti-FLAG M2 monoclonal antibody (Sigma), anti-RpoS monoclonal antibody 6A7-101, anti-OspC monoclonal antibody 1B2-105A, or anti-DbpA monoclonal antibody 6B3, respectively. Immunoblots were developed colorimetrically using 4-chloro-1-napthol as the substrate or by chemiluminescence using the ECL Plus Western Blotting Detection system (Amersham Biosciences). Images were documented by using a Fujifilm LAS-3000 Imager (Fujifilm), and semi-quantitative analyses were performed by using the MultiGauge V3.0 software (Fujifilm).

Results and Discussion

Complementation of the *rrp2* Point Mutation by using an IPTG-inducible *rrp2* Expression Construct

Previously, we introduced a G239C mutation into *rrp2* and created an *rrp2* point mutant OY01 (*rrp2*[G239C]) in *B. burgdorferi* [29]. The expression of *rpoS* and *ospC* is abolished in OY01. To confirm that the loss of *rpoS* and *ospC* expression in the *rrp2* point mutant was due solely to the mutation of *rrp2*, a *trans*-complementation approach was employed. To this end, an IPTG-inducible plasmid pRrp2 (Fig. 1A) was created by placing the expression of *rrp2* under the control of the IPTG-inducible PpQE30 promoter. By adjusting the amount of the inducer (IPTG) added to the medium, the expression of *rrp2* on pRrp2 could be controlled. pRrp2 was then introduced into the *rrp2* point mutant OY01, yielding the streptomycin-resistant strain OY160.

To induce the expression of wild-type Rrp2, OY160 was grown in BSK-II medium containing varying concentrations of IPTG. In this experiment and subsequently, bacterial growth and morphology were not affected when spirochetes were grown in media with indicated levels of IPTG. Cells were collected when bacterial growth reached early stationary phase (~10^8 cells/ml). As shown in Fig. 1C, when 10-, 20-, 50-, or 100-μM of IPTG was added into the media, synthesis of Rrp2 was enhanced in a dose-dependent manner in *B. burgdorferi*. Of note, when IPTG was not added into the medium, a band was also detected in immunoblot by using the Rrp2 antibody; this band represents the existing mutated protein Rrp2(G239C). The induction of Rrp2 also resulted in the synthesis of RpoS and OspC in OY160. As shown in Fig. 1B and Fig. 1C, when bacteria were grown in BSK-II containing 50- or 100-μM of IPTG, RpoS and OspC were readily detected in OY160, and the increased protein levels correlated well with the increased levels of Rrp2 in OY160. These data suggest that Rrp2 expressed from pRrp2 is capable of complementing the *rrp2*(G239C) point mutation, thereby activating the RpoN-RpoS pathway in *B. burgdorferi*.

Up-regulation of Rrp2 in *B. Burgdorferi* Induces the Expression of *RpoS* and *OspC*

The approach of gene overexpression has proven to be a highly valuable tool for examining gene functions, particularly for genes that cannot be inactivated [72]. This strategy was successfully employed previously to study the role of Rrp2 in *B. burgdorferi*, where it was reported that overexpression of Rrp2 in *B. burgdorferi* led to the induction of OspC [69]. However, it remained unknown whether RpoS synthesis was influenced by the overexpression of Rrp2 in *B. burgdorferi*. In this current study, the IPTG-inducible

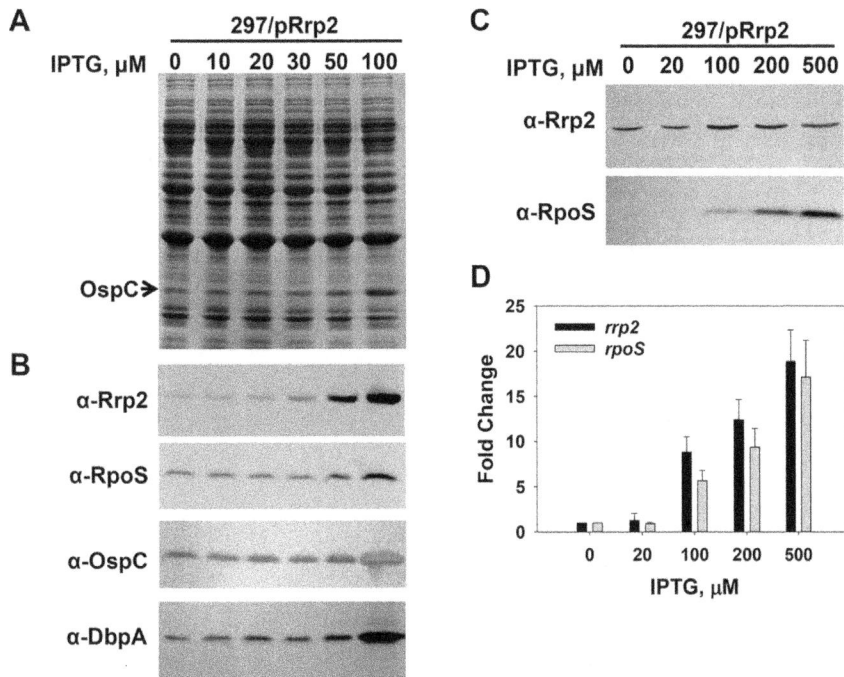

Figure 2. Up-regulation of Rrp2 in *B. burgdorferi* induces the expression of *rpoS*. Gene expression in OY159 was analyzed by SDS-PAGE (A), immunoblot (B, C), and qRT-PCR analyses (D). In (A) and (B), spirochetes grown in BSK-II media containing varying concentrations of IPTG were harvested when bacterial growth reached early stationary phase ($\sim 10^8$ cells per ml). In (C) and (D), spirochetes were grown in BSK-II medium. When bacterial growth reached mid-log phase ($\sim 10^7$ cells per ml), various amounts of IPTG were added into culture. Cells were collected at 9 h post-induction. In (A) and (C), concentrations of IPTG are indicated above the image. The arrow indicates OspC in (A). Specific antibodies, denoted as α-used in the immunoblot (B, C), are indicated on the left. In (D), data were collected from three independent experiments, and the bars represent the mean measurements ± standard deviation. The mean values between induced groups (100-, 200-, or 500 μM IPTG) and the uninduced group (0 μM IPTG) were compared using the Student's *t* test and are significantly different ($p<0.05$). For data normalization, the *B. burgdorferi* flaB gene was used as an internal control.

rrp2 expression construct pRrp2 was introduced into *B. burgdorferi* wild-type strain 297, yielding the merodiploid strain OY159. Gene expression in these spirochetes was measured through SDS-PAGE, semi-quantitative immunoblot, and quantitative RT-PCR analyses. In this experiment, two strategies employing shorter or longer induction times were used to induce Rrp2 in *B. burgdorferi*. For the longer induction time, bacteria were continuously grown in BSK-II containing various amounts of IPTG for about 7 days and harvested when growth reached the early stationary phase ($\sim 10^8$ cells/ml). As shown in Fig. 2B, synthesis of Rrp2 in OY159 was enhanced when bacteria were grown in media containing IPTG (compared with spirochetes cultivated in media without IPTG). Moreover, synthesis of RpoS and OspC was also found to be enhanced when IPTG was added into the medium (Fig. 2A, 2B). To exclude the possibility that gene induction might be an indirect effect resulting from prolonged exposure to IPTG, a shorter (9 h) induction period was also examined. When bacterial growth in BSK-II reached mid-log phase ($\sim 10^7$ cells/ml), varying amounts of IPTG were added into the media. After 9 h of induction, spirochetes were collected and gene expression was examined. As shown in Fig. 2C, immunoblot analyses showed that the syntheses of both Rrp2 and RpoS were induced by IPTG in a dose-dependent manner. Gene expression was also measured by using real-time quantitative RT-PCR (qRT-PCR) analyses. When gene transcription in spirochetes grown in BSK-II with 20-, 100-, 200-, or 500-μM of IPTG was compared with gene expression in bacteria grown without IPTG, transcription of *rrp2* was induced at 1.3-, 8.8-, 12.4-, or 18.9-fold, respectively; accordingly, *rpoS* transcription was induced at 0.9-, 5.6-, 9.4-, or 17.1-fold,

respectively. These combined data strongly suggest that Rrp2 is responsible for the induction of RpoS in *B. burgdorferi*.

Generation of an *rrp2* Conditional Mutant in *B. burgdorferi*

Given that many attempts to fully inactivate *rrp2* in *B. burgdorferi* have failed, and that *rrp2* thus seems to be essential for *B. burgdorferi in vitro* growth [69], we generated a *rrp2* conditional mutant in *B. burgdorferi* using a similar approach as described previously [69,73]. In this conditional lethal mutant, the wild-type chromosomal copy of *rrp2* is disrupted; *rrp2* is expressed from an IPTG-inducible shuttle plasmid. As a prelude to this approach, we first created another IPTG-inducible *rrp2* expression construct pRrp2-FLAG (Fig. 3A). From this shuttle plasmid, the expression of *rrp2* in *B. burgdorferi* is tightly controlled by IPTG added into the medium. To assist in the detection of Rrp2, a DNA fragment encoding the FLAG tag was fused to the 3′ of *rrp2*. Because adding a FLAG tag to Rrp2 could affect its general function, this plasmid pRrp2-FLAG was initially introduced into the *rrp2* point mutant OY01 (*rrp2*[G239C]) to create the strain OY173. OY173 was then used to test whether Rrp2-FLAG was functional in activating σ⁵⁴-dependent *rpoS* expression. OY173 bacteria were grown continuously in media with varying concentrations of IPTG, and collected when growth reached the early stationary phase. As shown in Fig. 3C, when IPTG was not added into the medium, Rrp2-FLAG was not detected. However, when 20-, 50-, 100-, or 200-μM of IPTG was added into the media, the synthesis of Rrp2-FLAG was enhanced in a dose-dependent manner. The produc-

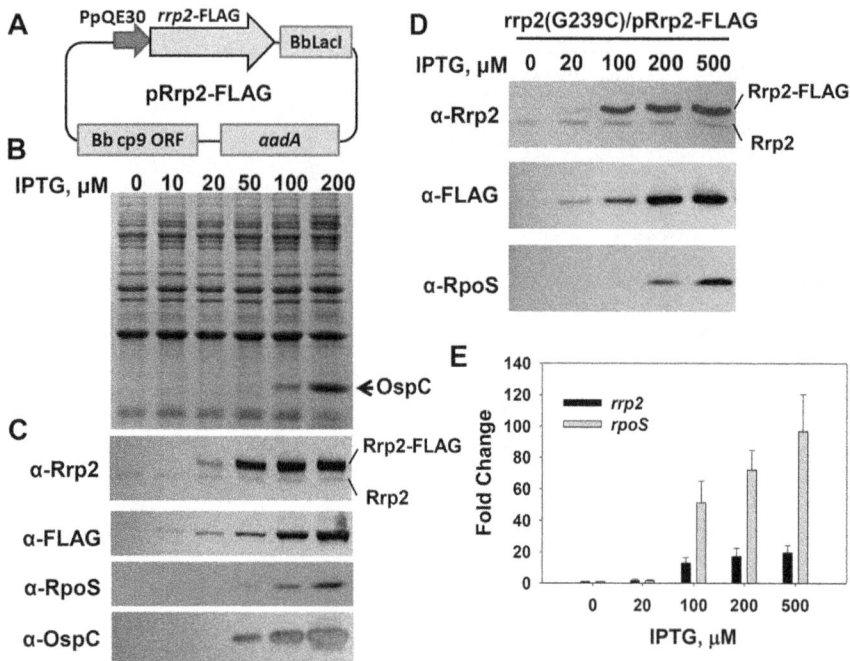

Figure 3. Gene expression in *B. burgdorferi* strain OY173. (A) Construction of an IPTG-inducible *rrp2*-FLAG expression shuttle plasmid. The plasmid pRrp2-FLAG pRrp2 was introduced into strain OY01 (*rrp2*[G239C]), yielding OY173. SDS-PAGE (B), immunoblot (C, D), and qRT-PCR analyses (E) were performed to analyze gene expression. In (B) and (C), spirochetes grown in BSK-II medium containing varying concentrations of IPTG were harvested when bacterial growth reached early stationary phase ($\sim 10^8$ cells per ml). In (D) and (E), spirochetes were grown in BSK-II medium. When bacterial growth reached mid-log phase ($\sim 10^7$ cells per ml), various amounts of IPTG were added into culture. Cells were collected at 9 h post-induction. In (B) and (D), concentrations of IPTG are indicated above the image. The arrow indicates OspC in (B). Specific antibodies, denoted as α- used in the immunoblot (C, D), are indicated on the left. In (E), the bars represent the mean measurements $\pm$ standard deviation. The mean values between induced groups (100-, 200-, or 500 μM IPTG) and the uninduced group (0 μM IPTG) were compared using the Student's *t* test and are significantly different (p<0.05). For data normalization, the *B. burgdorferi flaB* gene was used as an internal control.

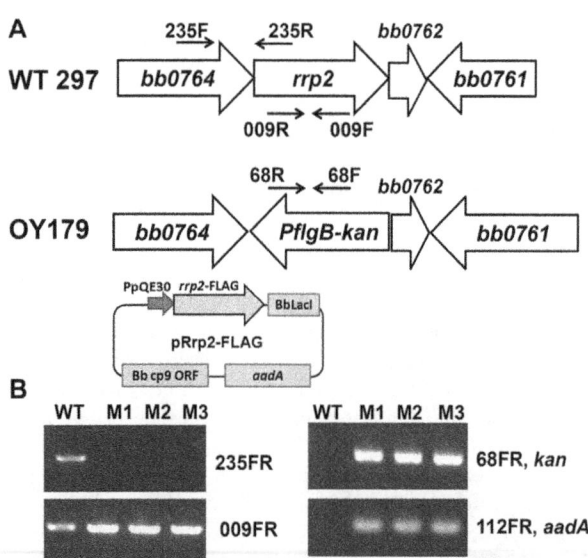

Figure 4. Generation of an *rrp2* conditional lethal mutant in *B. burgdorferi*. (A) Schematic representation of the *bb0764–bb0761* genes in the *B. burgdorferi* chromosome and the insertion of PflgB-kan cassette into *rrp2* by homologous recombination. Arrows indicate the approximate positions of the oligonucleotide primers used for subsequent analyses. (B) Analyses of the wild-type 297 and the *rrp2* conditional lethal mutant OY179 by PCR. The specific primer pairs are indicated on the right. Lanes WT, 297; lanes M1, M2, and M3, three clones of OY179.

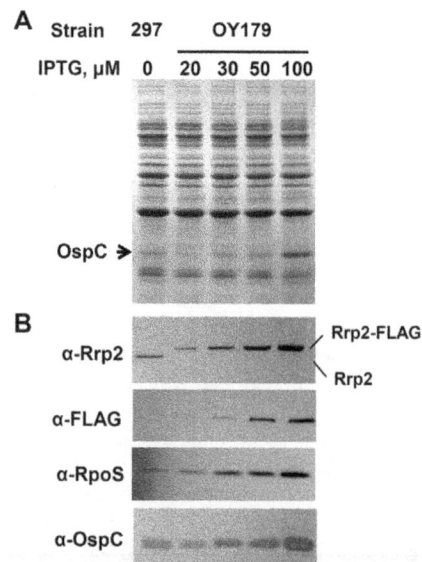

Figure 5. Gene expression in the *rrp2* conditional lethal mutant OY179. SDS-PAGE (A) and semi-quantitative immunoblot (B) analyses were performed to analyze gene expression. Bacteria were grown at 37°C in BSK-II medium with various concentrations of IPTG. When bacterial growth reached $\sim 10^8$ cells per ml, spirochetes were harvested. Approximately 4×10^7 spirochetes were loaded onto each lane of a 12.5% SDS-PAGE gel. Concentrations of IPTG are indicated above the image, and the arrow in (A) indicates OspC. Specific antibodies, denoted as α- used in the immunoblot (B), are indicated on the left.

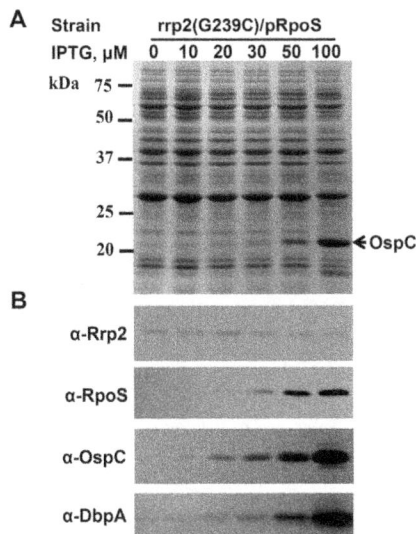

Figure 6. Rrp2 controls expression of *OspC* and *DbpA* via RpoS. (A, B) the *rrp2* point mutant *rrp2*(G239C) harboring the IPTG-inducible *rpoS* construct (pRpoS) was grown at 37°C with various concentrations of IPTG and gene expression was analyzed by SDS-PAGE (A) and immunoblot (B). The arrow in (A) indicates OspC. Specific antibodies, denoted as α-, used in the immunoblot (B) are indicated on the left.

tion of Rrp2-FLAG also resulted in the synthesis of RpoS and OspC in OY173. As shown in Fig. 3B and 3C, when bacteria were grown in BSK-II containing 50-, 100-, or 200-μM of IPTG, RpoS and OspC were readily detected in OY173, and the protein levels correlated closely with the increased levels of Rrp2-FLAG produced in *B. burgdorferi*. A 9-hr induction experiment was also carried out to examine gene induction in OY173. As shown in Fig. 3D, when bacteria at mid-log phase were exposed to 20-, 100-, 200-, or 500-μM of IPTG for 9 h, the synthesis of Rrp2-FLAG was readily detected in OY173. RpoS was also efficiently expressed in this strain (Fig. 3D). Moreover, qRT-PCR analyses revealed that exposing bacteria to IPTG for 9 h resulted in the concomitant induction of *rrp2* and *rpoS* (Fig. 3E). Taken together, these data suggested that Rrp2-FLAG expressed from pRrp2-FLAG was capable of functioning like native Rrp2 to activate the expression of σ^{54}-dependent *rpoS*.

Next, to fully inactivate *rrp2*, a suicide vector pOY202 that would target *rrp2* was first introduced into strain 297, but no transformants could be recovered. We then introduced pOY202 into strain OY173 (containing pRrp2-FLAG), and the desired transformants were selected by using streptomycin and kanamycin, together with IPTG (IPTG was used to induce *rrp2* expression from pRrp2-FLAG). Through allelic exchange, the chromosomal copy of *rrp2* was replaced by the PflgB-kan cassette, yielding strain OY179 (Fig. 4A). The inactivation of chromosomal *rrp2* was confirmed by PCR analyses. As shown in Fig. 4B, by using primers ZM235F and ZM235R, a 941-bp fragment spanning *bb0764* and *rrp2* was successfully amplified from WT strain 297 (lane 1), but not from the conditional mutant OY179 (lanes M1, M2, and M3). Also, PCR amplification revealed that strain OY179 contained the *kan* gene (conferring kanamycin resistance) and the *aadA* gene (conferring streptomycin resistance) (Fig. 4B).

The growth of the *rrp2* conditional mutant OY179 was assessed by cultivating the bacteria in BSK-II medium containing various concentrations of IPTG. OY179 grew in BSK-II with 20-, 30-, 50-, or 100-μM IPTG, but failed to grow in BSK-II without IPTG

(Fig. S1). Consistent with previous observations [69], we also conclude that *rrp2* is essential for *B. burgdorferi* growth *in vitro*.

Expression of *rrp2* Correlates with the Expression of *RpoS* and *OspC* in the *rrp2* Conditional Mutant

To assess whether the protein level of Rrp2 correlated with the expression of *rpoS*, gene expression in OY179 was measured via SDS-PAGE and semi-quantitative immunoblot analyses. To this end, OY179 was grown in BSK-II with varying concentrations of IPTG and cultures were harvested when cell growth reached early stationary phase. For bacteria grown under these conditions, no obvious differences were observed when spirochete morphology and motility were examined using dark-field microscopy. As shown in Fig. 5B, immunoblot analyses revealed that OY179 produced Rrp2-FLAG but not native Rrp2. Moreover, the levels of Rrp2-FLAG were dependent on the concentrations of the inducer IPTG. Specifically, when 20-μM IPTG was added into the medium, the level of Rrp2-FLAG produced in OY179 was ~3-fold lower than the level of Rrp2 observed in WT strain 297. When bacteria were cultivated in BSK-II containing 30-μM IPTG, Rrp2-FLAG was produced at a level commensurate with the level of Rrp2 in WT strain 297. When 50- or 100-μM IPTG was added into the medium, Rrp2-FLAG produced in OY179 was 3.5-fold or 10.3-fold higher, respectively, than the level of Rrp2 in strain 297. In addition, there was a close correlation between the levels of Rrp2-FLAG and the levels of RpoS and OspC (Fig. 5A and 5B). Taken together, the increased synthesis of RpoS was coincident with the IPTG-inducible production of Rrp2-FLAG, supporting that Rrp2 activates the RpoN-RpoS pathway in *B. burgdorferi*.

Rrp2 Indirectly Controls the Expression of *OspC* and *DbpA* via RpoS

Although the expression of *ospC* and *dbpA* was found to be lost in the *rrp2* point mutant [10,29,45], how Rrp2 ultimately controls the expression of these key lipoproteins has remained somewhat unclear. Given that (1) the expression of *ospC* and *dbpA* are directly regulated by RpoS through RpoS-specific promoters [74–76], and (2) *rpoS* transcription is abolished in the *rrp2* point mutant [10,29,45], we have hypothesized that Rrp2 likely regulates the expression of *rpoS* which, in turn, influences *ospC* and *dbpA* expression. To further test this hypothesis, we generated an IPTG-inducible *rpoS* expression shuttle construct pRpoS (i.e., pOY110) [30], in which *rpoS* expression is controlled solely by the IPTG-inducible PpQE30 promoter. This construct was introduced into the *rrp2* point mutant OY01, to investigate whether the IPTG-induced RpoS could restore *ospC* and *dbpA* expression in this mutant. As shown in Fig. 6A and 6B, when RpoS was induced from pRpoS by IPTG, production of OspC and DbpA was consequently rescued. As aforementioned, when the *rrp2* point mutation was complemented by the IPTG-inducible *rrp2* expression construct pRrp2, expression of *ospC* was restored in the complemented strain OY160 (Fig. 1B and 1C). Consistent with previous findings [69], however, when pRrp2 was introduced into a *B. burgdorferi rpoS* mutant AH206 (ΔrpoS), RpoS and OspC were not produced in this strain, despite the fact that Rrp2 synthesis was induced by IPTG (Fig. S2). These combined data suggest that controlled induction of RpoS can overcome the Rrp2 deficiency, which constitutes compelling evidence that Rrp2 indirectly controls *ospC* and *dbpA* expression via RpoS.

Conclusions

This is the first study, to our knowledge, to show that the level of Rrp2 produced in *B. burgdorferi* directly correlates with RpoS levels. Moreover, our study provides further validation for the application of an IPTG-inducible expression system [65,66] for assessing *B. burgdorferi* gene regulation. By using this system, an *rrp2* conditional lethal mutant was generated. Given the inability to inactivate *rrp2* via conventional deletion or insertion mutagenesis, and the fact that both σ^{54} and RpoS can be readily inactivated in *B. burgdorferi* [14,18,29,48,49], Rrp2 likely also controls the expression of genes independent of σ^{54} or RpoS control, which in turn suggests that Rrp2 may function as a unique bEBP (other characterized bEBPs activate only σ^{54}–dependent promoters). Our new conditional mutant thus provides an innovative way not only to define the direct control of Rrp2 over the central RpoN-RpoS pathway, but also to interrogate the overall regulatory role of Rrp2 in *B. burgdorferi* gene regulation. For example, by profiling gene expression in this new conditional mutant, it may be possible to identify novel Rrp2-controlled, σ^{54}- or RpoS-independent genes essential for *B. burgdorferi* growth or survival. Such studies will likely uncover new virulence-associated genes important for spirochetal pathogenesis, but also will provide the first definitive evidence that Rrp2 acts as an atypical bEBP to orchestrate virulence expression in *B. burgdorferi*. Although Rrp2 was presumed to be activated through phosphorylation [10,29,45], it still remains unanswered whether or how phosphorylated Rrp2 dynamically controls *rpoS* expression. Future work, such as pulse-chase analyses of the *rrp2* point mutant OY01 *trans*-complemented with *rrp2* variants, may help address these questions.

References

1. Burgdorfer W, Barbour AG, Hayes SF, Benach JL, Grunwaldt E, et al. (1982) Lyme disease-a tick-borne spirochetosis? Science 216: 1317–1319.
2. Steere AC (1993) Current understanding of Lyme disease. Hosp Pract (Off Ed) 28: 37–44.
3. Crother TR, Champion CI, Whitelegge JP, Aguilera R, Wu XY, et al. (2004) Temporal analysis of the antigenic composition of *Borrelia burgdorferi* during infection in rabbit skin. Infect Immun 72: 5063–5072.
4. de Silva AM, Telford SR 3rd, Brunet LR, Barthold SW, Fikrig E (1996) *Borrelia burgdorferi* OspA is an arthropod-specific transmission-blocking Lyme disease vaccine. J Exp Med 183: 271–275.
5. Liang FT, Nelson FK, Fikrig E (2002) Molecular adaptation of *Borrelia burgdorferi* in the murine host. J Exp Med 196: 275–280.
6. Montgomery RR, Malawista SE, Feen KJ, Bockenstedt LK (1996) Direct demonstration of antigenic substitution of *Borrelia burgdorferi* ex vivo: exploration of the paradox of the early immune response to outer surface proteins A and C in Lyme disease. J Exp Med 183: 261–269.
7. Ohnishi J, Piesman J, de Silva AM (2001) Antigenic and genetic heterogeneity of *Borrelia burgdorferi* populations transmitted by ticks. Proc Natl Acad Sci U S A 98: 670–675.
8. Schwan TG, Piesman J (2000) Temporal changes in outer surface proteins A and C of the lyme disease-associated spirochete, *Borrelia burgdorferi*, during the chain of infection in ticks and mice. J Clin Microbiol 38: 382–388.
9. Schwan TG, Piesman J, Golde WT, Dolan MC, Rosa PA (1995) Induction of an outer surface protein on *Borrelia burgdorferi* during tick feeding. Proc Natl Acad Sci U S A 92: 2909–2913.
10. Boardman BK, He M, Ouyang Z, Xu H, Pang X, et al. (2008) Essential role of the response regulator Rrp2 in the infectious cycle of *Borrelia burgdorferi*. Infect Immun 76: 3844–3853.
11. Boylan JA, Posey JE, Gherardini FC (2003) *Borrelia* oxidative stress response regulator, BosR: a distinctive Zn-dependent transcriptional activator. Proc Natl Acad Sci U S A 100: 11684–11689.
12. Burtnick MN, Downey JS, Brett PJ, Boylan JA, Frye JG, et al. (2007) Insights into the complex regulation of *rpoS* in *Borrelia burgdorferi*. Mol Microbiol 65: 277–293.
13. Dunham-Ems SM, Caimano MJ, Eggers CH, Radolf JD (2012) *Borrelia burgdorferi* requires the alternative sigma factor RpoS for dissemination within the vector during tick-to-mammal transmission. PLoS Pathog 8: e1002532.
14. Fisher MA, Grimm D, Henion AK, Elias AF, Stewart PE, et al. (2005) *Borrelia burgdorferi* σ^{54} is required for mammalian infection and vector transmission but not for tick colonization. Proc Natl Acad Sci U S A 102: 5162–5167.

Supporting Information

Figure S1   Growth of the *rrp2* conditional mutant OY179 *in vitro*. *B. burgdorferi* was inoculated into BSK-II medium with various concentrations of IPTG at 1000 spirochetes/ml. Spirochetes were enumerated using darkfield microscopy. Values are the means from three independent experiments. Error bars indicate standard deviations ($n = 3$).

Figure S2   Overexpression of Rrp2 does not restore expression of *OspC* and *DbpA* in the *RpoS* mutant. The *rpoS* point mutant AH206 harboring the IPTG-inducible *rrp2* construct (pRrp2) was grown at 37°C with various concentrations of IPTG and gene expression was analyzed by SDS-PAGE (A) and immunoblot (B). RpoS and OspC were not detected in this strain, which is consistent with previous findings [69]. Specific antibodies, denoted as α-, used in the immunoblot (B) are indicated on the left.

Table S1   Oligonucleotide primers used in this study.

Author Contributions

Conceived and designed the experiments: ZO MVN. Performed the experiments: ZO JZ. Analyzed the data: ZO MVN. Wrote the paper: ZO MVN.

15. Fraser CM, Casjens S, Huang WM, Sutton GG, Clayton R, et al. (1997) Genomic sequence of a Lyme disease spirochaete, *Borrelia burgdorferi*. Nature 390: 580–586.
16. Freedman JC, Rogers EA, Kostick JL, Zhang H, Iyer R, et al. (2010) Identification and molecular characterization of a cyclic-di-GMP effector protein, PlzA (BB0733): additional evidence for the existence of a functional cyclic-di-GMP regulatory network in the Lyme disease spirochete, *Borrelia burgdorferi*. FEMS Immunol Med Microbiol 58: 285–294.
17. He M, Ouyang Z, Troxell B, Xu H, Moh A, et al. (2011) Cyclic di-GMP is essential for the survival of the Lyme disease spirochete in ticks. PLoS Pathog 7: e1002133.
18. Hubner A, Yang X, Nolen DM, Popova TG, Cabello FC, et al. (2001) Expression of *Borrelia burgdorferi* OspC and DbpA is controlled by a RpoN-RpoS regulatory pathway. Proc Natl Acad Sci U S A 98: 12724–12729.
19. Hyde JA, Shaw DK, Smith Iii R, Trzeciakowski JP, Skare JT (2009) The BosR regulatory protein of *Borrelia burgdorferi* interfaces with the RpoS regulatory pathway and modulates both the oxidative stress response and pathogenic properties of the Lyme disease spirochete. Mol Microbiol 74: 1344–1355.
20. Jutras BL, Chenail AM, Carroll DW, Miller MC, Zhu H, et al. (2013) Bpur, the Lyme disease spirochete's PUR domain protein: identification as a transcriptional modulator and characterization of nucleic acid interactions. J Biol Chem 288: 26220–26234.
21. Jutras BL, Verma A, Adams CA, Brissette CA, Burns LH, et al. (2012) BpaB and EbfC DNA-binding proteins regulate production of the Lyme disease spirochete's infection-associated Erp surface proteins. J Bacteriol 194: 778–786.
22. Karna SL, Sanjuan E, Esteve-Gassent MD, Miller CL, Maruskova M, et al. (2011) CsrA modulates levels of lipoproteins and key regulators of gene expression critical for pathogenic mechanisms of *Borrelia burgdorferi*. Infect Immun 79: 732–744.
23. Katona LI, Tokarz R, Kuhlow CJ, Benach J, Benach JL (2004) The *fur* homologue in *Borrelia burgdorferi*. J Bacteriol 186: 6443–6456.
24. Kostick JL, Szkotnicki LT, Rogers EA, Bocci P, Raffaelli N, et al. (2011) The diguanylate cyclase, Rrp1, regulates critical steps in the enzootic cycle of the Lyme disease spirochetes. Mol Microbiol 81: 219–231.
25. Lybecker MC, Abel CA, Feig AL, Samuels DS (2010) Identification and function of the RNA chaperone Hfq in the Lyme disease spirochete *Borrelia burgdorferi*. Mol Microbiol 78: 622–635.
26. Lybecker MC, Samuels DS (2007) Temperature-induced regulation of RpoS by a small RNA in *Borrelia burgdorferi*. Mol Microbiol 64: 1075–1089.
27. Medrano MS, Policastro PF, Schwan TG, Coburn J (2010) Interaction of *Borrelia burgdorferi* Hbb with the p66 promoter. Nucleic Acids Res 38: 414–427.

28. Miller CL, Karna SL, Seshu J (2013) *Borrelia* host adaptation Regulator (BadR) regulates *rpoS* to modulate host adaptation and virulence factors in *Borrelia burgdorferi*. Mol Microbiol 88: 105–124.

29. Ouyang Z, Blevins JS, Norgard MV (2008) Transcriptional interplay among the regulators Rrp2, RpoN and RpoS in *Borrelia burgdorferi*. Microbiology 154: 2641–2658.

30. Ouyang Z, Deka RK, Norgard MV (2011) BosR (BB0647) controls the RpoN-RpoS regulatory pathway and virulence expression in *Borrelia burgdorferi* by a novel DNA-binding mechanism. PLoS Pathog 7: e1001272.

31. Ouyang Z, Kumar M, Kariu T, Haq S, Goldberg M, et al. (2009) BosR (BB0647) governs virulence expression in *Borrelia burgdorferi*. Mol Microbiol 74: 1331–1343.

32. Radolf JD, Caimano MJ, Stevenson B, Hu LT (2012) Of ticks, mice and men: understanding the dual-host lifestyle of Lyme disease spirochaetes. Nat Rev Microbiol 10: 87–99.

33. Rogers EA, Terekhova D, Zhang HM, Hovis KM, Schwartz I, et al. (2009) Rrp1, a cyclic-di-GMP-producing response regulator, is an important regulator of *Borrelia burgdorferi* core cellular functions. Mol Microbiol 71: 1551–1573.

34. Salman-Dilgimen A, Hardy PO, Dresser AR, Chaconas G (2011) HrpA, a DEAH-box RNA helicase, is involved in global gene regulation in the Lyme disease spirochete. PLoS One 6: e22168.

35. Samuels DS, Radolf JD (2009) Who is the BosR around here anyway? Mol Microbiol 74: 1295–1299.

36. Samuels DS (2011) Gene regulation in *Borrelia burgdorferi*. Annu Rev Microbiol 65: 479–499.

37. Samuels DS, Radolf JD (2010) *Borrelia*: molecular biology, host interaction and pathogenesis. Wymondham: Caister Academic. xii, 547 p.

38. Sanjuan E, Esteve-Gassent MD, Maruskova M, Seshu J (2009) Overexpression of CsrA (BB0184) alters the morphology and antigen profiles of *Borrelia burgdorferi*. Infect Immun 77: 5149–5162.

39. Scheckelhoff MR, Telford SR, Wesley M, Hu LT (2007) *Borrelia burgdorferi* intercepts host hormonal signals to regulate expression of outer surface protein A. Proc Natl Acad Sci U S A 104: 7247–7252.

40. Smith AH, Blevins JS, Bachlani GN, Yang XF, Norgard MV (2007) Evidence that RpoS (σ^S) in *Borrelia burgdorferi* is controlled directly by RpoN (σ^{54}/σ^N). J Bacteriol 189: 2139–2144.

41. Sultan SZ, Pitzer JE, Miller MR, Motaleb MA (2010) Analysis of a *Borrelia burgdorferi* phosphodiesterase demonstrates a role for cyclic-di-guanosine monophosphate in motility and virulence. Mol Microbiol 77: 128–142.

42. Sze CW, Li C (2011) Inactivation of *bb0184*, which encodes carbon storage regulator A, represses the infectivity of *Borrelia burgdorferi*. Infect Immun 79: 1270–1279.

43. Sze CW, Smith A, Choi YH, Yang X, Pal U, et al. (2013) Study of the response regulator Rrp1 reveals its regulatory role in chitobiose utilization and virulence of *Borrelia burgdorferi*. Infect Immun 81: 1775–1787.

44. Xu Q, Shi Y, Dadhwal P, Liang FT (2012) RpoS regulates essential virulence factors remaining to be identified in *Borrelia burgdorferi*. PLoS One 7: e53212.

45. Yang XF, Alani SM, Norgard MV (2003) The response regulator Rrp2 is essential for the expression of major membrane lipoproteins in *Borrelia burgdorferi*. Proc Natl Acad Sci U S A 100: 11001–11006.

46. Blevins JS, Hagman KE, Norgard MV (2008) Assessment of decorin-binding protein A to the infectivity of *Borrelia burgdorferi* in the murine models of needle and tick infection. BMC Microbiol 8: 82.

47. Caimano MJ, Eggers CH, Gonzalez CA, Radolf JD (2005) Alternate sigma factor RpoS is required for the *in vivo*-specific repression of *Borrelia burgdorferi* plasmid lp54-borne *ospA* and *lp6.6* genes. J Bacteriol 187: 7845–7852.

48. Caimano MJ, Eggers CH, Hazlett KR, Radolf JD (2004) RpoS is not central to the general stress response in *Borrelia burgdorferi* but does control expression of one or more essential virulence determinants. Infect Immun 72: 6433–6445.

49. Caimano MJ, Iyer R, Eggers CH, Gonzalez C, Morton EA, et al. (2007) Analysis of the RpoS regulon in *Borrelia burgdorferi* in response to mammalian host signals provides insight into RpoS function during the enzootic cycle. Mol Microbiol 65: 1193–1217.

50. Grimm D, Tilly K, Byram R, Stewart PE, Krum JG, et al. (2004) Outer-surface protein C of the Lyme disease spirochete: a protein induced in ticks for infection of mammals. Proc Natl Acad Sci U S A 101: 3142–3147.

51. Ouyang Z, Narasimhan S, Neelakanta G, Kumar M, Pal U, et al. (2012) Activation of the RpoN-RpoS regulatory pathway during the enzootic life cycle of *Borrelia burgdorferi*. BMC Microbiol 12: 44.

52. Pal U, de Silva AM, Montgomery RR, Fish D, Anguita J, et al. (2000) Attachment of *Borrelia burgdorferi* within *Ixodes scapularis* mediated by outer surface protein A. J Clin Invest 106: 561–569.

53. Pal U, Yang X, Chen M, Bockenstedt LK, Anderson JF, et al. (2004) OspC facilitates *Borrelia burgdorferi* invasion of *Ixodes scapularis* salivary glands. J Clin Invest 113: 220–230.

54. Shi Y, Xu Q, McShan K, Liang FT (2008) Both decorin-binding proteins A and B are critical for overall virulence of *Borrelia burgdorferi*. Infect Immun 76: 1239–1246.

55. Tilly K, Krum JG, Bestor A, Jewett MW, Grimm D, et al. (2006) *Borrelia burgdorferi* OspC protein required exclusively in a crucial early stage of mammalian infection. Infect Immun 74: 3554–3564.

56. Weening EH, Parveen N, Trzeciakowski JP, Leong JM, Hook M, et al. (2008) *Borrelia burgdorferi* lacking DbpBA exhibits an early survival defect during experimental infection. Infect Immun 76: 5694–5705.

57. Yang XF, Pal U, Alani SM, Fikrig E, Norgard MV (2004) Essential role for OspA/B in the life cycle of the Lyme disease spirochete. J Exp Med 199: 641–648.

58. Bush M, Dixon R (2012) The role of bacterial enhancer binding proteins as specialized activators of σ^{54}-dependent transcription. Microbiol Mol Biol Rev 76: 497–529.

59. Morett E, Segovia L (1993) The σ^{54} bacterial enhancer-binding protein family: mechanism of action and phylogenetic relationship of their functional domains. J Bacteriol 175: 6067–6074.

60. Rappas M, Bose D, Zhang X (2007) Bacterial enhancer-binding proteins: unlocking σ^{54}-dependent gene transcription. Curr Opin Struct Biol 17: 110–116.

61. Wigneshweraraj S, Bose D, Burrows PC, Joly N, Schumacher J, et al. (2008) *Modus operandi* of the bacterial RNA polymerase containing the σ^{54} promoter-specificity factor. Mol Microbiol 68: 538–546.

62. Ghosh T, Bose D, Zhang X (2010) Mechanisms for activating bacterial RNA polymerase. FEMS Microbiol Rev 34: 611–627.

63. Studholme DJ, Dixon R (2003) Domain architectures of σ^{54}-dependent transcriptional activators. J Bacteriol 185: 1757–1767.

64. He M, Oman T, Xu H, Blevins J, Norgard MV, et al. (2008) Abrogation of *ospAB* constitutively activates the Rrp2-RpoN-RpoS pathway (σ^N-σ^S cascade) in *Borrelia burgdorferi*. Mol Microbiol 70: 1453–1464.

65. Blevins JS, Revel AT, Smith AH, Bachlani GN, Norgard MV (2007) Adaptation of a luciferase gene reporter and *lac* expression system to *Borrelia burgdorferi*. Appl Environ Microbiol 73: 1501–1513.

66. Gilbert MA, Morton EA, Bundle SF, Samuels DS (2007) Artificial regulation of *ospC* expression in *Borrelia burgdorferi*. Mol Microbiol 63: 1259–1273.

67. Hughes CA, Kodner CB, Johnson RC (1992) DNA analysis of *Borrelia burgdorferi* NCH-1, the first northcentral U.S. human Lyme disease isolate. J Clin Microbiol 30: 698–703.

68. Pollack RJ, Telford SR 3rd, Spielman A (1993) Standardization of medium for culturing Lyme disease spirochetes. J Clin Microbiol 31: 1251–1255.

69. Groshong AM, Gibbons NE, Yang XF, Blevins JS (2012) Rrp2, a prokaryotic enhancer-like binding protein, is essential for viability of *Borrelia burgdorferi*. J Bacteriol 194: 3336–3342.

70. Samuels DS (1995) Electrotransformation of the spirochete *Borrelia burgdorferi*. Methods Mol Biol 47: 253–259.

71. Ouyang Z, He M, Oman T, Yang XF, Norgard MV (2009) A manganese transporter, BB0219 (BmtA), is required for virulence by the Lyme disease spirochete, *Borrelia burgdorferi*. Proc Natl Acad Sci U S A 106: 3449–3454.

72. Prelich G (2012) Gene overexpression: uses, mechanisms, and interpretation. Genetics 190: 841–854.

73. Lenhart TR, Akins DR (2010) *Borrelia burgdorferi* locus BB0795 encodes a BamA orthologue required for growth and efficient localization of outer membrane proteins. Mol Microbiol 75: 692–709.

74. Alverson J, Bundle SF, Sohaskey CD, Lybecker MC, Samuels DS (2003) Transcriptional regulation of the *ospAB* and *ospC* promoters from *Borrelia burgdorferi*. Mol Microbiol 48: 1665–1677.

75. Ouyang Z, Haq S, Norgard MV (2010) Analysis of the *dbpBA* upstream regulatory region controlled by RpoS in *Borrelia burgdorferi*. J Bacteriol 192: 1965–1974.

76. Yang XF, Lybecker MC, Pal U, Alani SM, Blevins J, et al. (2005) Analysis of the *ospC* regulatory element controlled by the RpoN-RpoS regulatory pathway in *Borrelia burgdorferi*. J Bacteriol 187: 4822–4829.

11

Brucella Cyclic β-1,2-Glucan Plays a Critical Role in the Induction of Splenomegaly in Mice

Mara S. Roset[1]*, **Andrés E. Ibañez**[2], **Job Alves de Souza Filho**[3], **Juan M. Spera**[1], **Leonardo Minatel**[1], **Sergio C. Oliveira**[3], **Guillermo H. Giambartolomei**[2], **Juliana Cassataro**[1], **Gabriel Briones**[1]*

1 Instituto de Investigaciones Biotecnológicas "Rodolfo Ugalde" - Instituto Tecnológico de Chascomús (IIB-INTECH), Universidad Nacional de San Martín (UNSAM), Consejo Nacional de Investigaciones Científicas y Técnicas (CONICET), Buenos Aires, Argentina, 2 Laboratorio de Inmunogenética, INIGEM-CONICET, Hospital de Clínicas "José de San Martín," Facultad de Medicina, Universidad de Buenos Aires (UBA), Buenos Aires, Argentina, 3 Department of Biochemistry and Immunology, Institute of Biological Sciences, Federal University of Minas Gerais, Belo Horizonte, Minas Gerais, Brazil

Abstract

Brucella, the etiological agent of animal and human brucellosis, is a bacterium with the capacity to modulate the inflammatory response. Cyclic β-1,2-glucan (CβG) is a virulence factor key for the pathogenesis of *Brucella* as it is involved in the intracellular life cycle of the bacteria. Using comparative studies with different CβG mutants of *Brucella*, *cgs* (CβG synthase), *cgt* (CβG transporter) and *cgm* (CβG modifier), we have identified different roles for this polysaccharide in *Brucella*. While anionic CβG is required for bacterial growth in low osmolarity conditions, the sole requirement for a successful *Brucella* interaction with mammalian host is its transport to periplasmic space. Our results uncover a new role for CβG in promoting splenomegaly in mice. We showed that CβG-dependent spleen inflammation is the consequence of massive cell recruitment (monocytes, dendritics cells and neutrophils) due to the induction of pro-inflammatory cytokines such as IL-12 and TNF-α and also that the reduced splenomegaly response observed with the *cgs* mutant is not the consequence of changes in expression levels of the characterized *Brucella* PAMPs LPS, flagellin or OMP16/19. Complementation of *cgs* mutant with purified CβG increased significantly spleen inflammation response suggesting a direct role for this polysaccharide.

Editor: Jean-Pierre Gorvel, Universite de la Mediterranee, France

Funding: This work was supported by grants from the Agencia Nacional de Promoción Científica y Técnológica, Buenos Aires, Argentina (PICT 2006-1208, PICT 2011-1772,), CONICET (PIP1142010010031401). The funders had no role in study design, data collection and analysis, decision to publish, or preparation of the manuscript.

* Email: gbriones@iibintech.com.ar (GB); mroset@iibintech.com.ar (MSR)

Introduction

Brucella is a bacterial pathogen that infects ruminants as the primary host and can be transmitted to humans by consumption of animal-derived contaminated products (e.g. unpasteurized dairy food) leading to brucellosis, a reticular endothelial disease, typically characterized by undulant fever [1]. *Brucella*, as the result of a longstanding association with the mammalian host, has evolved modified PAMPs (pathogen-associated molecular patterns molecules) such LPS and flagellin. These modified PAMPs limit host recognition by innate immune receptors (TLRs and NLRs) during the acute phase of the infectious process, reducing the induction of the inflammatory response known to be necessary for an efficient control of the infection [2–4]. In addition *Brucella* expresses virulence factors that actively interfere with the innate immune response such as the *Brucella* proteins Btp-A/TcpB and BtpB that down-regulate TLR2/TLR4 signaling [5–7]. Paradoxically, when the host is persistently infected at the chronic phase of brucellosis, inflammation becomes a dominant clinical sign that has been described to affect several organs producing symptoms such as arthritis, endocarditis, meningitis, epididymitis and splenomegaly [8].

To date few virulence mechanisms have been described in *Brucella* infection and different genomic studies have confirmed the absence of classical virulence factors like fimbriae, pilli, toxins or plasmids [9,10]. In our laboratory, we have identified and characterized a critical molecule for *Brucella* virulence: the cyclic β-1,2-glucan (CβG). CβG is composed by a family of cyclic polymers of 17 to 25 D-glucose molecules linked by β-1,2 linkages, synthesized by a membrane bound enzyme named Cgs (for cyclic glucan synthase) that utilizes UDP-glucose as the sugar donor [11]. Cgs initiates and elongates a linear chain of glucoses covalently bound to the Cgs which is subsequently cyclated and released to the bacterial cytoplasm [12]. A specific *Brucella* ABC-transporter Cgt (for Cyclic glucan transporter) translocates the CβG to the periplasmic space where they become chemically modified with the addition of succinyl groups, a reaction catalyzed by the membrane enzyme Cgm (for Cyclic glucan modifier) [13,14]. Although CβG is present within the periplasmic space, being this localization critical for hypo-osmotic adaptation, secretion of large amount of this polysaccharide to the supernatant by a non-characterized mechanism has been described in *Agrobacterium* and *Sinohizobium* [15]. Interestingly, *Brucella cgs* mutant strain presents a defect in intracellular trafficking in epithelial cells that can be complemented by the addition of purified CβG (or Cyclodextrins) to tissue culture medium [16]. This observation suggests that also in *Brucella*, CβG must be secreted within the host cell to exert its role in virulence. The proposed mechanism of action for CβG in *Brucella* infection is the sequestration of cholesterol from intracellular host membranes leading to lipid raft disorganization and modulation of intracellular trafficking [16].

We have previously observed that mice infected with a *cgs* mutant had a reduced spleen inflammatory response even though they had a high number of replicating bacteria within this organ [17]. Since inflammation is the consequence of the host recognition of microbial PAMPs that ultimately lead to the induction of an inflammatory process, we hypothesized that either the periplasmic CβG may be stabilizing the expression of *Brucella* PAMPs such as LPS, flagellin, and OMPs or that CβG could be recognized by the innate immune receptors in the context of *Brucella* infection triggering inflammation.

It has been shown recently, using *in vitro* studies and purified CβG, that this molecule acts directly as an agonist of the innate immune system mediated by TLR4 in a MyD88/TRIF dependent fashion (and in a CD14 independent manner) [18]. Here we describe the role of *Brucella* CβG in splenomegaly and inflammation.

Materials and Methods

Ethics statement

The protocol of this study (reference number 10/2011) was approved by the Committee on the Ethics of Animal Experiments of the Universidad Nacional de San Martín, which also approved protocol development under the recommendations in the Guide for the Care and Use of Laboratory Animals of the National Institutes of Health.

Growth of *B. abortus*

Bacterial strains used in this study were: *Brucella abortus* strain 2308, *Brucella abortus* strain S19, *Brucella abortus* cgs08 mutant, *Brucella abortus* cgs19 mutant [17]; *Brucella abortus* cgt19 mutant [13] and *Brucella abortus* cgm19 mutant [14]. All experiments involving live *B. abortus* were conducted in a biosafety level 3 (BSL3) facilities at the University of San Martín, Buenos Aires, Argentina. *B. abortus* strains were grown at 37°C on tryptic soy agar (TSA) and in tryptic soy broth (TSB) on a rotary shaker (250 rpm). If necessary, culture media were supplemented with the appropriate antibiotics at the following concentrations: kanamycin (Km), 50 μg/ml; ampicillin (Amp), 100 μg/ml; and nalidixic acid (Nal), 5 μg/ml.

CβG purification from *B. abortus*

For preparative purposes, 6-liter of stationary-phase cultures of *B. abortus* S19 grown for 48 h at 37°C (250 rpm) were harvested by centrifugation at 10,000×g for 10 min. The pellets were extracted with ethanol (70% ethanol; 1 h at 37°C). The ethanolic extracts were centrifuged and the supernatants were dried in a speed-vac centrifuge and subjected to Bio-Gel P4 chromatography and HPLC on a C18 silica column for further purification as described previously [14]. Purity of CβG was confirmed by NMR analysis [14]. Thin-layer chromatography (TLC) was performed as previously describe [12].

Western blot analysis

To monitor the expression levels of outer membrane proteins and LPS in *B. abortus* strains, bacteria were grown in TSB and harvested at stationary phase. Equivalent bacterial pellets were resuspended in Laemmli buffer and samples were subjected to SDS-PAGE. Proteins were transferred onto nitrocellulose mem-

branes using a semi-dry transfer procedure. Immunoblotting was performed using mouse monoclonal antibodies against Omp16 and Omp19 (kindly provided by Dr. Axel Cloeckaert) and mouse anti-O-polysaccharide monoclonal antibody M84 (kindly provided by Dr. K Nielsen). The correct O-antigen display on the membrane of *B. abortus* was confirmed by a *Brucella* phage sensitivity test (Tb[s], Rc[r]) and crystal violet staining [30].

In order to detect the expression of flagellin in *B. abortus* strains, bacteria were grown in 2YT medium [19] and harvested at early log phase (OD_{600} 0.2). Equivalent bacterial pellets were resuspended in Laemmli buffer and samples were subjected to 12% SDS-PAGE. Proteins were transferred as described above. Immunoblotting was performed using rabbit polyclonal antibodies against *Brucella* flagellin (kindly provided by Dr. J J Letteson).

Effect of the osmolarity on *B. abortus* growth

All strains were grown in regular LB (170 mM NaCl) until stationary phase. Cultures were harvested by centrifugation (12,500×g for 5 min) and washed twice with PBS. Cells were suspended to an OD_{600} of 0.9 and serially diluted in PBS, and 10 μl of the dilutions were spotted on LB agar or LB agar without the addition of NaCl. The plates were incubated at 37°C for 5 days before being read.

B. abortus virulence in mice

Virulence was determined by quantitating *Brucella* survival in mouse spleens at two weeks postinfection, as previously described [20]. Groups of 9-week-old female BALB/c mice were intraperitoneally (i.p) injected with different doses of *B. abortus* strains in 0.2 ml of sterile PBS. The animals were euthanized, and the spleens were removed, weighed, homogenized in PBS, serially diluted, and plated onto TSA with the appropriate antibiotics to determine the number of CFU per spleen.

Complementation of *B. abortus* cgs mutant with purified CβG

Mice were i.p inoculated with 1×10^6 CFU of *B. abortus* S19 or 1×10^6 CFU of its isogenic strain *B. abortus* cgs mutant. Sets of five mice inoculated with *B. abortus* cgs mutant were i.p injected with 15 μg of purified CβG during the first five days of infection. At two weeks postinfection, the animals were euthanized, and the spleens were removed, weighed, homogenized in PBS, serially diluted, and plated onto TSA with the appropriate antibiotics to determine the number of CFU per spleen.

Histological analysis of spleens infected with *B. abortus*

Group of mice were i.p infected with 1×10^6 CFU of *B. abortus* S19 or its isogenic *B. abortus* cgs mutant strain. At two weeks postinfection spleens were excised, fixed with 10% neutral buffered formalin and paraffin embedded. Finally, four micrometers thick longitudinal sections of spleens were obtained and stained with hematoxylin and eosin to assess the degree of spleen inflammation.

Preparation of single-cell suspensions of spleens infected with *B. abortus*

Spleens were aseptically removed and single cell suspensions were prepared by gently teasing through a sterile stainless steel screen. Erythrocytes were lysed in red blood cell lysis buffer and cells were washed twice in PBS solution.

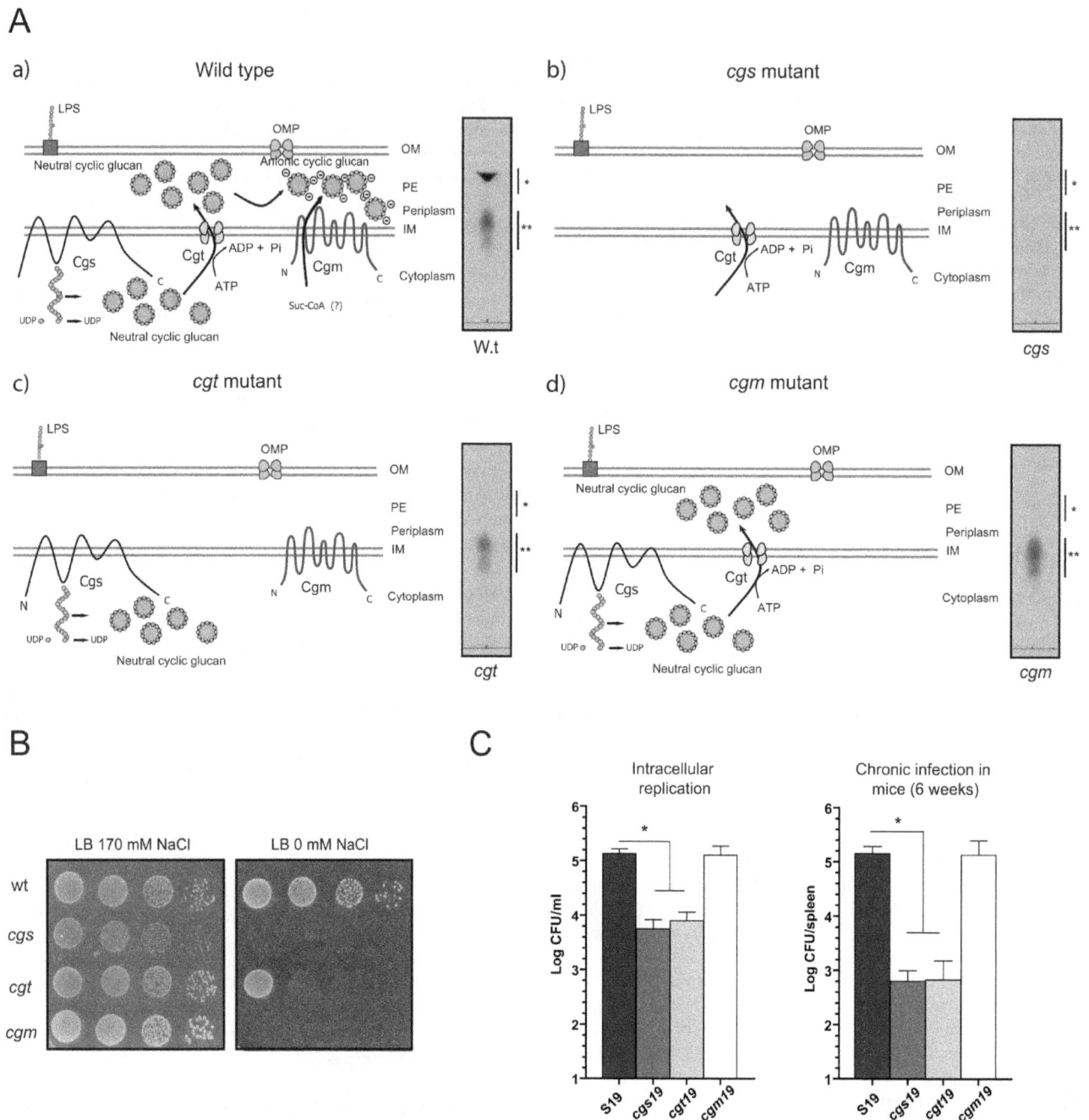

Figure 1. Different roles for *Brucella* CβG. (A) Schematic representation of *Brucella* CβG biosynthesis in the wild type strain and its isogenics CβG mutants strains, *cgs*, *cgt* and *cgm*. Inset pictures show TLC analysis of neutral (**) and anionic (*) CβG. Cgs: Cyclic β-1,2-glucan synthase, Cgt: Cyclic β-1,2-glucan transporter, Cgm: Cyclic β-1,2-glucan modifier. IM: Inner membrane, OM: outer membrane, OMP: outer membrane protein, UDP-⬤: uridine diphospho-glucose, LPS: Lipopolysaccharide. (B) Effect of osmolarity on growth of different strains of *B. abortus*. Dilutions were spotted on LB agar (170 mM NaCl) or LB agar without the addition of NaCl. The plates were incubated at 37°C for 5 days before determining CFU. (C) CβG mutant strains phenotype during *Brucella*-host interaction. Intracellular replication was monitored in HeLa cells at 48 h p.i and establishment of chronic infection of *Brucella* was evaluated in BALB/c mice by determining spleen CFU at six weeks p.i.

Determination of inflammatory cell recruitment in spleens infected with *B. abortus*

BALB/c mice were infected with 1×10^6 CFU of *B. abortus* S19 or its isogenic *B. abortus cgs* mutant strain. Spleens were obtained at two weeks postinfection and single cell suspensions were prepared. To assess recruitment of different cell subtypes after infection, splenocytes (2×10^6) were stained with anti-CD4 (FITC), -CD8 (PE-Cy5), -CD11c (PE), -CD11b (PE-Cy7), -Ly6G (PE), -Ly6C (PerCP-Cy5.5) and -B220 (PE) and analyzed by flow cytometry using FACSAriaII (BD Biosciences, San Jose, CA) and FlowJo software (Tree Star, Ashland, OR). Monoclonal antibodies were purchased from eBioscience (San Diego, CA) and BD Biosciences (San Jose, CA).

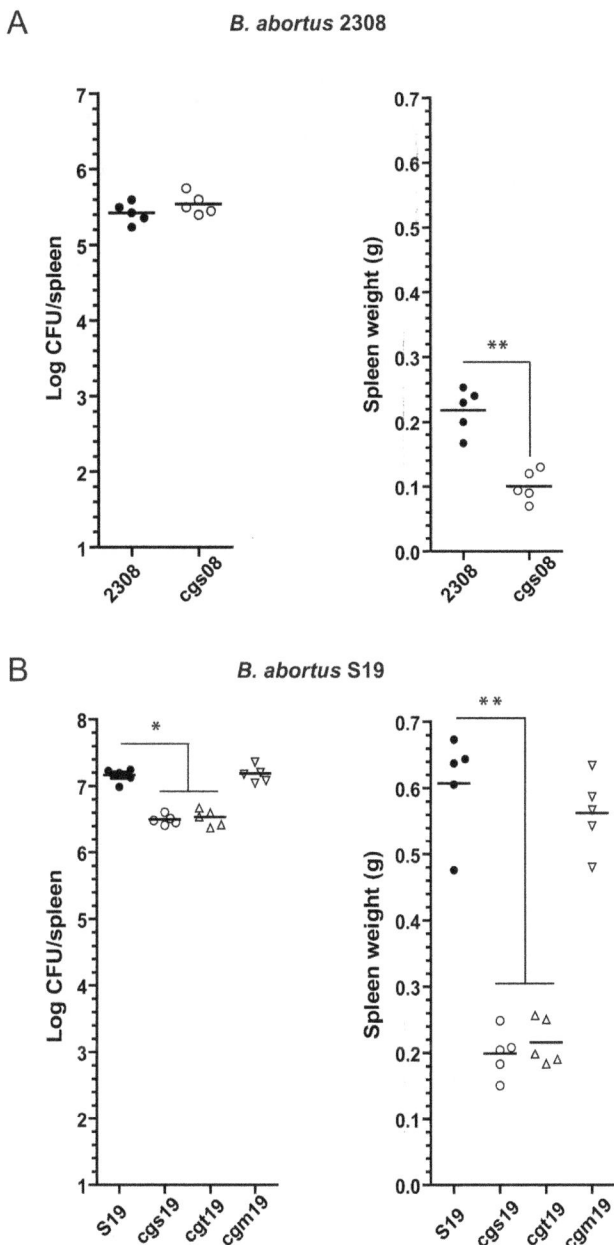

Figure 2. B. abortus cgs and cgt mutants elicit a reduced splenomegaly in mice. BALB/c mice were intraperitoneally infected (1×10^6 CFU) with (A) B. abortus 2308 or (B) B. abortus S19 and their isogenic CβG mutant strains. At two weeks postinfection, spleens were removed, weighed (right panel) and the numbers of CFU recovered were determined by serial dilutions and plating onto TSA (left panel). Five animals were used for each determination. *, $P<0.05$; **, $P<0.01$, Mann-Whitney test.

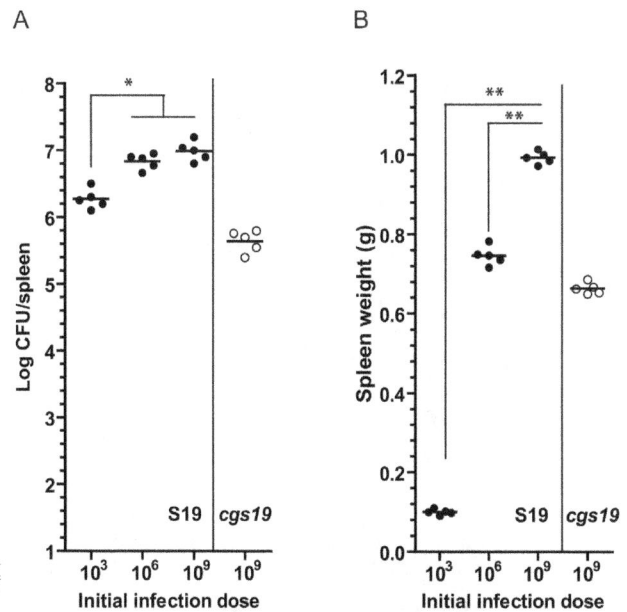

Figure 3. Splenomegaly induced by B. abortus is dependent on the initial dose of infection. BALB/c mice were intraperitoneally infected with different doses of B. abortus S19 or B. abortus cgs19 mutant. At two weeks postinfection, spleens were removed, weighed and the numbers of CFU recovered from spleens determined by serial dilutions and plating onto TSA. (A) Recovery of viable bacteria from spleens of mice. (B) Spleens weight of infected mice. Bacterial doses were calculated retrospectively by colony counting. *, $P<0.05$; **, $P<0.01$, Mann-Whitney test.

Determination of cytokines in bone-marrow derived macrophages (BMDM) infected with B. abortus strains

Macrophages were derived from bone marrow of C57BL/6, Mal/tirap, TLR4, TLR6 and TLR9 KO mice as follows. Each femur and tibia was flushed with 5 ml of Hank's balanced salt solution (HBSS). The resulting cell suspension was centrifuged, and the cells were resuspended in Dulbecco's modified Eagle's medium (DMEM; Gibco, Grand Island, NY) containing 10% fetal bovine Serum (FBS; Gibco), 1% penicillin/streptomycin (100 μg/ mL) and 10% L929 cell-conditioned medium (LCCM) as a source of macrophage colony-stimulating factor (M-CSF). The cells were distributed in 24-well plates and incubated at 37°C in a 5% CO_2 atmosphere. Three days after seeding, another 0.1 ml of LCCM was added. On the seventh day, the medium was renewed. On the 10th day of culture, the cells were completely differentiated into macrophages [21]. The BMDM were infected with B. abortus S19 or its isogenic B. abortus cgs mutant strain, corresponding to a multiplicity of infection (MOI) of 100:1. After 60 min incubation with the bacteria, wells were washed three times with phosphate-buffered saline (PBS) and incubated with fresh medium containing 50 μg ml−1 Gm and 100 μg ml−1 streptomycin to kill non-internalized bacteria After 24 h postinfection, the level of IL-12 (p40) and TNFα in the supernatants of BMDM were measured by ELISA Duoset kit (R&D, Minneapolis, MN). At 4, 24 and 48 hours postinfection, infected C57BL/6 BMDM were washed three times with PBS and lysed with 500 μl 0.1% Triton X-100 (Sigma-Aldrich Co.). The intracellular CFU was determined by plating serial dilutions on TSA with the appropriated antibiotic.

Results and Discussion

Transport of Brucella CβG to periplasm is required for bacterial-host interaction

As shown in Fig. 1A–a, Cgs, a 320 kDa membrane bound enzyme (the second largest protein in Brucella), is the enzyme responsible for the synthesis of cytoplasmic CβG, which is afterwards translocated to the periplasmic space by Cgt, a CβG-specific ABC transporter, and modified with succynil groups by the activity of Cgm (Fig. 1A–a). In order to determine if the different biosynthetic intermediate states of CβG play differential

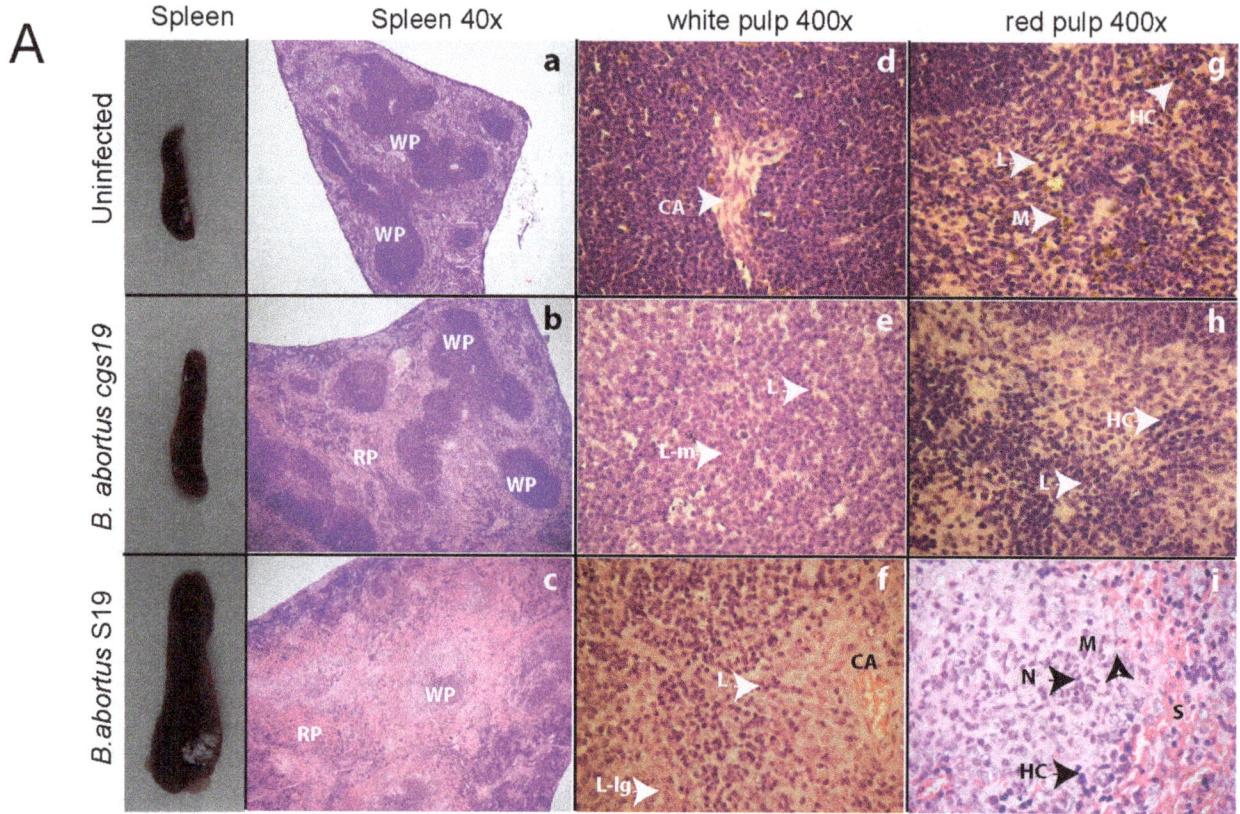

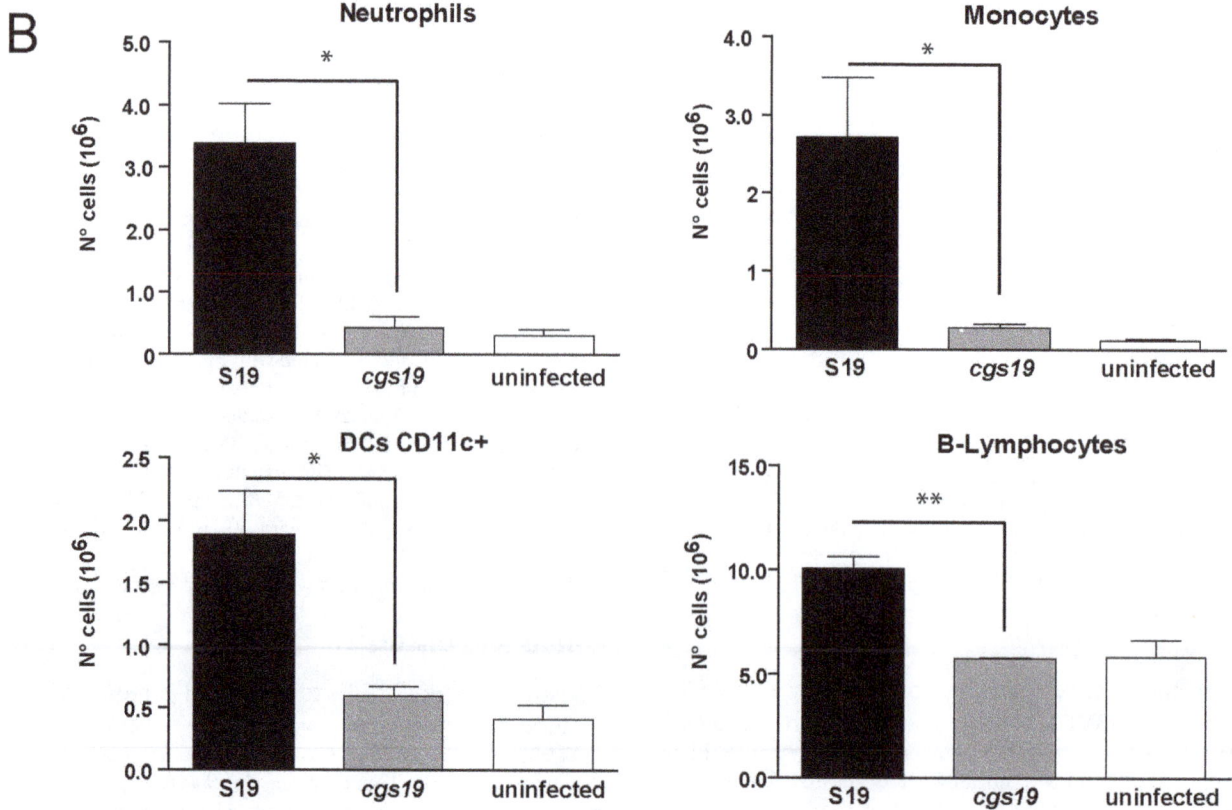

Figure 4. Reduced splenomegaly elicited by *cgs B. abortus* infection is a consequence of a lesser degree of cell recruitment. (A) BALB/c mice were intraperitoneally infected (1×10^6 CFU) with *B. abortus* S19, *B. abortus cgs19* mutant or PBS as a control. Mice were euthanized at two weeks postinfection and spleens were removed and examined by histological analysis. WP: white pulp; RP: Red pulp; CA: central spleen artery, L:

lymphocytes, L-m: medium lymphocytes; L-lg: large lymphocytes, HC: Hematopoietic cells, N: neutrophils, M: macrophages, S: dilated sinuses. (B) The number of differential counts of monocytes, neutrophils, dendritic cells (DCs) and B lymphocytes were determined as described in Materials and Methods. *, $P<0.05$; **, $P<0.01$, t test.

roles in the bacteria the phenotype of three different *B. abortus* CβG mutants were compared: *cgs* [17], *cgt* [13] and *cgm* mutant [14]. As shown in Fig. 1A–b, deletion of *cgs* gene abolished the presence of CβG but mutations in *cgt* or *cgm* lead to the production of neutral CβG (Fig. 1A–c and d) with a different cellular localization being cytoplasmic in *cgt* mutant and periplasmic in the case of *cgm*. From the comparison between *cgt* and *cgm* it became evident that CβG plays at least two roles: adaptation to hypo-osmotic growth conditions (Fig. 1B) and *Brucella*-host interaction (Fig. 1C). While the adaptation to hypo-osmotic conditions requires the presence of anionic periplasmic CβG (Fig. 1B), the sole requirement for host-pathogen interaction is the transport of CβG to the periplasmic space regardless of its anionic charge, as demonstrated by the *cgm* phenotype in intracellular replication and chronic infection in mice (Fig. 1C).

B. abortus mutants unable to produce CβG or to transport it to the periplasm elicit a reduced splenomegaly in mice

As we reported, deletion of *cgs* gene either in the *wild type* strain *B. abortus* 2308 or in the vaccine strain *B. abortus* S19 (which is virulent for humans) reduced their ability to infect mice and hampered their efficiency to reach the intracellular replication niche in epithelial cells [16,17]. Although attenuation of virulence occurs in both *Brucella* backgrounds, in the S19 strain the phenotype is already evidenced after 4 weeks postinfection while in the 2308 background, attenuation is observed after 12 weeks postinfection [17]. Interestingly, at two weeks postinfection when the infection is still not resolved and the number of bacteria in the spleens are equivalent (in the case of *B. abortus* cgs08 mutant, Fig. 2A left panel) or slightly reduced (in the case of *B. abortus* cgs19 mutant, Fig. 2B left panel) compared to their respective wild type parental strains, the splenomegaly was significantly reduced in mice infected with both *cgs* isogenic mutants (Fig. 2A and 2B, right panel). In an earlier report, Crawford et al. observed that splenomegaly elicited by *Brucella melitensis* infection (which peaks from two to three weeks postinfection) is dependent on the initial dose of infection rather than on the bacterial burden, concluding that very early events in the *Brucella* infection are the driving force controlling the severity of the inflammatory response in the spleen [22].

Since *cgs* mutants induced a reduced splenomegaly phenotype in both *B. abortus* 2308 and *B. abortus* S19 backgrounds, we decided to use the vaccine strain of *B. abortus* S19 for the next set of experiments as the wild type control because it elicited a significantly increased splenomegaly compared to the one induced by the *B. abortus* 2308 strain [23] allowing us to develop a more sensitive assay. Additionally, since *B. abortus* S19 presents an intracellular trafficking defect in epithelial cells [24], similar to the defect described for *B. abortus* cgs mutant strains, we reasoned that the use of S19 as our wild type control would reduce also the experimental variability due to differences in intracellular localization. Fig 2B (right panel) shows that the *B. abortus* S19 and *B. abortus* cgm19 mutant strains evoked a significantly increased inflammatory response in the spleens in comparison with the mutants *B. abortus* cgs19 and *B. abortus* cgt19, suggesting that the splenomegaly correlated with the presence of CβG within the periplasmic space. As mentioned above, CβG is likely secreted within the host cell and therefore periplasmic localization

requirement might be potentially a prerequisite for its delivery outside the bacteria.

Splenomegaly is dependent on the initial dose of *B. abortus* infection

To study the impact of the initial dose of infection on the intensity of the splenomegaly, BALB/c mice were intraperitoneally infected with different doses of *B. abortus* S19 (10^3, 10^6 and 10^9 CFU) (Fig. 3) and after two weeks postinfection spleens were removed, weighed, homogenized and the number of bacteria determined by serial dilution and plating to determine CFUs. Fig. 3 shows that, at two weeks postinfection although the numbers of replicating *B. abortus* S19 recovered from spleens were similar (about 10^6–10^7 CFU per spleen) (Fig. 3A), the splenomegaly varied from negligible (about 0.1 grams) to a massive one (1 gram) (Fig. 3B) depending on the initial dose of infection, confirming previous observations in *Brucella melitensis* [22]. To estimate the impact of *cgs* phenotype on spleen inflammation, we infected mice with different doses of *B. abortus* cgs19 mutant strain and at two weeks postinfection, splenomegaly was determined. The results showed that, to achieve an equivalent degree of splenomegaly elicited by 1×10^6 *B. abortus* S19, it was necessary to increase a thousand times the initial dose of the *B. abortus* cgs19 (1×10^9) (Fig. 3B).

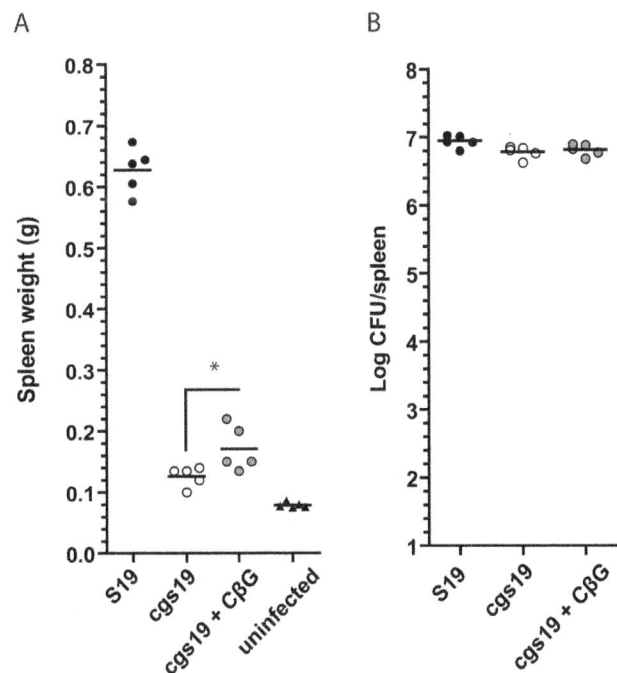

Figure 5. *B. abortus* CβG partially complement splenomegaly in mice. BALB/c mice were intraperitoneally infected with 1×10^6 CFU of *B. abortus* S19 or 1×10^6 CFU of *B. abortus* cgs19 mutant. Sets of five mice inoculated with *B. abortus* cgs mutant were injected intraperitoneally with 15 µg of purified CβG during the first five days of infection. At two weeks postinfection, mice were euthanized, and spleens were removed, weighed (A) and the number of CFU recovered determined by serial dilutions and plating onto TSA (B). *, $P<0.05$, Mann-Whitney test.

Reduced splenomegaly elicited by the *cgs* mutant is the consequence of a lesser degree of cell recruitment

At two weeks postinfection spleens from *Brucella* infected mice were processed for histological analysis and the results are shown in Fig 4A. While spleens from wild type strain infected mice showed an increase in the global cellularity with a massive infiltration of the red and white pulp (Fig. 4A–c), spleens from *cgs* infected mice shown a reduced degree of cellular infiltration (Fig. 4A–b) similar to what was observed in spleens of non-infected animals (Fig. 4A–a). Remarkably, at 400× magnification, the presence of neutrophils and macrophages within the red pulp of *B. abortus* S19 infected mice was observed (Fig. 4A–i) indicating

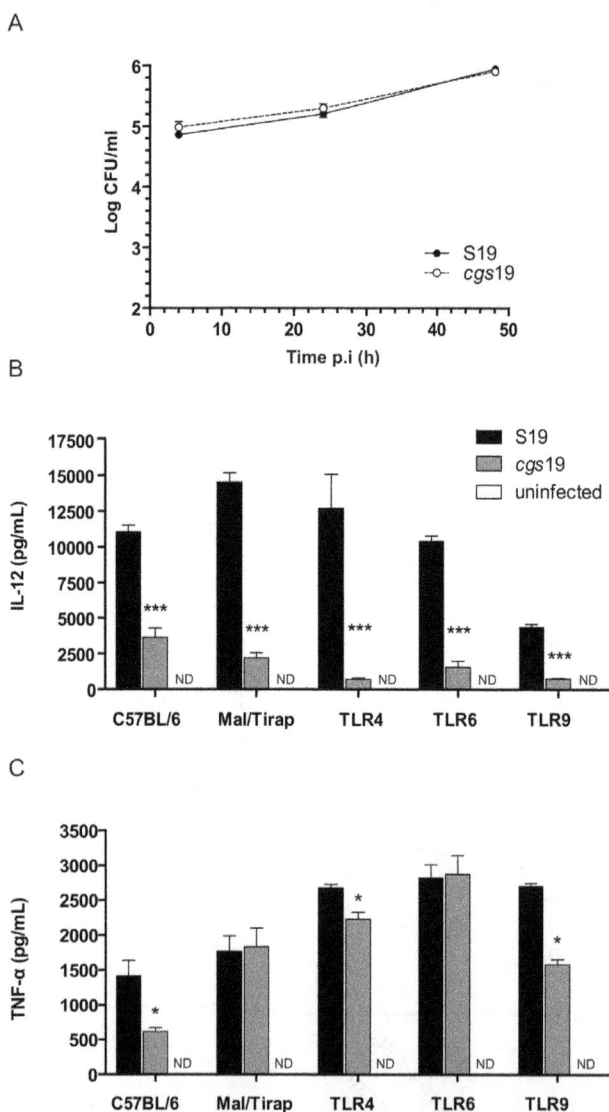

massive cell recruitment to the spleen. In addition, spleen white pulp of *B. abortus* S19 and *B. abortus cgs19* infected mice presented reactive lymphocytes (larger and medium lymphocytes) likely due to antigen stimulation (Fig. 4A–e and f).

To further characterize the spleen cell population, antibodies against specific markers were used to identified and quantify neutrophils, monocytes, B cells, T cells and dendritic cells by flow cytometry. As shown in Fig 4B spleens from *B. abortus* S19 infected mice presented eight times more neutrophils, ten times more monocytes, three times more dendritic cells and two times more B cells than spleens infected with *B. abortus cgs19* mutant strain. No difference was observed in T-cell recruitment (not shown).

One possible explanation for the reduced splenomegaly and cell recruitment associated with the lack of CβG biosynthesis or CβG transport to the periplasm could be that *cgs/cgt* strains have a differential expression of PAMPs that might lead to a less efficient engagement of the host innate immune receptors and consequently to a diminished inflammatory response.

B. abortus cgs mutant has normal expression of flagellin and Omps, displaying normal amounts of smooth LPS (S-LPS) on the membrane

It has been demonstrated that *Agrobacterium* and *Sinorhizobium cgs* mutants have a reduced expression of flagellin and a defect in flagella assembly that consequently leads to a non-motile phenotype [25] [26]. Since bacterial flagellins are powerful agonists of innate immune receptors, being recognized extracellularly by the surface receptor TLR-5 and intracellularlly by Nod-like receptors (such Ipaf or Naip5) [2], we explored if the *Brucella cgs* mutant has an altered expression of flagellin that might explain the diminished inflammatory response. In *Alphaproteobacteria* (including the *Brucella* genus) flagellin has a modification in the protein domain recognized by the TLR-5 innate immune receptor supporting the idea that *Brucella* flagellin has evolved to escape the host innate immune recognition [2]. In addition, in *Brucella*, flagellin expression is tightly controlled and only expressed under very strict culture conditions [19] and, because it has never been identified in any intracellular *Brucella* proteomic studies, this suggests that it is expressed poorly within the host cell [27,28]. In order to determine if the absence of periplasmic CβG affects flagellin expression, *B. abortus* S19 and its isogenic *cgs*, *cgt* and *cgm* mutants were grown in the conditional media to allow flagella assembly as described in Materials and Methods. Expression of *Brucella* flagellin was monitored by Western blot analysis, and the results, shown in Fig. S1A, indicates no changes in flagellin expression. Afterwards, we studied the expression of other critical *Brucella* PAMPs such as OMP16, OMP19 [29] or LPS [3,4][31] and no differences in expression were observed associated to any *B. abortus* CβG mutant strains (Fig. S1A and S1B).

Purified *Brucella* CβG partially complemented the splenomegaly defect in *cgs*-infected mice

As was mentioned above, purified CβG is capable to restore the intracellular replication deficiency of *cgs*-mutant strain in HeLa cells [16] and this observation suggests that CβG must be secreted within the host cell to exert its role in virulence. To evaluate the direct role of *Brucella* CβG in the inflammatory response we designed a trans-complementation experiment using purified CβG. For this, we added the purified carbohydrate to the bacterial initial inoculum and during the following five days postinfection by i.p injection. Splenomegaly determined at two weeks postinfection showed an increase in spleen weight in the CβG complemented mice compared to the *cgs* mutant although

Figure 6. *B. abortus cgs* mutant failed to induce IL-12 in BMDM. (A) Intracellular multiplication of *B. abortus* strains in Bone marrow cells derived from C57BL/6. Number of CFU of intracellular bacteria was determined after lysis of infected cells at the indicated times postinfection. Each determination was performed in duplicate, and values are shown as the means ± standard deviations and are representative of three independent experiments. (B) Bone marrow cells derived from C57BL/6, Mal/Tirap, TLR4, TLR6 or TLR9 KO mice were infected with *B. abortus* (MOI 1:100). IL-12 and TNF-α were measured by ELISA at 24 h postinfection. *, *P*<0.05; **, *P*<0.01, *** *P*<0.01 *t* test.

not to the degree of the *Brucella* wild type strain (Fig. 5A). Although significant, trans-complementation with purified CβG elicits a moderate increase on spleen enlargement compared with *cgs* mutant strain, an effect that can be explained as the result of the intrinsic limitations of this experimental approach. For instance, since injected CβG is diluted within the peritoneal cavity is not possible to know the effective CβG concentration at the *Brucella* intracellular replicative niche. However, these results suggest that *Brucella* CβG plays a direct role in the induction of the inflammatory response in the spleen. It has been recently proposed that purified *Brucella* CβG may act as a novel class of adjuvant that can induce, *in vitro*, the activation of mouse and human dendritic cells, enhancing T cell responses and CD4+ T cells memory immune responses in a TLR4 dependent/CD14 independent fashion [18]. Interestingly, it was described recently that inflammation far from be an undesirable byproduct of the bacterial infection can be a process that pathogens can actively promote to create favorable conditions for its own advantage. Thus enterobacterial pathogens can outgrow commensal bacteria or promote host release of compounds of nutritional relevance mediated by the activity of its virulence mechanisms [32,33]. In that manner it is conceivable to speculate that CβG dependent splenomegaly might work as an advantage for host colonization in *Brucella* infection.

B. abortus cgs mutant strain elicits a reduced inflammatory response in BMDM

In previous studies, Zhan *et al* demonstrated that the macrophage-synthesized cytokines IL-12 and TNF-α are required for an efficient control of *Brucella* infection. In addition, it was shown that depletion of both pro-inflammatory cytokines by antibody treatment abolished the development of splenomegaly in animals infected with *B. abortus* S19 at two weeks postinfection [34]. To understand if the reduced spleen inflammation observed in mice infected with *B. abortus cgs19* was due to a lower induction of IL-12 or TNF-α, we performed an *in vitro* infection experiment with naïve BMDM. Differently to the defective intracellular replication phenotype reported for *B. abortus cgs* mutant strains in HeLa cells [17], in BMDM *B. abortus cgs19* mutant strain showed no defect in intracellular replication in comparison with its parental wild type strain (Fig. 6A). The same phenotype was also reported for *cgs* mutant strain for intracellular replication in dendritic cells [5]. As shown in Fig. 6B and 6C, wild type BMDM infected with *B. abortus* S19, secreted higher levels of IL-12 (Fig. 6B) and TNF-α (Fig. 6C) to the supernatant than cells infected with *B. abortus cgs19* mutant strain, suggesting that CβG promotes the induction of proinflammatory cytokines from BMDM. To understand if this CβG-dependent IL-12/TNF-α induction is dependent on Toll-like receptor (TLR) recognition, an *in vitro* experiment with BMDM from Mal/Tirap (the TLR2/TLR4 adapter protein), TLR4, TLR6 and TLR9 KO mice was performed (Fig. 6B and C). As shown in Fig. 6B, CβG-dependent IL-12 induction was independent on the presence of TLR2, TLR4, TLR6 or TLR9 (Fig 6B) while CβG-dependent TNF-α induction was independent on the presence of TLR4 or TLR9 (Fig. 6C). In absence of TLR2 or TLR6, *B. abortus* S19 and its isogenic *cgs* mutant strain elicited

similar levels of TNF-α (Fig. 6C). These results suggest that TLR2 and TLR6 are potentially involved in the TNF-α induction elicited by CβG. It is interesting to notice that TLR2 and TLR6 are able to interact to form a heterodimer which is responsible for bacterial deacylated lipoproteins recognition [35].

Taken together all these results suggest that the reduced splenomegaly observed in *cgs* mutant strain infected mice is a consequence of a lower induction of proinflammatory cytokines that lead to a lesser cell recruitment to this organ.

Concluding Remarks

In the present study we describe the role of the *Brucella* cyclic β-1,2-glucan in promoting spleen enlargement during bacterial infection. Splenomegaly was the result of massive cell recruitment, mediated by the induction of pro-inflammatory cytokines. Since mutants deficient in CβG biosynthesis in the soil bacteria *Sinorhizobium* and *Agrobacterium* have shown to have membrane alterations that lead to non-motile phenotypes and an increased sensitivity to dyes and detergents, we evaluated if the low-inflammation phenotype observed with the *cgs/cgt* mutants was due to changes in expression of membrane bound complexes with inflammatory activity. No differences in flagellin, OMPs or LPS expression were evident and results suggested that CβG per se is responsible for the splenomegaly observed. The molecular mechanism underlying CβG induced splenomegaly remains to be identified and further studies will be performed to characterize this process.

Supporting Information

Figure S1 Western blot analysis of flagellin, outer membrane proteins (Omps) (A) and LPS (B) in *B. abortus* CβG mutant strains. Immunoblotting was performed using: rabbit polyclonal antibodies against *Brucella* flagellin, monoclonal antibodies against Omp16 and Omp19; and O-antigen specific monoclonal antibody (M84). SDS-PAGE and Western blot were carried out as described in Materials and Methods. The same amount of total protein extracts were loaded into the gels. The estimated molecular weight of each protein is shown.

Acknowledgments

We would like to dedicate this work to the memory of Dr. Rodolfo A. Ugalde.

We thank Luis Purón and Juan Ugalde for critical reading of the manuscript and useful suggestions. J.M.S. is a fellow of CONICET. M.S.R., G.H.G., J.C. and G.B. are members of the Research Career of CONICET.

Author Contributions

Conceived and designed the experiments: MSR GHG SCO JC GB. Performed the experiments: MSR AEI JASF JMS LM GB. Analyzed the data: MSR GB. Contributed reagents/materials/analysis tools: LM. Wrote the paper: MSR GB.

References

1. Pappas G, Papadimitriou P, Christou L, Akritidis N (2006) Future trends in human brucellosis treatment. Expert Opin Investig Drugs 15: 1141–1149.

2. Andersen-Nissen E, Smith KD, Strobe KL, Barrett SL, Cookson BT, et al. (2005) Evasion of Toll-like receptor 5 by flagellated bacteria. Proc Natl Acad Sci U S A 102: 9247–9252.

3. Lapaque N, Takeuchi O, Corrales F, Akira S, Moriyon I, et al. (2006) Differential inductions of TNF-alpha and IGTP, IIGP by structurally diverse classic and non-classic lipopolysaccharides. Cell Microbiol 8: 401–413.

4. Ferguson GP, Datta A, Baumgartner J, Roop RM 2nd, Carlson RW, et al. (2004) Similarity to peroxisomal-membrane protein family reveals that *Sinorhizobium* and *Brucella* BacA affect lipid-A fatty acids. Proc Natl Acad Sci U S A 101: 5012–5017.

5. Salcedo SP, Marchesini MI, Lelouard H, Fugier E, Jolly G, et al. (2008) *Brucella* control of dendritic cell maturation is dependent on the TIR-containing protein Btp1. PLoS Pathog 4: e21.

6. Newman RM, Salunkhe P, Godzik A, Reed JC (2006) Identification and characterization of a novel bacterial virulence factor that shares homology with mammalian Toll/interleukin-1 receptor family proteins. Infect Immun 74: 594–601.

7. Salcedo SP, Marchesini MI, Degos C, Terwagne M, Von Bargen K, et al. (2013) BtpB, a novel *Brucella* TIR-containing effector protein with immune modulatory functions. Front Cell Infect Microbiol 3: 28.

8. Young EJ (1995) An overview of human brucellosis. Clin Infect Dis 21: 283–289; quiz 290.

9. Moreno E, Moriyon I (2002) *Brucella melitensis*: a nasty bug with hidden credentials for virulence. Proc Natl Acad Sci U S A 99: 1–3.

10. DelVecchio VG, Kapatral V, Redkar RJ, Patra G, Mujer C, et al. (2002) The genome sequence of the facultative intracellular pathogen *Brucella melitensis*. Proc Natl Acad Sci U S A 99: 443–448.

11. Inon de Iannino N, Briones G, Tolmasky M, Ugalde RA (1998) Molecular cloning and characterization of cgs, the *Brucella abortus* cyclic beta(1–2) glucan synthetase gene: genetic complementation of *Rhizobium meliloti* ndvB and *Agrobacterium tumefaciens* chvB mutants. J Bacteriol 180: 4392–4400.

12. Briones G, Inon de Iannino N, Steinberg M, Ugalde RA (1997) Periplasmic cyclic 1,2-beta-glucan in *Brucella* spp. is not osmoregulated. Microbiology 143 (Pt 4): 1115–1124.

13. Roset MS, Ciocchini AE, Ugalde RA, Inon de Iannino N (2004) Molecular cloning and characterization of cgt, the *Brucella abortus* cyclic beta-1,2-glucan transporter gene, and its role in virulence. Infect Immun 72: 2263–2271.

14. Roset MS, Ciocchini AE, Ugalde RA, Inon de Iannino N (2006) The *Brucella abortus* cyclic beta-1,2-glucan virulence factor is substituted with O-ester-linked succinyl residues. J Bacteriol 188: 5003–5013.

15. Breedveld MW, Miller KJ (1994) Cyclic beta-glucans of members of the family Rhizobiaceae. Microbiol Rev 58: 145–161.

16. Arellano-Reynoso B, Lapaque N, Salcedo S, Briones G, Ciocchini AE, et al. (2005) Cyclic beta-1,2-glucan is a *Brucella* virulence factor required for intracellular survival. Nat Immunol 6: 618–625.

17. Briones G, Inon de Iannino N, Roset M, Vigliocco A, Paulo PS, et al. (2001) *Brucella abortus* cyclic beta-1,2-glucan mutants have reduced virulence in mice and are defective in intracellular replication in HeLa cells. Infect Immun 69: 4528–4535.

18. Martirosyan A, Perez-Gutierrez C, Banchereau R, Dutartre H, Lecine P, et al. (2012) *Brucella* beta 1,2 cyclic glucan is an activator of human and mouse dendritic cells. PLoS Pathog 8: e1002983.

19. Fretin D, Fauconnier A, Kohler S, Halling S, Leonard S, et al. (2005) The sheathed flagellum of *Brucella melitensis* is involved in persistence in a murine model of infection. Cell Microbiol 7: 687–698.

20. Montaraz JA, Winter AJ (1986) Comparison of living and nonliving vaccines for *Brucella abortus* in BALB/c mice. Infect Immun 53: 245–251.

21. Macedo GC, Magnani DM, Carvalho NB, Bruna-Romero O, Gazzinelli RT, et al. (2008) Central role of MyD88-dependent dendritic cell maturation and proinflammatory cytokine production to control *Brucella abortus* infection. J Immunol 180: 1080–1087.

22. Crawford RM, Van De Verg L, Yuan L, Hadfield TL, Warren RL, et al. (1996) Deletion of purE attenuates *Brucella melitensis* infection in mice. Infect Immun 64: 2188–2192.

23. Enright FM, Araya LN, Elzer PH, Rowe GE, Winter AJ (1990) Comparative histopathology in BALB/c mice infected with virulent and attenuated strains of *Brucella abortus*. Vet Immunol Immunopathol 26: 171–182.

24. Pizarro-Cerda J, Meresse S, Parton RG, van der Goot G, Sola-Landa A, et al. (1998) *Brucella abortus* transits through the autophagic pathway and replicates in the endoplasmic reticulum of nonprofessional phagocytes. Infect Immun 66: 5711–5724.

25. Douglas CJ, Halperin W, Nester EW (1982) *Agrobacterium tumefaciens* mutants affected in attachment to plant cells. J Bacteriol 152: 1265–1275.

26. Geremia RA, Cavaignac S, Zorreguieta A, Toro N, Olivares J, et al. (1987) A *Rhizobium meliloti* mutant that forms ineffective pseudonodules in alfalfa produces exopolysaccharide but fails to form beta-(1----2) glucan. J Bacteriol 169: 880–884.

27. Al Dahouk S, Jubier-Maurin V, Scholz HC, Tomaso H, Karges W, et al. (2008) Quantitative analysis of the intramacrophagic *Brucella suis* proteome reveals metabolic adaptation to late stage of cellular infection. Proteomics 8: 3862–3870.

28. Lamontagne J, Forest A, Marazzo E, Denis F, Butler H, et al. (2009) Intracellular adaptation of *Brucella abortus*. J Proteome Res 8: 1594–1609.

29. Giambartolomei GH, Zwerdling A, Cassataro J, Bruno L, Fossati CA, et al. (2004) Lipoproteins, not lipopolysaccharide, are the key mediators of the proinflammatory response elicited by heat-killed *Brucella abortus*. J Immunol 173: 4635–4642.

30. Alton GG, Jones LM, Angus RD, Verger JM (1988) Techniques for the brucellosis laboratory. (INRA, Paris, France).

31. Cheers C, Pavlov H, Riglar C, Madraso E (1980) Macrophage activation during experimental murine brucellosis. III. Do macrophages exert feedback control during brucellosis? Cell Immunol 49: 168–177.

32. Winter SE, Thiennimitr P, Winter MG, Butler BP, Huseby DL, et al. (2010) Gut inflammation provides a respiratory electron acceptor for *Salmonella*. Nature 467: 426–429.

33. Stecher B, Robbiani R, Walker AW, Westendorf AM, Barthel M, et al. (2007) *Salmonella enterica* serovar typhimurium exploits inflammation to compete with the intestinal microbiota. PLoS Biol 5: 2177–2189.

34. Zhan Y, Liu Z, Cheers C (1996) Tumor necrosis factor alpha and interleukin-12 contribute to resistance to the intracellular bacterium *Brucella abortus* by different mechanisms. Infect Immun 64: 2782–2786.

35. Kang JY, Nan X, Jin MS, Youn SJ, Ryu YH, et al. (2009) Recognition of lipopeptide patterns by Toll-like receptor 2-Toll-like receptor 6 heterodimer. Immunity 31: 873–884.

12

Chemical Interference with Iron Transport Systems to Suppress Bacterial Growth of *Streptococcus pneumoniae*

Xiao-Yan Yang[1,9], Bin Sun[2,9], Liang Zhang[1], Nan Li[1], Junlong Han[1], Jing Zhang[1], Xuesong Sun[1]*, Qing-Yu He[1]*

1 Key Laboratory of Functional Protein Research of Guangdong Higher Education Institutes, Institute of Life and Health Engineering, College of Life Science and Technology, Jinan University, Guangzhou, China, **2** School of Pharmaceutical Sciences, Southern Medical University, Guangzhou, China

Abstract

Iron is an essential nutrient for the growth of most bacteria. To obtain iron, bacteria have developed specific iron-transport systems located on the membrane surface to uptake iron and iron complexes such as ferrichrome. Interference with the iron-acquisition systems should be therefore an efficient strategy to suppress bacterial growth and infection. Based on the chemical similarity of iron and ruthenium, we used a Ru(II) complex R-825 to compete with ferrichrome for the ferrichrome-transport pathway in *Streptococcus pneumoniae*. R-825 inhibited the bacterial growth of *S. pneumoniae* and stimulated the expression of PiuA, the iron-binding protein in the ferrichrome-uptake system on the cell surface. R-825 treatment decreased the cellular content of iron, accompanying with the increase of Ru(II) level in the bacterium. When the *piuA* gene (SPD_0915) was deleted in the bacterium, the mutant strain became resistant to R-825 treatment, with decreased content of Ru(II). Addition of ferrichrome can rescue the bacterial growth that was suppressed by R-825. Fluorescence spectral quenching showed that R-825 can bind with PiuA in a similar pattern to the ferrichrome-PiuA interaction *in vitro*. These observations demonstrated that Ru(II) complex R-825 can compete with ferrichrome for the ferrichrome-transport system to enter *S. pneumoniae*, reduce the cellular iron supply, and thus suppress the bacterial growth. This finding suggests a novel antimicrobial approach by interfering with iron-uptake pathways, which is different from the mechanisms used by current antibiotics.

Editor: Roy Martin Roop II, East Carolina University School of Medicine, United States of America

Funding: This work was supported by the National Natural Science Foundation of China (21271086, to Q.-Y. H.; 31000373, to X. S.), Guangdong Natural Science Research Grant (32213027, to Q.-Y. H.; S2012010008685 to X. S.), the Fundamental Research Funds for the Central Universities (11610101 to Q.-Y. H.; 21611201, to X. S.), and the Pearl River Rising Star of Science and Technology of Guangzhou City (2011J2200003, to X. S.). The funders had no role in study design, data collection and analysis, decision to publish, or preparation of the manuscript.

Competing Interests: The authors have declared that no competing interests exist.

* Email: tqyhe@jnu.edu.cn (QYH); tsunxs@jnu.edu.cn (XS)

⑨ These authors contributed equally to this work.

Introduction

Iron is a critical nutrient for bacterial growth and survival, as a major determinant in the development of infection in host. However, the concentration of free iron in host is extremely low ($<10^{-18}$ M). In order to acquire enough iron from their host environments, bacteria have developed highly specific and effective iron-acquisition systems located on the membrane surface [1–3]. Blocking or interfering with the iron-acquisition systems could disrupt bacterial iron homeostasis and thus suppress bacterial growth.

Streptococcus pneumoniae is a dangerous bacterium responsible for various life-threatening diseases including otitis media, septicemia, pneumonia and meningitis in immuno-compromised individuals [4,5]. In *S. pneumoniae*, there are three ABC transporters as known iron-transport systems including PiaABC, PiuABC and PitABC, respectively responsible for the acquisition of heme, ferrichrome and ferric irons [6–9]. Heme, ferrichrome and ferric irons are firstly bound by lipoproteins PiaA, PiuA and PitA and transferred to the permeases PiaB, PiuB and PitB to go

across the cell membrane using energy provided by the ATP hydrolysis through PiaC, PiuC and PitC, respectively. Since the lipoproteins PiaA, PiuA and PitA work as iron-receptors located on the cellular surface, they are the valuable targets for the design of novel antibacterial agents. It must be pointed out that Cheng *et al* have characterized a ferrichorme-binding PiaA protein encoded by gene SP_1032 in *S. pneumnoniae* TIGR4 (corresponding to gene SPD_0915 in *S. pneumoniae* D39) [9], which is actually the protein PiuA through sequence alignments.

Earlier investigations have suggested that bacteria obtain iron mainly in the forms of complexes, heme-iron and ferrichrome-iron compounds as for *S. pneumoniae*. In this study, we selected our previously synthesized Ru(II) complex R-825 to chemically resemble the heme/ferrichrome compounds. Based on the chemical similarity between Fe(III) and Ru(II), we expected that R-825 could compete with heme/ferrichrome for binding to PiaA/PiuA in *S. pneumoniae*. Our experiments showed that R-825 indeed inhibited the bacterial growth by competing with ferrichrome for PiuA binding. We verified that complex R-825 can get access into the bacterium through ferrichrome-transport systems, reducing the iron supply and thus suppressing the bacterial growth.

Materials and Methods

Materials

Ru(II) complex R-825 was synthesized according to the procedure with minor modifications as described in our previous publication, where the complex was defined as 1a [10]. The structure of R-825 is shown in Figure 1.

Bacterial strains and growth conditions

Single gene deleted *piaA*- (SPD_1652), *piuA*- (SPD_0915) and *pitA*- (SPD_0226) mutant strains of *S. pneumoniae* D39 were constructed by long flanking homology-polymerase chain reaction (LFH-PCR) [11,12], mutants were made by replacing *piaA*, *piuA* and *pitA* genes of D39 with gene encoding resistance to erythromycin (erm)[13]. Briefly, the 500 bp region upstream of *piaA* was amplified using primers piaA–P1 and piaA–P3, while the 500 bp region downstream of *piaA* was amplified using primers piaA–P2 and piaA–P4. The erythromycin gene was amplified using primers erm-F and erm-R. The three PCR fragments generated were joined together by overlap extension PCR using primers piaA–P1 and piaA–P2 to form approximately 2-kb final linear DNA construct the deletion fragment. Then, the linear DNA construct was used for homologous recombination and transformed into *S. pneumoniae* D39. Transformants were selected with 0.25 µg/mL erythromycin selection agar plates, mutants were confirmed by PCR and Western blotting. The *piuA*- and *pitA*- mutant strains were constructed using a similar method. All mutations were stable after six sequential passages in Todd-Hewitt broth supplemented with 0.5% yeast extract (0.5% THY) without antibiotic selection.

These mutant strains together with wild-type *S. pneumoniae* D39 were cultured at 37°C and 5% CO_2 in 0.5% THY or on Columbia agar added with 5% sterile defibrinated sheep blood purchased from Ruite company (Guangzhou, China) in which all animal experiment procedures were conducted in strict accordance with the recommendations in the Guide for the Care and Use of Laboratory Animals of the National Institutes of Health. Disc susceptibility assays were performed as described by CLSI guidelines [14]. In short, filter paper disks (5 mm in diameter) containing 50 µg of R-825 were placed on the plates (10[7] bacteria/plate), and inhibition zones were examined after 24 h of incubation at 37°C. Tests using gentamicin as the control were carried out in parallel.

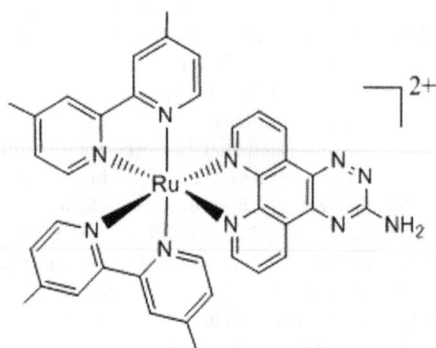

Figure 1. Structure of ruthenium(II) complex R-825 [10].

Minimal inhibitory concentration (MIC) assay

R-825 was dissolved in water to a concentration of 10 mg/mL (12 mM). The MIC assays were carried out in triplicate by using the standard micro-dilution method [14]. Different concentrations of R-825 with 7.5, 15, 30, and 60 µM were added to diluted media containing 10[6] bacteria/mL, followed by incubation in 24-well plates at 37°C for 24 h. The concentration of R-825 corresponding to the well with no visible bacteria growth ($OD_{600} < 0.1$) was taken as the MIC.

Cytotoxicity assay

The human alveolar epithelial cell line A549 was cultivated in DMEM media with 10% fetal bovine serum (FBS) at 37°C in a humidified atmosphere of 5% CO_2. The cytotoxicity of R-825 in A549 cells was assessed using LDH cytotoxicity assay kit by the following procedure as described in the manufacture's manual. Briefly, 5×10^3 A549 cells were seeded into a sterile 96-well plate in triplicate, incubated for 24 h without R-825, followed by a 48 h incubation in the dark with different amounts of R-825 (30, 60, 120, 240 and 480 µM). At the end of the incubation, the cell plate was centrifuged at 400 *g* for 5 min, then 120 µL of the media from each well was transferred to a new plate, and 60 µL of LDH mixture was added to the supernatant and incubated for 30 min in dark at room temperature. Cytotoxicity was assessed by monitoring the absorbance at 490 nm in a microplate reader. Whole cell lysate was used as a positive control.

ICP-MS analysis

Sample preparation for ICP-MS was conducted as a previously described method [15]. Briefly, bacteria were grown in 0.5% THY in the presence or absence of R-825 with sub-MIC concentrations at 37°C and then harvested when the optical density at 600 nm (OD_{600}) reached 0.5–0.6. Cells were pelleted by centrifugation at 8,000 *g* at 4°C for 10 min, and then washed three times with $1 \times PBS$ that had been treated with chelex-100 resin. Subsequently, the wet cell pellets were dried using a Scanvac Freeze Dryer (Labgene Scientific, Switzerland) and the dry weights were calculated. The dry cell mass was disrupted by resuspending pellets in 2 mL of 14% HNO_3, then heated to 95°C for 20 min. Samples were centrifuged at 13,200 *g* for 30 min, the supernatants were collected and an internal standard indium (In) was added to the samples. Metal contents of these samples were analyzed in an iCAP ICP-MS (Thermo Scientific, U.S.A.). Results were showed as ng of Fe and Ru per mg dry weight of cells. All data were evaluated with at least three independent biological experiments.

Real-time quantitative PCR (RT-qPCR)

Total RNA was extracted from *S. pneumoniae* D39 strain with and without R-825 (in sub-MIC) treatment by TRIZOL method according to the manufacturer's manual, and quantified by Nanodrop 2000 spectrophotometer. Any genomic DNA was removed using RNase-free DNase I. cDNA synthesis was performed using 1 µg of total RNA and iScript Reverse Transcriptase (Bio-Rad) according to the manufacture's instruction. RT-qPCR was carried out using EvaGreen Dye (Bio-Rad) in a Miniopticon RT-qPCR System (Bio-Rad). The cycle threshold (Ct) value was measured; relative quantification of specific gene expression was calculated using the $2^{-\Delta\Delta Ct}$ method, with the 16S rRNA as the reference gene. Genes with a two-fold or greater difference in expression level relative to control were considered significant. The primer sequences used for RT-qPCR are shown in Table 1.

Table 1. The primer sequences used for RT-qPCR experiments.

Primer	Sequence (5'–3')
16S rRNA-F	5'-CTGCGTTGTATTAGCTAGTTGGTG-3'
16S rRNA-R	5'-TCCGTCCATTGCCGAAGATTC-3'
PiaA-F	5'-TAGTCAGACAGAGACCAGT-3'
PiaA-R	5'-CTTTCATAGAACCAACATT-3'
PiuA-F	5'-ATTTGACGATTTGGATGGACTT-3'
PiuA-R	5'-GATTTGTATGCTGCTACAGGAG-3'
PitA-F	5'-ATGACTGTTGGTCTCTCTT-3'
PitA-R	5'-TTGTTTTAGCATTTTTACG-3'

Cloning and purification of PiuA protein

The *piuA* gene without the N-terminal lipoprotein signal sequence was PCR amplified from *S. pneumoniae* D39 genomic DNA with the forward primer 5'-CCGCCGGAGCTCTCTTC-TAATTCTGTTAAAAA-3' and the reverse primer 5'-GCCGCCGAATTCTTATTTCGCATTTTTGC-3', creating the SacI and EcoRI restriction sites (underlined). The PCR product was digested with SacI and EcoRI and ligated into the expression vector pBAD/HisA to generate pBAD-PiuA. The construct was transformed into *E. coli* TOP10 for high-level expression of recombinant protein. The transformants were incubated at 37°C with vigorous shaking in LB medium when the OD_{600} reached at 0.8, followed by induction with 0.05% L-arabinose for 6 h. Harvested cells were lysed by sonication for 30 min (5 s on/5 s off, on ice), the supernatant was collected by centrifugation at 4 °C, 10,000 g for 30 min and the protein was then isolated by using Ni-NTA His-bind Resin (1.5 mL, Qiagen). Fractions containing His6-PiuA protein were harvested and verified with SDS-PAGE and Western blotting. The His-tag was cleaved with entorokinase for 24 h at room temperature, and then removed by Ni-NTA to produce purified PiuA protein. The purity of the purified protein was examined by SDS-PAGE. The identity of the PiuA protein was further confirmed by using ABI-4800 plus MALDI TOF/TOF mass spectrometer according to a previously described method [16]. Proteins were identified by the MASCOT search engine (V2.1) against NCBI *S. pneumoniae* D39 protein database based on the MS and MS/MS spectra, protein identifications with Mascot scores C. I. % >95 were considered significant. PiaA protein (the forward primer piaA-F: 5'-GCGAGCTCGAGACCAGTAGCTCTGCTC-3', and the reverse primer piaA-R: 5'-CGCCGCGAATTCTTATTT-CAAAGCTTTTTG-3') and PitA protein (the forward primer pitA-F: 5'-GCGAGCTCATGACTGTTGGTCTCTCTTA-3', and the reverse primer pitA-R: 5'-CGCCGCGAATTCT-TACTGTTTAGATTGGATAT-3') were also cloned and purified using the similar procedure.

Immunization experiments and Western blotting

Purified His6-PiaA, His6-PiuA and His6-PitA proteins were used as antigens for the immunization experiments as previously described in the literature to generate multicolon antibodies [17]. The specificities of the antibodies were detected with Western blotting using purified proteins and whole-cell lysates of wild type and *piaA*-, *piuA*- and *pitA*- mutant D39 strains.

For Western blotting analysis, untreated and R-825-treated (sub-MIC) *S. pneumoniae* D39 were harvested by centrifugation at 6,000 g for 10 min at 4°C when the absorbance reading of 0.6 at 600 nm was reached. Then pellets were washed three times with $1\times$PBS and disrupted by sonication to extract proteins, the concentrations of the cellular proteins were measured by Bradford assay. The protein extracts were separated by 12% SDS–PAGE and then electroblotted onto polyvinylidene fluoride membranes. The protein expressions of PiaA, PiuA and PitA were detected with anti-PiaA, -PiuA and -PitA antibodies and quantified using ImageMaster 2D Platinum 6.0. Total proteins separated by SDS-PAGE and stained with Coomassie brilliant blue R250 were used as the loading control.

Fluorescence spectroscopy of apo-PiuA protein titrated with R-825

Fluorescence measurements were performed in a Hitachi F7000 spectrofluorophotometer. Fluorescence emission spectra were recorded from 290 to 450 nm after exciting at 280 nm. Both slit widths of excitation and emission beams were 5 nm. Spectra were acquired for 2 μM apo-PiuA (20 mM Tris-HCl, 100 mM NaCl, pH 7.4) with varying concentrations of R-825 (from 1.2 to 19.6 μM) and ferrichrome (from 0.4 to 4.0 μM). Vancomycin (from 1.2 to 19.2 μM) was also titrated to the apo-PiuA solution as a negative control. Relative changes in fluorescence emission (ΔF) at 343 nm during the titration verses the concentrations were fitted to a titration curve. The data were analyzed with Hill plot equation in Origin 8.5 to acquire the affinity constants (Ka).

Results

Ru(II) complex R-825 possesses antibacterial activity against *S. pneumoniae*

To investigate the antibacterial activity of R-825, we tested its effects on the growth of *S. pneumoniae* D39 strain in batch cultures, and determined its MIC value for *S. pneumoniae*. As shown in Figure 2A, R-825 can inhibit the bacterial growth in a concentration-dependent manner. At the concentration of 30 μM (MIC), R-825 completely suppressed the growth of *S. pneumoniae*. The sub-MIC, 15 μM, was selected to treat *S. pneumoniae* for the following real-time quantitative PCR and ICP-MS assays. MIC determination for the mutant strains was also performed and the results are listed in Table 2. Notably, *piuA*- mutant has a MIC = 60 μM, double of those for the wild-type and other mutant strains.

R-825 is not toxic to human A549 cells

The cytotoxic activity of R-825 against human A549 cells was evaluated by using LDH assay kit, with whole cell lysate as a positive control. As shown in Figure 2B, the viability of A549 cells had not significant changes under the 48 h treatment of R-825

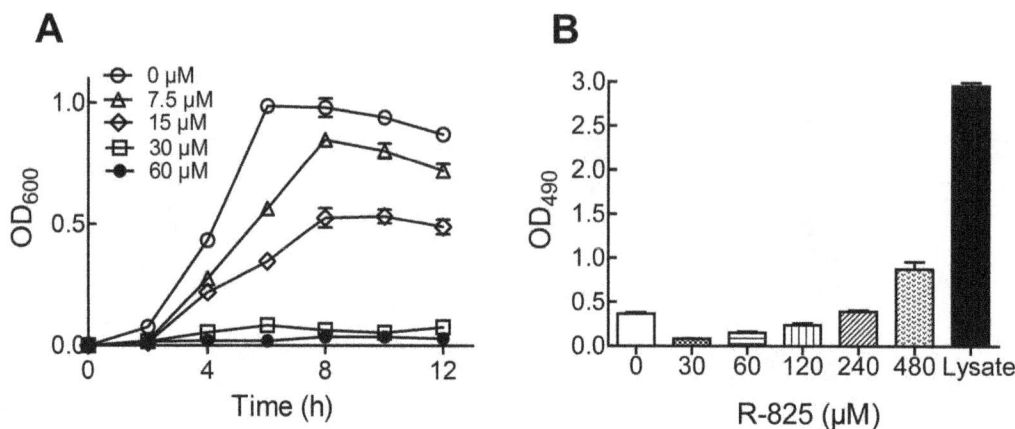

Figure 2. Effect of R-825 on *S. pneumoniae* and human A549 cell growth. (A) R-825 inhibited *S. pneumoniae* growth in a concentration-dependent manner. (B) R-825 showed little cytotoxicity to human cells. Determination of the cytotoxicity against A549 cell line was performed by incubating the cells with R-825 for 48 h using LDH kit, cell lysate was used as positive control. The data shown represent the mean of three experiments; error bars indicate SEM.

with up to 480 μM, a concentration substantially higher than the corresponding MIC value for the bacterium. Results demonstrated that R-825 has very low toxicity to human cells, implicating its high selective toxicity toward bacteria over human host.

R-825 reduces bacterial iron content and up-regulates PiaA and PiuA

To determine whether R-825 can compete with iron for the uptake system by *S. pneumoniae*, we measured the intracellular iron and ruthenium levels in the bacterium with and without R-825 treatment by using ICP-MS analysis. The results are shown in Figure 3A. As compared with control, the treatment with R-825 in its sub-MIC significantly decreased the iron concentration and correspondingly increased the ruthenium concentration in wild-type *S. pneumoniae*.

Previous studies have shown that many metal ion transporters are regulated by their substrates [18–20]. To investigate whether R-825 regulates iron-uptake proteins, we monitored the mRNA expression levels of lipoproteins PiaA, PiuA and PitA, the three surface iron chelates in the known iron-uptake ABC transporter systems in *S. pneumoniae* [6,7]. We observed that R-825 treatment up-regulated *piuA* and *piaA* gene expressions, but exhibited no significant impact on *pitA* expression, as shown in Figure 3B. Western blotting was also performed to measure the protein expression levels after R-825 treatment. As shown in Figure 3C, R-825 stimulated the expression of PiuA and PiaA but exerted no effect on PitA expression.

To confirm the interference of R-825 with the iron uptake system, we also investigated the protein expression level in the

bacteria cultured with iron replete and restricted medium. As shown in Figure S1 in File S1, upon iron starvation, the expression level of PiaA is significantly increased, and PiuA is slightly increased while PitA is unchanged in wild-type D39 strain. However, the expression levels of both PiuA and PitA are evidently up-regulated in *piaA-* mutant strain upon iron starvation. These results are basically consistent with the protein change tendency upon Ru-825 treatment. These consistent observations suggest that R-825 may enter *S. pneumoniae* via PiaABC or PiuABC iron-uptake systems.

R-825 is uptaken via ferrichrome-transport system

To determine whether R-825 would be taken up by *S. pneumoniae* via PiaABC (for heme uptake) or PiuABC (for ferrichrome uptake) system, we individually deleted *piaA* and *piuA* genes in the bacterium to construct the *piaA-* and *piuA-* single mutant strains, in which heme- and ferrichrome-binding ability was respectively impaired in the bacterium, and verified the effects using Western blotting (figure S2 in File S1). The sensitivity of the mutant strains to R-825 was compared to that of wild-type *S. pneumoniae* by measuring the growth inhibition zones and MIC values corresponding to R-825 treatment. As comparison, *pitA-* mutant was also constructed and tested (figure S2 in File S1).

Our experimental results are shown in Table 2. As compared with wild-type *S. pneumoniae*, both *piaA-* and *pitA-* mutants exhibited no significant difference in growth inhibition zones and MICs. In contrast, *piuA-* mutant strain displayed much smaller growth inhibition zone with a doubled MIC value, indicating that the *piuA-* strain is resistant to R-825 (Table 2). This suggests that

Table 2. Minimal Inhibitory Concentrations (MICs) and Diameter of Growth Inhibition Zones (mm) of R-825 against WT *S. pneumoniae* D39, *piaA-*, *piuA-* and *pitA-* mutant strains.

Bacteria	MICs (μM)	Diameter of Growth Inhibition Zones (mm)
WT *S. pneumoniae* D39	30	16
piaA- mutant	30	16
piuA- mutant	60	10
pitA- mutant	30	14

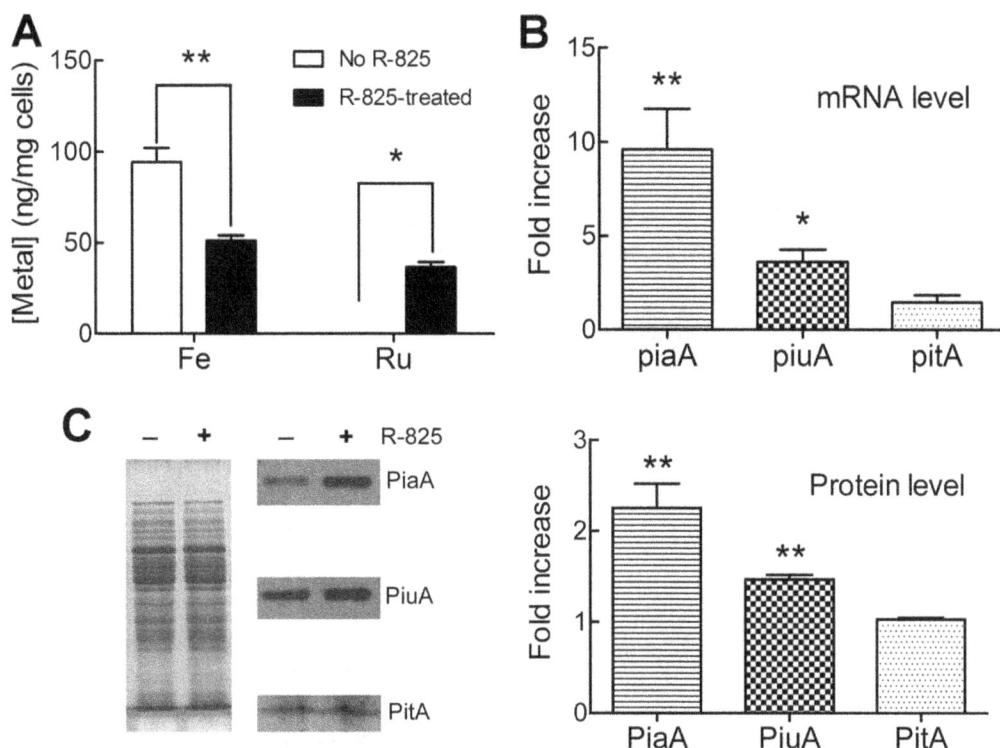

Figure 3. R-825 reduced iron uptake and stimulated the expressions of PiaA and PiuA in *S. pneumoniae*. (A) The cellular iron and ruthenium contents of the wild-type *S. pneumoniae* with and without R-825 treatment. (B) The increase of mRNA expression levels of *piaA*, *piuA* and *pitA* in the presence of 15 μM R-825, as compared with their expressions in the absence of R-825. Data were normalized with housekeeping gene 16S rRNA (mean ± SEM, n = 3, **$p < 0.01$, *$p < 0.05$ versus the untreated control). (C) The comparison of the protein expression levels in the bacterium with and without R-825 treatment. Whole cell proteins were used as loading control for the Western blotting.

the antibacterial activity of R-825 is largely related to PiuA but not PiaA and PitA; without PiuA, the uptake of R-825 may be impaired in the *piuA-* mutant strain. To confirm this hypothesis, we detected the intracellular ruthenium levels in the *piuA-* mutant and wild-type strains under R-825 treatment. As observed in Figure 4A, the level of intracellular ruthenium in *piuA-* mutant strain was substantially reduced to half content of the wild-type bacterium.

Ferrichrome but not iron ion and hemin rescues the bacterial growth suppressed by R-825

The above experiments indicated that R-825 may be taken up by *S. pneumoniae* through PiuABC, the ferrichrome transport system. Accordingly, ferrichrome should be able to compete with R-825 for the uptake and antagonize the antimicrobial activity of R-825. We therefore examined the antimicrobial activity of R-825 against *S. pneumoniae* in the presence of increasing concentrations of ferrichrome, hemin and FeCl₃. As shown in Figure 5,

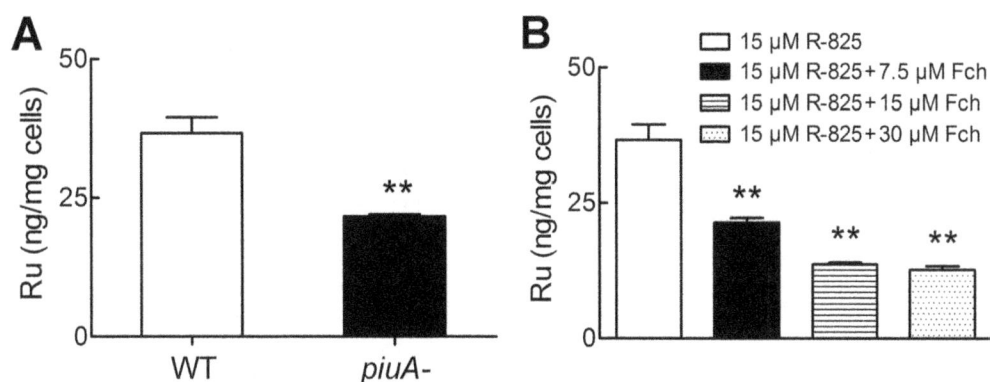

Figure 4. Cellular ruthenium concentrations in R-825 treated *S. pneumoniae* and with ferrichrome (Fch) competition as determined by ICP-MS. (A) Ruthenium contents in wild-type *S. pneumoniae* and in *piuA-* mutant as treated with 15 μM R-825. (B) Ruthenium contents in wild-type *S. pneumoniae* in the addition of increasing ferrichrome. Results are representative of the mean±SEM from three independent experiments (*$p < 0.01$).

ferrichrome addition indeed reversed the growth-inhibitory effects of R-825 in a dose-dependent manner with either sub-MIC or MIC treatments (Fig. 5A and 5B, respectively). When the molar ratio of ferrichrome to R-825 reached higher than 2:1, the maximum OD_{600} of bacterial growth could be restored to near normal level. In contrast, adding either hemin (Fig. 5C) or $FeCl_3$ (Fig. 5D) could not rescue the bacterial growth inhibited by R-825. The complete inhibition observed in Figure 5C with 30 μM hemin addition was due to the toxicity of hemin itself to the bacterium.

We also measured the bacterial ruthenium contents in response to the ferrichrome addition in Figure 5A. As shown in Figure 4B, ferrichrome addition gradually reduced the cellular concentration of ruthenium while rescuing the bacterial growth (Fig. 5A). This observation indicated an actual competition between ferrichrome and R-825 for the bacterial uptake. These results all together confirmed that R-825 enters *S. pneumoniae* indeed *via* ferrichrome-uptake pathway PiuABC, rather than PiaABC and PitABC transport systems.

R-825 can bind to PiuA protein *in vitro*

The binding between R-825 and PiuA protein was tested to validate the interaction of R-825 with PiuABC systems. For comparison, the binding between ferrichrome/vancomycin and PiuA was also determined. The quenching fluorescence spectra of apo-PiuA upon titration with the chemicals are shown in Figure 6; both R-825 and ferrichrome have a similar binding pattern with PiuA protein, while vancomycin as a negative control almost does not bind to PiuA protein. When the step-wise fluorescence quenching versus chemical concentration data were curve-fitted to Hill plot equation 1, the binding constants were determined to be $(0.30 \pm 0.18) \times 10^6$ M^{-1} and $(1.12 \pm 0.28) \times 10^6$ M^{-1} for R-825 and ferrichrome, respectively. These results revealed that both R-825 and ferrichrome could specifically bind to PiuA, with ferrichrome having a higher binding affinity than R-825 for PiuA.

Discussion

Bacterial resistance to antibiotics is becoming a significant threat to global public health [21]. There is an urgent need to develop novel antimicrobial drugs with an action mode different from the current antibiotics. The bacterial cell membrane demonstrates decreased permeability, serving as a barrier to limit the intracellular access of substances such as antibiotics by passive diffusion. Reduced membrane permeability is one of the main mechanisms for antibiotic resistance (reviewed in [22]). A strategy to circumvent the 'impermeability' resistance problem is to target the iron-transport systems for the drug delivery into the bacterial

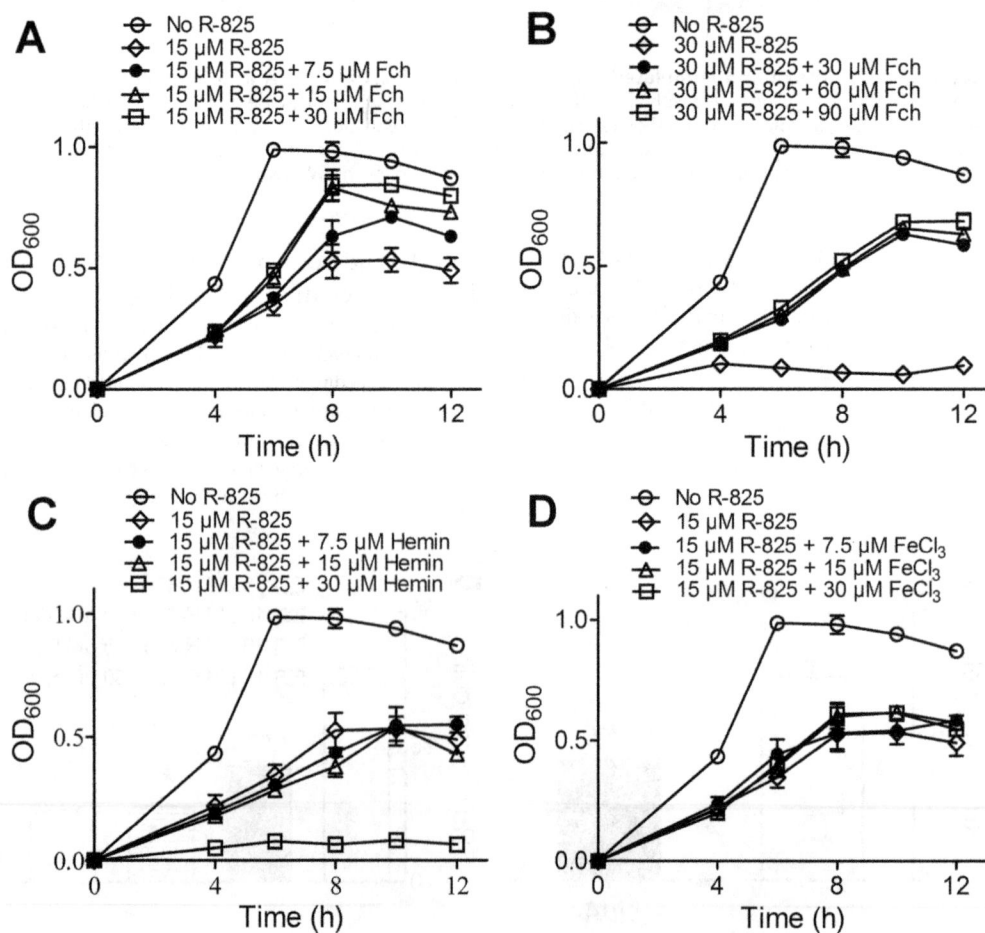

Figure 5. Ferrichrome, but not hemin or FeCl₃ rescued the bacterial growth suppressed by R-825. (A) Sub-MIC treated *S. pneumoniae* with increasing amounts of Fch. (B) MIC-treated *S. pneumoniae* with increasing amounts of Fch. (C) Sub-MIC treated *S. pneumoniae* with increasing amounts of hemin. Hemin itself at ≥30 μM has toxicity to the bacterium, causing the further inhibition of the bacterial growth. (D) Sub-MIC treated *S. pneumoniae* with increasing amounts of FeCl₃.

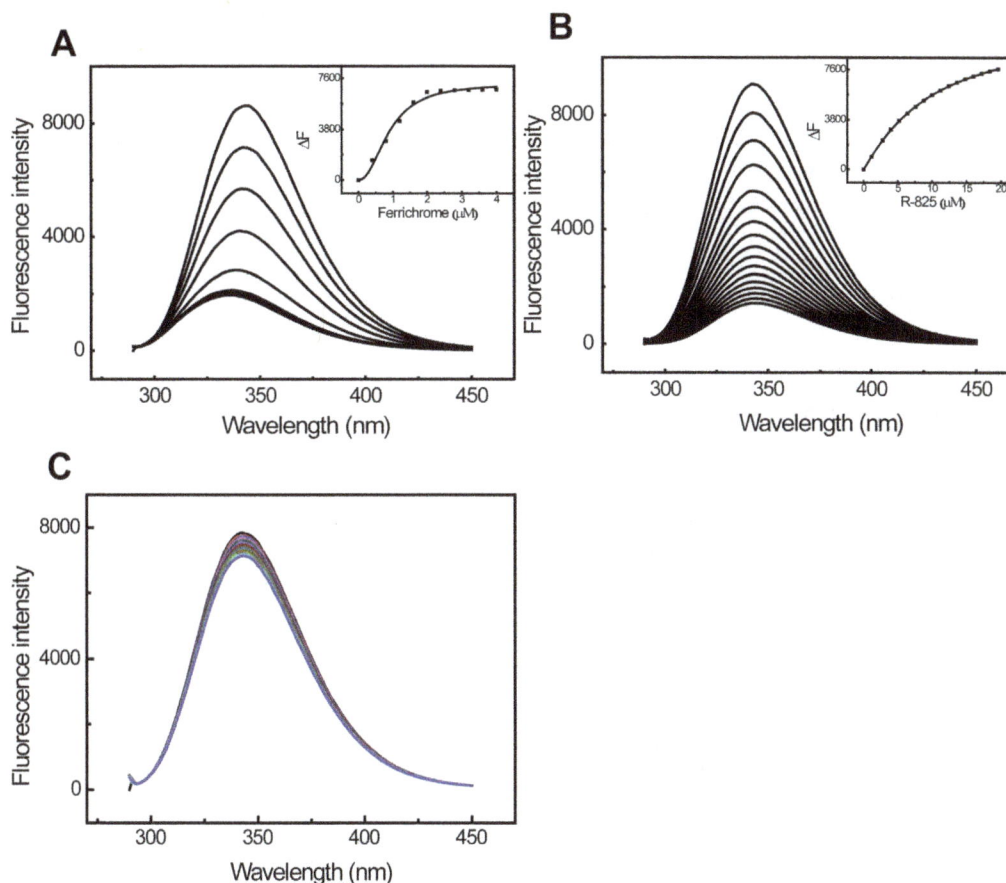

Figure 6. R-825 binds to PiuA protein *in vitro*. Fluorescence spectra of 2 μM apo-PiuA protein in 20 mM Tris-HCl (pH 7.4) titrated with aliquot ferrichrome (A), R-825 (B), or vancomycin (C). The inserted binding isotherm curves were built from the relative change values of fluorescence intensity (ΔF) at 343 nm versus concentrations. The inner panels are the curve-fitting analyses using Hill plot in the program Origin 8.5.

cells. Since the iron-transport systems are widespread among bacteria and work as active channels to transfer iron, some studies had explored antibacterial compounds including Ga(NO₃)₃ [23], DFO-Ga [24], non-iron metalloporphyrins (MPs) [25], and albomycin [8], enterobactin-cargo conjugates [26] as 'Trojan horses' to enter the bacteria through the iron-uptake systems.

In our long-term research on the mechanism of iron acquisition in bacteria, we understand that most iron obtained by bacteria is in the form of siderophores or heme since free iron ion is severely restricted in the host. Accordingly, blocking or interfering with either the heme- or siderophore-uptake pathway would be an effective approach to limit the acquisition of iron, one of the crucial elements for bacterial growth and survival. Here we selected our previously synthesized ruthenium(II) complex R-825 (Fig. 1) for the experiment, with an expectation that this compound can compete with either heme or ferrichrome to bind to the iron-receptors PiaA or PiuA, reduce the availability of iron to the bacteria and thus suppress the bacterial growth. Both ruthenium and iron are the members of VIII family in the periodic table, sharing chemical similarities in terms of coordination and binding with ligands. More importantly, Ru(II) is not toxic to the human body, making it an excellent candidate for antimicrobial drug development. Our current experiments demonstrated that R-825 indeed has selective antimicrobial activity against *S. pneumoniae*, with no evident toxic effects towards human A549

cells even in a concentration significantly greater than the corresponding MIC value (Fig. 2).

As expected, we detected that R-825 can be internalized by *S. pneumoniae*, accompanying with the decrease of bacterial iron uptake (Fig. 3A). Correspondingly, the expression of lipoproteins PiaA and PiuA in the bacterium was stimulated to compensate for the decreased iron availability under R-825 competition (Fig. 3B&C). Obviously, PiaA and/or PiuA iron-uptake systems were involved and thus we constructed *piaA*- and *piuA*- gene deletion mutant strains for further tests. Based on the MIC and growth inhibition zones assays (Table 2), we observed that only *piuA*- mutant showed an increased resistance to R-825, accompanying with a significant decrease in the content of cellular ruthenium in the mutant (Fig. 4A). This means that, without PiuA, less R-825 can be uptaken into cells and thus the bacterium became resistant to the drug. In other words, R-825 may be internalized by the bacterium mainly *via* PiuABC pathway.

Among the three iron-transport systems in *S. pneumoniae*, PiuABC system mainly conveys ferrichrome and its analogues into cells with the interaction between PiuA and ferrichrome [8]. Our experiments validated that R-825 can also interact with PiuA *in vitro*, in a binding pattern similar to that of PiuA-ferrichrome interaction (Fig. 6), suggesting that an active competition between R-825 and ferrichrome may occur for PiuA binding *in vivo*. Correspondingly, adding ferrichrome to the medium should compete with R-825 for the binding to PiuA, decrease the

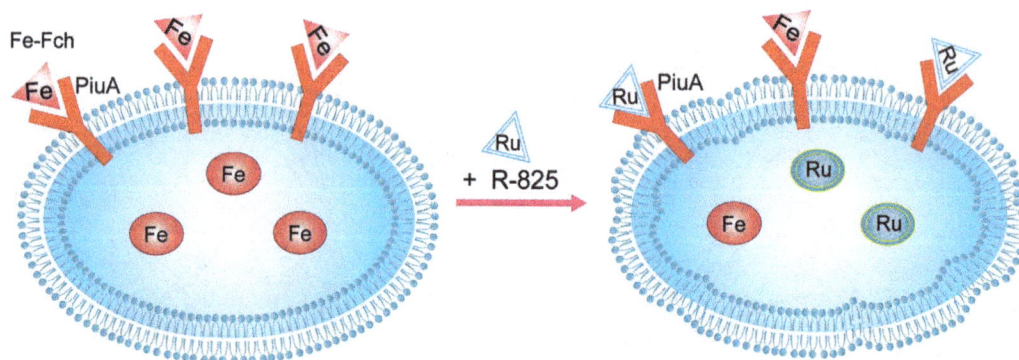

Figure 7. A model to summarize the proposed mechanism of Ru-complex action on the bacteria. R-825 is uptaken by *S. pneumoniae via* ferrichrome-transport pathway, reducing the competitive iron uptake into cells and thus suppressing the bacterial growth.

ruthenium uptake (Fig. 4B) and thus antagonize the antibacterial activity of R-825. Our competition experiments support this observation, in which adding ferrichrome can rescue the R-825-suppressed bacterial growth in a dose-dependent manner while adding hemin or FeCl$_3$ had no affect (Fig. 5). We can therefore conclude that R-825 interferes with the ferrichrome-uptake system by competing with ferrichrome for PiuA binding, reduces the iron internalization and thus suppresses the bacterial growth.

The fact that ruthenium compounds have antimicrobial activities has been recognized recently [27–31]. In particular, Keene and his co-workers characterized the susceptibility and cellular uptake of inert Ru(II) complexes [30,31] and demonstrated that di-nuclear Ru(II) complexes may enter eukaryotic cells by passive diffusion, while mononuclear complexes may have a different mode of action. This observation echoes our current finding that mononuclear Ru(II) complex R-825 is mostly uptaken into cells *via* the active ferrichrome-transport pathway. Certainly, this active transportation may not be the only way for R-825 to enter the bacterium, as attested by the fact that certain amounts of ruthenium were still detected in the *piuA*– mutant strain (Fig. 4A). On the other hand, our R-825 has a lower affinity than ferrichrome for PiuA binding (Fig. 6), suggesting that this compound can be further modified in terms of its structure and lipophilicity to enhance the ability to compete with ferrichrome and thus to optimize its antibacterial activity against *S. pneumoniae*.

In summary, we used a Ru(II) complex R-825 to chemically compete with iron compounds for binding to the iron-binding ligands in the iron-transport systems in *S. pneumoniae*. By performing various experiments, we demonstrated that R-825 can be transported into the bacterium *via* ferrichrome-transport pathway, competitively reducing iron uptake into cells and thus suppressing the bacterial growth (Fig. 7). Since the ferrichrome-uptake pathways are widely spread among bacteria, R-825 and derivatives may represent a new candidate class of antimicrobial agents for further development.

Supporting Information

File S1. Figure S1. The relative levels of PiaA, PiuA and PitA proteins in the iron replete or restricted conditions. **Figure S2.** The constructed *piaA*-, *piuA*- and *pitA*- mutant strains were verified by Western blotting and the purity of PiuA protein was verified by SDS-PAGE.

Author Contributions

Conceived and designed the experiments: XS QYH. Performed the experiments: XYY BS. Analyzed the data: LZ. Contributed reagents/materials/analysis tools: NL JH JZ. Contributed to the writing of the manuscript: XYY XS QYH.

References

1. Ratledge C, Dover LG (2000) Iron metabolism in pathogenic bacteria. Annu Rev Microbiol 54: 881–941.
2. Schaible UE, Kaufmann SH (2004) Iron and microbial infection. Nat Rev Microbiol 2: 946–953.
3. Cassat JE, Skaar EP (2013) Iron in infection and immunity. Cell Host Microbe 13: 509–519.
4. Mitchell TJ (2000) Virulence factors and the pathogenesis of disease caused by *Streptococcus pneumoniae*. Res Microbiol 151: 413–419.
5. Whalan RH, Funnell SG, Bowler LD, Hudson MJ, Robinson A, et al. (2006) Distribution and genetic diversity of the ABC transporter lipoproteins PiuA and PiaA within *Streptococcus pneumoniae* and related streptococci. J Bacteriol 188: 1031–1038.
6. Brown JS, Gilliland SM, Holden DW (2001) A *Streptococcus pneumoniae* pathogenicity island encoding an ABC transporter involved in iron uptake and virulence. Mol Microbiol 40: 572–585.
7. Brown JS, Gilliland SM, Ruiz-Albert J, Holden DW (2002) Characterization of pit, a *Streptococcus pneumoniae* iron uptake ABC transporter. Infect Immun 70: 4389–4398.
8. Pramanik A, Braun V (2006) Albomycin uptake via a ferric hydroxamate transport system of *Streptococcus pneumoniae* R6. J Bacteriol 188: 3878–3886.
9. Cheng W, Li Q, Jiang YL, Zhou CZ, Chen Y (2013) Structures of *Streptococcus pneumoniae* PiaA and its complex with ferrichrome reveal insights into the

substrate binding and release of high affinity iron transporters. PLoS One 8: e71451.
10. Sun B, Chu J, Chen Y, Gao F, Ji LN, et al. (2008) Synthesis, characterization, electrochemical and photophysical properties of ruthenium(II) complexes containing 3-amino-1,2,4-triazino[5,6-f]-1,10-phenanthroline. J Mol Struct 890: 203–208.
11. Wach A (1996) PCR-synthesis of marker cassettes with long flanking homology regions for gene disruptions in *S. cerevisiae*. Yeast 12: 259–265.
12. Ong CL, Potter AJ, Trappetti C, Walker MJ, Jennings MP, et al. (2013) Interplay between manganese and iron in pneumococcal pathogenesis: role of the orphan response regulator RitR. Infect Immun 81: 421–429.
13. Lanie JA, Ng WL, Kazmierczak KM, Andrzejewski TM, Davidsen TM, et al. (2007) Genome sequence of Avery's virulent serotype 2 strain D39 of *Streptococcus pneumoniae* and comparison with that of unencapsulated laboratory strain R6. J Bacteriol 189: 38–51.
14. Institute CaLS (2012) Performance Standards for Antimicrobial Susceptibility Testing: Twenty-second Informational Supplement M100-S22.CLSI, Wayne, PA, USA.
15. McDevitt CA, Ogunniyi AD, Valkov E, Lawrence MC, Kobe B, et al. (2011) A molecular mechanism for bacterial susceptibility to zinc. PLoS Pathog 7: e1002357.

16. Wang Y, Cheung YH, Yang Z, Chiu JF, Che CM, et al. (2006) Proteomic approach to study the cytotoxicity of dioscin (saponin). Proteomics 6: 2422–2432.

17. Brown JS, Ogunniyi AD, Woodrow MC, Holden DW, Paton JC (2001) Immunization with components of two iron uptake ABC transporters protects mice against systemic *Streptococcus pneumoniae* infection. Infect Immun 69: 6702–6706.

18. Whitby PW, Sim KE, Morton DJ, Patel JA, Stull TL (1997) Transcription of genes encoding iron and heme acquisition proteins of Haemophilus influenzae during acute otitis media. Infect Immun 65: 4696–4700.

19. Kehres DG, Zaharik ML, Finlay BB, Maguire ME (2000) The NRAMP proteins of *Salmonella typhimurium* and *Escherichia coli* are selective manganese transporters involved in the response to reactive oxygen. Mol Microbiol 36: 1085–1100.

20. Li C, Tao J, Mao D, He C (2011) A novel manganese efflux system, YebN, is required for virulence by *Xanthomonas oryzae* pv. *oryzae*. PLoS One 6: e21983.

21. Ho J, Tambyah PA, Paterson DL (2010) Multiresistant Gram-negative infections: a global perspective. Curr Opin Infect Dis 23: 546–553.

22. Pages JM, James CE, Winterhalter M (2008) The porin and the permeating antibiotic: a selective diffusion barrier in Gram-negative bacteria. Nat Rev Microbiol 6: 893–903.

23. Kaneko Y, Thoendel M, Olakanmi O, Britigan BE, Singh PK (2007) The transition metal gallium disrupts *Pseudomonas aeruginosa* iron metabolism and has antimicrobial and antibiofilm activity. J Clin Invest 117: 877–888.

24. Banin E, Lozinski A, Brady KM, Berenshtein E, Butterfield PW, et al. (2008) The potential of desferrioxamine-gallium as an anti-Pseudomonas therapeutic agent. Proc Natl Acad Sci U S A 105: 16761–16766.

25. Stojiljkovic I, Kumar V, Srinivasan N (1999) Non-iron metalloporphyrins: potent antibacterial compounds that exploit haem/Hb uptake systems of pathogenic bacteria. Mol Microbiol 31: 429–442.

26. Zheng T, Bullock JL, Nolan EM (2012) Siderophore-mediated cargo delivery to the cytoplasm of *Escherichia coli* and *Pseudomonas aeruginosa*: syntheses of monofunctionalized enterobactin scaffolds and evaluation of enterobactin-cargo conjugate uptake. J Am Chem Soc 134: 18388–18400.

27. Dwyer FP, Gyarfas EC, Rogers WP, Koch JH (1952) Biological activity of complex ions. Nature 170: 190–191.

28. Biersack B, Diestel R, Jagusch C, Sasse F, Schobert R (2009) Metal complexes of natural melophlins and their cytotoxic and antibiotic activities. J Inorg Biochem 103: 72–76.

29. Bolhuis A, Hand L, Marshall JE, Richards AD, Rodger A, et al. (2011) Antimicrobial activity of ruthenium-based intercalators. Eur J Pharm Sci 42: 313–317.

30. Li F, Mulyana Y, Feterl M, Warner JM, Collins JG, et al. (2011) The antimicrobial activity of inert oligonuclear polypyridylruthenium(II) complexes against pathogenic bacteria, including MRSA. Dalton Trans 40: 5032–5038.

31. Li F, Feterl M, Mulyana Y, Warner JM, Collins JG, et al. (2012) In vitro susceptibility and cellular uptake for a new class of antimicrobial agents: dinuclear ruthenium(II) complexes. J Antimicrob Chemother 67: 2686–2695.

13

Response of Bacterial Metabolic Activity to Riverine Dissolved Organic Carbon and Exogenous Viruses in Estuarine and Coastal Waters: Implications for CO_2 Emission

Jie Xu[1]*, Mingming Sun[2], Zhen Shi[1], Paul J. Harrison[2], Hongbin Liu[2]

1 State Key Laboratory of Tropical Oceanography, South China Sea Institute of Oceanology, Chinese Academy of Sciences, Guangzhou, China, **2** Division of Life Sciences, The Hong Kong University of Science and Technology, Clear Water Bay, Kowloon, Hong Kong

Abstract

A cross-transplant experiment between estuarine water and seawater was conducted to examine the response of bacterial metabolic activity to riverine dissolved organic carbon (DOC) input under virus-rich and virus-free conditions, as well as to exogenous viruses. Riverine DOC input increased bacterial production significantly, but not bacterial respiration (BR) because of its high lability. The bioavailable riverine DOC influenced bulk bacterial respiration in two contrasting ways; it enhanced the bulk BR by stimulating bacterial growth, but simultaneously reduced the cell-specific BR due to its high lability. As a result, there was little stimulation of the bulk BR by riverine DOC. This might be partly responsible for lower CO_2 degassing fluxes in estuaries receiving high sewage-DOC that is highly labile. Viruses restricted microbial decomposition of riverine DOC dramatically by repressing the growth of metabolically active bacteria. Bacterial carbon demand in the presence of viruses only accounted for 7–12% of that in the absence of viruses. Consequently, a large fraction of riverine DOC was likely transported offshore to the shelf. In addition, marine bacteria and estuarine bacteria responded distinctly to exogenous viruses. Marine viruses were able to infect estuarine bacteria, but not as efficiently as estuarine viruses, while estuarine viruses infected marine bacteria as efficiently as marine viruses. We speculate that the rapid changes in the viral community due to freshwater input destroyed the existing bacteria-virus relationship, which would change the bacterial community composition and affect the bacterial metabolic activity and carbon cycling in this estuary.

Editor: Douglas Andrew Campbell, Mount Allison University, Canada

Funding: Financial support for this research was provided by the Hundred Talented Program Startup Fund (Y35L041001) by South China Sea Institute of Oceanology, the Key Project of National Natural Science Fundation of China (No. 91128212) and the University Grants Council of Hong Kong AoE project (AoE/P-04/04-4-II). The funders had no role in study design, data collection and analysis, decision to publish, or preparation of the manuscript.

Competing Interests: The authors have declared that no competing interests exist.

* Email: xujie@scsio.ac.cn

Introduction

Estuaries are in general a significant source of CO_2 to the atmosphere, where CO_2 efflux is estimated to be 0.25 ± 0.25 Pg C y^{-1} at the global scale [1]. Low-latitude estuarine/coastal waters receive approximately two-thirds of the terrestrial organic carbon (OC) and have higher microbial decomposition rates due to higher temperatures [2–5]. Hence, it is speculated that more CO_2 is degassing in lower latitude estuarine and coastal waters [1].

The Pearl River is subtropical and the second largest river in China, with an annual average freshwater discharge of 10,524 $m^3 s^{-1}$, which was calculated by *Yin et al.* [6] from *Zhao* [7]. About 80% of the discharge occurs in the wet season (April–September). High dissolved organic carbon (DOC) concentrations up to 473 μmol C L^{-1} occur in the Pearl River estuary, most of which originates from sewage [8]. However, the CO_2 degassing flux is reported to be 6.92 mol C $m^2 y^{-1}$ in the entire Pearl River estuary [9], much lower than that (36.5–182.5 mol C $m^2 y^{-1}$) for the European estuaries [4,10]. Similarly, low CO_2 flux (only 8.1 mol C $m^2 y^{-1}$) has also been documented in the Hoogly estuary (NE India)

that receives a huge amount (~400 million tons yearly) of sewage [11]. One explanation for the surprisingly lower CO_2 flux in these low-latitude estuaries is due to the large freshwater discharge and short residence time of the discharged water [12,13]. To date, the underlying mechanism for the lower CO_2 flux remains unknown.

The riverine DOC is primarily mineralized by heterotrophic bacteria that produce a significant amount of new biomass from DOC and transform DOC into inorganic carbon [14–16]. High terrestrial organic carbon input exerts a great influence on bacterial metabolism and community composition [17,18]. Prior studies suggest that the lability of DOC plays a key role in regulating bacterial respiration rates [19]. Cell-specific bacterial respiration increases sharply in response to the decreasing quality of DOC since a larger percentage of energy is used for maintenance processes to safeguard metabolic flexibility at the cost of energetic efficiency [19,20]. A recent study indicates that bacteria grown on sewage-derived DOC have high production and low respiration rates, while those on marine-derived DOC at have high respiration rates [21]. Hence, the lability of DOC could potentially affect CO_2 degassing fluxes in high DOC estuaries.

In recent years, it has been recognized that viruses play an important role in regulating bacterial metabolic activities [22–24]. Bacterial production is reported to be reduced by up to 50% due to viral lysis [25]. However, little attention has been paid to interactions between bacteria and indigenous viruses and the effect of viruses on microbial decomposition of riverine DOC in estuaries, which may be important in understanding the fate of terrestrial organic carbon in the ocean. To our knowledge, only few studies have been conducted which suggested that the bacterial response to different sources of viruses may vary [26–28]. Therefore, a study on response of bacterial metabolic processes to riverine DOC and exogenous viruses is needed to determine the reasons for the lower CO_2 degassing fluxes in the Pearl River estuary. A cross-transplant experiment of estuarine water and seawater was conducted to examine the response of bacterial metabolic activity to riverine DOC input under virus-rich and virus-free conditions. This study will increase our understanding of the role of bacteria in influencing riverine organic carbon cycling in estuarine waters.

Materials and Methods

No specific permissions were required for sampling at stations PM7 and NM3 in Hong Kong waters. We confirmed that the field studies did not involve endangered or protected species at these two stations. Seawater samples were taken from 1 m depth in August (wet summer season) 2013 at two contrasting coastal sites, a relatively pristine coastal station (PM7) with no influence from sewage or freshwater discharge, and a eutrophic estuarine station near the Pearl River (NM3) (Fig. 1). The water depths of PM7 and NM3 were 17 and 15 m, respectively. Samples for nutrients, dissolved organic carbon (DOC), and bacterial and viral abundance were taken.

Experimental set-up

Water samples were passed through a 1 μm polycarbonate membrane and the filtrate was used as the bacterial inoculum. Virus-rich water was obtained by filtration of water samples through a 0.1 μm cartridge. Relatively virus-free water was prepared by filtration of the 0.1 μm filtrate through a 100 kDa cutoff polysulfone cartridge (Millipore) using tangential flow ultra-filtration. About 600 ml of the bacterial inoculum was mixed with 5.4 L of virus-rich and relatively virus-free water from each station in acid-washed and Milli-Q rinsed polycarbonate carboys and distributed to triplicate glass bottles (1 L) to obtain eight treatments (Fig. 2) as follows: 1) estuarine bacteria + virus-rich estuarine water (Be+Ve); 2) estuarine bacteria + virus-free estuarine water (Be-Ve); 3) estuarine bacteria + virus-rich seawater (Be+Vs); 4) estuarine bacteria + virus-free seawater (Be-Vs); 5) marine bacteria + virus-rich estuarine water (Bs+Ve); 6) marine bacteria + virus-free estuarine water (Bs-Ve); 7) marine bacteria + virus-rich seawater (Bs+Vs) and; 8) marine bacteria + virus-free seawater (Bs-Vs).

Samples were incubated in the dark for 24 h. Running seawater was used to maintain the surface *in situ* temperature. Samples for nutrients, bacterial abundance (BA), bacterial production (BP), bacterial respiration (BR), viral abundance (VA) were taken at the beginning and end of the incubations. DOC samples were taken at the beginning of the incubation.

Nutrients and DOC concentrations

Nutrient concentrations (NO_3, NO_2^-, NH_4, PO_4 and SiO_4) were determined colorimetrically with a SKALAR autoanalyser following the protocols described by *Strickland and Parsons* [29] and *Grasshoff et al.* [30]. Dissolved inorganic nitrogen (DIN) was the sum of NO_3, NO_2^- and NH_4. Samples for DOC were filtered through a 0.2 μm acetate cellulose membrane and analyzed with a high temperature combustion method using a Shimadzu TOC-5000 analyzer [31].

Bacterial and viral abundance, bacterial production, respiration and cell size

Samples for bacterial and viral abundance were taken in micro-centrifuge tubes, fixed with buffered paraformaldehyde (final concentration 0.5%), and then stored at $-80°C$ until analysis by a flow cytometer (Becton-Dickinson FACSCalibur). Bacterial and viral abundance was determined according to the methods described by *Marie et al.* [32,33] and *Xu et al.* [24], respectively. Bacteria were determined on a plot of green fluorescence vs. side scatter (SSC), which also provided information on the relative DNA content and cell size by the means of SYB-green fluorescence side scatter, respectively. The values for relative DNA content and cell size were calibrated with 1 μm sized beads.

The measurement of bacterial production (BP) and bacterial respiration (BR) was described previously [24]. BP was measured using ^{3}H-leucine following the JGOFS protocol [31] and the final concentration of ^{3}H-leucine was 35 nM. BP was calculated with the empirical conversion factor of 3 kg C mol leucine^{-1} [34]. Dissolved oxygen for bacterial respiration was titrated using an automated titration apparatus (716 DMS Titrino Metrohm) [35]. BR was presented in carbon units assuming that the respiratory quotient was 1 [36].

Bulk BP was calculated using the following equation:

$$BP = (BP_i + BP_f)/2 \times \Delta t$$

where BP_i and BP_f are bacterial production (μg C L^{-1} h^{-1}) measured at the beginning and the end of the experiment, Δt is the time interval between the beginning and the end of the experiment.

The cell-specific BP (sBP) and BR (sBR) were calculated according to the following equations, respectively:

$$sBP = 2 \times BP/(A_i + A_f);$$

$$sBR = 2 \times BR/(A_i + A_f)$$

where A_i and A_f are the initial and final abundance of bacteria, respectively.

The growth rates of bacteria and viruses were calculated using the following equation:

$$\mu = \frac{\ln(A_f/A_i)}{\Delta t}$$

where A_i and A_f are the initial and final abundance of bacteria or viruses, respectively. Δt is the time interval between the beginning and the end of the experiment.

Bacterial growth efficiency was calculated using the following equation:

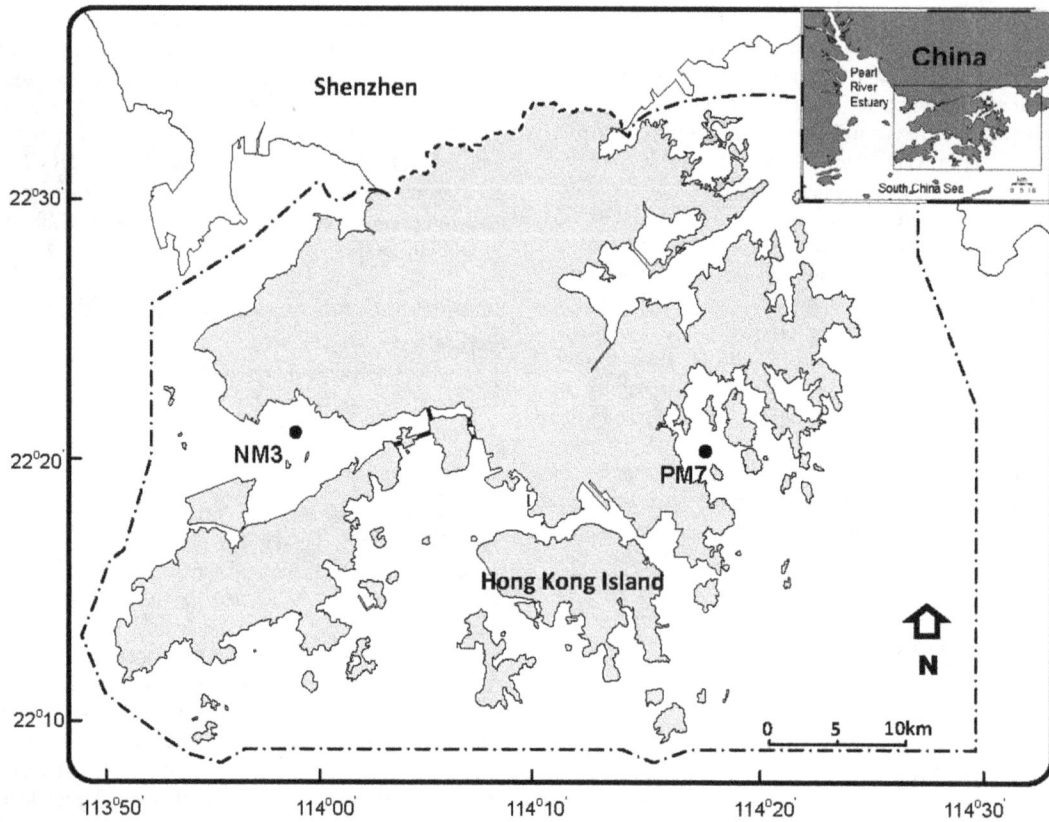

Figure 1. Location of the sampling stations in Hong Kong waters. These 2 stations are the same as the Environmental Protection Department of Hong Kong monitoring stations. The water depths for NM3 and PM7 are 15 and 17 m, respectively.

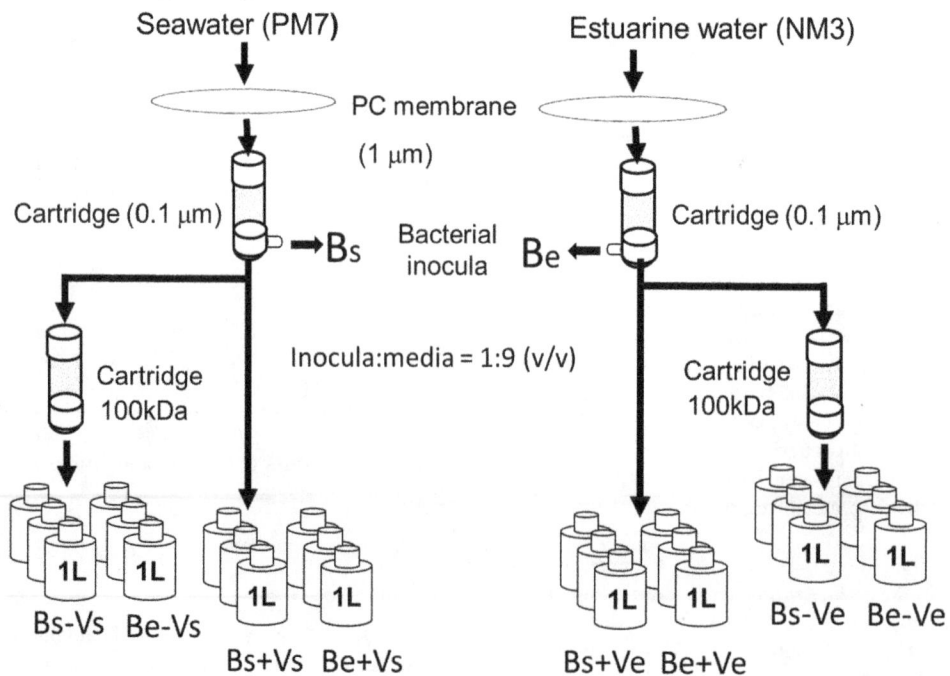

Figure 2. A schematic of the experimental setup. Bs and Be denote marine and estuarine bacteria, respectively. Vs and Ve denote marine and estuarine viruses, respectively.

$$BGE = BP/(BP + BR)$$

Statistical analyses

Statistical analyses were performed using SPSS software. A parametric ANOVA analysis with an LSD (least square difference) multiple comparison technique was conducted to determine any significant difference between treatments ($p < 0.05$). The error bars represent the standard error from triplicates for each treatment.

Results

Surface salinity (22.40) at estuarine station (NM3) was much lower than that (28.64) at coastal station (PM7) (Table 1). Nutrient and DOC concentrations were high at NM3 and low at PM7. However, bacterial and viral abundance (2.45×10^9 cells L^{-1} and 2.57×10^{10} particles L^{-1}, respectively) at PM7 was higher than those (4.28×10^8 cells L^{-1} and 9.40×10^9 particles L^{-1}, respectively) at NM3 (Table 1).

For the bioassay experiments with estuarine bacteria, viral abundance was the highest (1.95×10^{10} particles L^{-1}) in the Be+Vs treatment, moderate (1.00×10^{10} particles L^{-1}) in the Be+Ve treatment and the lowest ($\sim 2.20 \times 10^9$ particles L^{-1}) in the Be-Ve and Be-Vs treatments (Fig. 3). The abundance ratio of viruses to bacteria followed the pattern of viral abundance. Bacterial growth rates differed significantly among the four treatments, with low growth rates (2.9 and 2.6 d^{-1}) in the virus-rich treatment (i.e. Be+Ve and Be+Vs) and high (3.7 and 5.3 d^{-1}) in the virus-free treatment (i.e. Be-Ve and Be-Vs). The relative cell size and DNA content of bacteria followed the pattern of bacterial growth rate (Fig. 3). Bacterial production (535 and 118 $\mu g\ L^{-1}\ d^{-1}$) and cell-specific bacterial production (134 and 136 fg cell$^{-1}\ d^{-1}$) in the virus-free conditions (i.e. Be-Ve and Be-Vs) was significantly higher than the counterpart (42.7 and 40.6 $\mu g\ L^{-1}\ d^{-1}$) and (79.3 and 106 fg cell$^{-1}\ d^{-1}$) in the virus-rich conditions (i.e. Be+Ve and Be+Vs), respectively. Bacterial production did not differ significantly between the Be+Ve and Be+Vs treatment, while the cell-specific bacterial production in the Be+Vs treatment was significantly higher than that in the Be+Ve treatment. Similarly, bacterial respiration and cell-specific bacterial respiration followed the same patterns as bacterial production and cell-specific bacterial production, respectively (Fig. 4).

For the bioassay experiments with marine bacteria, viral abundance was the highest (2.10×10^{10} particles L^{-1}) in the Bs+Vs treatment, moderate (1.33×10^{10} particles L^{-1}) in the Bs+Ve treatment and the lowest ($\sim 3.30 \times 10^9$ particles L^{-1}) in the Bs-Ve and Bs-Vs treatments (Fig. 3). Bacterial growth rates differed significantly among four treatments, being low (1.1 and 1.0 d^{-1}) in the autochthonous DOC treatment (i.e. Bs+Vs and Bs-Vs) and high (1.7 and 3.2 d^{-1}) in the riverine DOC treatment (i.e. Bs+Ve and Bs-Ve). The relative cell size and DNA content of bacteria showed the same pattern (Fig. 3). Bacterial production (586 and 39.1 $\mu g\ L^{-1}\ d^{-1}$) and cell-specific bacterial production (178 and 84.4 fg cell$^{-1}\ d^{-1}$) in the virus-free conditions (i.e. Bs-Ve and Be-Vs) was significantly higher than the counterpart (56.0 and 19.5 $\mu g\ L^{-1}\ d^{-1}$) and (65.4 and 34.5 fg cell$^{-1}\ d^{-1}$) in the virus-rich conditions (i.e. Bs+Ve and Bs+Vs), respectively. Bacterial production and cell-specific bacterial production in the Bs+Ve treatment were significantly higher than those in the Bs+Vs treatment. Bacterial respiration (783 and 209 $\mu g\ L^{-1}\ d^{-1}$) in the virus-free conditions (i.e. Bs-Ve and Bs-Vs) was significantly higher than that (110 and 112 $\mu g\ L^{-1}\ d^{-1}$) in the virus-rich conditions (i.e. Bs+Ve and Bs+Vs), while bacterial respiration did not differ significantly between the Bs+Ve and Bs+Vs treatment. The cell-specific bacterial respiration (239 and 438 fg cell$^{-1}\ d^{-1}$) in the virus-free conditions (i.e. Bs-Ve and Bs-Vs) was significantly higher than that (129 and 197 fg cell$^{-1}\ d^{-1}$) in the virus-rich conditions (i.e. Bs+Ve and Bs+Vs) (Fig. 4).

For the bioassay experiments with marine bacteria, bacterial production and cell-specific bacterial production on the riverine DOC were significantly higher than those on the autochthonous coastal DOC (Stn PM7), irrespective of the presence and absence of viruses. Bacterial respiration on the riverine DOC was significantly higher than that on the autochthonous DOC in the virus-free conditions, but not in the virus-rich conditions. In contrast, the cell-specific bacterial respiration (129 and 239 fg cell$^{-1}\ d^{-1}$) on the riverine DOC was significantly lower than that (197 and 438 fg cell$^{-1}\ d^{-1}$) on autochthonous DOC, irrespective of the presence or absence of viruses. Bacterial growth efficiency (0.33 and 0.43) on the riverine DOC was significantly higher than that (0.15 and 0.17) on autochthonous DOC, irrespective of the presence or absence of viruses (Fig.4).

Discussion

Response of bacterial metabolic activity to riverine DOC input

The effects of DOC on bacterial metabolic activity are often assessed in the presence of viruses, which might bias the bacterial response to DOC supply alone [37] and underestimate bioavailability of DOC. In this study, a comparison of the bacterial response to riverine DOC between the presence or absence of

Table 1. Initial nutrient (DIN, PO$_4$ and Si(OH)$_4$) and DOC concentrations, bacterial abundance (BA), and viral abundance (VA) at two stations (PM7 and NM3; see map in Fig. 1). (DIN $= NO_3 + NO_2^- + NH_4$).

Parameters	PM7	NM3
Salinity	28.64	22.40
DIN (μM)	3.42	32.0
PO$_4$ (μM)	0.24	0.92
Si(OH)$_4$ (μM)	13.7	42.7
DOC (mg L^{-1})	1.12	3.66
BA (cells L^{-1})	2.45×10^9	4.28×10^8
VA (particles L^{-1})	2.57×10^{10}	9.40×10^9

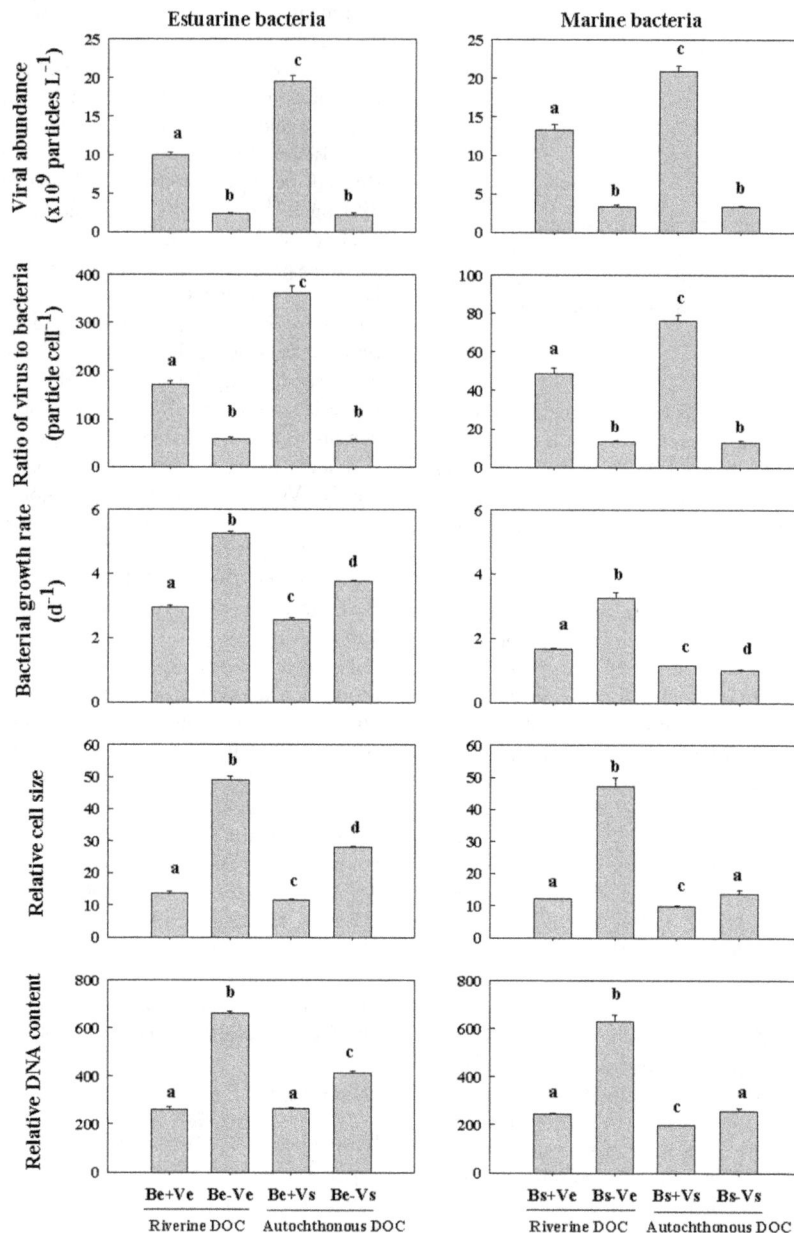

Figure 3. Viral abundance (VA) and virus to bacteria abundance ratio at the beginning of the incubation. Bacterial growth rate, relative cell size, and relative DNA content at the end of the incubation among eight treatments: 1) estuarine bacteria + virus-rich estuarine water (Be+Ve); 2) estuarine bacteria + virus-free estuarine water (Be-Ve); 3) estuarine bacteria + virus-rich seawater (Be+Vs); 4) estuarine bacteria + virus-free seawater (Be-Vs); 5) seawater bacteria + virus-rich estuarine water (Bs+Ve); 6) seawater bacteria + virus-free estuarine water (Bs-Ve); 7) seawater bacteria + virus-rich seawater (Bs+Vs); 8) seawater bacteria + virus-free seawater (Bs-Vs). Vertical bars indicate ± 1SE and n = 3. Different letters (a, b, c, d) denote that the treatment was significantly different ($p < 0.05$) and the same letters denote that the treatment was not significantly different ($p > 0.05$).

viruses provided insight into the bioavailability of riverine DOC and the role of viruses in regulating carbon cycling in the Pearl River estuary.

Our results showed that bacterial metabolic activity responded significantly to the riverine DOC input irrespective of the presence or absence of viruses. Generally, bacterial production, cell-specific BP, growth rate, relative cell size and DNA content were significantly higher in the riverine DOC treatments (e.g. Bs+Ve and Bs-Ve) than the counterpart in the autothchonous coastal DOC treatments (e.g. Bs+Vs and Bs-Vs), respectively. These results suggested that riverine DOC was highly bioavailable and

efficiently utilized, which favored the growth of large and metabolically active marine bacteria, especially under virus-free conditions. However, in the presence of viruses, the increased DOC utilization was mitigated considerably by viral lyses, since the large and metabolically active bacteria may have been preferentially killed by viruses [38] and less competitive bacterial species that were resistant to viral infection survived [39,40]. In addition, the viral shunt is partly responsible for low bacterial production in the presence of viruses [41]. It is estimated that up to 50% of bacterial mortality is due to viral lyses in marine environments [22,25]. A significant correlation between bacterial

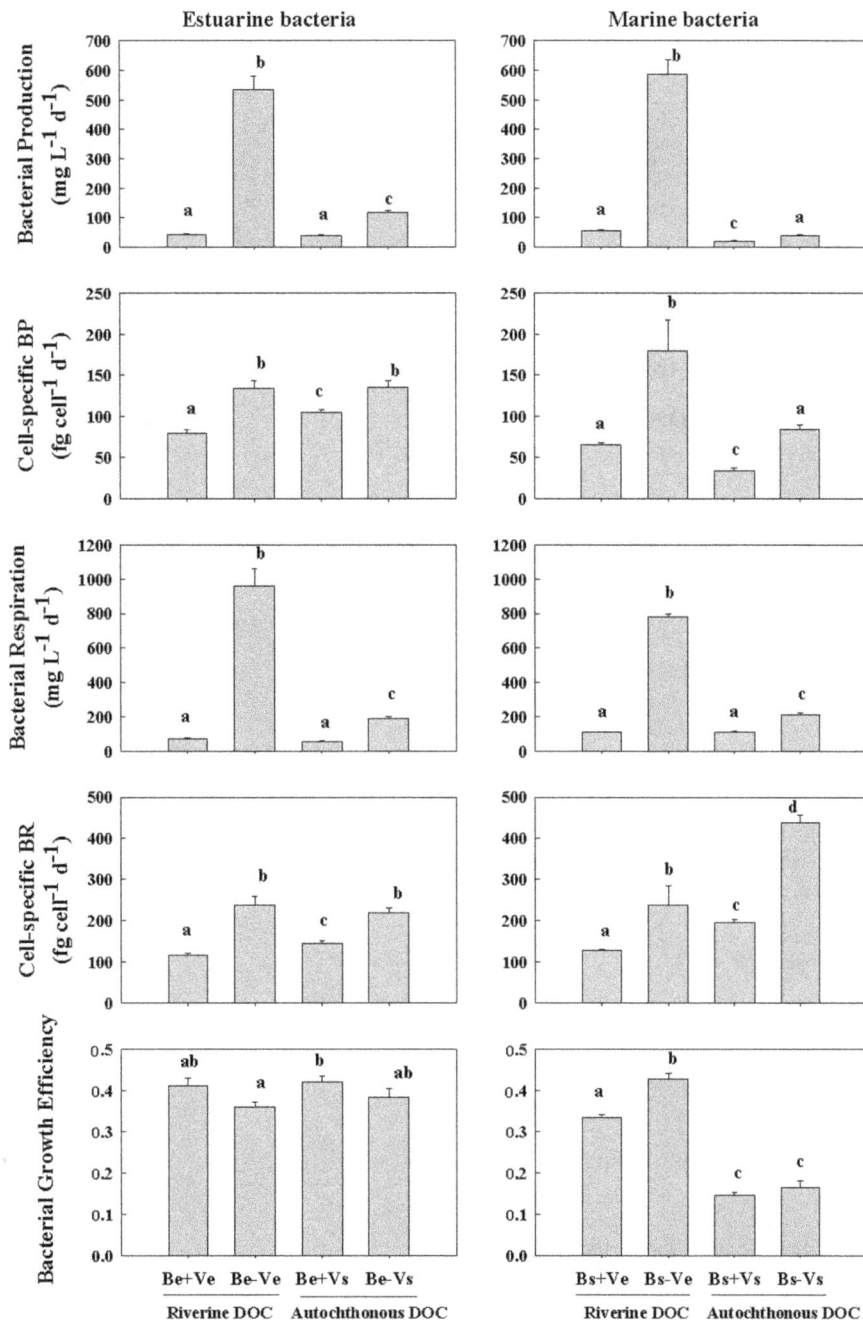

Figure 4. Bacterial production (BP), cell-specific bacterial production, bacterial respiration (BR), cell-specific bacterial respiration and bacterial growth efficiency at the end of the incubation among eight treatments. Error bars indicate ± 1SE and n = 3. Different letters (a, b, c, d) denote that the treatment was significantly different ($p < 0.05$) and the same letters denote that the treatment was not significantly different ($p > 0.05$). See Fig. 3 for treatment abbreviations.

production and the relative DNA content or cell size (Fig. 5) further confirmed that viruses played a major role in regulating bacterial production. The stimulation of bacterial production by riverine DOC in the presence of viruses implied that the addition of bioavailable DOC might, to some extent, weaken viral control of bacterial production by improving the capability of the metabolically active bacteria to resist viral lysis. This finding was consistent with previous reports [42,43].

In response to riverine DOC input, the bulk bacterial respiration was enhanced significantly in the absence of viruses,

but not in the presence of viruses, which was contradictory with previous reports that viruses enhanced the role of bacteria as oxidizers of organic matter and producers of CO_2 by stimulating bacterial respiration [26]. Viral infection has been reported to enhance BR likely due to the decomposition of the viral lysate with an energetic cost [44]. Interestingly, the cell-specific bacterial respiration in the riverine DOC treatments (i.e. Bs+Ve and Bs-Ve) was significantly lower than that in the autochthonous coastal DOC treatments (Bs+Vs and Bs-Vs) irrespective of the presence or absence of viruses, which was the opposite pattern for the cell-

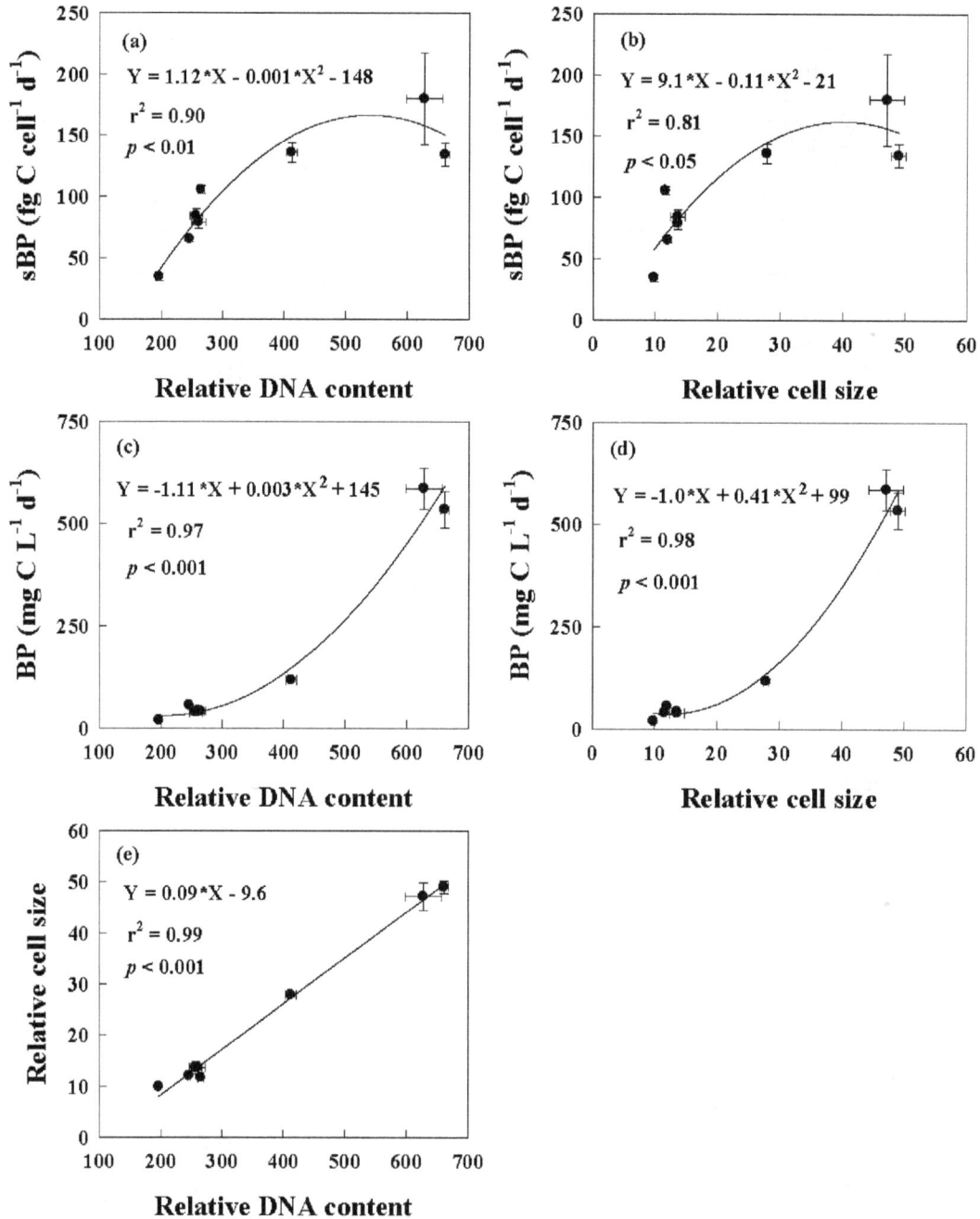

Figure 5. The relationship at the end of the incubation between: (a) cell-specific bacterial production (sBP) vs relative DNA content, (b) cell-specific bacterial production (sBP) vs relative cell size, (c) bacterial production (BP) vs relative DNA content, (d) bacterial production (BP) vs relative cell size, and (e) relative cell size vs relative DNA content for eight treatments.

specific bacterial production. The cell-specific bacterial respiration has been found to be sensitive to changes in substrates, which increases sharply with the deceasing quality and quantity of the substrates [20,24]. The high cell-specific bacterial production and low cell-specific bacterial respiration in the riverine DOC treatment further corroborated that riverine DOC was highly labile, resulting in higher bacterial growth efficiency. An early study showed that sewage effluents with high concentrations of carbohydrates and amino acids, which are among the most labile fraction of the bulk organic matter [45,46], were the major sources (32–54%) of riverine DOC in the Pearl River estuary [8]. Our

observations are in agreement with the finding from our previous sewage-seawater cross-transplant experiment that sewage effluents enhanced bacterial production, but not bacterial respiration, resulting in high bacterial growth efficiency [21]. The bioavailable riverine DOC input influenced the bulk bacterial respiration in two contrasting ways; it enhanced the bulk BR by stimulating bacterial growth and reduced the cell-specific BR due to its high lability. Riverine DOC input altered the partitioning between production and respiration within the natural bacterial community, with a larger proportion of carbon used for biomass synthesis and less for maintenance. The high supply of riverine DOC exposed bacteria to labile DOC in the Pearl River estuary with repaid discharge and a short residence time. As a result, there was little stimulation of the bulk BR by riverine DOC. This might partly explain the lower CO_2 emission in some estuaries that receive a large amount of riverine DOC with a high percentage of sewage-derived DOC. The relatively low bacterial production in the presence of viruses was more likely due to the repression of the metabolically active bacteria, rather than the viral shunt, since the viral shunt would increase bacterial respiration [26,47].

Viral infection dramatically decreased bacterial carbon demand. Bacterial carbon demand (= BP+BR) in the presence of viruses only accounted for 7–12% of that in the absence of viruses. The bioavailability of riverine DOC in the presence of viruses was greatly underestimated, since bacterial production was primarily regulated by viruses, rather than DOC [24]. In conventional microbial decomposition experiments, the biodegradable DOC is estimated as the difference between the DOC at the beginning and end of the incubation that could last for several days or up to one month. In this case, the estimated bioavailability of DOC is likely biased because a fraction of the bacterial biomass is returned to the DOC pool as semi-labile or recalcitrant DOC via the viral shunt [48], which might be substantial during a long-term incubation.

Response of bacterial metabolic activity to exogenous viruses

Shifts in the viral community composition are observed along a salinity gradient in estuaries [49], as well as viral infection rates [50]. To date, little attention has been paid to the composition and metabolic response of the natural bacterial assemblage to varying sources of viruses in estuaries [27]. In our study, bacterial mortality induced by exogenous viruses was assessed through a cross-transplant experiment of estuarine water and seawater. The relative cell size, DNA content, and cell-specific bacterial production in the Be+Vs treatment were significantly lower than that in the Be-Vs treatment, suggesting that marine viruses were able to repress the metabolically active estuarine bacteria. However, the cell-specific bacterial production in the Be+Ve treatment was significantly lower than that in the Be+Vs treatment, despite the high DOC availability in the Be+Ve treatment, implying that marine viruses were able to infect estuarine bacteria, but not as efficiently as estuarine viruses. This was inconsistent with a previous report that marine viruses may not be able to infect freshwater bacteria [27]. In contrast, a 2.7-fold difference in the cell-specific bacterial production for marine bacteria between in virus-free (Bs-Ve) and virus-rich (Bs+Ve) estuarine water was similar to the difference (2.4-fold) between in virus-free (Bs-Vs) and virus-rich (Bs+Vs) seawater. It revealed that estuarine viruses were able to efficiently infect marine bacteria. The same finding has been reported by *Bonilla-Findji et al.* [27]. The rapid changes in the viral community induced by freshwater input into estuaries might alter the existing relationship between bacteria and viruses, leading to changes in bacterial metabolic activity and community composition along a salinity gradient in estuarine environments, in addition to the salinity effect [51,52].

As the capability of utilizing the various DOC component varies with the type of microorganism [53], the effect of changes in the bacterial community composition and bacterial metabolic activity induced by exogenous viruses on the carbon cycle in estuaries should not be ignored. Further studies are needed to confirm the interaction of bacteria and exogenous viruses.

Evaluation of the Experimental Approach

Recent studies suggest that salinity may be an important factor regulating virus-host interactions [52,54]. In this study, estuarine water in the middle section of the Pearl River estuary served as the riverine DOC source instead of freshwater in the upper section of the estuary, in order to reduce the effect of low salinity on the bacterial metabolic activity in our experiments. Our earlier study has shown that large bacteria typically captured on a 1 μm polycarbonate membrane filter were negligible when the bacterial inoculum was prepared [21]. The empirical conversion factor of 3.0 kg C mol leucine^{-1} was adopted to calculate bacterial production and we cannot rule out the possibility that BP for bacteria grown on autochthonous DOC was overestimated, since DOC at coastal station PM7 was at the relatively low level and less labile. The DNA content is a useful proxy for bacterial activity [55]. The significant correlation between cell-specific bacterial production and relative DNA content (Fig. 5) indicated that the estimation of BP based on the commonly used empirical conversion factor was valid.

Conclusions

Riverine DOC input increased bacterial production significantly, but not bacterial respiration because of its high lability. The bioavailable riverine DOC input influenced the bulk bacterial respiration in two contrasting ways; it enhanced the bulk BR by stimulating bacterial growth and reduced the cell-specific BR due to its high lability. As a result, there was little stimulation of the bulk BR by riverine DOC. The lability of riverine DOC might be partly responsible for lower CO_2 degassing fluxes in some DOC-rich estuaries. Viruses reduced bacterial carbon demand dramatically by repressing the growth of metabolically active bacteria, which eventually restricted microbial decomposition of riverine DOC. Bacterial carbon demand in the presence of viruses only accounted for 7–12% of that in the absence of viruses. Hence, a significant fraction of riverine DOC was transported offshore to the shelf and slope ocean.

Acknowledgments

We thank Candy Lee for helping to count viruses with a flow cytometer.

Author Contributions

Conceived and designed the experiments: JX PH HL. Performed the experiments: JX MS ZS. Analyzed the data: JX. Contributed reagents/materials/analysis tools: JX MS ZS. Contributed to the writing of the manuscript: JX PH HL.

References

1. Cai WJ (2011) Estuarine and coastal ocean carbon paradox: CO_2 sinks or sites of terrestrial carbon incineration? Annu Rev Mar Sci 3: 123–145.

2. Laws EA, Falkowski PG, Smith WO Jr, Ducklow H, McCarthy JJ (2000) Temperature effects on export production in the open ocean. Glob Biogeochem Cycles 14: 1231–1246.

3. Price PB, Swoers T (2004) Temperature dependence of metabolic rates for microbial growth, maintenance, and survival. Proc Natl Acad Sci U S A 101: 4631–4636.

4. Borges AV, Dellile B, Frankignoulle M (2005) Budgeting sinks and sources of CO_2 in the coastal ocean: diversity of ecosystems counts. Geophys Res Lett 32: L14601.

5. Fuhrman JA, Steele JA, Hewson I, Schwalbach MS, Brown MV, et al. (2008) A latitudinal diversity gradient in planktonic marine bacteria. Proc Natl Acad Sci U S A 1005: 7774–7778.

6. Yin K, Qian PY, Chen JC, Hsieh DPH, Harrison PJ (2000) Dynamics of nutrients and phytoplankton biomass in the Pearl River estuary and adjacent

waters of Hong Kong during summer: preliminary evidence for phosphorus and silicon limitation. Mar Ecol Prog Ser 194: 295–305.

7. Zhao H (1990) Evolution of the Pearl River estuary. Ocean Press, Beijing (in Chinese).

8. He BY, Dai MH, Zhai WD, Wang L.F, Wang KJ, et al. (2010) Distribution, degradation and dynamics of dissolved organic carbon and its major compound classes in the Pearl River estuary, China. Mar Chem 119: 52–64.

9. Guo X, Dai M, Zhai W, Cai WJ, Chen B (2009) CO_2 flux and seasonal variability in a large subtropical estuarine system, the Pearl River estuary, China. J Geophys Res 114: G03013.

10. Zhai WD, Dai MH, Guo XH (2007) Carbonate system and CO_2 degassing fluxes in the inner estuary of Changjiang (Yangtze) River, China. Mar Chem 107: 342–356.

11. Mukhopadhyay SK, Biswas H, De TK, Sen S, Jana TK (2002) Seasonal effects on the air-water carbon dioxide exchange in the Hooghly estuary, NE coast of Bay of Bengal, India. J Environ Monit 41: 549–552.

12. Shetye SR (1999) Propagation of tides in the Mandovi and Zuri estuaries. Sadhana 24: 5–16.

13. Sundar D, Shetye SR (2005) Tides in the Mandovi and Zuari estuaries, Goa, west coast of India. J Earth Syst Sci 114: 493–503.

14. Williams PJL (1981) Microbial contribution to overall marine plankton metabolism: direct measurement of respiration. Oceanologica Acta 4: 359–364.

15. Pomeroy LR, Wiebe WJ, Deibel D, Thompson JR, Rowe GT, et al. (1991) Bacterial responses to temperature and substrate concentration during the Newfoundland spring bloom. Mar Ecol Prog Ser 75: 143–159.

16. Fuhrman JA (1992) Bacterioplankton roles in cycling of organic matter: the microbial food web, p 361–383. In Woodhead PGF (ed), Primary Productivity and Biogeochemical Cycles in the Sea, Plenum Press, New York.

17. Stepanauskas R, Leonardson L, Tranvik LJ (1999) Bioavailability of wetland-derived DON to freshwater and marine bacterioplankton. Limnol Oceanogr 44: 1477–1485.

18. Langenheder S, Kisand V, Lindström ES, Wikner J, Tranvik LJ (2004) Growth dynamics within bacterial communities in riverine and estuarine batch cultures. Aquat Microb Ecol 37: 137–148.

19. Carlson CA, del Giorgio PA, Herndl GJ (2007) Microbes and the dissipation of energy and respiration: from cells to ecosystems. Oceanography 20: 89–100.

20. del Giorgio PA, Cole JJ (2000) Bacterial energetics and growth efficiency, p 289–325. In Kirchman KL (ed), Microbial ecology of the ocean, Wiley-Liss Inc., New York.

21. Xu J, Jing H, Kong L, Sun M, Harrison PJ, et al. (2013) Effect of seawater-sewage cross-transplants on bacterial metabolism and diversity. Microb Ecol 66: 60–72.

22. Bouvier T, del Giorgio PA (2007) Key role of selective virus-induced mortality in determining marine bacterioplankton composition. Environ Microbiol 9: 287–297.

23. Campbell B, Yu L, Heidelberg JF, Kirchman DL (2011) Activity of abundant and rare bacteria in a coastal ocean. Proc Nat Acad Sci U S A 108: 12776–12781.

24. Xu J, Jing H, Sun M, Harrison PJ, Liu H (2013) Regulation of bacterial metabolic activity by dissolved organic carbon and viruses. J Geophys Res Biogeosci 118: 1573–1583.

25. Fuhrman JA, Schwalbach M (2003) Viral influence on aquatic bacterial communities. Biol Bull 204: 192–195.

26. Bonilla-Findji O, Malits A, Lefèvre D, Rochelle-Newall E, Lemée R, et al. (2008) Viral effects on bacterial respiration, production and growth efficiency: Consistent trends in the Southern Ocean and the Mediterranean Sea. Deep-Sea Res II 55: 790–800.

27. Bonilla-Findji O, Rochelle-Newall E, Weinbauer MG, Pizay MD, Kerros ME, et al. (2009) Effect of seawater-freshwater cross-transplantations on viral dynamics and bacterial diversity and production. Aquat Microb Ecol 54: 1–11.

28. Rochelle-Newall EJ, Pizay MD, Middelburg JJ, Boschker HTS, Gattuso JP (2004) Degradation of riverine dissolved organic matter by seawater bacteria. Aquat Microb Ecol 37: 9–22.

29. Strickland JDH, Parsons TR (1968) Determination of reactive nitrate. In: A practical handbook of seawater analysis. Bull Fish Res Board Can167: 71–75.

30. Grasshoff KM, Ehrhardt M, Kremling K (1983) Methods of seawater analysis. Weinheim, Verlag Chemie.

31. Knap A, Michaels A, Close A, Ducklow H, Dickson A (1996) Protocols for the Joint Global Ocean Flux Study (JGOFS) core measurement, Rep. JGOFS Nr. 19, IOC Manual and Guide, No 29, UNESCO.

32. Marie D, Partensky F, Jacquet S, Vaulot D (1997) Enumeration and cell cycle analysis of natural populations of marine picoplankton by flow cytometry using the nucleic acid stain SYBR Green I. Appl Environ Microbiol 63: 186–193.

33. Marie D, Brussaard CPD, Thyrhaug R, Bratbak G, Vaulot D (1999) Enumeration of marine viruses in culture and natural samples by flow cytometry. Appl Environ Microbiol 65: 45–52.

34. Pedrós-Alió C, Calderón-Paz J, Guixa-Boixereu N, Estrada M, Gasol JM (1999) Bacterioplankton and phytoplankton biomass and production during summer stratification in the northwestern Mediterranean Sea. Deep Sea Res I 46: 985–1019.

35. Outdot CR, Gerard R, Morin P, Gningue I (1988) Precise shipboard determination of dissolved oxygen (Winkler procedure) for productivity studies with a commercial system. Limnol Oceanogr 33: 146–150.

36. Hopkinson CS Jr (1985) Shallow-water benthic and pelagic metabolism: Evidence of heterotrophy in the nearshore Georgia Bight. Mar Biol 87: 19–32.

37. Zhang R, Weinbauer MG, Tam YK, Qian PY (2013) Response of bacterioplankton to a glucose gradient in the absence of lysis and grazing. FEMS Microbiol Ecol 85: 443–451.

38. Thingstad TF, Lignell R (1997) Theoretical models for the control of bacterial growth rate, abundance, diversity and carbon demand. Aquat Microb Ecol 13: 19–27.

39. Hewson I, Vargo GA, Fuhrman JA (2003) Bacterial diversity in shallow oligotrophic marine benthos and overlying waters: effects of virus infection, containment, and nutrient enrichment. Microb Ecol 46: 322–336.

40. Winter C, Smit A, Herndl GJ, Weinbauer MG (2005) Linking bacterial richness with viral abundance and prokaryotic activity. Limnol Oceanogr 50: 968–977.

41. Motegi C, Nagata T, Miki T, Weinbauer MG, Legendre L, et al. (2009) Viral control of bacterial growth efficiency in marine pelagic environments. Limnol Oceanor 54: 1901–1910.

42. Motegi C, Nagata T (2009) Addition of monomeric and polymeric organic substrates alleviates lytic pressure on bacterial communities in coastal seawaters. Aquat Microb Ecol 57: 343–350.

43. Zhang R, Weinbauer MG, Qian PY (2007) Viruses and flagellates sustain apparent richness and reduce biomass accumulation of bacterioplankton in coastal marine waters. Environ Microbiol 9: 3008–3018.

44. Middelboe M, Lyck PG (2002) Regeneration of dissolved organic matter by viral lysis in marine microbial communities. Aquat Microb Ecol 27: 187–194.

45. Benner R, Pakulski JD, McCarthy M, Hedges JI, Hatcher PG (1992) Bulk chemical characteristics of dissolved organic matter in the ocean. Science 255: 1561–1564.

46. Middelboe M, Borch NH, Kirchman DL (1995) Bacterial utilization of dissolved free amino acids, dissolved combined amino acids and ammonium in the Delaware Bay estuary: effects of carbon and nitrogen limitation. Mar Ecol Prog Ser 128: 109–120.

47. Eissler Y, Sahlsten E, Quiñones RA (2003) Effects of virus infection on respiration rates of marine phytoplankton and microplankton communities. Mar Ecol Prog Ser 262: 71–80.

48. Jiao N, Herndl GJ, Hansell DA, Benner R, Kattner G, et al. (2010) Microbial production of recalcitrant dissolved organic matter: long-term carbon storage in the global ocean. Nat Rev Microbiol 8: 593–599.

49. Wommack KE, Ravel J, Hill RT, Chun J, Clowell RR (1999) Population dynamics of Chesapeake Bay virioplankton: total-community analysis by pulse-field gel electrophoresis. Appl Environ Microbiol 65: 231–240.

50. Almeida MA, Cunda MA, Alcantara F (2001) Loss of estuarine bacteria by viral infection and predation in microcosm conditions. Microb Ecol 42: 562–571.

51. Lozupone CA, Knight R (2007) Global patterns in bacterial diversity. Proc Natl Acad Sci U S A 104: 11436–11440.

52. Combe M, Bouvier T, Pringault O, Rochelle-Newal E, Bouvier C, et al. (2013) Freshwater prokaryote and virus communities can adapt to a controlled increase in salinity through changes in their structure and interactions. Estuar Coast Shelf Sci 133: 58–66.

53. Carlson CA, Giovannoni SJ, Hansell DA, Goldberg SJ, Parsons R, et al. (2004) Interactions among dissolved organic carbon, microbial processes, and community structure in the mesopelagic zone of the northwestern Sargasso Sea. Limnol Oceanogr 49: 1073–1083.

54. Kukkaro P, Bamford DH (2009) Virus–Host interactions in environments with a wide range of ionic strengths. Environ Microbiol Reports 1: 71–77.

55. Jellet JF, Li WKW, Dickie PM, Boraie A, Kepkay PE (1996) Metabolic activity of bacterioplankton communities assessed by flow cytometry and single carbon substrate utilization. Mar Ecol Prog Ser 136: 213–225.

14

Amoebal Endosymbiont *Neochlamydia* Genome Sequence Illuminates the Bacterial Role in the Defense of the Host Amoebae against *Legionella pneumophila*

Kasumi Ishida[1], Tsuyoshi Sekizuka[2], Kyoko Hayashida[3], Junji Matsuo[1], Fumihiko Takeuchi[2], Makoto Kuroda[2], Shinji Nakamura[4], Tomohiro Yamazaki[1], Mitsutaka Yoshida[5], Kaori Takahashi[5], Hiroki Nagai[6], Chihiro Sugimoto[3], Hiroyuki Yamaguchi[1]*

1 Department of Medical Laboratory Science, Faculty of Health Sciences, Hokkaido University, Sapporo, Hokkaido, Japan, **2** Pathogen Genomics Center, National Institute of Infectious Diseases, Tokyo, Japan, **3** Research Center for Zoonosis Control, Hokkaido University, Sapporo, Hokkaido, Japan, **4** Division of Biomedical Imaging Research, Juntendo University Graduate School of Medicine, Tokyo, Japan, **5** Division of Ultrastructural Research, Juntendo University Graduate School of Medicine, Tokyo, Japan, **6** Research Institute for Microbial Diseases, Osaka University, Osaka, Japan

Abstract

Previous work has shown that the obligate intracellular amoebal endosymbiont *Neochlamydia* S13, an environmental chlamydia strain, has an amoebal infection rate of 100%, but does not cause amoebal lysis and lacks transferability to other host amoebae. The underlying mechanism for these observations remains unknown. In this study, we found that the host amoeba could completely evade *Legionella* infection. The draft genome sequence of *Neochlamydia* S13 revealed several defects in essential metabolic pathways, as well as unique molecules with leucine-rich repeats (LRRs) and ankyrin domains, responsible for protein-protein interaction. *Neochlamydia* S13 lacked an intact tricarboxylic acid cycle and had an incomplete respiratory chain. ADP/ATP translocases, ATP-binding cassette transporters, and secretion systems (types II and III) were well conserved, but no type IV secretion system was found. The number of outer membrane proteins (OmcB, PomS, 76-kDa protein, and OmpW) was limited. Interestingly, genes predicting unique proteins with LRRs (30 genes) or ankyrin domains (one gene) were identified. Furthermore, 33 transposases were found, possibly explaining the drastic genome modification. Taken together, the genomic features of *Neochlamydia* S13 explain the intimate interaction with the host amoeba to compensate for bacterial metabolic defects, and illuminate the role of the endosymbiont in the defense of the host amoebae against *Legionella* infection.

Editor: Matthias Horn, University of Vienna, Austria

Funding: This study was supported by grants-in-aid for scientific research from KAKENHI, grant numbers 21590474, 24659194, and 24117501, "Innovation Areas (Matryoshika-type evolution)". The funders had no role in study design, data collection and analysis, decision to publish, or preparation of the manuscript.

Competing Interests: The authors declare that they have no competing interests.

* E-mail: hiroyuki@med.hokudai.ac.jp

Introduction

Obligate intracellular chlamydiae have evolved into two groups since the divergence of ancient chlamydiae 0.7–1.4 billion years ago. Pathogenic chlamydiae species (e.g. *Chlamydia trachomatis*) have adapted with their vertebrate hosts, whereas environmental chlamydiae (e.g. *Neochlamydia* species) have evolved as endosymbionts of lower eukaryotes, such as free-living amoebae (*Acanthamoeba*) [1–4]. Both types of chlamydiae have unique intracellular developmental cycles, defined by two distinct stages: the elementary body (EB), which is the form that is infectious to host cells, and the reticulate body (RB), which is the replicative form in the cells [4]. Interestingly, pathogenic chlamydiae have evolved through a decrease in genome size, with genomes of approximately 1.0–1.2 Mb, which may be a strategy to evade the host immune network, resulting in a shift to parasitic energy and metabolic requirements [1–4]. Meanwhile, the genome of the representative environmental chlamydia, *Protochlamydia* UWE25, is not decreasing and has stabilized at 2.4 Mb [4], implying that environmental chlamydiae still possess certain genes that pathogenic chlamydiae have lost. Therefore, environmental chlamydiae are useful tools for elucidating chlamydial evolution and obligate intracellular parasitism.

Recently, we isolated several environmental amoebae harboring endosymbiotic environmental chlamydiae from Sapporo, Hokkaido, Japan [5]. Of these, the amoebal endosymbiont *Neochlamydia* S13 was particularly interesting because its rate of amoebal infection was always 100%, but no amoebal lysis or transfer to other host amoebae was observed [5,6]. This suggested an intimate mutualistic interaction of *Neochlamydia* S13 with its host amoebae, which is possibly a unique genomic feature. The reason why amoebae continually feed the endosymbiotic bacteria remains unknown, although the endosymbiotic bacteria may protect the host against *Legionella*, which also grow in and kill amoebae [7–9]. Therefore, in this study we evaluated the interaction of *Neochlamydia* S13 with the host amoebae, including its protective role against *Legionella*, through analysis of a draft genome of *Neochlamydia* S13.

Results and Discussion

Neochlamydia S13 intimately interacts with host amoebae and plays a significant role in the amoebal protection system against *Legionella pneumophila* infection

Transmission electron microscopic (TEM) analysis revealed a wide distribution of RBs in the amoebal cytoplasm, but no EBs were observed, suggesting persistent infection and an intimate interaction between the bacteria and the host amoeba (Figure 1).

Why the amoebae allow *Neochlamydia* to persist within the cells remains unknown. We therefore assessed whether the amoebae harboring *Neochlamydia* S13 could resist infection by *L. pneumophila*, which can kill amoebae in natural environments [7–9]. In contrast to the extensive growth observed in the aposymbiotic strain of amoeba (S13RFP: treatment with rifampicin), *L. pneumophila* failed

to replicate in amoebae harboring *Neochlamydia* S13 wild-type (WT) (Figure 2A). Another amoebal strain, harboring *Protochlamydia* R18 (R18WT) [5], also allowed intracellular growth of *L. pneumophila*, as did the aposymbiotic amoeba R18DOX (treatment with doxycycline) and the reference C3 amoebal strain, which lacks any endosymbiotic bacteria (Figure 2B and C). Gimenez staining showed that *L. pneumophila* failed to grow in amoebae harboring *Neochlamydia* S13 (Figure 2D, top (arrows, *Neochlamydia* S13)). Thus, the results strongly suggested that *Neochlamydia* S13 confers a survival advantage on the host by providing resistance to *L. pneumophila* infection in amoebal environments such as biofilms [10,11].

Phylogenetic analysis of 16S rRNA sequences showed that *Neochlamydia* S13 belonged to its own cluster, sequestered from other chlamydiae (Figure S1, arrow). Thus, the phylogenetic data suggest unique genomic features in *Neochlamydia* S13.

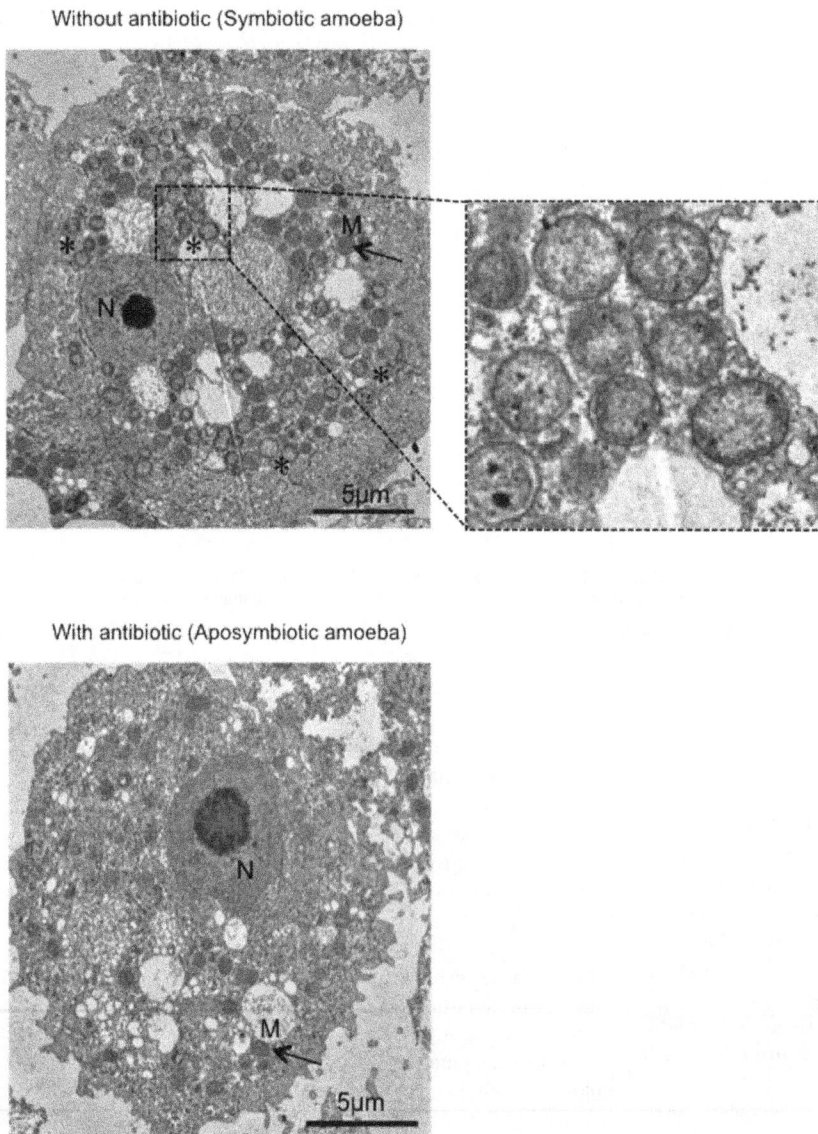

Without antibiotic (Symbiotic amoeba)

With antibiotic (Aposymbiotic amoeba)

Figure 1. Morphological traits of *Neochlamydia* S13 inside host amoebae. Representative TEM images showing symbiotic amoebae harboring *Neochlamydia* S13 (up and right (enlarged)) and aposymbiotic amoebae constructed by treatment with antibiotics (down). Enlarged image in the square with a dotted line shows the bacterial reticular body (no elementary body was observed in the amoebae). *, bacteria. M, mitochondria (arrows). N, nucleus.

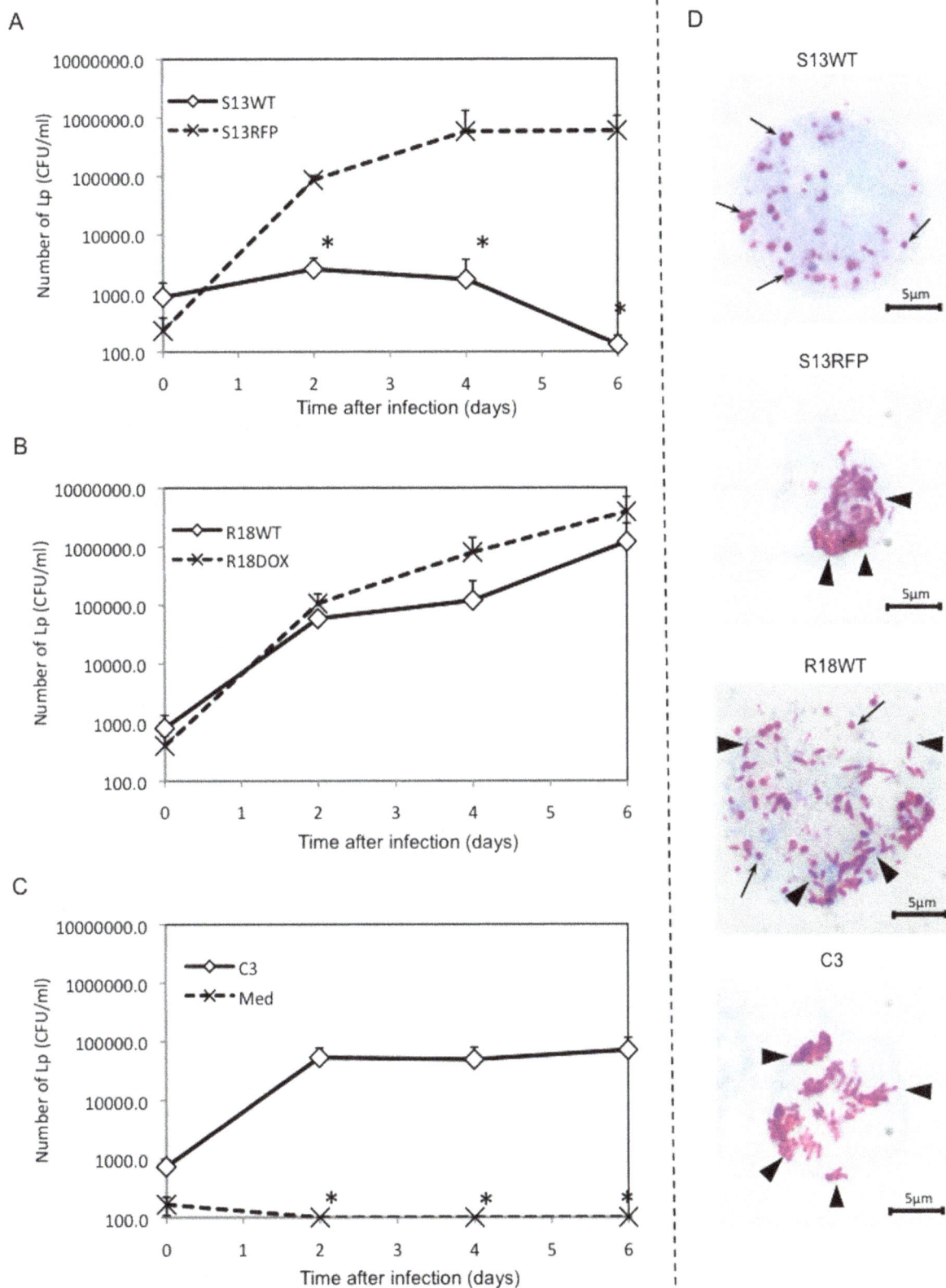

Figure 2. Amoebae harboring *Neochlamydia* S13 are completely protected from *L. pneumophila* infection. (A) *L. pneumophila* survival in *Neochlamydia* S13-infected amoebae (S13WT) and aposymbiotic amoebae (S13RFP). (B) *L. pneumophila* growth in *Protochlamydia* R18-infected amoebae (R18WT) and aposymbiotic amoebae (R18DOX). (C) *L. pneumophila* survival in C3 reference strain amoebae (C3), and PYG medium only (Med). Data are the means ± SD from at least three independent experiments performed in triplicate. * denotes *p*<0.05 at each of the time points. (D) Representative Gimenez staining images showing *L. pneumophila* replication into amoebae at 5 days post-infection. Arrows, *Neochlamydia* S13 (S13WT) or *Protochlamydia* R18 (R18WT). Arrowheads, replicated *L. pneumophila*.

General genome features and validity

A draft genome of *Neochlamydia* S13 was determined by Illumina GAIIx, assembled with ABySS-pe, and then annotated by the RAST server with manual local BLAST analysis. The draft genome of *Neochlamydia* S13 contained 3,206,086 bp (total contig size) with a GC content of 38.2% in 1,317 scaffold contigs (DDBJ accession number: BASK01000001–BASK01001342). The genome contains 2,832 protein-coding sequences (CDSs), 43 tRNAs, and six ribosomal RNAs (Table S1). About half of the CDSs (1,577; 57%) were classified as hypothetical proteins, including 1,030 unknown proteins (36.4%) that had no significant similarity to those of other chlamydiae (*Protochlamydia amoebophila* UWE25 (NC_005861.1), *Parachlamydia acanthamoebae* UV-7 (NC_015702.1), *Simkania negevensis* Z (NC_015713.1), *Waddlia chondrophila* WSU 86-1044 (NC_014225.1), *Chlamydia trachomatis* D/UW-3/CX (NC_000117.1), *C. trachomatis* L2 434/Bu (NC_010287.1), *Chlamydia pneumoniae* CWL029 (NC_000922.1), *C. pneumoniae* TW183 (NC_005043.1)) in public databases (E-value $> 1e^{-10}$) (Figure S2). This suggested that *Neochlamydia* S13 contains unique genomic features, which may provide hints for discovering the intimate mutual relationship underlying symbiosis.

Because of the presence of so many unknown genes with repeat sequences in the predicted genome, we were unable to fill all of the gaps to complete the genome of *Neochlamydia* S13. The validity of our system, including scaffold contig assembly and gene annotation, was confirmed by comparing the reference genome of *Protochlamydia* UWE25 (NC_005861.1) with the draft genome of its related amoebal endosymbiont, *Protochlamydia* R18 (originally isolated from a river in Sapporo City, Japan [5]), assembled using our system for this study. The draft genome of *Protochlamydia* R18 contained 2,727,392 bp with a GC content of 38.8% in 770 scaffold contigs (DDBJ accession number: BASL01000001–BASL01000795). An ORF annotation coverage of 87.6% was observed.

Glycolytic pathway, tricarboxylic acid cycle, and respiratory chain are incomplete

As mentioned above, the previous findings of an amoebal infection rate of 100% and absence of amoebal lysis and transferability to other host amoebae suggest a defective energy reserve system in *Neochlamydia* S13. We therefore used KEGG-module analysis to determine whether *Neochlamydia* S13 contained complete metabolic pathways. This showed that the *Neochlamydia* S13 modules that mapped onto the metabolic pathways differed significantly from those of other chlamydiae (*Protochlamydia* UWE25 and *C. trachomatis* L2) (Figure 3). While the glycolytic pathway from fructose 1,6-bisphosphate to pyruvate was complete in *Neochlamydia* S13, hexokinase and 6-phosphfructokinase were missing, indicating a truncated Embden-Meyerhof-Parnas pathway (Figure S3). Analysis also confirmed that while the Entner-Doudoroff pathway was truncated, the pentose phosphate pathway was intact, suggesting ribulose-5-phosphate synthesis with folate metabolic activity (Figure S3). Surprisingly, the tricarboxylic acid (TCA) cycle, which oxidizes acetyl-CoA to CO_2, was almost entirely missing, except for malate dehydrogenase, the dihydrolipoamide succinyltransferase component (E2) of the 2-oxoglutarate dehydrogenase complex, and dihydrolipoamide dehydrogenase of 2-oxoglutarate dehydrogenase (Figure S3). As pathogenic chlamydiae are still viable when carrying at least half of the TCA cycle genes [2,12], and all previously reported environmental chlamydiae possess complete TCA cycles [4,13,14], the lack of the cycle in the *Neochlamydia* S13 genome is unique. We also found a defective respiratory chain, equipped with only the NADH dehydrogenase complex, cytochrome *c* oxidase complex, and V-type ATPase units; the succinate dehydrogenase complex, cytochrome *c* reductase complex, and F-type ATPase units were completely missing (Figure S4). Thus, these data show that the central metabolic pathway of *Neochlamydia* S13 is drastically truncated, even when compared with pathogenic chlamydiae, indicating a strong dependence on host amoebae.

Fatty acid biosynthesis pathways are conserved

As the genes of the fatty acid initiation and biosynthesis pathways were conserved, as in other chlamydiae (Figure S5), it seemed likely that *Neochlamydia* S13 could produce fatty acid. However, synthesis pathways for CoA, which is a starting material for the acetyl-CoA required for fatty acid initiation [15,16], and biotin, which is a coenzyme required for fatty acid construction [17,18] were lacking in *Neochlamydia* S13. These findings therefore suggest that both CoA and biotin may be transported from the amoebal cytoplasm into the bacteria by unknown transporters.

ATP/ADP translocases, ATP-binding cassette (ABC) transporters, the Sec-dependent type II secretion system, and the type III secretion systems, but not the type IV secretion system, are well conserved

Similar to pathogenic chlamydiae, *Neochlamydia* S13 lacked many key enzymes in the purine and pyrimidine metabolic pathways that are directly connected to nucleotide biosynthesis. It therefore seemed likely that *Neochlamydia* S13 might obtain ATP from the host amoebal cytoplasm via a number of ATP/ADP translocases. As expected, three translocases (NTT1–NTT3) similar to those of pathogenic chlamydiae (Figure S6A) were identified in the genome, although environmental chlamydia strain UWE25 contains five ATP/ADP translocases [19,20]. We also found several ABC transport systems (spermidine/putrescine, zinc, mannan, lipopolysaccharide, and lipoprotein) in the annotated genome (Figure S6B). These findings suggest significant roles for the ABC transporters in compensating for the defective metabolic systems of the bacteria, possibly explaining the intimate symbiotic interaction and strong host dependency. Meanwhile, the number of ABC transporters identified in the draft genome was limited, although these transporters are generally widespread among living organisms and are highly conserved in all genera. They are responsible for essential biological processes such as material transport, translation elongation, and DNA repair [21–23].

We next assessed whether secretion systems (Sec-dependent type II, type III, and type IV) were conserved in the *Neochlamydia* S13 genome. As shown in Figure S7, the type III secretion system, which is widely distributed among chlamydiae [24–26], was well conserved in the *Neochlamydia* S13 genome, and the Sec-dependent type II secretion system was nearly completely conserved (Figure S8). These findings suggested that both systems aid *Neochlamydia* S13 survival in the host amoebae. Interestingly, in contrast to previously reported environmental chlamydiae [4,13,27,28], no gene cluster encoding the type IV secretion system was found, similar to pathogenic chlamydiae, although the *Protochlamydia* R18 genome contained a complete type IV gene cluster (Figure S9, *Protochlamydia* UWE25 versus *Protochlamydia* R18). Recent works have strongly suggested that bacterial type IV secretion systems might induce inflammasome or caspase activation, resulting in bacterial elimination via accumulation of professional effector cells [29–31]. It is therefore possible that the type IV secretion system is harmful to the symbiotic interaction in host cells, as well as persistent infection that generally occurs in mammalian cells.

vs. *Protochlamydia* UWE25

vs. *Chlamydia trachomatis* L2 434 Bu

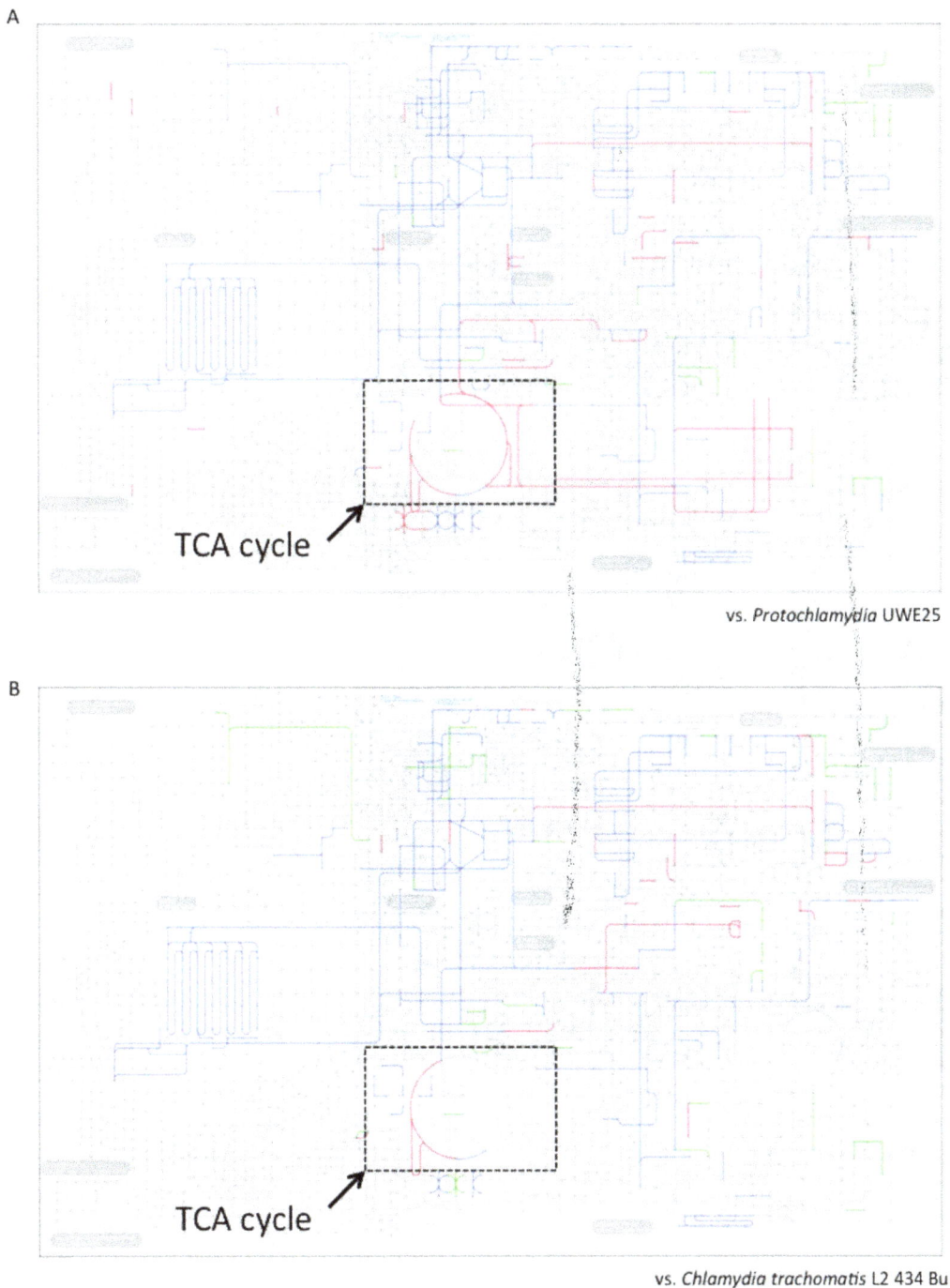

Figure 3. Comparison of metabolic pathways among representative chlamydiae. (A) *Neochlamydia* S13 (this study) versus *Protochlamydia amoebophila* UWE25 (NC005861.1). (B) *Neochlamydia* S13 versus *C. trachomatis* L2 434 Bu (NC010287.1). Green lines, unique in the *Neochlamydia* active modules. Blue lines, shared modules. Red lines; modules specific for *Protochlamydia* or *C. trachomatis*. Squares surrounded with dotted line show TCA cycle.

Predicted outer membrane proteins were truncated

In contrast to *Protochlamydia* UWE25 [13,32,33], *Neochlamydia* S13 contained fewer annotated genes encoding outer membrane proteins, which presumably localize to the outer leaflet membrane and periplasmic space. These genes, *pomS*, the 76-kDa protein gene (*Protochlamydia* UWE25, pc0004), *ompW*, and *omcB* (Figure S10), indicate successful adaptation to the host amoebal cytoplasm through loss of redundant molecules. The predicted 3D model of

PomS (NEOS13_1146), constructed using MMDB (see "Methods"), showed a porin with a β-barrel structure and a channel (Figure 4), suggesting an active transporter.

Predicted proteins with leucine-rich repeats (LRRs) or ankyrin domains

As mentioned above, we found that amoebae harboring *Neochlamydia* S13 were never infected with *L. pneumophila*, which

Worms   Wire

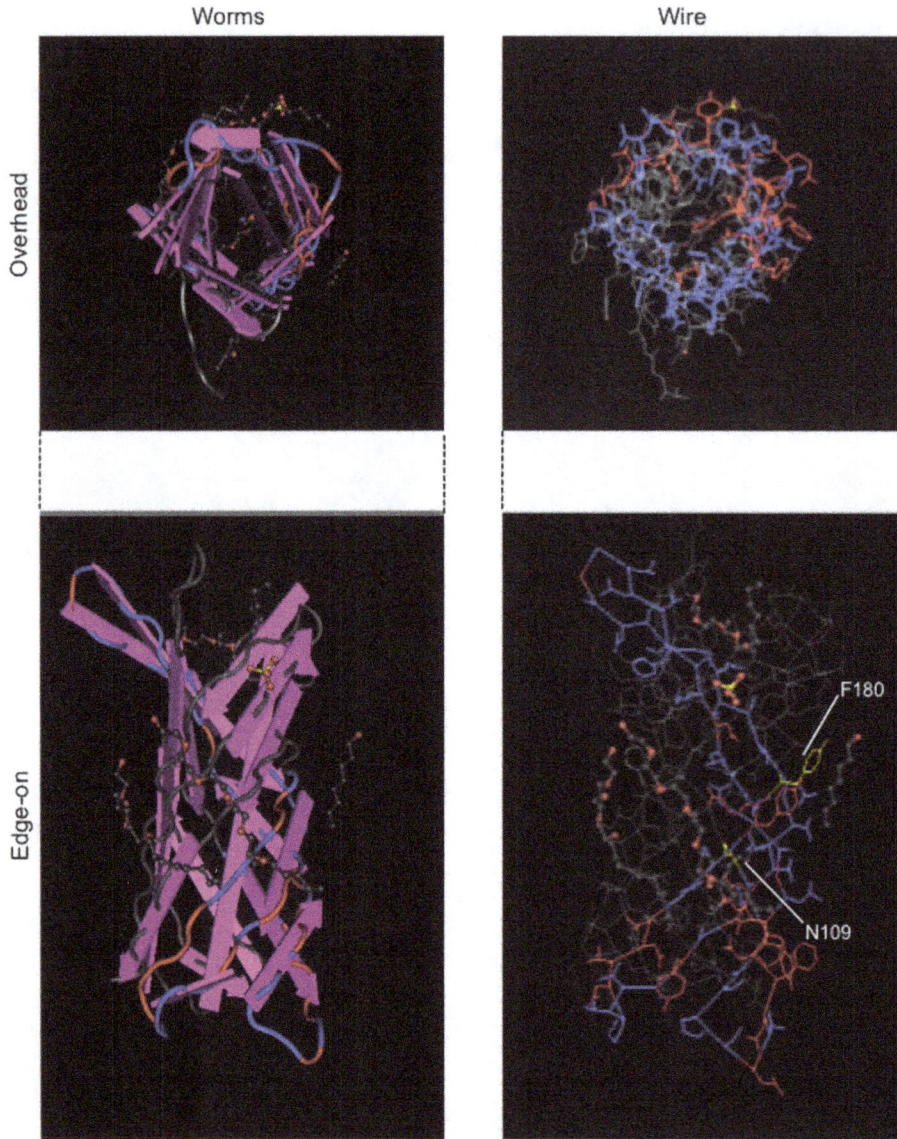

*Aligned to Chain A, Crystal Structure Of Neisserial Surface Protein A (NspA) (pdb|1PTA)

Figure 4. Predicted three-dimensional structure of PomS (NEOS13_1146), which presumably localizes to the outer membrane as a porin. The structure was constructed by alignment with the crystal structure of chain A of *Neisseria* surface protein A (NspA) (pdb:1PTA). N109 (yellow), amino acid aligned at start position. F180 (yellow), amino acid aligned at end position.

is a natural killer of amoebae [7–9,10,11]. We hypothesized that *Neochlamydia* S13 effector molecules secreted into the amoeba might be associated with protection against *L. pneumophila* infection. Recent studies have intriguingly revealed that pathogenic bacteria have evolved effector proteins with LRR or ankyrin domains that may mimic host signaling molecules when injected into host cells [34–36]. Therefore, we searched for unique molecules with LRRs [37,38] or ankyrin domains [39–41], which may be responsible for protein-protein interaction and possibly for controlling *L. pneumophila* infection in host amoebae, in the *Neochlamydia* S13 genome. We identified 199 genes encoding predicted candidate molecules with LRRs, 30 of which were unique, showing no homology with other environmental chlamydiae (Table S2). This suggests possible expansion of these genes from a small number of ancestral genes containing LRRs, although the mechanism of expansion remains unknown. How-

ever, it is possible that *L. pneumophila* infection could stimulate the expansion of the *Neochlamydia* genes encoding LRR domains. Among these genes, 15 were well conserved with those of *Micromonas* (algae) and *Nostoc punctiforme* (a nitrogen-fixing cyanobacterium), with 45–74% identity (Table S2). These results suggest horizontal gene transfer between *Neochlamydia* S13 and such plant-related microbes, allowing us to hypothesize an ancestral relationship between chlamydiae and algae or cyanobacteria [42–44].

As it is well known that molecules with ankyrin domains play a critical role in protein-protein interaction, we also searched for these genes in the *Neochlamydia* S13 genome. RAST analysis with manual local BLAST analysis predicted eight genes (NEOS13_0151, NEOS13_0209, NEOS13_0435, NEOS13_0856, NEOS13_1517, NEOS13_1563, NEOS13_2364, NEOS13_2796) that encode molecules with ankyrin domains.

Interestingly, NEOS13_0151 had a unique coiled-coil structure that was not similar to other chlamydial proteins (Table S1, Figure S11). Meanwhile, phylogenetic analysis of NEOS13_0151 revealed close similarity with functional molecules found in eukaryotes (Figure S12), presumably associated with host cell modification or cellular functions. Recent work has shown that the *L. pneumophila* (strain AA100/130b) F-box ankyrin effector is involved in eukaryotic host cell exploitation, allowing intracellular growth [45]. Thus, we suggest that *Neochlamydia* S13 possesses unique genes encoding ankyrin domains, possibly responsible for resisting *L. pneumophila* infection via host amoebae, although the underlying mechanism remains to be determined.

Presence of transposases implies drastic genome modification

We found 33 genes encoding transposases in the *Neochlamydia* S13 genome, as annotated by RAST analysis with manual local BLAST analysis (Table S1). It has been reported that *Chlamydia suis* possesses a novel insertion element (IScs605) encoding two predicted transposases [46], and that *Protochlamydia* UWE25 contains 82 transposases [47]. Thus, the features of the *Neochlamydia* S13 genome were unique, without genome reduction, but with specified genes for controlling host-parasite interaction, resulting in successful adaptation to the host amoeba. Although the reason why the *Neochlamydia* S13 genome size has not reduced remains unknown, such transposases may be responsible for genome modification without genome reduction.

Conclusions

We determined a draft genome sequence of *Neochlamydia* S13, which provided hints as to why the mutualistic interaction between the bacteria and the host amoebae is maintained, and how the bacteria manipulate the host amoebae. Such unique genome features of *Neochlamydia* S13 strongly indicate an intimate dependency on the host amoebae to compensate for lost bacterial metabolic activity, and a possible role for the bacterial endosymbiont in defense against *L. pneumophila*. These findings provide new insight into not only the extraordinary diversity between chlamydiae, but also why symbiosis occurred between the amoebae and environmental chlamydiae.

Materials and Methods

Amoebae

As described previously [5], two different amoebal (*Acanthamoeba*) strains persistently infected with *Protochlamydia* R18 or *Neochlamydia* S13 were isolated from a river water sample and a soil sample, respectively. The prevalence of amoebae with endosymbionts, as defined by 4′,6-diamidino-2-phenylindole staining, was always approximately 100% [6]. *Acanthamoeba castellanii* C3 (ATCC 50739) was purchased from the American Type Culture Collection (ATCC) and used as a reference strain. Aposymbiotic amoebae derived from *Protochlamydia* R18-infected amoebae (designated R18DOX, established by treatment with doxycycline (64 μg/ml)) and *Neochlamydia* S13-infected amoebae (designated S13RFP, established by treatment with rifampicin (64 μg/ml)) [6] were also used for this study. All amoebae were maintained in PYG broth (0.75% (w/v) peptone, 0.75% (w/v) yeast extract, and 1.5% (w/v) glucose) at 30°C [5].

Bacteria

Human isolate *L. pneumophila* Philadelphia I (JR32), equipped with a complete *dot/icm* gene set encoding a type IV secretion system, which is required for intracellular amoebal growth [48], was kindly provided by Dr. Masaki Miyake of the University of Shizuoka, Japan. *L. pneumophila* was cultured on BCYE agar (OXOID, Hampshire, UK) at 37°C for 2 days.

Infection of amoebae with *L. pneumophila*

Amoebae (5×10^5 cells) were infected with *L. pneumophila* (5×10^5 colony-forming units [CFU]) at a multiplicity of infection of one for 2 h at 30°C, and then uninfected bacteria were killed by the addition of 50 μg/ml gentamycin. After washing with PYG medium, the infected amoebae were incubated for up to 6 days. The infected amoebae were collected every other day, and bacterial CFUs were estimated by serial dilution on BCYE agar.

Bacterial purification and genomic DNA extraction

Both *Neochlamydia* S13- and *Protochlamydia* R18-infected amoebae were collected by centrifugation at $1,500\times g$ for 30 min. The resulting pellets were suspended in PYG medium. Each of the amoebae were disrupted by bead-beating for 5 min according to a previously described method [49], and then centrifuged at $150\times g$ for 5 min to remove unbroken cells and nuclei. The supernatant including intact bacteria was incubated with DNase (Sigma) for 30 min at room temperature, and then the bacteria were washed and suspended in 10 mM HEPES buffer containing 145 mM NaCl. The suspension was carefully overlayed onto 30% Percoll. Following centrifugation at $30,000\times g$ for 30 min, the bacteria were collected from the lower layer. Finally, bacterial pellets were stored at −20°C until use. DNA was extracted with a phenol-chloroform method.

Genome sequencing, annotation and prediction of metabolic pathway modules, and genome comparison

Bacterial 1 kb insert DNA libraries (Purified *Neochlamydia* S13 and *Protochlamydia* R18) were prepared using a genomic DNA Sample Prep Kit (Illumina, San Diego, CA). DNA clusters were generated on a slide using a Cluster Generation Kit (ver. 4) on an Illumina Cluster Station, according to the manufacturer's instructions. Sequencing runs for 81-mer paired-end sequence were performed using an Illumina Genome Analyzer IIx (GA IIx). The 81-mer paired-end reads were assembled (parameters k64, n51, c32.1373) using ABySS-pe v1.2.0 [50]. Annotation of genes from the draft genome sequences was performed using Rapid Annotation using Subsystem Technology (RAST: http://rast.nmpdr.org/) [51] with a local manual BLASTp search. Metabolic pathway modules were predicted using the Kyoto Encyclopedia of Genes and Genomes (KEGG: http://www.genome.jp/kegg/) [52]. Genome comparison was performed using RAST, and then manually visualized by GenomeMatcher 1.69 (http://www.ige.tohoku.ac.jp/joho/gmProject/gmhome.html) [53].

Prediction of 3D structure for annotated genes

Three-dimensional structures of annotated protein sequences of interest were predicted using a web program, protein BLAST with the Molecular Modeling Database (MMDB) (http://www.ncbi.nlm.nih.gov/Structure/MMDB/mmdb.shtml) [54]. Cn3D 4.3 (http://www.ncbi.nlm.nih.gov/Structure/CN3D/cn3dmac.shtml) was used to display the predicted structure [55].

Phylogenetic analysis

Phylogenetic analyses of all nucleotide sequences were conducted using the neighbor-joining method with 1,000 bootstrap replicates in ClustalW2 (http://blast.ncbi.nlm.nih.gov/Blast.cgi) [55]. The website viewer was also used to display the generated tree for Figure S12. Other tree in supplementary figure (Figure S1) was visualized using TreeViewX (version 0.5.0) [56].

TEM

For TEM analysis, amoebal cells were immersed in a fixative containing 3% glutaraldehyde in 0.1 M phosphate buffered saline (PBS), pH 7.4, for 24 h at 4°C. Following a brief wash with PBS, cells were processed by alcohol dehydration and embedding in Epon 813. Ultrathin cell sections were stained with lead citrate and uranium acetate prior to visualization by electron microscopy (Hitachi H7100; Hitachi, Tokyo, Japan) as described previously [57].

Statistical analysis

Data were compared using a Student's t-test. A P-value of less than 0.05 was considered significant.

Contig sequence accession numbers

The draft genome sequences for the *Neochlamydia* S13 and *Protochlamydia* R18 strains have been deposited in the DNA Data Bank of Japan [DDBJ accession numbers: BASK01000001–BASK01001342 (*Neochlamydia* S13), BASL01000001–BASL01000795 (*Protochlamydia* R18)].

Supporting Information

Figure S1   Phylogenetic analysis of chlamydial 16S rRNA sequences. Bacterial names follow the accession numbers. Arrow denotes *Neochlamydia* 16S rRNA. The gene accession numbers are as follows: *Chlamydia trachomatis* D/UW_3/CX, NC_000117.1; *Chlamydia trachomatis* L2/434/Bu, NC_010287.1; *Chlamydia pneumoniae* TW-183, NC_005043.1; *Simkania* Z gsn131, NC_015713.1; *Parachlamydia acanthamoebae* UV7, NC_015702.1; endosymbiont *Acanthamoeba* UWC22, AF083616.1; endosymbiont *Acanthamoeba* TUME1, AF098330.1; *Neochlamydia hartmannellae*, AF177275.1; *Parachlamydia acanthamoebae* Bn₉, NR_026357.1; *Parachlamydia* Hall's coccus, AF366365.1; *Parachlamydia acanthamoebae* Seine, DQ309029.1; *Protochlamydia naegleriophila* KNic, DQ632609.1.

Figure S2   Venn diagram showing the numbers of common and unique proteins among three chlamydiae. Red, *Chlamydia trachomatis* D/UW_3/CX (NC000117.1). Green, *Protochlamydia* R18 (this study). Blue, *Neochlamydia* S13 (this study).

Figure S3   Predicted genes annotated as glycolytic pathway and TCA cycle with pentose phosphate pathway and Entner-Doudoroff pathway. Black lines with arrows show predicted active modules. Gray lines show incomplete modules. Red names with numbers indicate *Neochlamydia* S13 gene IDs.

Figure S4   Predicted genes annotated as oxidative phosphorylation pathway. Solid lines with arrows show predicted active modules. Gray lines show incomplete modules. Red names with numbers indicate *Neochlamydia* S13 gene IDs.

Figure S5   Predicted genes annotated as fatty acid initiation and elongation. Black lines with arrows show predicted active modules. Red names with numbers indicate *Neochlamydia* S13 gene IDs.

Figure S6   Predicted genes annotated as ATP/ADP translocases (NTTs) (A) and ABC transporters (B). Black lines with arrows show predicted active modules. Gray lines show incomplete modules. Red names with numbers indicate *Neochlamydia* S13 gene IDs.

Figure S7   Comparison of genes encoding a type III secretion system from *Neochlamydia* S13 and *Protochlamydia* UWE25. The type III operon structures of the two chlamydiae are compared. Top panel, T3SS-T1; second panel, T3SS-A1; third panel, T3SS-A2; bottom panel, T3SS-A3. Right scale values show % identity estimated by BLASTp. Each of the gene cluster sequences in *Protochlamydia amoebophila* UWE25 (NC_005861.1) was obtained from NCBI (http:www.ncbi.nlm.nih.gov/genome).

Figure S8   Predicted genes annotated as Sec-dependent type II secretion machinery. Black line with arrow shows predicted active module. Gray boxes show incomplete molecules. Red names with numbers indicate *Neochlamydia* S13 gene IDs.

Figure S9   Comparative analysis of genes encoding type IV secretion machinery from *Protochlamydia* UWE25 and *Protochlamydia* R18. No annotated type IV genes were found in the *Neochlamydia* S13 genome. Blue boxes indicate individual coding regions of the type IV cluster.

Figure S10   Predicted outer membrane structures. Blue molecules were predicted to be active. Gray molecules are absent. Red names with numbers indicate *Neochlamydia* S13 gene IDs. This figure depicts the predicted outer membrane structure based on a findings described by Heinz et al. [34] and previous findings published by Aistleitner et al. [33].

Figure S11   Characterization of a unique molecule with ankyrin domains (NEOS13_0151). (A) Detection of ankyrin domains in the molecule encoded by NEOS13_0151. (B) Alignment scores and 3D prediction. The scores and prediction were performed using the web program, protein BLAST with the MMDB (http://www.ncbi.nlm.nih.gov/Structure/MMDB/docs/mmdb_search.html).

Figure S12   Phylogenetic comparison of the predicted protein sequence encoded by NEOS13_0151 with other eukaryotic proteins. The predicted protein sequence encoded by NEOS13_0151 was phylogenetically compared with previously reported sequences obtained from the GenBank database using ClustalW2. The phylogenetic trees generated from the aligned sequences were constructed by neighbor-joining in ClustalW2, and then visualized with the website viewer.

Table S1   *Neochlamydia* S13 gene IDs with features.

Table S2   Homologs of eukaryote genes in the *Neochlamydia* S13 genome encoding predicted LRR-molecules.

Acknowledgments

We thank the staff at the Department of Medical Laboratory Science, Faculty of Health Sciences, Hokkaido University, for their assistance throughout this study.

References

1. Greub G, Raoult D (2003) History of the ADP/ATP-translocase-encoding gene, a parasitism gene transferred from a *Chlamydiales* ancestor to plants 1 billion years ago. Appl Environ Microbiol 69: 5530–5535.
2. Stephens RS, Kalman S, Lammel C, Fan J, Marathe R, et al. (1998) Genome sequence of an obligate intracellular pathogen of humans: *Chlamydia trachomatis*. Science 282: 754–759.
3. Kalman S, Mitchell W, Marathe R, Lammel C, Fan J, et al. (1999) Comparative genomes of *Chlamydia pneumoniae* and *C. trachomatis*. Nat Genet 21: 385–389.
4. Horn M, Collingro A, Schmitz-Esser S, Beier CL, Purkhold U, et al. (2004) Illuminating the evolutionary history of chlamydiae. Science 304: 728–730.
5. Matsuo J, Kawaguchi K, Nakamura S, Hayashi Y, Yoshida M, et al. (2010) Survival and transfer ability of phylogenetically diverse bacterial endosymbionts in environmental *Acanthamoeba* isolates. Environ Microbiol Rep 2: 524–533.
6. Okude M, Matsuo J, Nakamura S, Kawaguchi K, Hayashi Y, et al. (2012) Environmental chlamydiae alter the growth speed and motility of host acanthamoebae. Microbes Environ 27: 423–429.
7. Lau HY, Ashbolt NJ (2009) The role of biofilms and protozoa in *Legionella* pathogenesis: implications for drinking water. J Appl Microbiol 107: 368–378.
8. Shin S, Roy CR (2008) Host cell processes that influence the intracellular survival of *Legionella pneumophila*. Cell Microbiol 10: 1209–1220.
9. Jules M, Buchrieser C (2007) *Legionella pneumophila* adaptation to intracellular life and the host response: clues from genomics and transcriptomics. FEBS Lett 2581: 2829–2838.
10. Rodríguez-Zaragoza S (1994) Ecology of free-living amoebae. Crit Rev Microbiol 20: 225–241.
11. Declerck P (2010) Biofilms: the environmental playground of *Legionella pneumophila*. Environ Microbiol 12: 557–566.
12. Nunes A, Borrego MJ, Gomes JP (2013) Genomic features beyond *Chlamydia trachomatis* phenotypes: What do we think we know? Infect Genet Evol 16: 392–400.
13. Sixt BS, Heinz C, Pichler P, Heinz E, Montanaro J, et al. (2011) Proteomic analysis reveals a virtually complete set of proteins for translation and energy generation in elementary bodies of the amoeba symbiont *Protochlamydia amoebophila*. Proteomics 11: 1868–1892.
14. Horn M (2008) Chlamydiae as symbionts in eukaryotes. Annu Rev Microbiol 62: 113–131.
15. Heath RJ, Rock CO (1996) Regulation of fatty acid elongation and initiation by acyl-acyl carrier protein in *Escherichia coli*. J Biol Chem 271: 1833–1836.
16. Heath RJ, Rock CO (1996) Inhibition of beta-ketoacyl-acyl carrier protein synthase III (FabH) by acyl-acyl carrier protein in *Escherichia coli*. J Biol Chem 271: 10996–11000.
17. Zempleni J, Wijeratne SS, Hassan YI (2009) Biotin. Biofactors 35: 36–46.
18. Purushothaman S, Annamalai K, Tyagi AK, Surolia A (2011) Diversity in functional organization of class I and class II biotin protein ligase. PLoS One 6: e16850.
19. Haferkamp I, Schmitz-Esser S, Linka N, Urbany C, Collingro A, et al. (2004) A candidate NAD+ transporter in an intracellular bacterial symbiont related to Chlamydiae. Nature 432: 622–6265.
20. Schmitz-Esser S, Linka N, Collingro A, Beier CL, Neuhaus HE, et al. (2004) ATP/ADP translocases: a common feature of obligate intracellular amoebal symbionts related to Chlamydiae and Rickettsiae. J Bacteriol 186: 683–691.
21. Moraes TF, Reithmeier RA (2012) Membrane transport metabolons. Biochim Biophys Acta 1818: 2687–2706.
22. Hinz A, Tampé R (2012) ABC transporters and immunity: mechanism of self-defense. Biochemistry 51:4981–4989.
23. Erkens GB, Majsnerowska M, ter Beek J, Slotboom DJ (2012) Energy coupling factor-type ABC transporters for vitamin uptake in prokaryotes. Biochemistry 51: 4390–4396.
24. Peters J, Wilson DP, Myers G, Timms P, Bavoil PM (2007) Type III secretion à la Chlamydia. Trends Microbiol 15: 241–251.
25. Betts-Hampikian HJ, Fields KA (2010) The Chlamydial Type III Secretion Mechanism: Revealing Cracks in a Tough Nut. Front Microbiol 1:114.
26. Dean P (2011) Functional domains and motifs of bacterial type III effector proteins and their roles in infection. FEMS Microbiol Rev 35: 1100–1125.
27. Greub G, Collyn F, Guy L, Roten CA (2004) A genomic island present along the bacterial chromosome of the *Parachlamydiaceae* UWE25, an obligate amoebal endosymbiont, encodes a potentially functional F-like conjugative DNA transfer system. BMC Microbiol 4: 48.
28. Eugster M, Roten CA, Greub G (2007) Analyses of six homologous proteins of *Protochlamydia amoebophila* UWE25 encoded by large GC-rich genes (lgr): a model of evolution and concatenation of leucine-rich repeats. BMC Evol Biol 7: 231.
29. Casson CN, Copenhaver AM, Zwack EE, Nguyen HT, Strowig T, et al. (2013) Caspase-11 activation in response to bacterial secretion systems that access the host cytosol. PLoS Pathog 9: e1003400.
30. Luo ZQ (2011) Striking a balance: modulation of host cell death pathways by *Legionella pneumophila*. Front Microbiol 2: 36.
31. Arlehamn CS, Evans TJ (2011) *Pseudomonas aeruginosa* pilin activates the inflammasome. Cell Microbiol 13: 388–401.
32. Aistleitner K, Heinz C, Hörmann A, Heinz E, Montanaro J, et al. (2013) Identification and characterization of a novel porin family highlights a major difference in the outer membrane of chlamydial symbionts and pathogens. PLoS One 8: e55010.
33. Heinz E, Tischler P, Rattei T, Myers G, Wagner M, et al. (2009) Comprehensive in silico prediction and analysis of chlamydial outer membrane proteins reflects evolution and life style of the Chlamydiae. BMC Genomics 10: 634.
34. Singer AU, Schulze S, Skarina T, Xu X, Cui H, et al. (2013) A pathogen type III effector with a novel E3 ubiquitin ligase architecture. PLoS Pathog 2013 9: e1003121, 2013s.
35. Zhu Y, Li H, Hu L, Wang J, Zhou Y, et al. (2008) Structure of a *Shigella* effector reveals a new class of ubiquitin ligases. Nat Struct Mol Biol 15:1302–1308, 2008.
36. Zhou JM, Chai J (2008) Plant pathogenic bacterial type III effectors subdue host responses. Curr Opin Microbiol, 11:179–185, 2008.
37. Bierne H, Sabet C, Personnic N, Cossart P (2007) Internalins: a complex family of leucine-rich repeat-containing proteins in *Listeria monocytogenes*. Microbes Infect 9: 1156–1166.
38. Kobe B, Kajava AV (2001) The leucine-rich repeat as a protein recognition motif. Curr Opin Struct Biol 11: 725–732.
39. Rikihisa Y, Lin M, Niu H (2010) Type IV secretion in the obligatory intracellular bacterium *Anaplasma phagocytophilum*. Cell Microbiol 12: 1213–1221.
40. Rikihisa Y, Lin M (2010) *Anaplasma phagocytophilum* and *Ehrlichia chaffeensis* type IV secretion and Ank proteins. Curr Opin Microbiol 13: 59–66.
41. Al-Khodor S, Price CT, Kalia A, Abu Kwaik Y (2010) Functional diversity of ankyrin repeats in microbial proteins. Trends Microbiol 18:132–139.
42. Huang J, Gogarten JP (2007) Did an ancient chlamydial endosymbiosis facilitate the establishment of primary plastids? Genome Biol 8: R99.
43. Moustafa A, Reyes-Prieto A, Bhattacharya D (2008) Chlamydiae has contributed at least 55 genes to Plantae with predominantly plastid functions. PLoS One 3: e2205.
44. Becker B, Hoef-Emden K, Melkonian M (2008) Chlamydial genes shed light on the evolution of photoautotrophic eukaryotes. BMC Evol Biol 8: 203.
45. Price CT, Al-Quadan T, Santic M, Jones SC, Abu Kwaik Y (2010) Exploitation of conserved eukaryotic host cell farnesylation machinery by an F-box effector of *Legionella pneumophila*. J Exp Med 207: 1713–1726.
46. Dugan J, Andersen AA, Rockey DD (2007) Functional characterization of IScs605, an insertion element carried by tetracycline-resistant *Chlamydia suis*. Microbiology 153(Pt 1): 71–79.
47. Greub G, Collyn F, Guy L, Roten CA (2004) A genomic island present along the bacterial chromosome of the *Parachlamydiaceae* UWE25, an obligate amoebal endosymbiont, encodes a potentially functional F-like conjugative DNA transfer system. BMC Microbiol 4: 48.
48. Harada T, Tanikawa T, Iwasaki Y, Yamada M, Imai Y, et al. (2012) Phagocytic entry of *Legionella pneumophila* into macrophages through phosphatidylinositol 3,4,5-trisphosphate-independent pathway. Biol Pharm Bull 35: 1460–1468.
49. Matsuo J, Oguri S, Nakamura S, Hanawa T, Fukumoto T, et al. (2010) Ciliates rapidly enhance the frequency of conjugation between *Escherichia coli* strains through bacterial accumulation in vesicles. Res Microbiol 161: 711–719.
50. Sekizuka T, Yamamoto A, Komiya T, Kenri T, Takeuchi F, et al. (2012) *Corynebacterium ulcerans* 0102 carries the gene encoding diphtheria toxin on a prophage different from the *C. diphtheriae* NCTC 13129 prophage. BMC Microbiol 12: 72.
51. Aziz RK, Bartels D, Best AA, DeJongh M, Disz T, et al. (2008) The RAST Server: rapid annotations using subsystems technology. BMC Genomics 9: 75.
52. Kanehisa M, Goto S (2000) KEGG: kyoto encyclopedia of genes and genomes. Nucleic Acids Res 28: 27–30.
53. Ohtsubo Y, Ikeda-Ohtsubo W, Nagata Y, Tsuda M (2008) GenomeMatcher: a graphical user interface for DNA sequence comparison. BMC Bioinformatics 9: 376.
54. Madej T, Addess KJ, Fong JH, Geer LY, Geer RC, et al. (2012) MMDB: 3D structures and macromolecular interactions. Nucleic Acids Res 40(Database issue): D461–464.
55. Larkin MA, Blackshields G, Brown NP, Chenna R, McGettigan PA, et al. (2007) Clustal W and Clustal X version 2.0. Bioinformatics 23: 2947–2948.
56. Peterson MW, Colosimo ME (2007) TreeViewJ: An application for viewing and analyzing phylogenetic trees. Source Code Biol Med 2: 7.
57. Matsuo J, Hayashi Y, Nakamura S, Sato M, Mizutani Y, et al. (2008) Novel *Parachlamydia acanthamoebae* quantification method based on coculture with amoebae. Appl Environ Microbiol 74: 6397–6404.

Author Contributions

Conceived and designed the experiments: HY. Performed the experiments: JM TY KI SN MY KT. Analyzed the data: KI MK HY TS FT HN KH CS. Wrote the paper: HY.

15

Characteristics of *Salmonella enterica* Serovar 4,[5],12:i:- as a Monophasic Variant of Serovar Typhimurium

Noriko Ido[1], Ken-ichi Lee[2], Kaori Iwabuchi[3], Hidemasa Izumiya[4], Ikuo Uchida[5], Masahiro Kusumoto[2], Taketoshi Iwata[2], Makoto Ohnishi[4], Masato Akiba[2,6]*

1 Iwate Prefecture Central Livestock Hygiene Service Center, Iwate, Japan, **2** Bacterial and Parasitic Disease Research Division, National Institute of Animal Health, National Agriculture and Food Research Organization, Ibaraki, Japan, **3** Research Institute for Environmental Sciences and Public Health of Iwate Prefecture, Iwate, Japan, **4** Department of Bacteriology, National Institute of Infectious Diseases, Tokyo, Japan, **5** Hokkaido Research Station, National Institute of Animal Health, National Agriculture and Food Research Organization, Hokkaido, Japan, **6** Graduate School of Life and Environmental Sciences, Osaka Prefecture University, Osaka, Japan

Abstract

Salmonella enterica subspecies *enterica* serovar 4,[5],12:i:- (*S.* 4,[5]12:i:-) is believed to be a monophasic variant of *S. enterica* serovar Typhimurium (*S.* Typhimurium). This study was conducted to corroborate this hypothesis and to identify the molecular and phenotypic characteristics of the *S.* 4,[5]12:i:- isolates in Japan. A total of 51 *S.* 4,[5]12:i:- isolates derived from humans, cattle, swine, chickens, birds, meat (pork), and river water in 15 prefectures in Japan between 2000 and 2010 were analyzed. All the *S.* 4,[5],12:i:- isolates were identified as *S.* Typhimurium by two different polymerase chain reactions (PCR) for identification of *S.* Typhimurium. Of the 51 *S.* 4,[5],12:i:- isolates, 39 (76.5%) harbored a 94-kb virulence plasmid, which is known to be specific for *S.* Typhimurium. These data suggest that the *S.* 4,[5],12:i:- isolates are monophasic variants of *S.* Typhimurium. The flagellar phase variation is induced by three adjacent genes (*fljA*, *fljB*, and *hin*) in the chromosome. The results of PCR mapping of this region and comparative genomic hybridization analysis suggested that the deletion of the *fljAB* operon and its flanking region was the major genetic basis of the monophasic phenotype of *S.* 4,[5],12:i:-. The *fljAB* operon and *hin* gene were detectable in eight of the *S.* 4,[5],12:i:- isolates with common amino acid substitutions of A46T in FljA and R140L in Hin. The introduction of these mutations into *S.* Typhimurium isolates led to the loss of selectability of isolates expressing the phase 2 H antigen. These data suggested that a point mutation was the genetic basis, at least in part, of the *S.* 4,[5],12:i:- isolates. The results of phenotypic analysis suggested that the *S.* 4,[5],12:i:- isolates in Japan consist of multiple distinct clones. This is the first detailed characterization of the *S.* 4,[5],12:i:- isolates derived from various sources across Japan.

Editor: Wolfgang Köster, University of Saskatchewan, Canada

Funding: This work was supported by grants-in-aid from the Ministry of Agriculture, Forestry and Fisheries of Japan (Research project for ensuring food safety from farm to table 24-7304), and the Ministry of Health, Labour, and Welfare of Japan (H24-Shokuhin-Ippan-008). The funders had no role in study design, data collection and analysis, decision to publish, or preparation of the manuscript.

Competing Interests: The authors have declared that no competing interests exist.

* Email: akiba@affrc.go.jp

Introduction

Nontyphoidal salmonellae are one of the most common cause of bacterial gastroenteritis in humans as well as salmonellosis in domestic and wild animals worldwide [1]. Serotyping is widely used as an epidemiological typing method to subdivide *Salmonella* species [2]. Each serovar is identified by the combination of lipopolysaccharide moieties on the cell surface (O antigens) and one or two different flagellar proteins (H antigens). Many serovars have the ability to express two different flagellin proteins, although individual cells can express one of the two flagellins [3]. According to the White-Kaufmann-Le Minor scheme, more than 2500 serovars to date have been recognized in the genus *Salmonella* [4].

The incidence of human salmonellosis caused by *Salmonella enterica* subspecies *enterica* serovar 4,[5],12:i:- (*S.* 4,[5],12:i:-) has been increasing in Europe, North and South America, and Asia since the mid-1990s [5–9]. This serovar is currently among the 10 most common serovars responsible for human infections in a variety of countries, including the second and sixth most prevalent serovar in Germany [10] and the United States [11], respectively. *S.* 4,[5],12:i:- was also reported as the fourth most common

serovar in slaughtered pigs in the EU [10]. In Japan, the rate of distribution of this serovar was more than 2% (ninth most prevalent) in 2009 for the first time, and then remained relatively high to date [12]. Larger outbreaks caused by this serovar have been reported in the United States and Luxemburg [13,14].

S. 4,[5],12:i:- does not appear in the White-Kaufmann-Le Minor scheme [4] and appears to be a monophasic variant of other biphasic serovars, which have lost phase 2 flagellin or the necessary switching mechanism of phase variation. Seven serovars of *S. enterica* subsp. *enterica* with same O and phase 1 H antigens are possible ancestors of this serovar, including Typhimurium (*S.* Typhimurium), Lagos, Agama, Farsta, Tsevie, Gloucester, and Tumodi [4]. Among these, *S.* Typhimurium is commonly isolated from humans, animals, and the environment, whereas the others are rarely isolated.

S. Typhimurium is believed to be an ancestor of *S.* 4,[5],12:i:- based on the following evidence. *S.* Typhimurium-specific sequences have been detected in *S.* 4,[5],12:i:- [15]. Some of the *S.* 4,[5],12:i:- isolates showed the same lysogenic patterns as those of *S.* Typhimurium by phage typing and displayed pulsed-field gel electrophoresis patterns identical or similar to those of *S.*

Typhimurium isolates [15–18]. Different deletions and mutations can be responsible for the lack of phase 2 flagellin expression among the S. 4,[5],12:i:- isolates. Specifically, some isolates from Spain lack 16 genes, including the fljAB operon and flanking genes, which encode phase 2 flagellin expression-related proteins [19]. Some of the S. 4,[5],12:i:- isolates from the United States appeared to have smaller deletions or point mutations in the fljAB operon and/or flanking genes not identified by deoxyribonucleic acid (DNA) probes [18]. To date, no specific point mutations affecting the phase 2 flagellin expression have been identified.

In the present study, we characterized the S. 4,[5],12:i:- isolates derived from various sources in Japan for the following purpose: (i) to corroborate the hypothesis that the S. 4,[5],12:i:- isolates are monophasic variants of S. Typhimurium, (ii) to elucidate the genetic basis of the monophasic phenotype of the S. 4,[5],12:i:- isolates, (iii) to identify the molecular and phenotypic characteristics of the S. 4,[5],12:i:- isolates in Japan because only limited information is currently available.

Materials and Methods

Bacterial isolation, identification, and typing

The S. 4,[5],12:i:- isolates used in this study are listed in Table 1. A total of 51 isolates were derived from humans, cattle, swine, chickens, birds, meat (pork), and river water from 15 prefectures in Japan between 2000 and 2010. The isolates from humans and cattle were obtained from fecal samples of patients or affected animals with different sporadic infections. The swine isolates were obtained from fecal samples of healthy or affected animals. The isolates from chickens and crows were obtained from fecal samples or organs of healthy birds. The isolates from a penguin and a parrot were obtained from organs of diseased birds. The isolates from pork and river water were obtained from a previous monitoring study. S. Typhimurium strain LT2 [20] was used as a positive control for polymerase chain reaction (PCR) analysis or as a reference for comparative genomic hybridization analysis, and strains L-3900 and L-3287 were used to introduce point mutations identified among the S. 4,[5],12:i:- isolates by gene replacement. In Japan, L-3900 was isolated from cattle in 2010, whereas L-3287 was isolated from chicken in 2001. Both these strains were susceptible to kanamycin. The isolation of S. enterica was performed by the staff of the local Institute of Public Health or local Animal Hygiene Service Centers for diagnostic or monitoring purposes. Patient information was anonymized and de-identified prior to analysis. The approval from the Institutional Animal Care and Use Committee is not required in case of isolation for diagnostic purpose. Isolates H1–5 and C1–10 were the same as H1–5 and C1–10, respectively, as described in a previous report [7]. Salmonella spp. were identified based on colony morphology on selective media and biochemical properties, as previously described [21]. Serovar identification was performed by microtiter and slide agglutination methods according to the latest version of the White-Kaufmann-Le Minor scheme [4] using antiserum (Denka Seiken Co., Ltd., Tokyo, Japan). Phage typing was performed using S. Typhimurium typing phages according to the methods and schemes previously described by Anderson et al. [22]. All isolates were maintained at −80°C in Luria–Bertani (LB) broth (Becton, Dickenson and Company, Sparks, MD, USA) containing 25% (v/v) glycerol.

Antimicrobial susceptibility testing

The Kirby–Bauer disc diffusion test was performed using Mueller–Hinton agar plates (Becton, Dickenson and Company) according to Clinical and Laboratory Standard Institute standards [23] using the following antimicrobials: ampicillin (10 µg), cefazolin (30 µg), kanamycin (30 µg), streptomycin (10 µg), tetracycline (30 µg), chloramphenicol (30 µg), fosfomycin (50 µg), colistin (10 µg), sulfamethizole (250 µg), and nalidixic acid (30 µg) (Becton, Dickinson and Company).

Plasmid isolation

Plasmid DNA was isolated by the method described by Kado and Liu [24] followed by phenol–chloroform extraction. The Bac-Tracker Supercoiled DNA ladder (Epicentre Biotechnologies, Madison, WI, USA) and a 94-kb plasmid from S. Typhimurium LT2 were used as size markers.

PCR and sequencing

All primers used in this study were purchased from Hokkaido System Science Co., Ltd. (Hokkaido, Japan) and are listed in Table S1. A single colony of each bacterial isolate was suspended in 50 µL of 25 mM NaOH and boiled for 5 min. After addition of 4 µL of 1 M Tris-HCl (pH 8.0), the suspension was centrifuged and the supernatant was used as a template DNA. Amplification was performed using an iCycler apparatus (Bio-Rad Laboratories, Hercules, CA, USA). Takara Ex Taq (Takara Bio Inc., Shiga, Japan) was used as DNA polymerase for each monoplex PCR. The Salmonella serovar Typhimurium Identification Kit (Takara Bio Inc.) was used to detect S. Typhimurium-related genes, including STM0292, STM2235, and STM4493, by multiplex PCR (m-PCR) as previously described [25]. Some of the PCR products were purified using the illustra ExoStar Kit for Enzymatic PCR and Sequencing Clean-up (GE Healthcare UK Ltd., Buckinghamshire, UK). Nucleotide sequences were determined on both strands using an Applied Biosystems 3130 xl genetic analyzer with the BigDye Terminator cycle sequencing kit (version 3.1; Applied Biosystems, Foster City, CA, USA). The sequences were assembled with Sequencher version 4 (Hitachi Solutions, Kanagawa, Japan) and the DNA alignments and deduced amino acid sequences were examined using the Basic Local Alignment Search Tool (http://blast.ncbi.nlm.nih.gov/Blast.cgi).

Comparative genomic hybridization (CGH) analysis

Copy number analysis of the selected S. 4,[5],12:i:- isolates was performed using the whole genomic CGH array (Roche Nimble-Gen, Inc., Madison, WI, USA) at their facility in Iceland according to previously published methods with some modifications [26]. In brief, a tiling array was designed with a mean probe density of 1 probe/10 bp, 50–75-mer length using the S. Typhimurium strain LT2 sequences of the chromosome (AE006468) and pSLT plasmid (AE006471). Labeling was performed using the NimbleGen Dual Color Labeling Kit according to the manufacturer's protocols. In brief, each DNA sample (1 µg) was denatured at 98°C in the presence of one optical density of 5′-Cy3- or 5′-Cy5-labeled random nonamer. The denatured sample was chilled on ice and then incubated with 100 U of (exo-) Klenow fragment and dNTP mix for 2 h at 37°C. Reactions were terminated by addition of 0.5 M ethylenediaminetetraacetic acid (pH 8.0) and the end products were precipitated with isopropanol and resuspended in water. The Cy-labeled test and reference samples (Cy3 and Cy5, respectively) were combined (31 µg each) and dried down by vacuum centrifugation. Each sample was rehydrated in 5.6 µL of PCR grade water included in the NimbleGen Sample Tracking Control Kit and added to the hybridization buffers included in the NimbleGen Hybridization Buffer Kit, denatured at 95°C for 5 min, and then cooled to 42°C. Hybridizations were conducted for 40–72 h at 42°C using the

Table 1. *Salmonella enterica* serovar 4,[5],12:i:- isolates used in this study.

Isolates	Source	Year	PCR results[a]								94 kb plasmid[b]	Phage type[c]	LDC[d]	Resistance profile[e]
			m-PCR	IS*200*	up-*fljA*	*fljA-fljB*	*fljB-hin*	*hin*-down	*fin*	*spvB*				
H1~4	Human	2006	+	+	-	-	-	+	-	+	+	193	+	-
H5	Human	2007	+	+	-	-	-	-	-	-	-	193	+	ASSu
H6	Human	2008	+	+	-	-	-	+	-	+	+	RDNC-a	+	-
H7	Human	2003	+	+	-	-	-	+	-	+	+	193	+	-
H8	Human	2007	+	+	+	+	+	+	-	+	+	26	+	-
H9~11	Human	2007	+	+	-	-	-	+	-	+	+	RDNC-a	+	-
H12	Human	2004	+	+	-	-	-	+	-	-	-	RDNC-c	+	-
H13	Human	2007	+	+	-	-	-	-	-	-	-	193	+	SSuT
H14	Human	2002	+	+	-	-	-	-	-	+	+	UT	+	ASuT
C1	Cattle	2003	+	+	-	-	-	+	-	+	+	RDNC-a	-	-
C2	Cattle	2005	+	+	-	-	-	+	-	+	+	RDNC-a	-	-
C3~4	Cattle	2007	+	+	-	-	-	+	-	+	+	RDNC-a	-	-
C5~8	Cattle	2008	+	+	-	-	-	+	-	+	+	RDNC-a	-	-
C9~10	Cattle	2008	+	+	-	-	-	-	-	+	+	RDNC-a	+	A
C11	Cattle	2004	+	+	-	-	-	+	-	+	+	RDNC-a	+	-
C12	Cattle	2005	+	+	-	-	-	+	-	+	+	120	+	-
C13	Cattle	2005	+	+	+	+	+	+	-	+	+	RDNC-b	+	-
C14	Cattle	2008	+	+	-	-	-	-	-	-	-	UT	+	ASuT
C15	Cattle	2007	+	+	-	-	-	+	-	+	+	RDNC	+	-
C16	Cattle	2010	+	+	-	-	-	+	-	+	+	RDNC-a	+	-
C17	Cattle	2010	+	+	-	-	-	+	-	+	+	RDNC-b	+	A
S1	Swine	2008	+	+	-	-	-	-	-	-	-	UT	+	ASSuT
S2	Swine	2009	+	+	-	-	-	-	-	-	-	UT	+	ASSu
S3	Swine	2002	+	+	-	-	-	-	-	-	-	RDNC-d	+	Ssu
S4	Swine	2003	+	+	-	-	-	-	-	-	-	RDNC-d	+	SSuT
S5	Swine	2008	+	+	-	-	-	+	-	-	-	193	+	SSuT
S6	Swine	2009	+	+	-	-	-	-	+	+	+	27	+	ASSuT
K1	Chicken	2001	+	+	-	-	-	+	-	+	+	RDNC-b	+	-
K2	Chicken	2004	+	+	-	-	-	+	-	-	-	RDNC	+	-
K3	Chicken	2005	+	+	-	-	-	+	-	-	-	RDNC-c	+	-
K4	Chicken	2006	+	+	-	-	-	+	-	-	-	RDNC-c	+	-
K5	Chicken	2010	+	+	-	-	-	+	-	+	+	RDNC	+	ASuT
B1	Penguin	2009	+	+	+	+	+	+	-	+	+	RDNC	+	-

Table 1. Cont.

Isolates	Source	Year	PCR results[a]								94 kb plasmid[b]	Phage type[c]	LDC[d]	Resistance profile[e]
			m-PCR	IS200	up-fljA	fljA-fljB	fljB-hin	hin-down	fin	spvB				
B2~3	Crow	2000	+	+	+	+	+	+	-	+	+	RDNC-e	+	-
B4	Parrot	2005	+	+	+	+	+	+	-	+	+	RDNC-e	+	-
M1	Pork	2005	+	+	-	-	-	+	-	+	+	RDNC-a	+	-
M2	Pork	2007	+	+	-	-	-	+	-	+	+	RDNC-a	+	-
R1	River water	2007	+	+	+	+	+	+	-	+	+	26	+	-
R2	River water	2007	+	+	-	-	-	+	-	+	+	RDNC-a	+	ASu
R3	River water	2007	+	+	+	+	+	+	-	+	+	26	+	-

a m-PCR, multiplex PCR to identify S. Typhimurium [25]; IS200, PCR to identify S. Typhimurium [15]; up-fljA, boundary region of fljA and its upstream intergenic region; fljA-fljB, boundary region of fljA and fljB; fljB-hin, boundary region of fljB and hin; hin-down, boundary region of hin and its downstream intergenic region; +, presence; -, absence.
b +, presence; -, absence.
c RDNC, reacted but did not conform; RDNC-a-e, same letter indicates the same lysogenic patterns among RDNC isolates.
d LDC, lysine decarboxylase; +, positive; -, negative.
e A, ampicillin; S, streptomycin; Su, sulfamethizole; T, tetracycline; -, pansusceptible.

NimbleGen Hybridization System. The arrays were washed using the NimbleGen Wash Buffer Kit and immediately dried down by centrifugation. Arrays were scanned at a resolution of 5 μm using the GenePix4000B scanner (Axon Instruments, Molecular Devices Corp., Sunnyvale, CA, USA). Data were extracted from scanned images using NimbleScan software (Roche NimbleGen, Inc.), which allows for automated grid alignment, extraction, and generation of data files.

Gene replacement in S. Typhimurium

The primer pair GR1F and GR1R was used to amplify a region containing a point mutation in fljA, and the primer pair GR2F and GR2R was used to amplify a region containing a point mutation in hin from the S. 4,[5],12:i:- isolates by PCR (Fig. 1, Table S1). After digestion with the XbaI and HindIII restriction enzymes, the resulting fragment was cloned to the temperature-sensitive vector pTH18ks1 [27] and used as a vector for gene replacement. S. Typhimurium strains L-3900 and L-3287 were transformed with one of each vector by electroporation. The cells were spread on LB agar (Becton, Dickinson and Co.) supplemented with kanamycin and incubated at 28°C for 18 h. The colonies were then streaked on the same agar plates pre-warmed to 42°C and incubated at 42°C for 18 h. The single crossover strains were purified under the same conditions and then passaged at 28°C several times. The double crossover strains were screened by allele-specific PCR using the primer pair SNP1 and GR1R for detection of a point mutation in fljA and the primer pair SNP2 and GR2R for detection of a point mutation in hin (Fig. 1, Table S1). The introduced point mutation was verified by PCR and sequencing using appropriate primers. S. Typhimurium derivatives with the point mutation in hin were transformed with the vector for the fljA gene replacement. Double crossover strains were selected using the abovementioned procedure to obtain double mutants.

Estimation of phase variation frequency

Phase variation frequency was estimated by the method described by Stocker [28] with minor modifications. In brief, the tested strains expressing the H-i antigen (phase 1) were serially passaged in LB broth media until the estimated number of generations reached 110. The culture was diluted to yield approximately 100 colonies on LB agar in 90-mm petri dishes and spread with a plastic spreader. The plates were incubated at 30°C for 18 h until the colonies grew to 1 mm in diameter. The plates were cooled in a refrigerator, and then 7 mL of semi-solid agar containing 0.35% Bacto Agar (Becton, Dickinson and

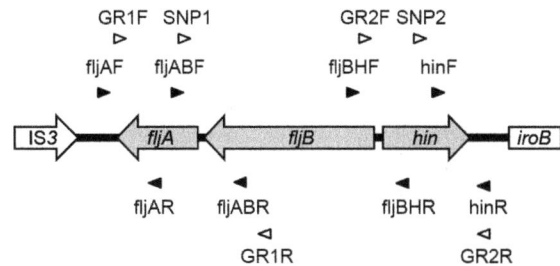

Figure 1. Schematic view of genetic organization of the chromosomal region related to flagellar phase variation of S. Typhimurium. Gray arrows indicate gene related to phase variation. Closed triangles indicate the primer locations for polymerase chain reaction mapping. Open triangles indicate the primer locations for mutant construction.

Table 2. Chromosomal genes that lacks in the CGH tested isolates.

NC_00319 tag number	Position (start-end)	Size (bp)	Description	Presence/Absence[a]		
				C1	C9	C13
STM0276–STM0279	316895–319135	2241	putative cytoplasmic/periplasmic proteins	+	+	-
STM0893–STM0929	962638–1005280	42643	Fels-1 prophage	-	-	-
STM1011–STM1019	1104868–1109917	5050	part of Gifsy-2 prophage	+	+	-
STM1555–STM1557	1632448–1635078	2631	putative Na$^+$/H$^+$ antiporter and others	+	+	-
STM2585–STM2636	2730851–2776671	45821	Gifsy-1 prophage	-	-	-[b]
STM2694–STM2739	2844326–2877883	33558	Fels-2 prophage	-	-	-
STM2740–STM2771	2877884–2914231	36348	fljAB and upstream genes	-	-	+
STM2951	3094339–3096696	2358	ygcF	+	+	-
STM3113	3271613–3272493	881	nupG	-	-	+
STM3255–STM3260	3425101–3430141	5041	putative phosphotransferase system and others	-	-	+

[a]Chromosomal genes with log2 ratios < -0.5 were identified as absent genes.
[b]Five genes (7782 bp) of the Gifsy-1 prophage were absent.

Company) was pipetted on top. This semi-solid agar contained 0.7% (v/v) anti-H-i serum (Denka Seiken Co.). Once solidified, the plates were incubated at 37°C for 1–2 h. Colonies expressing the H-1,2 (phase 2) antigen were found to be surrounded by a wide zone of opacity with an indefinite edge, indicating that the organisms were swarming out into the semi-solid agar. However, colonies expressing the phase 1 antigen were surrounded by a narrow dense zone of opacity with a clear-cut edge. The frequencies of swarming colonies among ca. 7000 colonies were calculated, and then the frequency was divided by the number of generations to determine the frequency of phase variation per generation. Each experiment was performed thrice. Differences in the results were tested using the two-tailed unpaired Student's *t* test.

Results and Discussion

S. 4,[5],12:i:- is very likely a monophasic variant of S. Typhimurium

To prove the hypothesis that S. 4,[5],12:i:- is a monophasic variant of S. Typhimurium, several molecular characteristics of the S. 4,[5],12:i:- isolates were investigated. In the m-PCR to identify S. Typhimurium, three serovar-related genomic regions were successfully amplified from all the S. 4,[5],12:i:- isolates tested in this study (Table 1). No false positives were observed using 117 *Salmonella* serovars, with the exception of S. 4,[5],12:i:- [25], which strongly suggested that S. 4,[5],12:i:- originated from S. Typhimurium. The results of PCR analysis to detect the *fliA–fliB* intergenic region also support this statement. The location of IS*200* between the genes *fliA* and *fliB* can be used as a specific marker for S. Typhimurium [29]. The amplicon sizes from the *fliA–fliB* intergenic regions from S. Typhimurium and other serovars were expected to be 1000 and 250 bp, respectively [15]. A 1000-bp amplicon was successfully detected in all the S. 4,[5],12:i:- isolates. These data suggest that S. 4,[5],12:i:- is a monophasic variant of S. Typhimurium. In other words, the S. Typhimurium-specific m-PCR, and PCR might be useful to verify

that the tested S. 4,[5],12:i:- isolate is a monophasic variant of S. Typhimurium.

As shown in Table 1, 39 (76.5%) of the 51 S. 4,[5],12:i:- isolates harbored a 94-kb plasmid with or without other plasmids. The *spvB* gene, which is a marker of the S. *enterica* virulence plasmid [30], was detected by PCR in all of the isolates with the 94-kb plasmid, thereby supporting the possibility that these isolates originated from S. Typhimurium. However, the lack of this plasmid does not contradict the possibility that the isolate is S. Typhimurium. The prevalence of the virulence plasmid in the S. Typhimurium isolates obtained from swine with systemic infections was 92%, whereas less than 20% in isolates from diarrhea samples and animals without any symptoms [31].

Deletion is a major basis of the monophasic phenotype of the S. 4,[5],12:i:- isolates

Most S. *enterica* serovars possess two different flagellin proteins, including FliC (phase 1) and FljB (phase 2), which are encoded by the genes *fliC* and *fljB*, respectively. Flagellar phase variation is induced by inversion of the genetic region called the H segment, which contains the *hin* gene encoding for DNA invertase and the promoter for the *fljB* gene. The *fljB* constitutes an operon with the *fljA* gene, which encodes a negative regulator of *fliC* expression. FljA binds to the operator region of FliC mRNA and inhibits its translation, leading to the rapid degradation of FliC mRNA. When the H segment is in the "on" state, both *fljB* and *fljA* are transcribed, resulting in synthesis of phase 2 flagellin and inhibition of phase 1 flagellin. However, when the H segment is switched to the "off" state, neither *fljB* nor *fljA* are transcribed, resulting in the synthesis of phase 1 flagellin only (Fig. S1) [32,33].

To determine whether the S. 4,[5],12:i:- isolates maintained the genetic structure of the *fljAB–hin* region, PCR mapping [7] to detect the boundary region of each gene was performed using the S. Typhimurium LT2 DNA sequence as a reference. The amplification targets were as follows; up-*fljA*, the boundary region of *fljA* and its upstream intergenic region; *fljA–fljB*, the boundary region of *fljA* and *fljB*; *fljB–hin*, the boundary region of *fljB* and *hin*; and *hin*-down, the boundary region of *hin* and its downstream

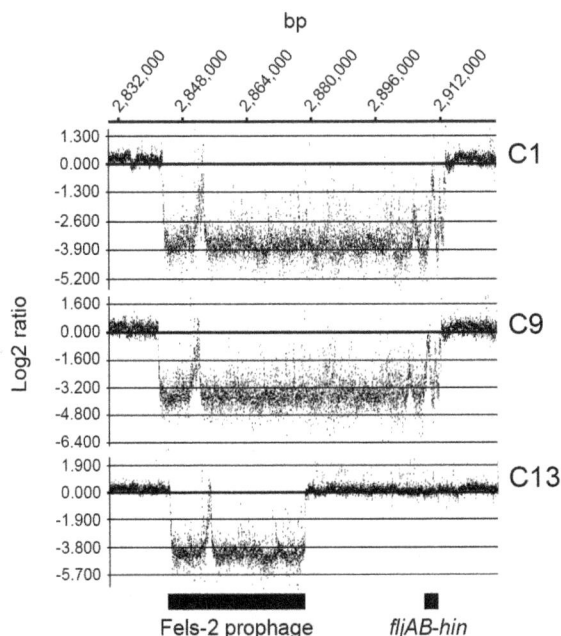

Figure 2. Partial quantitative data from the comparative genomic hybridization of the S. 4,[5],12:i:- isolates. The ruler indicates the nucleotide number of S. Typhimurium LT2 chromosome (AE006468). The vertical scale indicates the log2 ratio of the signal intensities. C1, C9, and C13 indicate the name of the isolates listed in Table 1. The underlying bold lines indicate the location of specific genetic structures in the chromosome.

intergenic region (Fig. 1). As shown in Table 1, three amplification patterns were observed: positive for only *hin*-down (31 isolates), all negative (12 isolates), and all positive (eight isolates). In eight of the positive isolates, the whole *fljAB–hin* structure was detectable with two common amino-acid substitutions: A46T in FljA and R140L in Hin. These amino-acid substitutions were not observed in six S. Typhimurium wild-type strains isolated in Japan including L-3900 and L-3287 (data not shown).

To compare the whole genome sequences of the S. 4,[5],12:i:- isolates with that of S. Typhimurium strain LT2, one of the each

representative isolate from the three amplification patterns determined by PCR mapping was analyzed by CGH. As shown in Table 2 and Fig. S2, whole sequences of prophages Fels-1 and Fels-2 were not detectable among the three isolates. In strains C1 and C9, an additional 36-kb sequence downstream of Fels-2 containing the *fljAB* operon was not detectable (Fig. 2). In addition, the whole sequence of the Gifsy-1 prophage was not detectable in strains C1 and C9, whereas part of the Gifsy-1 prophage sequence was not detected in strain C13 (Table 2 and Fig. S2). Among these, a broad scale deletion event stretching from the Fels-2 prophage to the *fljAB–hin* region was determined as the genetic basis of the monophasic phenotype of the S. 4,[5],12:i:- isolates C1 and C9.

S. Typhimurium strain LT2 contains the DNA invertase gene *fin*, which contributes to the phase induction of H antigens other than *hin*. The *fin* gene is located in the Fels-2 prophage of S. Typhimurium LT2 [34]. All the S. 4,[5],12:i:- isolates were PCR-negative for *fin*, except for one isolate obtained from a swine, suggesting that the Fels-2 prophage was not distributed among most of the S. 4,[5],12:i:- isolates, as indicated by the CGH analysis of representative isolates.

In total, the presence/absence patterns of isolates C1 and C9 were identical, whereas strain C13 was different from other strains. Approximately 80% or more of the total length of the absence region corresponded to the Fels-1, Fels-2, Gifsy-1, and Gifsy-2 prophages (Table 2). Garaizar et al. [19] reported the deletion of most of the Fels-1 and Fels-2 sequences and a partial sequence of Gifsy-1 among the S. 4,[5],12:i:- isolates in Spain. Prophage sequences may be selectively neutral for this serovar.

Point mutations reduce the phase variation frequency of S. Typhimurium

To manifest the effect of the amino acid substitutions, A46T in FljA and R140L in Hin observed in the eight S. 4,[5],12:i:- isolates by phase variation frequency analysis, these point mutations were independently or simultaneously introduced to S. Typhimurium strains L-3900 and L-3287. As shown in Table 3, phase variation was successfully observed in both parental strains and the frequency from phase 1 to phase 2 of L-3900 was 1.84×10^{-4}, whereas that of L-3287 was less than that of the detection limit (10^{-6}). The phase variation frequency of the L-3900 *fljA* mutant was significantly lower ($p = 0.04$) than that of the parental strain.

Table 3. Expression of phase 2 antigen and phase variation frequency.

Parental strain	Genotype[a]		Selectability of phase 2[b]	Phase variation frequency[c]
	fljA	*hin*		
L-3900	WT	WT	+	1.84×10^{-4}
	A46T	WT	+	$9.11 \times 10^{-5*}$
	WT	R140L	-	$<10^{-6}$
	A46T	R140L	-	$<10^{-6}$
L-3287	WT	WT	+	$<10^{-6}$
	A46T	WT	-	ND
	WT	R140L	-	ND
	A46T	R140L	-	ND

[a]WT, wild type; A46T and R140L, amino acid substitutions.
[b]+, selectable; -, not detected.
[c]ND, not done;
*, significantly lower than parental strain ($p = 0.04$).

An isolate expressing the phase 2 H antigen was successfully selected from this mutant, but not from any other mutants. These data suggest that the point mutations reduced the phase variation frequency and may be the genetic basis of the monophasic phenotype of all *fljAB–hin* detectable isolates.

Hin invertase catalyzes DNA inversion of the H segment. This site-specific recombination event controls the alternate expression of two flagellin genes by reversing the orientation of the *fljB* promoter [32,33]. The interaction between Hin invertase and the target DNA has been fully elucidated. The 52 carboxyl-terminal residues is the DNA-binding domain of the Hin invertase. Particularly, the sequence G^{139}-R^{140}-P^{141}-R^{142} is essential to maintain DNA binding ability. The deletion of residues G^{139} and R^{140} abolished the sequence-specific binding to DNA [35]. The *fin* gene located in the Fels-2 prophage region encodes an invertase that can support inversion of the H segment without Hin invertase. As both the strains L-3900 and L-3287 were found to be negative for the *fin* gene by PCR, R140L in Hin may diminish the DNA-binding ability; thus, resulting in a reduction in the phase variation frequency. No information is available regarding the effect of the A46T substitution in FljA on the phase variation frequency to date.

S. 4,5,12:i:- isolates in Japan consist of multiple distinct clones

As shown in Table 1, phage typing of the 51 S. 4,[5],12:i:- isolates examined using the S. Typhimurium typing phages identified four phage types: DT193 (eight isolates), DT26 (three isolates), DT27 (one isolate), and DT120 (one isolate). The remaining 34 isolates were RDNC (reacts with phages but does not confirm to a recognized pattern) and four isolates were UT (untypable). Among the 34 RDNC isolates, five lysogenic patterns were observed in more than one isolate, which were named RDNC-a–e. All eight isolates of lysine decarboxylase-negative S. 4,[5],12:i:- belonged to RDNC-a. In Germany and Switzerland, DT193 was the most prevalent definitive phage type among the S. 4,[5],12:i:- isolates [6,17]. Most of the S. 4,[5],12:i:- isolates from Spain belonged to the definitive phage-type U302 [16]. The S. 4,[5],12:i:- isolates from Japan appeared to consist of multiple clones with greater variation than those from European countries.

Antimicrobial susceptibility testing using nine antimicrobials showed that 14 (27.5%) of the 51 S. 4,[5],12:i:- isolates were resistant to one or more antimicrobials among ampicillin, streptomycin, sulfamethizole, and tetracycline. The remaining 37 isolates were pan-susceptible (Table 1). According to the data published to date, most of the S. 4,[5],12:i:- isolates from Europe (i.e., Spain, Germany, and Switzerland) appeared to have a multidrug resistance phenotype, whereas most of the S. 4,[5],12:i:-

isolates from North and South American countries (i.e., USA and Brazil) appeared to be pan-susceptible or resistant to only a few antimicrobials [6,9,16,17]. In Japan, pan-susceptible isolates were dominant, although multidrug resistance isolates with resistance to up to four antimicrobials were detected. All of the six isolates obtained from swines exhibited resistance to multiple antimicrobials. This may reflect the extensive use of antibiotics as feed additives in the pig industry in Japan [36].

Conclusions

The results of the molecular characterization of the 51 S. 4,[5],12:i:- isolates derived from various sources in Japan suggested that these isolates were very likely monophasic variants of S. Typhimurium. Deletion and point mutations were the bases of the monophasic phenotype of the S. 4,[5],12:i:- isolates. The results of phenotypic characterization suggested these isolates consisted of multiple distinct clones.

Supporting Information

Figure S1 A model for the molecular mechanism of phase variation in *Salmonella* cited from Yamamoto and Kutsukake [33] with slight modifications. This system consists of two major parts: (i) the switching mechanisms of *fljB* promoter orientation by inversion of H segments and (ii) the FljA-mediated translational repression of *fliC* mRNA, leading to the rapid degradation of the mRNA. IR, inverted repeat; *fljBp*, *fljB* promoter; *fliCp*, *fliC* promoter; OP, operator region.

Figure S2 Quantitative data from the comparative genomic hybridization of the S. 4,[5],12:i:- isolates. The ruler indicates the nucleotide number of S. Typhimurium LT2 chromosome (AE006468). The vertical scale indicates the log2 ratio of the signal intensities. C1, C9, and C13 indicate the name of the isolates listed in Table 1. The underlying bold lines indicate the locations of prophages in the chromosome.

Table S1 Primers used in this study.

Author Contributions

Conceived and designed the experiments: NI MA. Performed the experiments: NI KL HI MA. Analyzed the data: NI KL MK TI MA. Contributed reagents/materials/analysis tools: NI KI IU MO MA. Wrote the paper: NI MA.

References

1. Majowicz SE, Musto J, Scallan E, Angulo FJ, Kirk M, et al. (2010) The global burden of nontyphoidal *Salmonella* gastroenteritis. Clin Infect Dis 50: 882–889.
2. Grimont PAD, Grimont F, Bouvet P (2000) Taxonomy of the Genus *Salmonella*. In: Wray C, Wray, A., etidor. *Salmonella* in Domestic Animals. Wallingford: CABI Publishing. pp. 1–17.
3. Iino T (1969) Genetics and chemistry of bacterial flagella. Bacteriol Rev 33: 454–475.
4. Grimont P, Weil F (2007) Antigenic formulae of the *Salmonella* serovars, 9th ed. World Health Organization Centre for Reference and Research on *Salmonella*. Paris, France: Pasteur Institute.
5. Dionisi AM, Graziani C, Lucarelli C, Filetici E, Villa L, et al. (2009) Molecular characterization of multidrug-resistant strains of *Salmonella enterica* serotype Typhimurium and Monophasic variant (S. 4,[5],12:i:-) isolated from human infections in Italy. Foodborne Pathog Dis 6: 711–717.
6. Gallati C, Stephan R, Hachler H, Malorny B, Schroeter A, et al. (2013) Characterization of *Salmonella enterica* subsp. enterica serovar 4,[5],12:i:-

clones isolated from human and other sources in Switzerland between 2007 and 2011. Foodborne Pathog Dis 10: 549–554.
7. Ido N, Kudo T, Sasaki K, Motokawa M, Iwabuchi K, et al. (2011) Molecular and phenotypic characteristics of *Salmonella enterica* serovar 4,5,12:i:- isolated from cattle and humans in Iwate Prefecture, Japan. J Vet Med Sci 73: 241–244.
8. Majtan V, Majtanova L, Majtan J (2011) Phenotypic and molecular characterization of human *Salmonella enterica* serovar 4,[5],12:i:- isolates in Slovakia. Curr Microbiol 63: 491–495.
9. Switt AI, Soyer Y, Warnick LD, Wiedmann M (2009) Emergence, distribution, and molecular and phenotypic characteristics of *Salmonella enterica* serotype 4,5,12:i. Foodborne Pathog Dis 6: 407–415.
10. European Food Safety Authority (2008) Report of the Task Force on Zoonoses Data Collection on the analysis of the baseline survey on the prevalence of *Salmonella* in slaughter pigs, in the EU, 2006–2007. EFSA Journal 135: 1–111.
11. Centers for Disease Control and Prevention (2008) *Salmonella* surveillance: annual summary, 2006.

12. National Institute of Infectious Diseases (2014) Infectious Agents Surveillance Report. Available: http://idsc.nih.go.jp/iasr/index.html. Accessed 2014 Jun 23.

13. Centers for Disease Control and Prevention (2007) Investigation of outbreaks of human infections caused by *Salmonella* serotype I 4,[5],12:i:-.

14. Mossong J, Marques P, Ragimbeau C, Huberty-Krau P, Losch S, et al. (2007) Outbreaks of monophasic *Salmonella enterica* serovar 4,[5],12:i:- in Luxembourg, 2006. Euro Surveill 12: E11–12.

15. Echeita MA, Herrera S, Usera MA (2001) Atypical, *fljB*-negative *Salmonella enterica* subsp. *enterica* strain of serovar 4,5,12:i:- appears to be a monophasic variant of serovar Typhimurium. J Clin Microbiol 39: 2981–2983.

16. de la Torre E, Zapata D, Tello M, Mejia W, Frias N, et al. (2003) Several *Salmonella enterica* subsp. *enterica* serotype 4,5,12:i:- phage types isolated from swine samples originate from serotype Typhimurium DT U302. J Clin Microbiol 41: 2395–2400.

17. Hauser E, Tietze E, Helmuth R, Junker E, Blank K, et al. (2010) Pork contaminated with *Salmonella enterica* serovar 4,[5],12:i:-, an emerging health risk for humans. Appl Environ Microbiol 76: 4601–4610.

18. Zamperini K, Soni V, Waltman D, Sanchez S, Theriault EC, et al. (2007) Molecular characterization reveals *Salmonella enterica* serovar 4,[5],12:i:- from poultry is a variant Typhimurium serovar. Avian Dis 51: 958–964.

19. Garaizar J, Porwollik S, Echeita A, Rementeria A, Herrera S, et al. (2002) DNA microarray-based typing of an atypical monophasic *Salmonella enterica* serovar. J Clin Microbiol 40: 2074–2078.

20. McClelland M, Sanderson KE, Spieth J, Clifton SW, Latreille P, et al. (2001) Complete genome sequence of *Salmonella enterica* serovar Typhimurium LT2. Nature 413: 852–856.

21. Edwards PR, Ewing WH (1986). Edwards and Ewing's identification of Enterobacteriaceae, 4th ed. New York, NY: Elsevier Science Publishing Co., Inc.

22. Anderson ES, Ward LR, Saxe MJ, de Sa JD (1977) Bacteriophage-typing designations of *Salmonella typhimurium*. J Hyg (Lond) 78: 297–300.

23. Clinical and Laboratory Standards Institute (2008) Performance Standards for Antimicrobial Disk and Dilution Susceptibility Test for Bacteria Isolated from Animals-Third Edition: Approved Standatrd M31-A3. Wayne, PA. Wayne, PA.

24. Kado CI, Liu ST (1981) Rapid procedure for detection and isolation of large and small plasmids. J Bacteriol 145: 1365–1373.

25. Akiba M, Kusumoto M, Iwata T (2011) Rapid identification of *Salmonella enterica* serovars, Typhimurium, Choleraesuis, Infantis, Hadar, Enteritidis, Dublin and Gallinarum, by multiplex PCR. J Microbiol Methods 85: 9–15.

26. Selzer RR, Richmond TA, Pofahl NJ, Green RD, Eis PS, et al. (2005) Analysis of chromosome breakpoints in neuroblastoma at sub-kilobase resolution using fine-tiling oligonucleotide array CGH. Genes Chromosomes Cancer 44: 305–319.

27. Hashimoto-Gotoh T, Yamaguchi M, Yasojima K, Tsujimura A, Wakabayashi Y, et al. (2000) A set of temperature sensitive-replication/-segregation and temperature resistant plasmid vectors with different copy numbers and in an isogenic background (chloramphenicol, kanamycin, *lacZ, repA, par, polA*). Gene 241: 185–191.

28. Stocker BA (1949) Measurements of rate of mutation of flagellar antigenic phase in *Salmonella typhi-murium*. J Hyg (Lond) 47: 398–413.

29. Burnens AP, Stanley J, Sack R, Hunziker P, Brodard I, et al. (1997) The flagellin N-methylase gene *fliB* and an adjacent serovar-specific IS*200* element in *Salmonella* Typhimurium. Microbiology 143: 1539–1547.

30. Gulig PA, Danbara H, Guiney DG, Lax AJ, Norel F, et al. (1993) Molecular analysis of *spv* virulence genes of the *Salmonella* virulence plasmids. Mol Microbiol 7: 825–830.

31. Namimatsu T, Asai T, Osumi T, Imai Y, Sato S (2006) Prevalence of the virulence plasmid in *Salmonella* Typhimurium isolates from pigs. J Vet Med Sci 68: 187–188.

32. Simon M, Zieg J, Silverman M, Mandel G, Doolittle R (1980) Phase variation: evolution of a controlling element. Science 209: 1370–1374.

33. Yamamoto S, Kutsukake K (2006) FljA-mediated posttranscriptional control of phase 1 flagellin expression in flagellar phase variation of *Salmonella enterica* serovar Typhimurium. J Bacteriol 188: 958–967.

34. Kutsukake K, Nakashima H, Tominaga A, Abo T (2006) Two DNA invertases contribute to flagellar phase variation in *Salmonella enterica* serovar Typhimurium strain LT2. J Bacteriol 188: 950–957.

35. Feng JA, Johnson RC, Dickerson RE (1994) Hin recombinase bound to DNA: the origin of specificity in major and minor groove interactions. Science 263: 348–355.

36. Takahashi T, Asai T, Kojima A, Harada K, Ishihara K, et al. (2006) Present situation of national surveillance of antimicrobial resistance in bacteria isolated from farm animals in Japan and correspondence to the issue. J Jap Assoc Infect Dis 80: 185–195 (*in Japanese with English summary*).

16

ComQXPA Quorum Sensing Systems May Not Be Unique to *Bacillus subtilis*: A Census in Prokaryotic Genomes

Iztok Dogsa[1,○], Kumari Sonal Choudhary[2,○], Ziva Marsetic[1], Sanjarbek Hudaiberdiev[2], Roberto Vera[2,3], Sándor Pongor[2,3]*, Ines Mandic-Mulec[1]*

1 Department of Food Science and Technology, Biotechnical Faculty, University of Ljubljana, Ljubljana, Slovenia, **2** Group of Protein Structure and Bioinformatics, International Centre for Genetic Engineering and Biotechnology, Trieste, Italy, **3** Faculty of Information Technology and Bionics, Pázmány Péter Catholic University, Budapest, Hungary

Abstract

The *comQXPA* locus of *Bacillus subtilis* encodes a quorum sensing (QS) system typical of Gram positive bacteria. It encodes four proteins, the ComQ isoprenyl transferase, the ComX pre-peptide signal, the ComP histidine kinase, and the ComA response regulator. These are encoded by four adjacent genes all situated on the same chromosome strand. Here we present results of a comprehensive census of *comQXPA*-like gene arrangements in 2620 complete and 6970 draft prokaryotic genomes (sequenced by the end of 2013). After manually checking the data for false-positive and false-negative hits, we found 39 novel *com*-like predictions. The census data show that in addition to *B. subtilis* and close relatives, 20 *comQXPA*-like loci are predicted to occur outside the *B. subtilis* clade. These include some species of Clostridiales order, but none outside the phylum Firmicutes. Characteristic gene-overlap patterns were observed in *comQXPA* loci, which were different for the *B. subtilis*-like and non-*B. subtilis*-like clades. Pronounced sequence variability associated with the ComX peptide in *B. subtilis* clade is evident also in the non-*B. subtilis* clade suggesting grossly similar evolutionary constraints in the underlying quorum sensing systems.

Editor: Raymond Schuch, Rockefeller University, United States of America

Funding: ID and IMM were supported by the Slovenian Research Agency grants J4-3631 and JP4-116 awarded to IMM. ZM was supported by a 3 month grant from the Erasmus Placement Program 2012/2013. Work at Pázmány Péter Catholic University, Budapest was partially supported by Hungarian National Innovation Office grants TÉT 10-1-2011-0058, TÁMOP-4.2.1.B-11/2/KMR-2011-0002, and TÁMOP-4.2.2/B-10/1-2010-0014. Funding for open access charge and for the services of a scientific writer/editor was provided by the University of Ljubljana and ICGEB, respectively. The funders had no role in study design, data collection and analysis, decision to publish, or preparation of the manuscript.

Competing Interests: The authors have declared that no competing interests exist.

* E-mail: pongor@icgeb.org (SP); ines.mandic@bf.uni-lj.si (IMM)

○ These authors contributed equally to this work.

Introduction

B. subtilis is one of the most studied prokaryotes and a frequently used model organism for Gram positive bacteria. It is capable of secreting a wide variety of molecules, including ribosomally and non-ribosomally produced peptides, polyketides, etc (for a recent review see [1,2]). Some of these play the role of quorum sensing (QS) molecules that coordinate social response of various bacterial populations in a cell density dependent manner [3]. Here we focus on the major *B. subtilis* quorum sensing *comQXPA* locus that operates through the ribosomally synthesized signaling peptide ComX. This peptide, which is modified by the ComQ isoprenyl transferase, specifically binds to the membrane receptor ComP, which then through phosphorylation of the response regulator ComA, elicits the QS response genes [4–10]. Note that we use the term "quorum sensing" with no implication of the relative importance of diffusion, flux, evolutionary and ecological roles (for recent reviews see [2,11–13]). We also use the term "signal" without reference to evolutionary adaptation, but simply to denote a chemical compound that alters the functioning of a bacterial cell.

ComX was first identified 20 years ago as a pheromone molecule present in the supernatant of high density cultures of *B. subtilis* [9,14]. Upon reaching a critical concentration it activates a large number of cellular responses including competence development, surfactin production, biofilm formation and extracellular DNA release [13,15–17]. Current research shows that ComX plays a pivotal role in activating cellular differentiation of *B. subtilis* in that ComX producing cells activate quorum sensing response only in a subpopulation of cells [4,18–20]. The QS response itself is regulated by a negative feedback loop that is based on production of ComX and sensed privately by the producing cell [19]. Another direction of studying subpopulations is the comparison of *B. subtilis* undomesticated strains isolated from the environment. These show high rates of polymorphisms of the ComQXPA quorum sensing system. The polymorphisms are reflected in functional diversification of *B. subtilis* strains into four different phenotype groups. These groups consist of strains among which effective communication is possible, but where communication across pherotypes is impaired [7,18,21,22]. Most recently, it was also found that pherotype diversification is present even within

2.5 mm^3 soil samples [23] and correlates, although imperfectly, with ecotype diversification [24].

The functional importance of the ComQXPA quorum sensing system and its interspecific polymorphisms make it a very interesting model system to study. The functioning of the ComQXPA quorum sensing system in *B. subtilis* relies on four proteins, ComQ, ComX, ComP, and ComA that are encoded by four adjacent reading frames situated on the same strand of the *B. subtilis* chromosome [7]. There is one common promoter in front of *comQ* and two putative promoters - one in front of *comX* and a second in front of *comA* [4]. We briefly overview the functional properties of the four proteins in the following paragraphs.

ComQ

A notable property of the ComQXPA QS system is that it requires an enzyme for postranslational modification of the ComX prepeptide. This is achieved by ComQ, which in *B. subtilis* is 286 to 309 amino acids in length (average length 299). It functions as an isoprenyl transferase and has the unique ability to attach the isoprenyl units to the tryptophan residue of ComX signal peptide. Isoprenylation occurs via an unusual ring-like structure formed upon addition of a farnesyl or geranyl group [25]. The sequence of the peptide pheromone and type of isoprenylation can vary from strain to strain. For instance, *B. subtilis* RO-E-2 uses an active pheromone of only five residues in length containing a geranylated tryptophan [25,26].

All *B. subtilis* ComQ protein sequences bear similarities to isoprenyl phosphate synthetases in general [4]. In domain databases, ComQ is reported to be a single-domain protein (e.g., PFAM: PF00348.12), with some databases (UNIPROT) indicating a putative membrane-binding segment of the protein. Consistent with such observations, some sequence-based predictions indicate that ComQ may have a membrane bound segment [8]. However, these predictions are apparently not unequivocal, and there is experimental evidence showing that the purified ComQ protein is enzymatically active *in vitro* even in the absence of a membrane

[27]. Therefore the condition for ComQ being a membrane protein was not included in our survey criteria. It is not known how ComX, once post-translationally modified, leaves the cell. Nevertheless, the expression of ComX and ComQ in *E. coli* is sufficient to reconstitute the synthesis of an active ComX [7].

ComX

In all *B. subtilis* strains known thus far, the *comX* gene codes for a 52 to 73 residue-long precursor protein. The sequence of ComX is extremely variable (PFAM identifier: PF05952), but it contains a tryptophan within the 5 to 10 amino acids of the C terminus [18].

ComP

Within the receptor ComP the two conserved intracellular motifs were included into the HMM model: histidine kinase domain (PFAM: PF07730) and an ATP-binding domain (PFAM: PF02518). The rest of the protein consists of a series of transmembrane helices separated by intra and extracellular loops, (**Figure 1**, also see the section on non-*B. subtilis*-like loci, below). The ComX binding site is located on the extracellular part of its N-terminal membrane-associated domain [6].

ComA

ComA is polypeptide of 196 to 245 residues in length (average length 214 residues). It has a typical structure of the transcriptional response regulator. The N-terminal domain (PFAM: PF00072.19) carries the phosphorylation domain involving an Asp residue. The C-terminal DNA-binding domain (PFAM: PF00196.14) is responsible for DNA binding [28,29].

In this work we searched bacterial genomes for candidate loci similar to the *B. subtilis* ComQXPA locus, by combining Hidden Markov Model (HMM) recognizers with filtering criteria based on the structural and organizational properties of ComQXPA QS systems. Here we show that the *comQXPA*-like gene arrangements are present in Firmicutes, outside the *B. subtilis* species. The

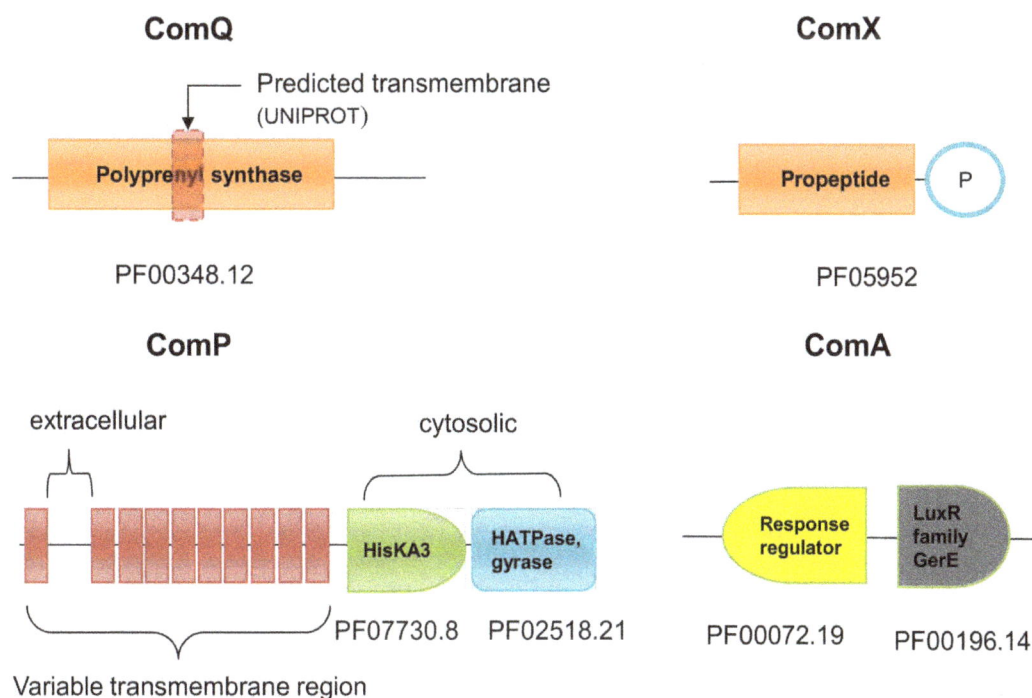

Figure 1. Domain structure of the Com proteins. The codes denote the PFAM (PF) domain identifiers.

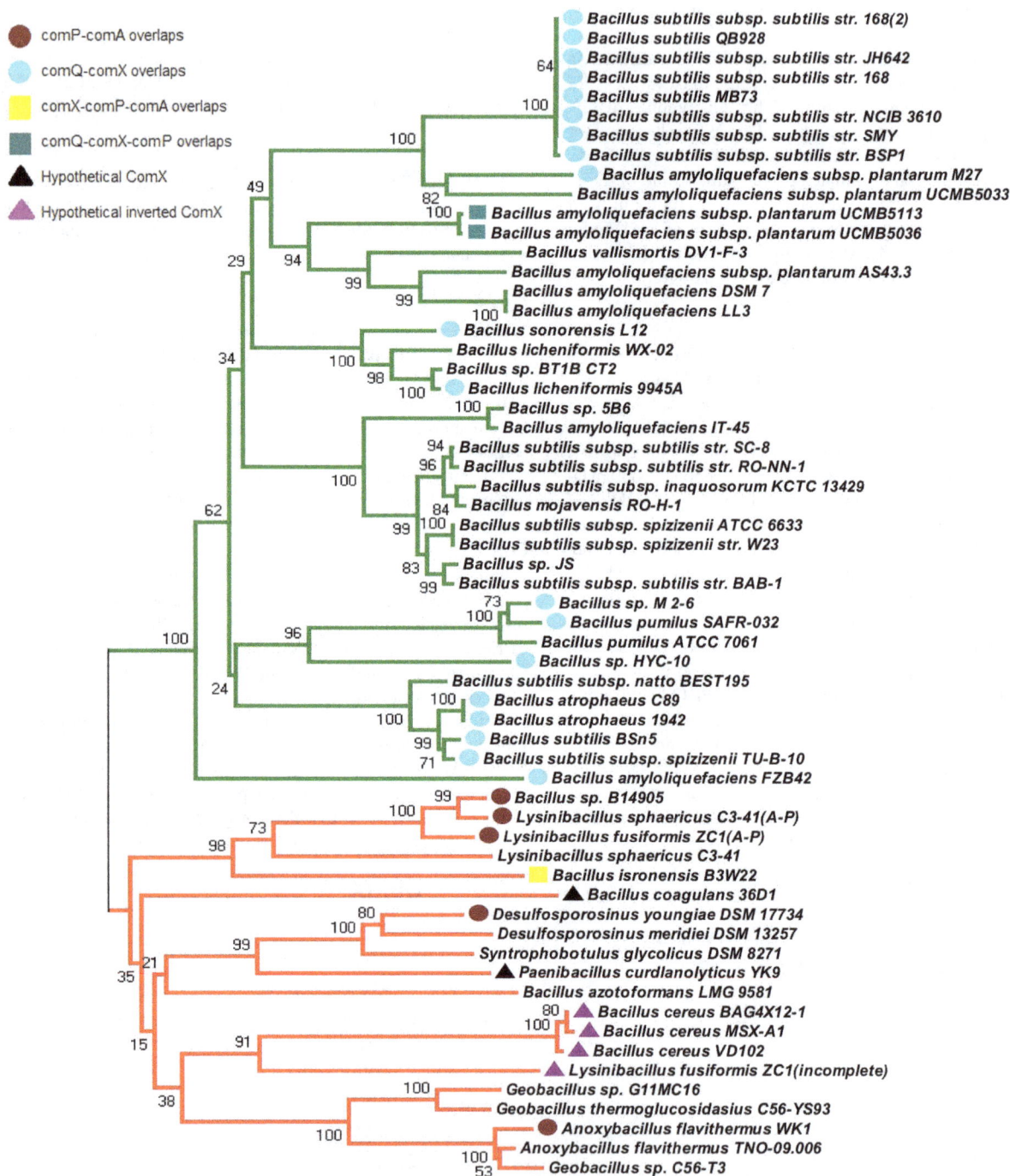

Figure 2. A similarity cladogram of ComQ sequences. The larger clade (green) that we termed "*B. subtilis*-like" contains the *B. subtilis* and a few other species from the *Bacillus* genus that are closely related. The other, visibly more varied one (red) contains non-*B. subtilis* sequences. The overlaps are indicated as highlighted bullets before each species.

candidate loci are characterized by similar sequence diversity profiles, but notable differences were also revealed in the overlap patterns found between *comQXPA* genes.

Results and Discussion

Census of *comQXPA* loci in prokaryotic genomes.

The complex nature of the ComQXPA quorum sensing system justifies questioning whether or not related circuit architectures

occur outside the *B. subtilis* species. We tried to answer this question by scanning all prokaryotic genomes (2620 complete and 6970 draft genomes having 644474 annotated and 505155 unannotated contigs, and a total of over 4.7 million protein sequences) for ComQXPA-related proteins, using Hidden Markov recognizers and additional filtering as described in Methods section. This census revealed that in addition to the 21 occurrences explicitly mentioned in the literature and/or in the databases, there are 39 new occurrences in which one, more or all of the

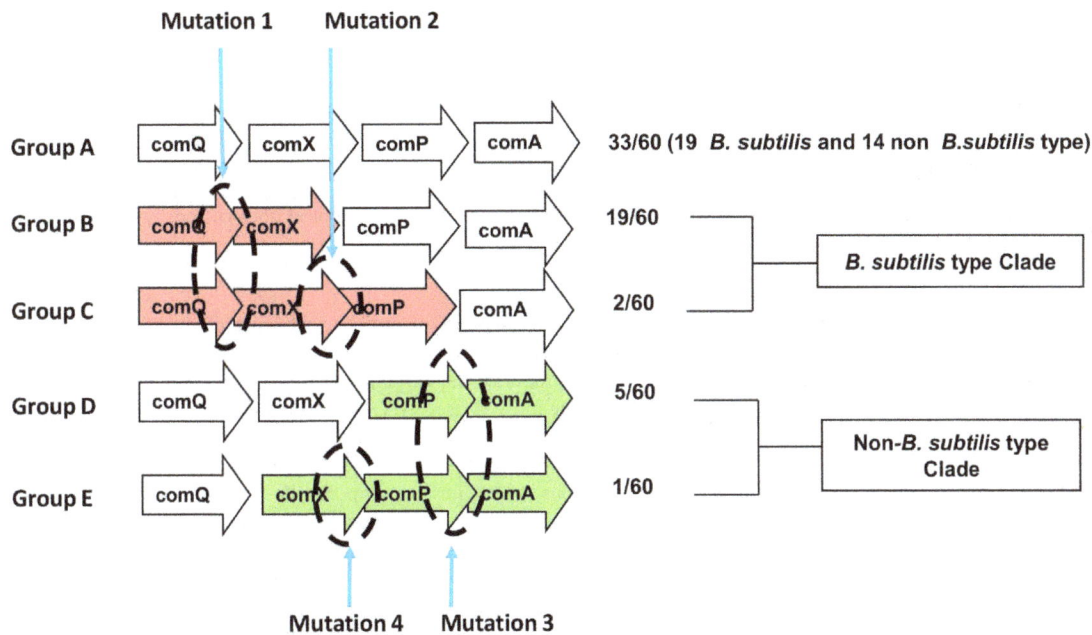

Figure 3. The gene overlaps within the comQXPA loci in *B. subtilis* **type and non-** *B. subtilis* **type clade, overlaps are shown in color.**
For the definition of the clades, see phylogenetic analysis. The numbers indicate the frequency of the type/the number total occurrences (i.e. 60).

functions were indicated as hypothetical. The complete list of the species is shown in **Table S1**.

Among the species in Table S1 we find a number of occurrences in which Com proteins have not yet been described. The similarity cladograms of all four ComQXPA sequences showed two large clades. **Figure 2** shows the ComQ tree as an example, while the trees of ComX, ComP and ComA are shown in Figures S1–S3. The larger clade, which we termed "*B. subtilis*-like", contains *B. subtilis* and a few other species from the *Bacillus subtilis-lichenoformis* group. We termed the other, visibly more varied clade as "non-*B. subtilis*-like" since it contains no *B. subtilis* sequences. This clade contains a few species from the *Bacillus* genus (*B. isronensis*, *B. coagulans*, *B. azotoformansname* and *B. cereus*), but also species from *Lysinbacillus*, *Geobacillus*, *Anoxybacillus*, *Desulfosporosinus* and more distantly related organisms such as *Clostridiales*. (Note that the same clades appear on the ComP and ComA trees, shown in Figures S2 and S3). To our knowledge, *comQXPA*- like loci in the members of this clade have not been reported before, so we carried out a detailed comparison of the two clades in terms of gene arrangements and sequence diversity.

Local arrangement of the *comQXPA* genes

The chromosomal arrangements of the loci are schematically shown in **Figure 3** where the reading frames are represented as arrows pointing to the putative direction of transcription. The types of overlaps were named A through E, and the species are given in Table S1. The majority of the loci contain no overlapping reading frames (A in Figure 3). A substantial number of the found loci contain overlapping reading frames that are spread in the two clades. In the *B. subtilis*-like clade the overlap types are dominated by an apparent mutation of the *comQ* stop codon. This results in a 13–18 amino acids long C-terminal extension, giving rise to the *comQ-comX* overlaps. In two loci within the same clade (*B. amyloliquefaciens subsp. plantarum* UCMB5113, *B. amyloliquefaciens subsp. plantarum* UCMB5036), there is a second mutation causing a 14 amino acids long N-terminal extension to ComP, leading to the

comQ-comX-comP overlaps. ii) In the non-*B. subtilis*-like clade the most frequent overlaps (six out of 14 species in the clade) are between ComP and ComA of which five are caused by a C-terminal extension of ComP. In contrast, in *Anoxybacillus flavithermus* WK1 the overlap is due to a 22 amino acids long N-terminal extension of ComA. In one locus (*B. isronensis* B3W22) there is an additional mutation that gives rise to a three amino acids long N-terminal extension of ComP, leading to a *comX-comP-comA* overlap.

In general, overlapping reading frames are not rare in bacteria [30] most likely because bacterial genes are frequently (>70%) located on one strand [31]. However, the fact that we see two coherent sets of mutations (Mutations 1–2, and Mutations 3–4, respectively) that are confined to two separate clades makes us believe that these mutations may have some logical reason for existing. In fact, the estimated probability of finding these mutations in two separate groups by chance is low. More precisely, a clear differentiation in the overlap type groups is evident, as both F-tests and P-tests [32] are highly significant (for F-test: $p < 0.0002$; for P-test: $p < 0.001$) which is further supported by the high bootstrap values obtained between the clades (Figure 2 and Figures S1 and S2). It has been suggested that overlapping reading frames may result in more efficient transcriptional control and reduce the need for more complex regulatory pathways [30]. Similarly, overlapping genes are often found in regulatory operons and indeed, the primary role of *comQXPA* loci is to control gene expression [5]. We thus speculate that the expression of *comQXPA* genes may be different in the two clades. In other words, the fact that different kinds of mutations are accepted in the two clades makes us hypothesize that the *com* loci of two clades may differ in terms of how the transcription/translation of the genes is coupled – a statement that would be worth testing by experimental methods in the future.

Unusual com-like loci

A few *com*-like arrangements were found outside the *Bacillus* genus in which the *comX* sequence was of the right length and in

A (4/60)   **B (2/60)**

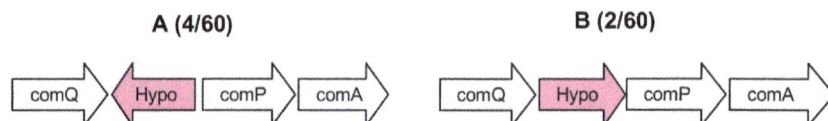

Figure 4. Unusual comQXPA-like loci. (A) Non-canonical unusual *com* system is present in *B. cereus* VD102, *B. cereus* BAG4X12 1, *B. cereus* MSX A1 and *L. fusiformis* ZC1. Note that *Lysinibacillus* is the only genus whose two species have two *com* loci each: *L. sphaericus* C3 41 has two canonical loci while in *L. fusiformis* ZC1 one locus is canonical and the other one is non-canonical, shown here. (B) Canonical unusual *com* system present in *Paenibacillus curdlanolyticus* YK9 and *B. coagulans* 36D1.

the right position between *comQ* and *comP*, but showed no appreciable homology with the known *comX* sequences, except the tryptophan residue in the C-terminus. We term these reading frames as hypothetical *comX*, (**Figure 4**). In one of the groups the hypothetical *comX* gene is on the opposite strand (4A). Three strains (*B. cereus* VD102, *B. cereus* BAG4X12 1, *B. cereus* MSX A1) and *L. fusiformis* ZC1 have this arrangement. We note that *Lysinibacillus* is the only genus whose two species have two *comQXPA* loci each: *L. sphaericus* C3 41 has two canonical loci while in *L. fusiformis* ZC1 one locus is canonical and the other has the hypothetical *comX* gene on the opposite strand, shown in Figure 4A.

The other arrangement (Figure 4B) is canonical in terms of the serial order of the genes. Two species *Paenibacillus curdlanolyticus* YK9 and *B. coagulans* 36D1 have this kind of arrangement. We note that we have no reason to believe that these *com*-like loci are functional or that they function in the same way as in *B. subtilis*. Nevertheless, the fact that they are at least in part conserved in relatively distinct species makes them interesting subjects for further experiments.

Sequence variability

As the non-*B. subtilis*-like clade contained interesting and novel *comQXPA* loci, which are known to be highly polymorphic within *B. subtilis* species [7,18,21,23,24], we compared the variability of the sequences within this clade and the more established *B. subtilis*-like clade.

The variability (π) of the ComQXPA proteins (Table 1) was calculated as the number of amino acid differences per site and by averaging over all sequence pairs using the MEGA program [33]. The non-*B. subtilis*-like clade was consistently more diverse than the *B. subtilis*-like which is in accordance with the branch lengths of the cladograms (see **Figure 2** and Figures S1–S3). In both clades the conservation of the genes followed the same order: ComA> ComP>ComQ>ComX. ComX showed the highest within-clade variability. Moreover, the diversity of the *com* sequences is substantially above the variability of the example housekeeping gene product RpoB (Table 1).

The distribution of amino acid variability along the protein sequences was investigated by calculating the similarity scores for each position using the Plotcon program [34]. Here, high values indicate more conserved regions (**Figure 5**). It is conspicuous that the general course of the variability curves is similar between the two clades, i.e. the peaks and the valleys are at similar sequence positions, which further supports the suggestion that the proteins in both clades are homologous and are under broadly similar evolutionary constraints. It is also conspicuous that the proteins of the non-*B. subtilis*-like clade are reasonably more variable than their counterparts in the *B.subtilis*-like clade, i.e. the blue curves run under the red curves. In other words, we see the same general tendency as in Table 1. The highest variability is seen at the C terminus of ComX that encodes the active peptide signal, whereas the rest of the propeptide is apparently more conserved. This pattern is evident for both clades.

A substantial variability was also detected in the N-terminal domain of the ComP protein that is predicted to interact with ComX [6]. The variability of the N-terminal domain is especially pronounced in comparison to the conserved cytoplasmic domain. Although earlier studies suggested that the N-terminal domain of ComP contains 8 transmembrane helices [6], the prediction methods used in this work suggest that the N-terminal domain contains 10 transmembrane helices (48 out of 60 occurrences). However, there are proteins in both clades in which the N-terminal transmembrane domain is missing (9 out of 60 occurrences, uniformly distributed in the two clades) meaning that the loop region of the intact protein becomes N-terminal (Table S3). Another interesting example of sequence variability is an unusual, C-terminal extension of ComX, that occurs only in the B. subtilis-like clade (*B. subtilis* natto, *B. subtilis* subsp. Spizienii TU-B-10, *B. subtilis* BSn5, *Bacillus atrophaeus* 1942 and *Bacillus atrophaeus* C89)

Even though the residue conservation follows the same general trends in both clades, there are a few interesting points that need to be discussed. It is not known which part in ComQ is responsible for the specificity of interactions between ComQ and ComX. However, we speculate that the two valleys (positioned at around

Table 1. Sequence diversity[1] of ComQXPA proteins.

	B. subtilis **clade**	**Non-*B.subtilis* clade**	**Clades combined**
ComA	0.13±0.01	0.57±0.02	0.41±0.02
ComP	0.29±0.01	0.64±0.01	0.51±0.01
ComQ	0.46±0.02	0.60±0.01	0.57±0.02
ComX	0.63±0.04	0.69±0.03	0.70±0.03
RpoB[2]	0.017±0.002	0.15±0.01	0.091±0.005

[1] π diversity values calculated with the MEGA program [33] from the multiple alignments used to construct the cladograms. High values indicate high diversity.
[2] RNA polymerase beta subunit, taken as a example of a conserved housekeeping protein.

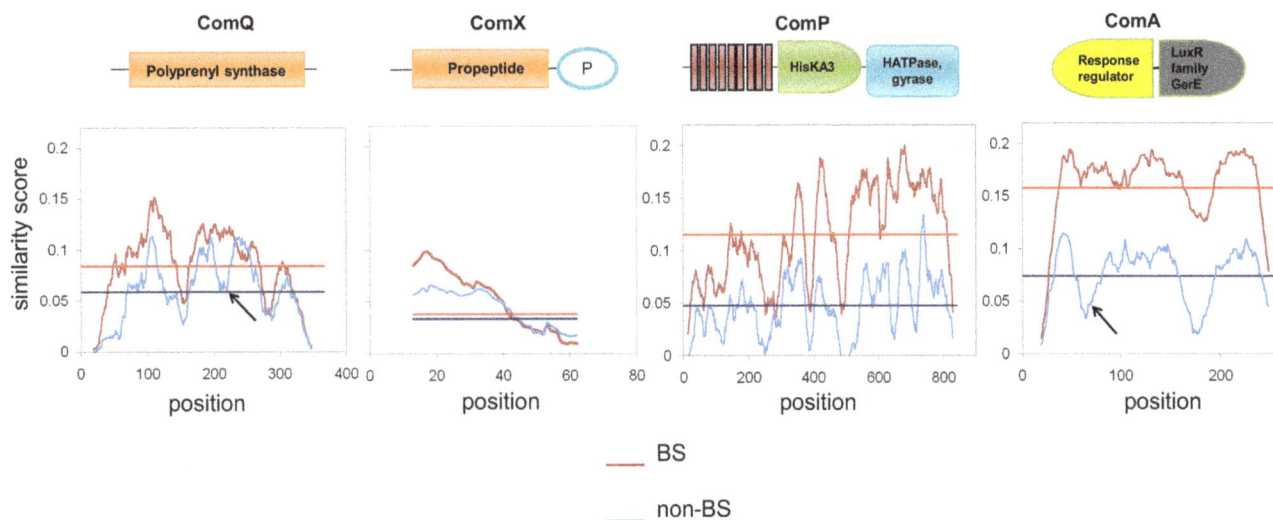

Figure 5. Variability of Com protein sequences in the *B. subtilis*-**like (red) and non-**B. subtilis*-**like (blue) proteins.** The average variability of the two clades is plotted for each protein respectively [32], the straight lines represent the average value of the plots. Note that the overall course of the variability is similar in both clades, i.e. peaks and valley are at the same position, and the non-*B. subtilis*-like group are more variable than the *B. subtilis*-like proteins. A conspicuous difference between the two clades is found in ComA protein where the response regulator domain in the non-*B. subtilis*-like clade has a variable region (valley indicated by arrow).

150 and 280, respectively) indicating a high degree of polymorphism might be responsible for these interactions. Although this prediction is highly speculative it should be interesting to address this problem by experimental methods in the future. ComA shows lowest overall diversity and there is one noteworthy difference between the plots of ComA proteins. The region of high variability is present around position 80 in the non-*B. subtilis* like clade, but it is absent in the *B. subtilis* like clade. The function of this variable region is not known but we speculate that it might be involved in interactions with additional, as yet unidentified modulators. Again, we point to this region as it might be of interest for future experimental studies.

Common traits of ComQXPA and AHL based QS systems.

It is interesting to compare the general features of *comQXPA* loci to the AHL regulatory system of Gram negative bacteria [35–37]. At first sight, the AHL system is very different in many respects. Firstly, its signal molecules, *N*-acyl homoserine lactones, are secondary metabolites, as opposed to the post-translationally modified, ribosomally synthesized ComX. Secondly, the core AHL system is simpler as it contains only two proteins, the signal synthetase and a LuxR-like response regulator. Most AHL systems contain well-defined additional negative feedback regulatory components (co-expressed repressors, RNA interference etc. [35,37]). Recently, it was shown that inhibition is also part of the ComQXPA locus; ComQ was found to provide negative feedback that modulates the QS response of the signal producer [19]. In addition, the overall architecture of the com-specific response regulator ComA and all AHL LuxR proteins bear similarities in as much as the C-terminal DNA-binding domain belongs to the same domain type (LuxR family GerE), and the signal-mediating domain is at the N-terminal in both proteins. The underlying signaling mechanisms are naturally different since LuxR binds the AHL autoinducer molecule while the N-terminal domain of ComA is phosphorylated by the histidine kinase receptor ComP. Both systems contain conserved gene overlaps (40% of the systems in AHL and 45% in com In terms of sequence similarity, AHL proteins in different topological arrangements are

typical orthologs that are closer to their QS-linked homologs in other species [35–37], than to the related proteins in the same genome. The same tendency holds for the Com proteins, even though the analysis could only be done for ComP proteins which have clear-cut non-QS homologues in most bacterial genomes. Here we see a similar tendency (Figure S4 and explanations in figure legend): QS-linked ComP proteins cluster together with other ComPs from different genomes rather than clustering with other histidine kinases of the same genome. This shows that ComP has diverged from the other, non-QS linked histidine kinases before the modern bacterial species appeared, i.e. the emergence of ComQXPA QS systems is a not a recent evolutionary event.

Conclusions

In this work, we presented a census of *comQXPA* locus proteins involved in the quorum sensing regulation of competence and other late growth adaptive traits in *B. subtilis*. We found 31 new occurrences, many of them outside the *Bacillus* genus, with some from different orders (for instance order: Clostridiales, family: *Peptococcaceae*, genera: *Desulfosporosinus, Syntrophobotulus*). The local arrangement of the genes was quite conserved in all *com*-like occurrences. The only variability of gene topology was found in putative *com*-related operons whose function may however, not be necessarily linked to quorum sensing, or may be based on a different type of peptide signal. We found two conserved classes of overlap patterns in *B. subtilis*-like and non-*B subtilis* like *comQXPA* loci, respectively, which may be due to hitherto unknown transcriptional/translational differences.

In summary, a wholesale scan of current databases showed a number of novel occurrences of com QS regulatory locus originally identified in *B. subtilis*. It was found that this locus is phylogenetically more widespread than previously thought, and its organization has some commonalities with unrelated QS systems such as conserved gene overlaps.

Methods

Data

Data relating to 2620 complete genomes and 6970 drafts genomes having 644474 annotated and 505155 unannotated contigs were downloaded from NCBI FTP, ftp://ftp.ncbi.nlm.nih.gov/genomes/Bacteria/ and ftp://ftp.ncbi.nlm.nih.gov/genomes/Bacteria_DRAFT respectively. Additional protein sequences were retrieved from the UNIPROT database (http://www.uniprot.org/). Data retrieval was completed on August 16, 2013.

QS gene detection

com genes were determined using Hidden Markov Model (HMM) recognizers built with HMMER 3.0, http://hmmer.janelia.org/). Briefly, a core set of protein sequences was taken for ComA, ComP, ComX and ComQ sequences from Uniprot database (see Table S2). The CLUSTAL program (accessed via the EBI Webportal, http://www.ebi.ac.uk/Tools/msa/clustalw2/) was used for constructing the multiple sequence alignments which were then processed by the HMMBBUILD program to give HMM recognizers that were in turn used to scan the protein sequence data (over 4.7 million protein sequences). Sequences giving an E-value below 0.1 were manually checked for length, alignment coverage and residue conservation. It was also determined whether the loci were full or incomplete. We proceeded with the following steps:

1.) The presence of transmembrane helices in ComP candidates was checked by TMHMM server v. 2.0 (http://www.cbs.dtu.dk/services/TMHMM/).

2.) The presence of a tryptophan residue at the end of each probable ComX was manually checked.

3.) Only those proteins were accepted where the length was within the range of protein sequences used for building the HMM recognizers. The hypothetical proteins in place of ComX also fell within this range.

Cladograms of ComA, ComP, ComX and ComQ sequence groups were built using the guide tree of the CLUSTAL program and visualized using the MEGA 5 program package installed from http://www.megasoftware.net. In the trees, the numerical value at each node indicates the bootstrap value supporting every split in the lineage (out of 1,000 bootstrap replicates).

The complete list of the species having *com* loci is shown in **Table S1**.

Statistical analysis

For the distribution of overlaps, the P-test of significance [32] was performed using MacClade 4.08 [38]. The P-test seeks for the co-variation of the sequence phylogeny with the distribution of the overlap-types among the groups. One way to study this is to use parsimony [39], which provides the minimum number of changes (i.e. switch from one type of overlap to another) necessary to explain the observed distribution of sequences among the groups, assuming that the group type (overlap type) evolved according to the sequence based phylogenetic tree [32]. The minimal number of changes between a given phylogeny and random phylogenies can be determined and significance values can then be calculated. F-test was performed on ComQ amino acids sequences as described by Martin [32]. Statistical significance was evaluated by randomly assigning sequences to group types and calculating F-values for 10000 permutations. The F-test assesses the degree of differentiation between the groups by comparing the total genetic diversity within each group to the total genetic diversity of the groups combined [40]. If the members within the groups (overlap type) are genetically very similar, but the groups themselves are very different from each other, the F value will be significantly higher than zero, which is an indicator of a correlation between the genetic sequences and the group type (i.e. overlap type) [41]. Other statistical tests were carried out as described in [42,43].

Supporting Information

Figure S1   A similarity cladogram of ComA sequences.

Figure S2   A similarity cladogram of ComP sequences.

Figure S3   A similarity cladogram of ComX sequences.

Figure S4   Cladogram of ComP and other Histidine Kinase protein sequences in 60 genomes in which *comQXPA* locus was identified.

Table S1   List of the species having *comQXPA* loci.

Table S2   List of protein sequence sets used for building HMM recognizers.

Table S3   Transmembrane domain architecture in ComP proteins.

Acknowledgments

The authors would like to thank Dr. Polonca Stefanic (University of Ljubljana) for her advice and helpful discussions and Dr. Max Bingham (Freelance Editor, Rotterdam, The Netherlands) for editorial assistance prior to submission.

Author Contributions

Conceived and designed the experiments: ID KSC IMM SP. Performed the experiments: KSC ZM SH RV. Analyzed the data: KSC ID RV SP IMM. Wrote the paper: ID KSC SP IMM.

References

1. Mongkolthanaruk W (2012) Classification of Bacillus beneficial substances related to plants, humans and animals. J Microbiol Biotechnol 22: 1597–1604.
2. Thoendel M, Horswill AR (2010) Biosynthesis of peptide signals in gram-positive bacteria. In: Gadd, Geoffrey; Sariaslani S, editor. Advances in applied microbiology. Iowa City, Iowa: Elsevier Inc., Vol. 71. pp. 91–112.
3. Fuqua WC, Winans SC, Greenberg EP (1994) Quorum sensing in bacteria: the LuxR-LuxI family of cell density-responsive transcriptional regulators. J Bacteriol 176: 269–275.
4. Bacon Schneider K, Palmer TM, Grossman AD (2002) Characterization of comQ and comX, two genes required for production of ComX pheromone in Bacillus subtilis. J Bacteriol 184: 410–419.
5. Comella N, Grossman AD (2005) Conservation of genes and processes controlled by the quorum response in bacteria: characterization of genes controlled by the quorum-sensing transcription factor ComA in Bacillus subtilis. Mol Microbiol 57: 1159–1174.
6. Piazza F, Tortosa P, Dubnau D (1999) Mutational analysis and membrane topology of ComP, a quorum-sensing histidine kinase of Bacillus subtilis controlling competence development. J Bacteriol 181: 4540–4548.
7. Tortosa P, Logsdon L, Kraigher B, Itoh Y, Mandic-Mulec I, et al. (2001) Specificity and genetic polymorphism of the Bacillus competence quorum-sensing system. J Bacteriol 183: 451–460. doi:10.1128/JB.183.2.451-460.2001

8. Weinrauch Y, Msadek T, Kunst F, Dubnau D (1991) Sequence and properties of comQ, a new competence regulatory gene of Bacillus subtilis. J Bacteriol 173: 5685–5693.

9. Magnuson R, Solomon J, Grossman AD (1994) Biochemical and genetic characterization of a competence pheromone from B. subtilis. Cell 77: 207–216.

10. Solomon JM, Magnuson R, Srivastava A, Grossman AD (1995) Convergent sensing pathways mediate response to two extracellular competence factors in Bacillus subtilis. Genes Dev 9: 547–558.

11. Ryan RP, Dow JM (2008) Diffusible signals and interspecies communication in bacteria. Microbiology 154: 1845–1858. doi:10.1099/mic.0.2008/017871-0

12. Taga ME, Bassler BL (2003) Chemical communication among bacteria. Proc Natl Acad Sci U S A 100 Suppl: 14549–14554. doi:10.1073/pnas.1934514100

13. López D, Vlamakis H, Losick R, Kolter R (2009) Paracrine signaling in a bacterium. Genes Dev 23: 1631–1638. doi:10.1101/gad.1813709

14. Shank E, Kolter R (2011) Extracellular signaling and multicellularity in Bacillus subtilis. Curr Opin Microbiol 14: 741–747. doi:10.1016/j.mib.2011.09.016.Extracellular

15. Dubnau D (1991) Genetic competence in Bacillus subtilis. Microbiol Rev 55: 395–424.

16. Nakano MM, Zuber P (1991) The primary role of comA in establishment of the competent state in Bacillus subtilis is to activate expression of srfA. J Bacteriol 173: 7269–7274.

17. Zafra O, Lamprecht-Grandío M, de Figueras CG, González-Pastor JE (2012) Extracellular DNA release by undomesticated Bacillus subtilis is regulated by early competence. PLoS One 7: e48716. doi:10.1371/journal.pone.0048716

18. Ansaldi M, Marolt D, Stebe T, Mandic-Mulec I, Dubnau D (2002) Specific activation of the Bacillus quorum-sensing systems by isoprenylated pheromone variants. Mol Microbiol 44: 1561–1573.

19. Oslizlo A, Stefanic P, Dogsa I, Mandic-Mulec I (2014) Private link between signal and response in Bacillus subtilis quorum sensing. Proc Natl Acad Sci U S A 2013: 1–6. doi:10.1073/pnas.1316283111

20. Maamar H, Raj A, Dubnau D (2007) Noise in gene expression determines cell fate in Bacillus subtilis. Science 317: 526–529. doi:10.1126/science.1140818

21. Tran LS, Nagai T, Itoh Y (2000) Divergent structure of the ComQXPA quorum-sensing components: molecular basis of strain-specific communication mechanism in Bacillus subtilis. Mol Microbiol 37: 1159–1171.

22. Mandic-Mulec I, Kraigher B, Cepon U, Mahne I (2003) Variability of the Quorum Sensing System in Natural Isolates of Bacillus sp. FOOD Technol Biotechnol 41: 23–28.

23. Stefanic P, Mandic-Mulec I (2009) Social interactions and distribution of Bacillus subtilis pherotypes at microscale. J Bacteriol 191: 1756–1764. doi:10.1128/JB.01290-08

24. Stefanic P, Decorosi F, Viti C, Petito J, Cohan FM, et al. (2012) The quorum sensing diversity within and between ecotypes of Bacillus subtilis. Environ Microbiol 14: 1378–1389. doi:10.1111/j.1462-2920.2012.02717.x

25. Okada M, Yamaguchi H, Sato I, Tsuji F, Dubnau D, et al. (2008) Chemical structure of posttranslational modification with a farnesyl group on tryptophan. Biosci Biotechnol Biochem 72: 914–918.

26. Okada M, Sato I, Cho SJ, Iwata H, Nishio T, et al. (2005) Structure of the Bacillus subtilis quorum-sensing peptide pheromone ComX. Nat Chem Biol 1: 23–24. doi:10.1038/nchembio709

27. Tsuji F, Ishihara A, Kurata K, Nakagawa A, Okada M, et al. (2012) Geranyl modification on the tryptophan residue of ComXRO-E-2 pheromone by a cell-free system. FEBS Lett 586: 174–179. doi:10.1016/j.febslet.2011.12.012

28. Weinrauch Y, Penchev R, Dubnau E, Smith I, Dubnau D (1990) A Bacillus subtilis regulatory gene product for genetic competence and sporulation resembles sensor protein members of the bacterial two-component signal-transduction systems. Genes Dev 4: 860–872.

29. Roggiani M, Dubnau D (1993) ComA, a phosphorylated response regulator protein of Bacillus subtilis, binds to the promoter region of srfA. J Bacteriol 175: 3182–3187.

30. Johnson ZI, Chisholm SW (2004) Properties of overlapping genes are conserved across microbial genomes. Genome Res 14: 2268–2272. doi:10.1101/gr.2433104

31. Fukuda Y, Nakayama Y, Tomita M (2003) On dynamics of overlapping genes in bacterial genomes. Gene 323: 181–187.

32. Martin A (2002) Phylogenetic approaches for describing and comparing the diversity of microbial communities. Appl Environ Microbiol 68: 3673–3682. doi:10.1128/AEM.68.8.3673

33. Tamura K, Peterson D, Peterson N, Stecher G, Nei M, et al. (2011) MEGA5: molecular evolutionary genetics analysis using maximum likelihood, evolutionary distance, and maximum parsimony methods. Mol Biol Evol 28: 2731–2739. doi:10.1093/molbev/msr121

34. Rice P, Longden I, Bleasby A (2000) EMBOSS: the European Molecular Biology Open Software Suite. Trends Genet 16: 276–277.

35. Choudhary KS, Hudaiberdiev S, Gelencsér Z, Gonçalves Coutinho B, Venturi V, et al. (2013) The Organization of the Quorum Sensing luxI/R Family Genes in Burkholderia. Int J Mol Sci 14: 13727–13747. doi:10.3390/ijms140713727

36. Gelencsér Z, Galbáts B, Gonzalez JF, Choudhary KS, Hudaiberdiev S, et al. (2012) Chromosomal arrangement of AHL.driven quorum sensing circuits in Pseudomonas. ISRN Microbiol 2012: 1–6.

37. Gelencsér Z, Choudhary KS, Coutinho BG, Hudaiberdiev S, Galbáts B, et al. (2012) Classifying the topology of AHL-driven quorum sensing circuits in proteobacterial genomes. Sensors 12: 5432–5444.

38. Maddison WP, Maddison DR (1989) Interactive analysis of phylogeny and character evolution using the computer program MacClade. Folia Primatol (Basel) 53: 190–202.

39. Slatkin WPM (1991) Null Models for the Number of Evolutionary Steps in a Character on a Phylogenetic Tree. Evolution (N Y) 45: 1184–1197.

40. Edwards S (1993) Mitochondrial gene genealogy and gene flow among island and mainland populations of a sedentary songbird, the grey-crowned babbler (Pomatostomus temporalis). Evolution (N Y) 47: 1118–1137.

41. Edwards S (1993) Mitochondrial gene genealogy and gene flow among island and mainland populations of a sedentary songbird, the grey-crowned babbler (Pomatostomus temporalis). Evolution (N Y) 47: 1118–1137.

42. Sonego P, Pacurar M, Dhir S, Kertész-Farkas A, Kocsor A, et al. (2007) A Protein Classification Benchmark collection for machine learning. Nucleic Acids Res 35: D232–6. doi:10.1093/nar/gkl812

43. Pintar A, Carugo O, Pongor S (2003) Atom depth in protein structure and function. Trends Biochem Sci 28: 593–597.

17

Sorting through the Wealth of Options: Comparative Evaluation of Two Ultraviolet Disinfection Systems

Michelle M. Nerandzic[1]*, **Christopher W. Fisher**[3], **Curtis J. Donskey**[1,2]

1 Research Service, Louis Stokes Cleveland Veterans Affairs Medical Center, Cleveland, Ohio, United States of America, **2** Geriatric Research, Education and Clinical Center, Cleveland Veterans Affairs Medical Center, Cleveland, Ohio, United States of America, **3** STERIS Corporation, Healthcare Group, Mentor, Ohio, United States of America

Abstract

Background: Environmental surfaces play an important role in the transmission of healthcare-associated pathogens. Because environmental cleaning is often suboptimal, there is a growing demand for safe, rapid, and automated disinfection technologies, which has lead to a wealth of novel disinfection options available on the market. Specifically, automated ultraviolet-C (UV-C) devices have grown in number due to the documented efficacy of UV-C for reducing healthcare-acquired pathogens in hospital rooms. Here, we assessed and compared the impact of pathogen concentration, organic load, distance, and radiant dose on the killing efficacy of two analogous UV-C devices.

Principal Findings: The devices performed equivalently for each impact factor assessed. Irradiation delivered for 41 minutes at 4 feet from the devices consistently reduced *C. difficile* spores by ~ 3 $\log_{10}CFU/cm^2$, MRSA by >4 $\log_{10}CFU/cm^2$, and VRE by >5 $\log_{10}CFU/cm^2$. Pathogen concentration did not significantly impact the killing efficacy of the devices. However, both a light and heavy organic load had a significant negative impacted on the killing efficacy of the devices. Additionally, increasing the distance to 10 feet from the devices reduced the killing efficacy to ≤3 $\log_{10}CFU/cm^2$ for MRSA and VRE and <2 $\log_{10}CFU/cm^2$ for *C.difficile* spores. Delivery of reduced timed doses of irradiation particularly impacted the ability of the devices to kill *C. difficile* spores. MRSA and VRE were reduced by >3 $\log_{10}CFU/cm^2$ after only 10 minutes of irradiation, while *C. difficile* spores required 40 minutes of irradiation to achieve a similar reduction.

Conclusions: The UV-C devices were equally effective for killing *C. difficile* spores, MRSA, and VRE. While neither device would be recommended as a stand-alone disinfection procedure, either device would be a useful adjunctive measure to routine cleaning in healthcare facilities.

Editor: Peter Setlow, University of Connecticut, United States of America

Funding: This work was supported by a Merit Review grant from the Department of Veterans Affairs to C.J.D. The funders had no role in study design, data collection and analysis, decision to publish, or preparation of the manuscript.

Competing Interests: C.W.F. is an employee of STERIS Corporation. The other authors declare that no competing interests exist.

* Email: michellenerandzi@aim.com

Introduction

Environmental surfaces may play an important role in transmission of healthcare-associated pathogens such as *Clostridium difficile*, methicillin-resistant *Staphylococcus aureus* (MRSA), and vancomycin-resistant *Enterococcus* (VRE) [1–6]. Pathogens are shed onto environmental surfaces and will remain for several days, or possibly months, if the surfaces are not effectively disinfected [1–6]. Unfortunately, several recent studies have demonstrated that environmental cleaning is often suboptimal in healthcare facilities [5–8]. Interventions such as education of housekeeping staff or use of fluorescent markers to provide feedback to housekeepers may result in improved cleaning [5–8]. Yet, despite the promise of improvement in routine cleaning, there remains a demand for novel, automated technologies that are effective against hard to kill *Clostridium difficile* spores, but are also safe and rapid. As a consequence, there has been an upsurge in automated disinfection technologies on the market, many of which have yet to be rigorously evaluated.

Novel ultraviolet disinfection devices are currently on the forefront of burgeoning automated technologies due to the well documented efficacy of ultraviolet-C (UV-C) irradiation for killing bacteria, viruses, and persistent spores [9–15]. The mechanism of killing of microorganisms by UV-C is primarily due to inactivation of DNA and RNA through absorption of photons resulting in formation of pyrimidine dimers from thymine and cytosine [9,12,15]. We previously demonstrated that an automated room disinfection device that utilizes low pressure mercury lamps for emitting UV-C radiation was effective for significantly reducing *C. difficile*, MRSA, and VRE contamination in hospital rooms (Tru-D Rapid Room Disinfection, Lumalier, Memphis, TN, USA) [16]. Similarly, Rutala et al. evaluated the Tru-D device and concluded that it was an efficacious and environmentally friendly method for disinfecting surfaces in healthcare facilities [17]. Here, we performed a side-by-side comparative evaluation of a homologous automated UV-C room disinfection device (Pathogon UV Disinfection System, Steris Corporation, Mentor, Ohio, USA) against the previously tested Tru-D device, in the laboratory setting. For each device, the impact of pathogen concentration, organic load, distance, and radiant dose on killing efficacy was assessed.

Materials and Methods

C. difficile, MRSA, and VRE Strains

Two clinical isolates of C. difficile, MRSA and VRE were studied. The MRSA strains were a pulsed-field gel electrophoresis (PFGE) type USA300 (community-associated) and USA800 (hospital-associated). The VRE strains were a VanA-type isolate (C37) and a VanB-type isolate (C68). The C. difficile strains were VA 17, a restriction endonuclease analysis (REA) type BI strain, and VA 11, an REA type J strain.

Preparation of C. difficile Spores

Spores were prepared by growth on brain-heart infusion agar (Becton Dickinson, Cockeysville, MD) supplemented with yeast extract (5 mg/ml) and L-cysteine (0.1%) at 37°C under anaerobic conditions as previously described [18]. Spores were stored at 4°C in sterile distilled water until use. Prior to testing, spore preps were confirmed by phase contrast microscopy and malachite green staining to be > 99% dormant, bright-phase spores.

Microbiology

For VRE, MRSA, and C. difficile cultures, media included Enterococcosel agar (Becton Dickinson, Cockeysville, MD) containing 20 μg/mL of vancomycin, CHROMagar (CHROMagar, Paris, France) containing 6 μg/mL of cefoxitin, and cycloserine-cefoxitin-brucella agar containing 0.1% taurocholic acid and lysozyme 5 mg/L (CDBA), respectively [19]. Plates containing MRSA or VRE were incubated aerobically at 37°C for 48 hours. C. difficile plates were incubated in a Whitley Workstation MG1000 anaerobic chamber (Microbiology International, Frederick, MD) at 37°C for 48 hours.

The UV-C Disinfection Devices

Figure 1A/B is a photograph of the devices. The Pathogon device (1A) is 28 inches wide, 31 inches long, and stands 67 inches tall. The system is a wheeled mobile unit that is controlled remotely by a Windows-based tablet controller. It is placed in the center of the room and commonly touched surfaces are arranged close to the device for optimal exposure to UV-C radiation (i.e. bedrails pulled up, call buttons placed on the bed, tables placed near the device). The device contains motion and heat sensor that are connected to a safety rated relay, aborting the UV-C cycle if someone enters the room during use. The unit has 24, 45 inch low pressure mercury bulbs. Once the operator has exited the room, a pre-programed germicidal dose is chosen based on the dimensions of the room. UV-C radiation penetrates all areas of the room that receive light, but the highest exposure occurs for areas that are in direct line of exposure to the output of the device; areas that are not in direct line of exposure to UV-C may receive radiation that is reflected from the walls and ceiling or from other surfaces in the room.

The Tru-D device (1B) is 72 inches tall and measures 24 inches at the widest portion of the base. It is a wheeled mobile device that is placed strategically in the center of the room just as described above for the Pathogon device. The operator exits the room, closes the door, and places a door sensor on the frame of the door. Continuous monitoring during operation is not required because the sensor triggers automatic discontinuation of the cycle if the door is opened. A handheld remote is used to select either a vegetative cycle that is effective for killing of non-spore forming

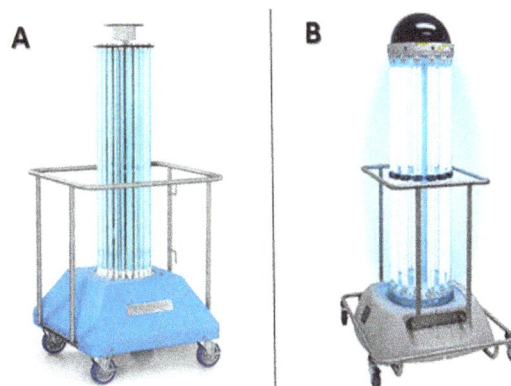

Figure 1. Photographs of the Pathogon (A) and Tru-D (B) devices.

organisms or a spore cycle that is effective in killing spores. The unit has 28, 36 inch low pressure mercury bulbs. The device contains eight sensors spaced at equal distances on a ring at the top of the device. The sensors measure the amount of UV-C light reflected back to the device. The device automatically ends the cycle when the area reflecting the lowest level of UV-C back to the sensors (i.e. shaded areas in the room) has received an adequate dose.

Efficacy of the UV-C Devices for the Reduction of Pathogens on Carriers

The Effect of Pathogen Concentration and Organic Load. Initial experiments were conducted to determine whether pathogen concentration (i.e. concentration of organisms per cm^2) or organic load influenced the disinfection efficacy of the UV-C devices. For pathogen concentration experiments, ten μl aliquots of two strains of C. difficile spores, MRSA, and VRE were suspended in sterile phosphate buffered saline (PBS), inoculated onto stainless steel carriers, and then spread to cover a 1 cm^2 area. Organisms were allowed to desiccate onto the carriers under ambient room conditions. For C. difficile spores, the inoculum applied to the carriers ranged from 2 to >5 $\log_{10}$CFU/cm^2. Previous experiments demonstrated that a reduction in vegetative organisms (MRSA and VRE) was observed after initial desiccation onto the carriers, however no further reduction was observed within the duration of treatment time (author's unpublished data). For each vegetative pathogen, the inoculum applied to the slide was adjusted such that 2 to >5 $\log_{10}$CFU/cm^2 were recovered from the positive control specimens after desiccation. For organic load experiments, two strains of C. difficile spores, MRSA, and VRE were suspended in either sterile PBS, light organic load (5% fetal calf serum), or heavy organic load (5% fetal calf serum and 5% tryptone) and inoculated onto stainless steel carriers as described above. However, the inoculum was altered such that each carrier yielded 6 $\log_{10}$ CFU at baseline.

The carriers were placed on a laboratory bench top 4 feet from the UV-C device, within the direct field of radiation. Baseline slides were left untreated outside of the room (i.e., positive controls). The room dimensions were approximately 10×10 feet. Based on these dimensions, the UV-C devices were run for 41 minutes, as suggested by the manufacturer of the Pathogon device to deliver a spore-killing dose of UV-C within a 10×10 foot range.

To quantify viable organisms, the carriers were submersed in 10 mL of sterile PBS, vortexed vigorously, and dilutions of the

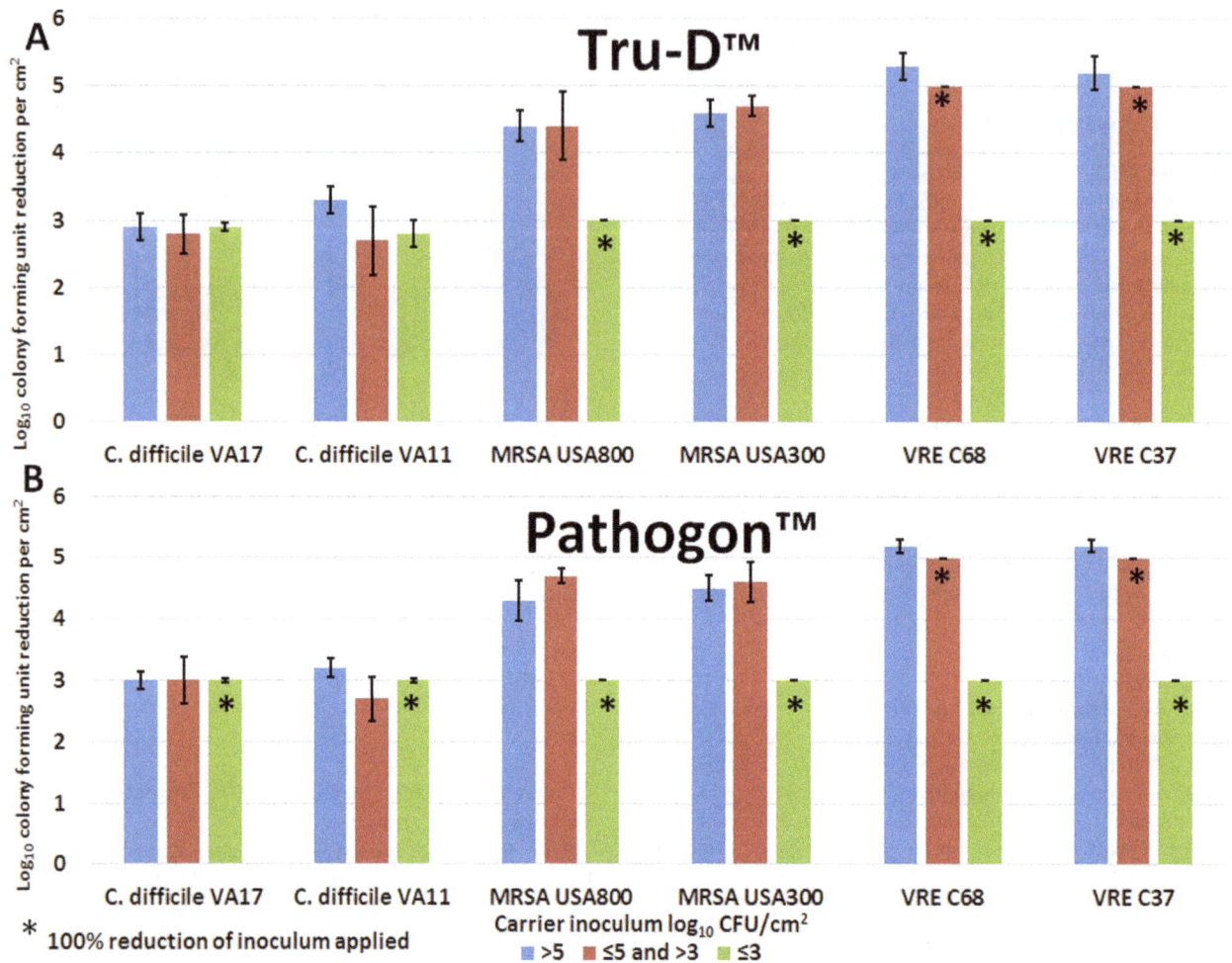

Figure 2. The effect of pathogen concentration on the efficacy of the UV-C devices. The log_{10}CFU reduction/cm^2 of two strains of *C. difficile* spores, MRSA, and VRE inoculated onto carriers. Carriers contained either >5, ≤5 and >3, or ≤3 log_{10}CFU of each pathogen. The carriers were irradiated for 41 minutes at a distance of 4 feet from the Tru-D (2A) or Pathogon (2B) device. The means of the data from experiments conducted in triplicate are presented. Error bars indicate standard error.

suspensions were plated onto selective media as described in *Microbiology*. Following 48 hours of incubation, log_{10} colony forming unit (CFU) reductions were calculated by comparing the log_{10}CFU recovered from carriers post UV-C disinfection to untreated controls. All experiments were repeated three times.

The Effect of Distance and Indirect Irradiation. The killing efficacies of the UV-C devices were evaluated at increasing distances and shaded from the direct field of radiation. Carriers were prepared and processed as described above in *The Effect of Pathogen Concentration and Organic Load*, however, the organisms were suspended in PBS and altered such that each carrier yielded 6 log_{10} CFU at baseline. Additionally, carriers were placed 6 inches, 4 feet and 10 feet within the direct field of radiation, and also 4 feet shaded from direct radiation. The UV-C devices were run for 41 minutes.

The Effect of Radiant Dose. The effect of radiant dose on the killing efficacies of the UV-C devices was determined for vegetative organism (MRSA and VRE) and spores (*C. difficile*). Carriers were prepared and processed as described above in *The Effect of Pathogen Concentration and Organic Load*, however, the organism were suspended in PBS and altered such that each carrier yielded 6 log_{10} CFU at baseline. Carriers were placed 4

feet from the device in the direct field of UV-C and irradiated for either 10, 20, or 40 minutes.

Statistical Analysis

Data were analyzed using STATA 9.0 software (StataCorp, College Station, TX). Continuous data were analyzed using paired *t* tests.

Results

Efficacy of the UV-C Devices for the Reduction of Pathogens on Carriers

The Effect of Pathogen Concentration and Organic Load. Figures 2A and 2B show the mean log_{10}CFU/cm^2 reductions of two strains of *C. difficile*, MRSA, and VRE on carriers after the use of the Tru-D and Pathogon devices, respectively. There was no significant differences between the log_{10}CFU reductions of the two strains of each pathogen tested. Therefore, in subsequent experiments, the two strains were calculated collectively in the mean. The concentration of pathogens on a surface (≤3 to>5 log_{10}CFU) did not have an impact on the killing efficacy of the UV-C devices. Furthermore,

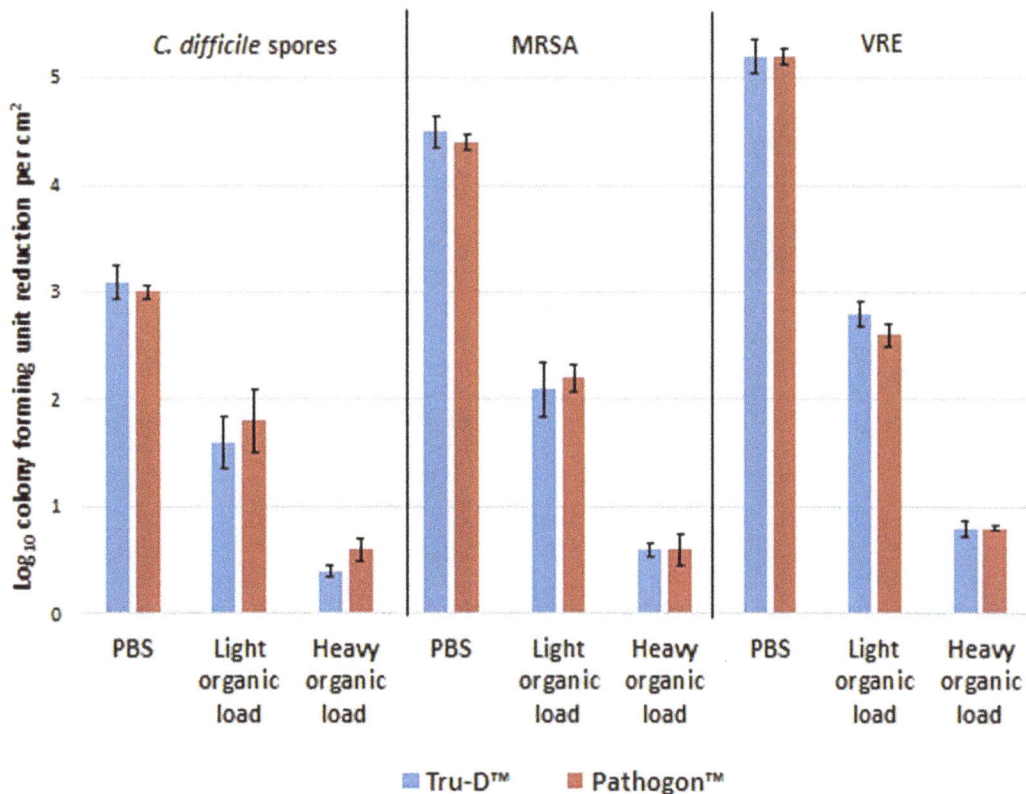

Figure 3. The effect of organic load on the efficacy of the UV-C devices. The $\log_{10}$CFU reduction/cm^2 of *C. difficile* spores, MRSA, and VRE suspended in phosphate-buffered saline (PBS), light organic load (5% fetal calf serum), or heavy organic load (5% fetal calf serum, 5% tryptone). Carriers contained 6 $\log_{10}$CFU of each pathogen. The carriers were irradiated for 41 minutes at a distance of 4 feet from the Tru-D or Pathogon device. The means of the data from experiments conducted in triplicate are presented. Error bars indicate standard error.

the UV-C devices were equally effective for reducing pathogens. There was no significant difference observed in the $\log_{10}$CFU reductions achieved by the Tru-D or Pathogon device for the two strains of each pathogen assessed (Tru-D vs. Pathogon: $P = 0.57$ (*C.difficile*), $P = 1.0$ (MRSA) and $P = 0.97$ (VRE)). Irradiation delivered by the Tru-D and Pathogon devices for 41 minutes (spore killing dose) consistently reduced *C. difficile* spores by ∼ 3 $\log_{10}$CFU/cm^2, MRSA by >4 $\log_{10}$CFU/cm^2, and VRE by > 5 $\log_{10}$CFU/cm^2.

Figure 3 shows the effects of a light and heavy organic load on the killing efficacy of the Tru-D and Pathogon device. Both the light (5% fetal calf serum) and heavy organic load (5% fetal calf serum, 5% tryptone) had a significant deleterious impact on the efficacy of the devices. The light organic load decreased the log reductions achieved by the devices to <2 $\log_{10}$CFU/cm^2 for *C.difficile* spores, <2.5 $\log_{10}$CFU/cm^2 for MRSA, and < 3 $\log_{10}$CFU/cm^2 for VRE. The heavy organic load had a more dramatic effect, decreasing the log reduction to <1 $\log_{10}$CFU/cm^2 for each pathogen assessed.

The Effect of Distance and Indirect Irradiation. The germicidal efficacy of UV-C light as a function of distance follows an inverse relationship, as shown in Figure 4. The Tru-D and Pathogon devices achieved analogous log reductions for each distance assessed. Six inches away from the device, vegetative organisms (MRSA and VRE) were completely eliminated (≥ 6 $\log_{10}$CFU/cm^2) and *C. difficile* spores were reduced by > 4 $\log_{10}$CFU/cm^2. As the distance from the device was increased to 4 feet, the log reduction decreased to ≤5 $\log_{10}$CFU/cm^2 for vegetative organisms and ≤3 $\log_{10}$CFU/cm^2 for *C. difficile*

spores. Shading the organisms from the direct field of radiation did not have a significant impact on the killing efficacy of the devices. Ten feet from the devices, the log reductions decreased further to ≤3 $\log_{10}$CFU/cm^2 for vegetative organisms and < 2 $\log_{10}$CFU/cm^2 for *C.difficile* spores.

The Effect of Radiant Dose. Figure 5 shows the effect of radiant dose on the killing efficacy of the UV-C devices. There was no significant difference between the killing efficacies of the Tru-D or Pathogon device for each of the timed doses of irradiation delivered. Killing achieved by the UV-C devices was directly proportional to the dose of irradiation delivered. MRSA and VRE were reduced by >3 $\log_{10}$CFU/cm^2 after only 10 minutes of irradiation, while the hardier *C. difficile* spores required 40 minutes of irradiation to achieve a >3 $\log_{10}$CFU/cm^2 reduction.

Discussion

We found that the Tru-D and Pathogon devices were equally effective for killing *C. difficile* spores, MRSA, and VRE in a laboratory setting. Surfaces in a real-world setting contain variable levels of contamination, and in our experience, yield between 4 to <1 $\log_{10}$CFU when cultured (author's unpublished data). Here, we determined that the concentration of pathogens on a surface did not have a significant impact on the killing efficacy of the UV-C devices. Conversely, organic load did significantly reduce the killing efficacy of both devices. These findings are inconsistent with previously published data showing that organic load did not impact the killing efficacy of the Tru-D device [16]. However, the organic load used in the current study was much more

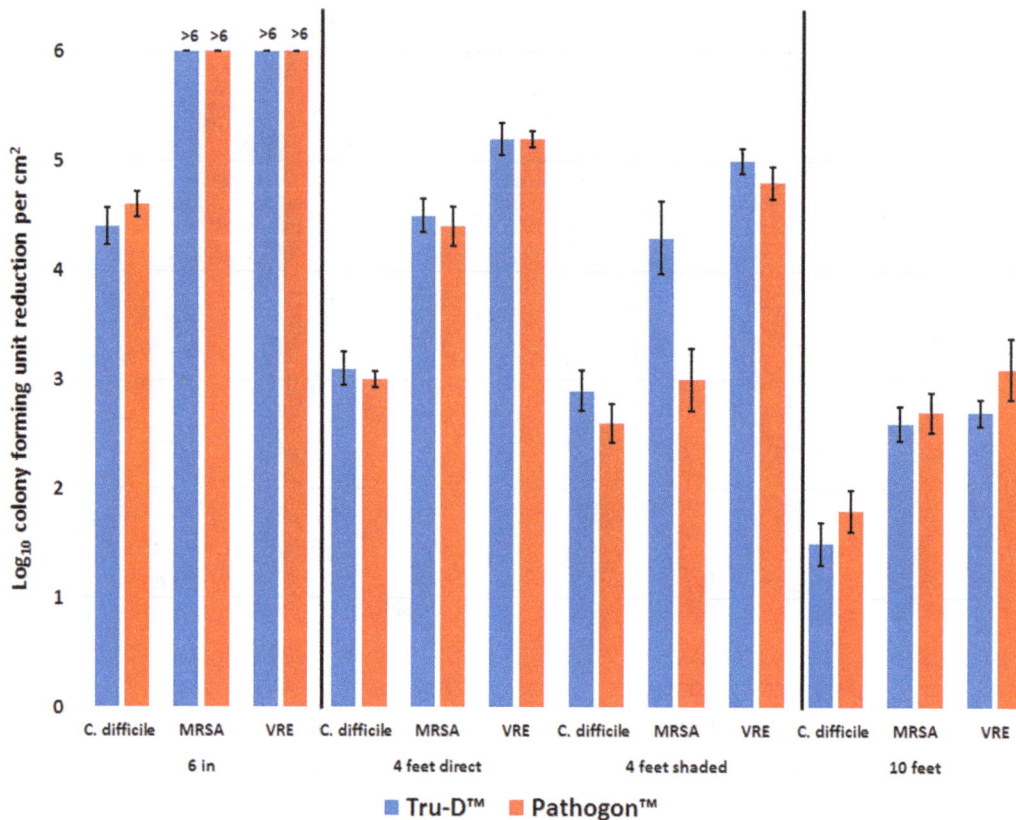

Figure 4. The effect of distance and indirect irradiation on the efficacy of the UV-C devices. The log$_{10}$CFU reduction/cm^2 of *C.difficile* spores, MRSA, and VRE at increasing distances and shaded from the direct field of radiation delivered by the UV-C device. Carriers contained 6 log$_{10}$CFU of each pathogen. The carriers were irradiated for 41 minutes at a distance of 6 in, 4 feet, 4 feet shaded, and 10 ft from the Tru-D or Pathogon device. The means of the data from experiments conducted in triplicate are presented. Error bars indicate standard error.

concentrated (5% fetal calf serum or 5% fetal calf serum plus 5% tryptone) than in the previously published study (1% bovine serum albumin). The current study demonstrates that as the matrix of organic load increased from light to heavy, the killing efficacy of the devices decreased, suggesting that UV-C light does not penetrate heavy soils, but may break through lighter organic loads.

Other factors known to impact the delivery of lethal doses of UV-C irradiation are distance from the device and time of radiant exposure [16–17]. The efficacy of the Tru-D and Pathogon devices significantly decreased as distance from the devices increased. We have previously demonstrated that shading from the direct field of irradiation inhibited the lethal effects of the Tru-D device (assessed at 10 feet from the device) [16]. Conversely, here we found that at 4 feet from the devices, shading did not have a significant impact on the killing efficacy of UV-C. For killing of *C. difficile* spores, time of UV-C exposure was of particular importance. While vegetative organisms were reduced by 3 log$_{10}$CFU after only 10 minutes of exposure, it took 40 minutes to achieve the same level of reduction for *C. difficile* spores. These results suggest that the Tru-D and Pathogon devices are similarly effective at delivering lethal doses of UV-C irradiation under analogous conditions. And as previously demonstrated for the Tru-D device, the Pathogon device may be a promising new environmental disinfection technology that could be a useful adjunct to routine cleaning measures in healthcare facilities.

UV-C devices have important advantages over other disinfection strategies that are effective against *C. difficile* spores. Sodium hypochlorite has corrosive effects on various materials, may irritate

the eyes and respiratory tracts of cleaning staff and patients, and the efficacy is dependent on correct application by housekeeping staff [20]. Hydrogen peroxide vapor and hydrogen peroxide dry-mist have been shown to be highly effective in elimination of *C. difficile* spores [20–22]. However, these systems are relatively expensive to operate, a dedicated staff is required, and up to several hours may be required to complete room disinfection [20–22]. In contrast, after the initial purchase of the UV-C device, the cost of operating and maintaining them is minimal (i.e., electricity and annual bulb replacement of ~ $20 each), a dedicated staff is not essential, and a 3 log$_{10}$CFU reduction in *C. difficile* spores can be achieved in less than an hour. Additionally, UV-C may be less damaging to surfaces than bleach and does not produce emissions that are harmful or irritating to operators.

The Tru-D and Pathogon devices do have some potential limitations. First, because spores require a minimum of 40 minutes of irradiation to achieve significant reductions, it may not be feasible to use the devices in circumstances where rapid turn-over of rooms is required. Second, surface properties and organic debris may potentially inhibit lethal doses of UV-C from killing pathogens. For example, UV-C does not penetrate porous surfaces such as sheets, upholstery and curtains [15]. In our current study, lethal doses of UV-C irradiation were significantly or completely inhibited by organic matrices. Finally, the efficacy of the UV-C devices was reduced at sites further from the devices. Therefore, it is recommended that commonly touched surfaces (e.g., bedside table, call button, telephone) be arranged close to the device for optimal exposure to UV-C radiation.

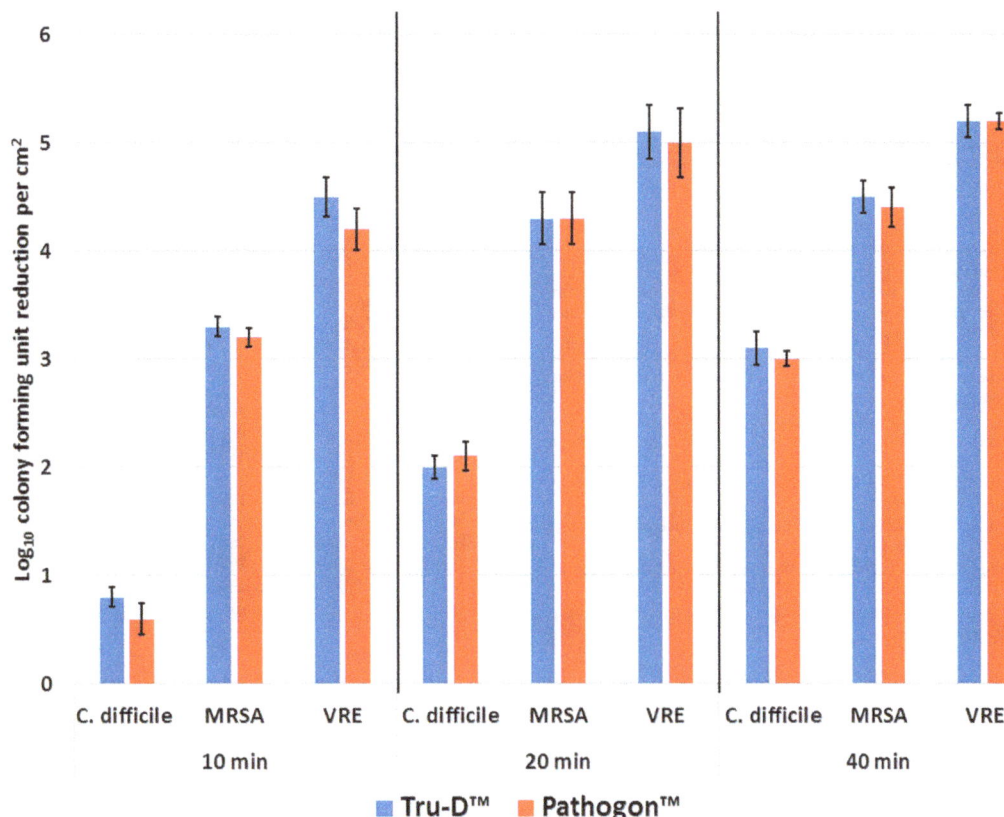

Figure 5. The effect of radiant dose on the efficacy of the UV-C devices. The $\log_{10}$CFU reduction/cm^2 of *C.difficile* spores, MRSA, and VRE after receiving increasing timed doses of irradiation delivered by the UV-C devices. Carriers contained 6 $\log_{10}$CFU of each pathogen. The carriers were irradiated for 10, 20, 40 minutes at a distance of 4 feet from the Tru-D or Pathogon device. The means of the data from experiments conducted in triplicate are presented. Error bars indicate standard error.

Both UV-C disinfection systems performed similarly in a laboratory setting, however, each system has certain advantages and limitations. The Tru-D device is unique in that it uses UV-C light reflected from the walls, ceilings, floors, and items in the room to calculate the amount of irradiation required to deliver a programmed lethal dose for either vegetative or spore-forming pathogens. This feature is advantageous because it delivers a customized dose of irradiation to each room based on the areas of the room that are hardest for light to penetrate. However, this advantage translates into longer cycle times for rooms that inhibit the reflection of light, often increasing the run time to greater than 50 minutes for a 10×10 foot room. On the other hand, the Pathogon device delivers a pre-programmed dose of irradiation that is configured based on the size of the room and type of pathogen contamination suspected, therefore, cycle time does not fluctuate. However, shaded areas of the room may not receive sufficient lethal doses of irradiation. In our opinion, the Pathogon control interface is more user-friendly, however, the next generation of the Tru-D device has been updated with an iPad interface. Last, the Pathogon device is significantly less expensive than the Tru-D device, however after the initial cost, both units require equivalent care and maintenance.

Our study does have some limitations. This study was not designed to address the impact of the UV-C devices on native pathogens found on surfaces in hospital rooms. Nevertheless, the two devices performed equivalently in a laboratory setting, and because the Tru-D device has been previously reported as effective for significantly reducing *C. difficile*, MRSA, and VRE contamination in hospital rooms it can be inferred that the Pathogon device may perform similarly. Further studies are needed to determine whether reductions achieved by these devices translates to reduced rates of infection.

Acknowledgments

We thank Steris Corporation, Mentor, OH for providing the Pathogon unit used in this study free of charge. Steris Corporation did not provide funding for the study, however C.W.F contributed to the study design and editing of the manuscript.

Author Contributions

Conceived and designed the experiments: MMN CWF CJD. Performed the experiments: MMN CWF. Analyzed the data: MMN CWF CJD. Contributed reagents/materials/analysis tools: CWF CJD. Wrote the paper: MMN CJD.

References

1. Goodman ER, Platt R, Bass R, Onderdonk AB, Yokoe DS, et al. (2008) Impact of an environmental cleaning intervention on the presence of methicillin-resistant *Staphylococcus aureus* and vancomycin-resistant enterococci on surfaces in intensive care unit rooms. Infect Control Hosp Epidemiol 29: 593–99.

2. Weber DJ, Anderson D, Rutala WA (2013) The role of the surface environment in healthcare associated infections. Curr Opin Infect Dis 26(4): 338–44.

3. Sitzlar B, Deshpande A, Fertelli D, Kundrapu S, Sethi AK, et al. (2013) An environmental disinfection odyssey: evaluation of sequential interventions to

improve disinfection of Clostridium difficile isolation rooms. Infect Control Hosp Epidemiol 34(5): 459–65.

4. Boyce JM, Potter-Bynoe G, Chenevert C, King T (1997) Environmental contamination due to methicillin-resistant *Staphylococcus aureus* (MRSA): possible infection control implications. Infect Control Hosp Epidemiol 18: 622–7.

5. Eckstein BC, Adams DA, Eckstein EC, Rao A, Sethi AK, et al. (2007) Reduction of *Clostridium difficile* and vancomycin-resistant Enterococcus contamination of environmental surfaces after an intervention to improve cleaning methods. BMC Infect. Dis 7: 61.

6. Hayden MK, Bonten JM, Blom DW, Lyle EA, van de Vijver D, et al. (2006) Reduction in acquisition of vancomycin-resistant *Enterococcus* after enforcement of routine environmental cleaning measures. Clin Infect Dis 42: 1552–60.

7. Hacek DM, Ogle AM, Fisher A, Robicsek A, Peterson LR (2010) Significant impact of terminal room cleaning with bleach on reducing nosocomial *Clostridium difficile*. Am J Infect Control 38(5): 350–3.

8. Carling PC, Parry MF, Bruno-Murtha LA, Dick B (2010) Improving environmental hygiene in 27 intensive care units to decrease multidrug-resistant bacterial transmission. Crit Care Med 38: 1212–4.

9. Conner-Kerr TA, Sullivan PK, Gaillard J, Jones RM (1998) The effects of ultraviolet radiation on antibiotic-resistant bacteria *in vitro*. Ostomy Wound Manage 44: 50–6.

10. Griego VM, Spence KD (1977) Inactivation of *Bacillus thuringiensis* spores by ultraviolet and visible light. Appl Env Microbiol 35: 906–10.

11. Hercik F (1936) Action of ultraviolet light on spores and vegetative forms of *Bacillus megatherium* sp. J Gen Physiol 20: 589–94.

12. Setlow P (2006) Spores of *Bacillus subtilis*: their resistance to and killing by radiation, heat and chemicals. J Appl Microbiol 101: 514–25.

13. Setlow P (2001) Resistance of spores of *Bacillus* species to ultraviolet light. Environ Mol Mutagen 38: 97–104.

14. Thai TP, Keast DH, Campbell KE, Woodbury MG, Houghton PE (2005) Effect of ultraviolet light C on bacterial colonization in chronic wounds. Ostomy Wound Manage 51: 32–45.

15. Owens MU, Deal DR, Shoemaker MO, Knudson GB, Meszaros JE, et al. (2005) High-dose ultraviolet C light inactivates spores of *Bacillus subtilis* var. niger and *Bacillus anthracis* Sterne on non-reflective surfaces. Appl Biosafety 10: 240–7.

16. Nerandzic MM, Cadnum JL, Pultz MJ, Donskey CJ (2010) Evaluation of an automated ultraviolet radiation device for decontamination of *Clostridium difficile* and other healthcare-associated pathogens in hospital rooms. BMC Infect Dis 10: 197.

17. Rutala WA, Gergen MF, Weber DJ (2010) Room decontamination with UV radiation. Infect Control Hosp Epidemiol 31(10): 1025–9.

18. Sorg JA, Sonenshein AL (2008) Bile salts and glycine as cogerminants for *Clostridium difficile* spores. J Bacteriol 190(7): 2505–2512.

19. Nerandzic MM, Donskey CJ (2009) Effective and reduced-cost modified selective medium for isolation of *Clostridium difficile*. J Clin Microbiol 47(2): 397–400.

20. Barbut F, Menuet D, Verachten M, Girou E (2009) Comparison of the efficacy of a hydrogen peroxide dry-mist disinfection system and sodium hypochlorite solution for reduction of *Clostridium difficile* spores. Infect Control Hosp Epidemiol 30: 507–14.

21. Boyce JM, Havill NL, Otter JA, McDonald LC, Adams NM, et al. (2008) Impact of hydrogen peroxide vapor room decontamination on *Clostridium difficile* environmental contamination and transmission in a healthcare setting. Infect Control Hosp Epidemiol 29: 723–9.

22. Boyce JM (2009) New approaches to decontamination of rooms after patients are discharged. Infect Control Hosp Epidemiol 30: 515–7.

18

In Vitro Antibacterial Activity of a Novel Resin-Based Pulp Capping Material Containing the Quaternary Ammonium Salt MAE-DB and Portland Cement

Yanwei Yang[1]**◕**, **Li Huang**[2]*◕, **Yan Dong**[1]◕, **Hongchen Zhang**[3], **Wei Zhou**[1], **Jinghao Ban**[1], **Jingjing Wei**[1], **Yan Liu**[1], **Jing Gao**[1], **Jihua Chen**[1]*

1 State Key Laboratory of Military Stomatology, Department of Prosthodontics, School of Stomatology, Fourth Military Medical University, Xi'an, China, **2** State Key Laboratory of Military Stomatology, Department of General Dentistry and Emergency, School of Stomatology, Fourth Military Medical University, Xi'an, China, **3** Department of Clinical Nursing, School of Nursing, Fourth Military Medical University, Xi'an, China

Abstract

Background: Vital pulp preservation in the treatment of deep caries is challenging due to bacterial infection. The objectives of this study were to synthesize a novel, light-cured composite material containing bioactive calcium-silicate (Portland cement, PC) and the antimicrobial quaternary ammonium salt monomer 2-methacryloxylethyl dodecyl methyl ammonium bromide (MAE-DB) and to evaluate its effects on *Streptococcus mutans* growth in vitro.

Methods: The experimental material was prepared from a 2:1 ratio of PC mixed with a resin of 2-hydroxyethylmethacrylate, bisphenol glycerolate dimethacrylate, and triethylene glycol dimethacrylate (4:3:1) containing 5 wt% MAE-DB. Cured resin containing 5% MAE-DB without PC served as the positive control material, and resin without MAE-DB or PC served as the negative control material. Mineral trioxide aggregate (MTA) and calcium hydroxide (Dycal) served as commercial controls. *S. mutans* biofilm formation on material surfaces and growth in the culture medium were tested according to colony-forming units (CFUs) and metabolic activity after 24 h incubation over freshly prepared samples or samples aged in water for 6 months. Biofilm formation was also assessed by Live/Dead staining and scanning electron microscopy.

Results: *S. mutans* biofilm formation on the experimental material was significantly inhibited, with CFU counts, metabolic activity, viability staining, and morphology similar to those of biofilms on the positive control material. None of the materials affected bacterial growth in solution. Contact-inhibition of biofilm formation was retained by the aged experimental material. Significant biofilm formation was observed on MTA and Dycal.

Conclusion: The synthesized material containing HEMA-BisGMA-TEGDMA resin with MAE-DB as the antimicrobial agent and PC to support mineralized tissue formation inhibited *S. mutans* biofilm formation even after aging in water for 6 months, but had no inhibitory effect on bacteria in solution. Therefore, this material shows promise as a pulp capping material for vital pulp preservation in the treatment of deep caries.

Editor: Zezhang Wen, LSU Health Sciences Center School of Dentistry, United States of America

Funding: This research was financially supported by grants from the National Natural Science Foundation of China (No. 81200816, No. 81130078 and No. 81200823) and by the Program for Changjiang Scholars and Innovative Research Team in University (No. IRT13051). The funders had no role in study design, data collection and analysis, decision to publish, or preparation of the manuscript.

Competing Interests: The authors have declared that no competing interests exist.

* Email: 519huang@fmmu.edu.cn (LH); jhchen@fmmu.edu.cn (JHC)

◕ These authors contributed equally to this work.

Introduction

Pulpal vitality is critical to the maintenance of the structural integrity and normal physiological function of teeth. As our understanding of the importance of pulp in tooth health increases, methods for preserving pulp vitality during caries treatment even after exposure during caries removal are in great demand [1]. Currently, pulp capping is the primary method for preserving vital pulp, but the success rate of this approach during the treatment of deep caries is low at only 33% [2]. The presence of bacteria is the major reason for failure [3]. Bacteria located in deep caries can

induce severe inflammatory reactions in the pulp and even cause pulp necrosis [4]. Therefore, the prevention of bacterial infections is an important objective for improving pulp capping methods in the treatment of deep caries.

In general, an ideal pulp capping material should possess both excellent antibacterial properties and the ability to induce mineralized tissue formation [5]. Currently, the most common pulp capping materials used clinically include various formulations of calcium hydroxide [$Ca(OH)_2$] and mineral trioxide aggregate (MTA). MTA has been shown to induce less pulp inflammation and greater dentin bridge formation as well as offer superior structural qualities compared to Ca(OH)$_2$ [6–8]. The better performance of MTA compared to Ca(OH)$_2$ may be due to continued dissolution of Ca(OH)$_2$ paste, which has a prolonged irritant effect (release of basic ions) on pulp tissues [6]. Although both Ca(OH)$_2$ and MTA promote the formation of mineralized tissue, they lack good antibacterial properties [9–11], and thus, cannot prevent the bacterial infection that commonly leads to treatment failure in cases of deep caries. Therefore, a novel pulp capping material that offers a combination of excellent antibacterial properties and the ability to induce mineralized tissue formation is highly desired.

Dental resins modified with a quaternary ammonium salt (QAS) have been shown to have excellent antibacterial properties [12–19]. QAS monomers such as 12-methacryloyloxydodecylpyridinium bromide (MDPB) and other antibacterial monomers can be copolymerized with dental resins to form antibacterial polymer matrices that effectively inhibit bacterial growth [12–19]. Our research group developed a novel QAS monomer, 2-methacryloxylethyl dodecyl methyl ammonium bromide (MAE-DB), which contains two polymerizable methacrylate groups that facilitate its facile polymerization with dental resin monomers and other MAE-DB monomers [20]. MAE-DB exhibits strong bactericidal action against oral bacteria [20] and can be copolymerized with dental resin monomers to form an antibacterial composite resin that effectively inhibits bacterial growth even after a 6-month aging process [21].

The ability of MTA to induce mineralized tissue formation has been attributed to its components [6–8,22]. According to the manufacturer, MTA is formed by mechanically mixing three powder ingredients: Portland cement (PC, 75%), bismuth oxide (20%), and gypsum (5%) [23,24]. Thus, PC is the major component of MTA [24,25], and several studies have demonstrated that PC shares the same physical and chemical properties with MTA [26–28] as well as the same antimicrobial activity [11], biocompatibility [29,30], and pulp capping effectiveness [22].

To create a novel pulp capping material that can both prevent bacterial infection and support mineralized tissue formation, we synthesized a composite resin containing MAE-DB and PC. We then investigated the immediate and long-term antibacterial effects of this new light-cured pulp capping material against *Streptococcus mutans* in vitro.

Materials and Methods

Specimen preparation and aging treatment

The structure of the QAS monomer MAE-DB is presented in Figure 1. The resin matrix of the experimental light-curable material evaluated in this study was composed primarily of 2-hydroxyethylmethacrylate (HEMA, Sigma–Aldrich, St. Louis, MO, USA), bisphenol glycerolate dimethacrylate (BisGMA, Esstech, Essington, PA, USA), and triethylene glycol dimethacrylate (TEGDMA, Esstech) with a mass ratio of 4:3:1 (Table 1). The photoinitiator camphorquinone (CQ, Sigma-Aldrich) and coinitiator ethyl 4-(dimethylamino)benzoate (EDMAB, Sigma-Aldrich) were added at concentrations of 0.5 wt% of the resin matrix each. The compositions of the experimental (HEMA-BisGMA-TEGDMA resin with MAE-DB and PC), positive control (HEMA-BisGMA-TEGDMA resin with MAE-DB only), and negative control (HEMA-BisGMA-TEGDMA resin without MAE-DB or PC) materials are listed in Table 1. In the experimental material, MAE-DB monomer was added as an immobilized bactericide at 5 wt% in the HEMA-BisGMA-TEGDMA resin, and white PC (P. W. 52.5, Aalborg, Anqing,

China) was then added at a PC:HEMA-BisGMA-TEGDMA resin mass ratio of 2:1. HEMA-BisGMA-TEGDMA resin containing 5 wt% MAE-DB without PC served as the positive control. HEMA-BisGMA-TEGDMA resin without MAE-DB or PC served as the negative control. Two common pulp capping materials, white ProRoot MTA (Dentsply, Tulsa, OK, USA) and Dycal (Dentsply, Milford, DE, USA), were used for comparisons to commercially available materials.

For sample preparation, the experimental, positive control, and negative control materials were placed into disk-shaped organic glass molds (inner diameter of 10 mm and depth of 1.5 mm). The top and bottom surfaces were covered with a Mylar strip and a microscope slide, which was slightly pressed to remove excess material. The resins were photo-cured for 60 s on each side with a light activation unit (QHL75, Dentsply). After the resin disks were removed from the mold, they were submersed in distilled water with agitation for 1 h for the removal of any uncured monomer [31]. According to the manufacturer's instructions, MTA was mixed with sterile water (powder:liquid ratio of 3:1), and Dycal was prepared by mixing equal amounts of catalyst paste and base paste. The same organic glass molds and procedures (except for light irradiation) were used to prepare disks of these materials. All of the prepared disks were sterilized with ethylene oxide, followed by degassing in the fuming cupboard for more than 48 h before testing [17,31].

For aging, specimens of each group were placed in wells of a 24-well plate containing 1 ml deionized water, which was changed every 48 h. After aging for 6 months at 37°C, specimens were retrieved, sterilized, and subjected to the following experiments [17].

Bacterial strain and culture conditions

Streptococcus mutans UA159 (State Key Laboratory of Military Stomatology, School of Stomatology, Fourth Military Medical University, Xi'an, China) was cultured overnight at 37°C in brain-heart infusion (BHI) broth (Difco, Detroit, MI, USA) in an anaerobic atmosphere enriched with 5% CO$_2$. The resulting bacterial suspension was adjusted to an optical density (OD) of 0.5 at 600 nm and then diluted 1:100 with fresh BHI for further use [31].

Bacterial growth on material surfaces and in culture medium

The sterile disks prepared for testing with or without aging were placed in wells of a 24-well plate with 2 ml BHI broth. Then 20 µl of the diluted *S. mutans* suspension was added to each well. After 24 h in anaerobic culture, biofilm formation on the disk and the planktonic bacteria in the culture medium were assessed using the following experimental techniques [31].

The total number of viable bacteria was evaluated according to the number of colony-forming units (CFU) both on the disk surface and in the culture medium over each disk. After bacteria within biofilms are properly dispersed and diluted, each viable bacterium results in a single, countable colony on an agar plate. After 24 h in culture to allow biofilm growth, disks were washed twice with PBS

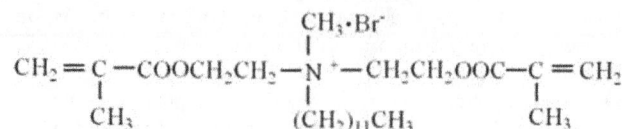

Figure 1. Structure of the QAS monomer MAE-DB.

Table 1. Compositions of experimental resin composites.

Group	HEMA-BisGMA-TEGDMA %	CQ %	EDMAB %	MAE-DB %	Portland cement %
Negative control	99	0.5	0.5	0	0
Positive control	94	0.5	0.5	5	0
Experimental material	31.33	0.17	0.17	1.67	66.7

Data are given in mass percentages.
HEMA-BisGMA-TEGDMA refers to a HEMA, BisGMA, and TEGDMA mixture at a 4:3:1 ratio.
Note that 5% MAE-DB in HEMA-BisGMA-TEGDMA resin equals 1.67% of the total content upon the addition of PC.

and then transferred into tubes (15-ml sterile centrifuge tubes, Nest, China) with 2 ml fresh BHI. Biofilms on individual disks were harvested by sonication (3510R, Branson, Danbury, CT, USA) for 3 min and vortex mixing at maximum speed for 20 s using a vortex mixer (Fisher Scientific, Pittsburgh, PA, USA), for removing and dispersing the bacteria [19].

Once the disks had been removed from the wells for biofilm harvesting, planktonic bacteria in the original medium samples were mixed thoroughly by repeated pipetting to achieve a homogeneous bacterial suspension. The bacterial suspensions from both the biofilms on the disks and the planktonic bacteria in the medium were serially diluted, spread onto BHI agar plates, and incubated for 1 day at 5% CO_2 and 37°C for CFU analysis (n = 6), following previously reported methods [19,31].

Bacterial metabolic activity on material surfaces and in culture medium

The bacterial suspensions obtained from biofilms formed on the disks and planktonic bacteria in the medium were prepared as described in Section 2.3. After brief mixing via repeated pipetting, 200-μl aliquots of the bacterial suspensions were transferred to wells of a 96-well plate, and then 20 μl of Cell Counting Kit-8 (CCK-8) dye solution was added to each well and incubated at 37°C in 5% CO_2 for 2 h. Instead of 3-(4,5-dimethylthiazol-2-yl)-2,5-diphenyltetrazolium bromide (known as MTT), the CCK-8 assay uses WST-8(2-(2-methoxy-4-nitrophenyl)-3-(4-nitrophenyl)-5-(2,4-disulfophenyl)-2H-tetrazolium, monosodium salt, which produces a yellow, water-soluble formazan production upon reduction by dehydrogenases in metabolically active bacteria. The absorbance at 450 nm of the resulting solution in each well was measured using a microplate reader (SpectraMax M5, Molecular Devices, Sunnyvale, CA). Each sample was assayed in triplicate, and an average value was calculated for each sample. A higher absorbance value indicates a higher formazan concentration, which in turn indicates the presence of more metabolically active bacteria in the sample.

Live/dead staining for visualization of S. mutans viability on material surfaces

Biofilm formation on sample disks during 24 h in culture was achieved as described in Section 2.3. The disks coated with biofilms were washed three times with sterile saline to remove loose bacteria, and then the remaining bacteria were stained using the Live/Dead BacLight Bacterial Viability Kit L13152 (Molecular Probes, Invitrogen, Eugene, OR, USA) with a 15-min incubation in the dark at room temperature. With this staining kit, live bacteria produce green fluorescence upon staining with Syto 9, and bacteria with compromised membranes produce red fluorescence upon staining with propidium iodide [21,32]. After

incubation with the fluorescent dyes, the samples were rinsed gently with distilled water and observed by confocal laser scanning microscopy (CLSM, FluoView FV1000, Olympus, Tokyo, Japan). Excitation with a 488-nm laser revealed the green fluorescence emission of live bacteria, and excitation with a 543-nm laser revealed the red fluorescence emission of bacteria with damaged membranes [21]. Three disks were used for each condition (type of material and aging status). Four images were collected at random locations on each disk, yielding 12 images per condition.

Scanning electron microscopy (SEM) of S. mutans on the tested material surfaces

Biofilm formation on sample disks during 24 h in culture was achieved as described in Section 2.3. Then the disks coated with biofilms were gently rinsed with PBS, soaked in 3% glutaraldehyde at 4°C overnight, washed twice with PBS, dehydrated in a graded series of ethanol solutions, and then dried in a critical-point drier [21]. After sputter coating of the samples with gold using an ion sputter (JFC-1100E, JEOL, Tokyo, Japan), all specimens were observed by field emission SEM (FESEM; S-4800; Hitachi Ltd, Tokyo, Japan).

Statistical analysis

One-way and two-way analyses of variance (ANOVAs) were performed to detect the significant effects of the variables (material type and aging status) on CFU count and bacterial metabolic activity. Tamhane multiple comparison test was used to compare differences between any two groups, with significance assumed at a p-value of 0.05. Standard deviation (SD) values serve as estimates for the standard uncertainty associated with particular measurements.

Results

CFU counts of S. mutans on the surfaces of the tested materials and in the culture medium away from the surfaces

Table 2 shows CFU counts of S. mutans on the surfaces of the tested materials with different aging treatments. Two-way ANOVA showed that only material type had significant effect on the CFU count ($P<0.05$). Differences among all the subgroups for both fresh and aged materials were assessed by one-way ANOVA. For each material, aging had no significant effect on the CFU count ($P>0.05$). The numbers of CFUs from S. mutans biofilms on the experimental material and positive control material were significantly less by about an order of magnitude than that for the negative control material ($P<0.05$). No significant difference was observed between the experimental and positive control groups ($P>0.05$). In contrast, the CFU counts from S. mutans biofilms on MTA and Dycal were significantly greater by about one order of magnitude compared to that for the negative control material ($P<0.05$) and by about two orders of magnitude compared to those for the experimental and positive control materials ($P<0.05$ for both). No significant difference was found between MTA and Dycal ($P>0.05$). Together these results indicate that the addition of PC at a 2:1 mass ratio to HEMA-BisGMA-TEGDMA resin containing 5% MAE-DB did not diminish the ability of the antibacterial resin to inhibit *S. mutans* growth on its surface even after 6 months of aging. Conversely, the growth of *S. mutans* on the surfaces of MTA and Dycal was even greater than that on HEMA-BisGMA-TEGDMA resin without MAE-DB or PC.

Table 3 lists the CFU counts of *S. mutans* from the culture medium away from the surfaces of the tested materials with different aging conditions. Two-way ANOVA showed that both material type and aging had no significant effect on the CFU count (all $P>0.05$). In addition, one-way ANOVA revealed no significant differences among all subgroups with different aging conditions (all $P>0.05$). These results indicate that the *S. mutans* growth in the culture medium was not inhibited by the experimental material, the positive control material, or the two commercial materials in comparison to that in the medium away from the negative control material.

Metabolic activity of *S. mutans* on the surfaces of the tested materials and in the culture medium away from the surfaces

The metabolic activity data for *S. mutans* biofilms on the material surfaces are plotted in Figure 2. Two-way ANOVA showed that only material type had a significant effect on the metabolic activity of the bacteria ($P<0.05$). According to one-way ANOVA, for each material, aging had no significant effect on the metabolic activity ($P>0.05$ for all materials). Regardless of the aging condition, the greatest absorbance values and thus highest levels of metabolic activity were observed for *S. mutans* biofilms formed on MTA and Dycal ($P<0.05$ compared to all other material types). Conversely, the lowest levels of metabolic activity were observed for *S. mutans* biofilms on the experimental and positive control materials, indicating these materials had the strongest antibacterial activity ($P<0.05$ compared to all other material types). These results are consistent with those for bacterial growth in *S. mutans* biofilms on the tested material surfaces based on CFU counts.

The metabolic activity data for *S. mutans* in the culture medium away from the disk samples are plotted in Figure 3. Two-way ANOVA showed that neither material type nor aging condition

significantly affected the metabolic activity of bacteria in the culture medium (all $P>0.05$). In addition, one-way ANOVA revealed the lack of significant differences in *S. mutans* metabolic activity in the culture medium among all material types and aging conditions (all $P>0.05$). These results also are consistent with those for bacterial growth in *S. mutans* in the culture medium over the surfaces based on CFU counts.

Viability of *S. mutans* on the tested material surfaces

Figure 4 shows representative CLSM images of Live/Dead-stained *S. mutans* biofilms after 24 h of anaerobic growth on the material surfaces. Both fresh and aged negative control surfaces (Fig. 4A and F) were covered with primarily live bacteria. By contrast, both fresh and aged positive control surfaces (Fig. 4B and G) showed more dead bacteria, compared to the negative control surfaces. Compared to that on the control surfaces, the total amount of bacteria was greatly increased on both fresh and aged experimental surfaces (Fig. 4C and H), and these surfaces were covered primarily with dead bacteria. The amounts of bacteria on fresh and aged MTA surfaces (Fig. 4D and I) also were greater than on control surfaces, but the MTA surfaces were covered primarily with live bacteria. The staining results on fresh and aged Dycal surfaces (Fig. 4E and J) were qualitatively similar to those on MTA surfaces. For each material, the Live/Dead staining results were qualitatively similar between the fresh and aged surfaces. These results are consistent with those for *S. mutans* growth according to CFU count and metabolic activity in biofilms on the tested material surfaces.

SEM imaging of *S. mutans* on the tested material surfaces

Figure 5 shows representative SEM images of *S. mutans* biofilms after 24 h of anaerobic growth on the tested material surfaces. Both fresh and aged negative control surfaces (A and F) were covered primarily with live bacteria with intact membranes. Both fresh and aged positive control surfaces (B and G) were covered with considerably more dead bacteria with compromised membranes, compared to the negative control surfaces. The total amounts of bacteria on both fresh and aged experimental surfaces (C and H) were greater than those on the control surfaces, and the experimental surfaces were covered primarily with dead bacteria. The total amounts of bacteria on fresh and aged MTA surfaces (D and I) also were greater than those on control surfaces, but the MTA surfaces were covered primarily with live bacteria. The total amount and viability of bacteria on fresh and aged Dycal surfaces according to SEM imaging (E and J) were qualitatively similar to those on MTA surfaces. For each material, the SEM observations were qualitatively similar between fresh and aged samples. These

Table 2. CFU counts from *S. mutans* biofilms on material surfaces.

Material	Biofilm CFU (per disk)	
	Without aging	**With aging**
Negative control	$5.95(0.98) \times 10^{7A}$	$6.12(1.01) \times 10^{7A}$
Positive control	$3.53(0.91) \times 10^{6B}$	$3.72(0.86) \times 10^{6B}$
Experimental material	$6.14(1.13) \times 10^{6B}$	$6.27(1.02) \times 10^{6B}$
MTA	$4.71(0.76) \times 10^{8C}$	$4.64(0.81) \times 10^{8C}$
Dycal	$4.55(0.88) \times 10^{8C}$	$4.65(0.97) \times 10^{8C}$

CFU values represent the mean (SD) of six replicates, and data were analyzed with one-way and two-way ANOVA at a significance level of 0.05. Values with dissimilar superscript letters are significantly different from each other ($p<0.05$). Values with the same superscript letter are not significantly different ($p>0.05$).

Table 3. CFU counts from *S. mutans* in culture medium away from the material surfaces.

Medium sample	*S. mutans* CFU (per mL)	
	Without aging	**With aging**
Culture medium of negative control	$1.52(0.16) \times 10^{9A}$	$1.57(0.13) \times 10^{9A}$
Culture medium of positive control	$1.50(0.28) \times 10^{9A}$	$1.43(0.20) \times 10^{9A}$
Culture medium of experimental material	$1.53(0.21) \times 10^{9A}$	$1.61(0.19) \times 10^{9A}$
Culture medium of MTA	$1.41(0.23) \times 10^{9A}$	$1.49(0.16) \times 10^{9A}$
Culture medium of Dycal	$1.42(0.23) \times 10^{9A}$	$1.51(0.19) \times 10^{9A}$

CFU values represent the mean (SD) of six replicates, and data were analyzed with one-way and two-way ANOVA at a significance level of 0.05. Values with the same superscript letter are not significantly different ($p > 0.05$).

results for *S. mutans* biofilms on the tested material surfaces are consistent with those obtained by counting CFUs, measuring metabolic activity, and labeling cells with the Live/Dead fluorescent staining kit.

Discussion

S. mutans is a major pathogen causing human dental caries [21] and was therefore chosen for evaluation of the antibacterial effects of the materials prepared in this study. The present study investigated the immediate and long-term antibacterial activity of a novel light-cured pulp capping composite resin containing PC filler and MAE-DB monomer on *S. mutans* both on its surface and in solution around its surface. The results showed that this new material inhibited *S. mutans* biofilm growth and metabolic activity on its surface but had no effect on bacteria in solution. Even after 6 months of aging in water, the experimental material retained its antibacterial activity at a level similar to that observed without aging.

Bacterial infection is an important reason for failure of pulp capping procedures, especially in the treatment of caries [22,33].

Previous studies have reported that the most common pulp capping materials used clinically, such as Dycal and MTA, lack sufficient antibacterial activity [10,11]. Currently, studies on resin-modified pulp capping materials are popular. For example, Formosa *et al.* incorporated MTA into light- and chemical-cured resins to improve the ability of the resins to induce formation of mineralized tissue and to shorten the curing time of MTA [34]. Gandolfi *et al.* [35,36] incorporated calcium-silicate PC-derived (MTA-like) fillers into light-curable resins to improve their mechanical properties and reduce the curing time of the MTA-like material, and their innovative materials were shown to promote the formation of bone-like carbonated-apatite on demineralized dentin. However, these novel materials showed no improvement in antibacterial activity due to the absence of an effective antimicrobial agent. Therefore, we attempted to apply a QAS antibacterial resin in the synthesis of a pulp capping material with antibacterial properties in addition to appropriate mechanical properties and the ability to induce mineralized tissue formation toward the goal of improving success rates for pulp capping treatment.

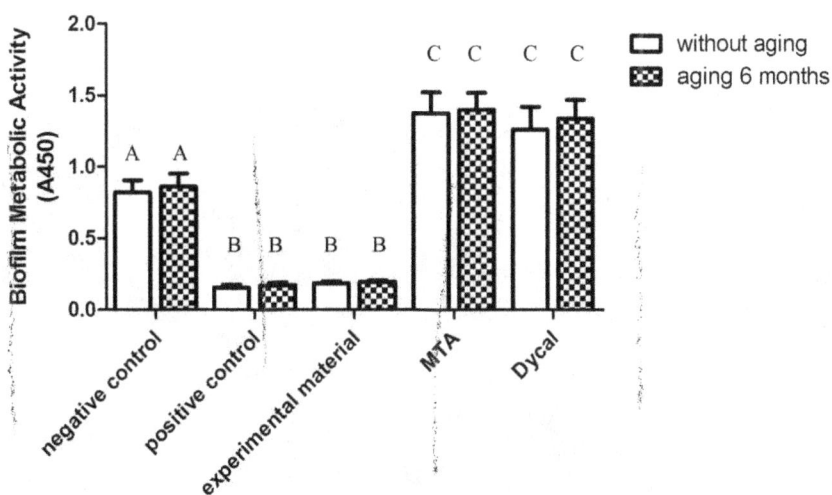

Figure 2. Metabolic activity of *S. mutans* in biofilms on material surfaces. Metabolic activity of *S. mutans* in biofilms on material surfaces (mean±SD; n = 6) for negative control, positive control, and experimental materials as well as MTA, Dycal, and the corresponding aged samples for each material. Absorbance values were analyzed with one-way and two-way ANOVA at a significance level of 0.05. Values with dissimilar letters are significantly different ($p < 0.05$). Values with the same letter are not significantly different ($p > 0.05$).

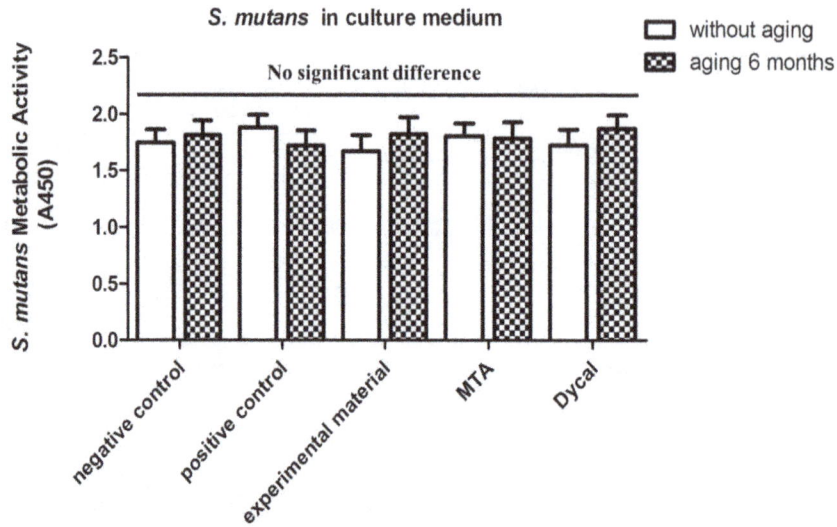

Figure 3. Metabolic activity of *S. mutans* in culture medium away from the surfaces. Metabolic activity of *S. mutans* in culture medium over the surfaces (mean±SD; n=6) of the negative control, positive control, and experimental materials as well as MTA, Dycal, and the corresponding aged samples for each material. Absorbance values were analyzed with one-way and two-way ANOVA at a significance level of 0.05, and no significant differences were observed between any conditions (p>0.05).

To achieve a material that promotes mineralized tissue formation, PC was incorporated into the novel material. We selected PC because it is the major component of MTA [24,25] and also has been judged to have the same physical, chemical, and biological properties as MTA, such as alkalinization, calcium ion leaching, curing mechanism [26–28], biocompatibility [29,30], and pulp capping effectiveness [22]. Moreover, studies using hydrophilic resins as direct pulp capping materials have reported promising results in animal models [37,38]. Optimal compatibility of the QAS antibacterial resin with pulp tissue was also confirmed by direct pulp capping experiments in dogs [39]. Therefore, both PC and a QAS antibacterial hydrophilic resin to develop are reasonable choices for the design of a new pulp capping composite material.

Previous studies investigated antibacterial resins containing a QAS monomer that inhibits bacteria on contact, such as MDPB-containing materials and adhesives incorporating DMAE-CB

[12,13,16,19,40,41]. MAE-DB, a QAS monomer developed by our research group, has two reactive groups located on either end of the molecule. QAS monomers with two reactive groups have been shown to have little effect on the mechanical properties of the resin [31] and are expected to result in minimal monomer leaching, compared with other QAS monomers based on monomethacylates [19,40]. The integration of MAE-DB in a resin matrix has also proved to be effective for providing chemically stable and long-lasting contact-inhibition of bacterial growth [20]. The antibacterial mechanism of MAE-DB involves the induction of bacterial lysis upon penetration and disruption of the cell membrane by the compound, which causes cytoplasmic leakage [21,42]. When the negatively charged bacterial membrane contacts the positively charged (N^+) sites of the QAS material, the electric balance of the cell membrane can be disturbed, and the bacterium may rupture as a result of the change in osmotic pressure [43]. Therefore, an antibacterial resin containing MAE-

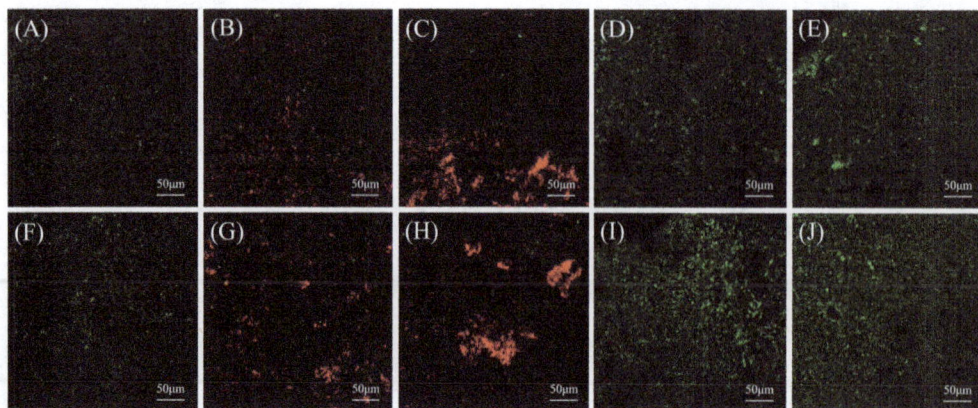

Figure 4. Representative CLSM images of Live/Dead-stained biofilms on material surfaces. Representative CLSM images of Live/Dead-stained biofilms after 24 h of anaerobic growth on the tested material surfaces: (A) negative control material, (B) positive control material, (C) experimental material, (D) MTA, and (E) Dycal. Biofilms on the corresponding aged samples are shown in (F)–(J). Live bacteria exhibited green fluorescence, and bacteria with compromised membranes exhibited red fluorescence. Scale bars, 50 μm.

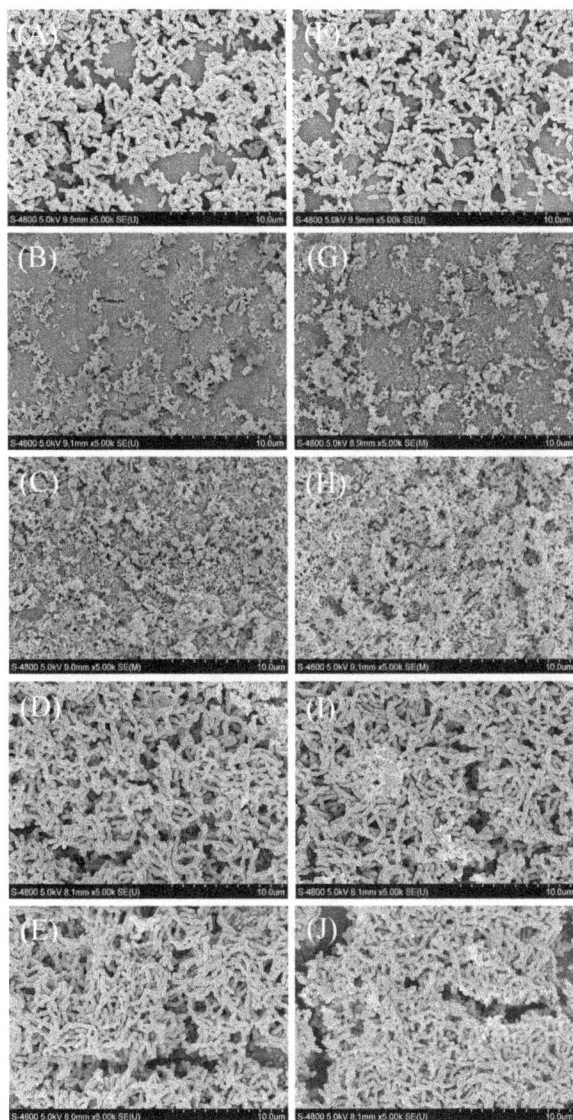

Figure 5. Representative SEM images of *S. mutans* biofilms on material surfaces. Representative SEM images of *S. mutans* biofilms after 24 h of anaerobic growth on the tested material surfaces: (A) negative control material, (B) positive control material, (C) experimental material, (D) MTA, and (E) Dycal. Biofilms on the corresponding aged samples are shown in (F)–(J).

DB may offer a pulp capping material with antibacterial activity, and this hypothesis was confirmed by the results obtained in the present study.

The present study compared the antibacterial activity of the new pulp capping material with that of two currently commercially available pulp capping materials. The results for *S. mutans* growth and metabolic activity consistently demonstrated that the experimental material exhibited stronger antibacterial activity against *S. mutans* on its surface than did MTA and Dycal, whereas none of the materials affected bacterial growth or metabolic activity in the solution surrounding the samples. These findings are consistent with the contact-inhibition characteristic of QAS resins in general [17,18,21,31]. The results of Live/Dead staining and SEM observation provide insights into the number and viability of

bacteria on the material surfaces, and these results were consistent with those for metabolic activity and CFU formation. Interestingly, compared to bacterial behavior observed on the negative control material, the CFU-based results showed that MTA and Dycal displayed greatly increased the adherence of *S. mutans* biofilms on the surfaces by about an order of magnitude ($P<0.05$, Table 2). The results obtained via Live/Dead staining, SEM observation, and metabolic activity assays were all consistent with the CFU results. The increase in biofilm formation on these surfaces may be due to the release of calcium ions from MTA and Dycal upon reaction with phosphate ions in the BHI broth to produce calcium phosphate deposits on the material surfaces [36], which increases the surface roughness and thus promotes bacterial adhesion. Moreover, although CFU counts and metabolic activity results showed no significant differences between the experimental and positive control material surfaces regarding *S. mutans* biofilm formation, the results of Live/Dead staining and SEM observation showed that more bacteria adhered to the experimental material surfaces (and consequently exhibited loss of membrane integrity) than to the positive control surfaces. This may be because experimental material could also release calcium ions to produce calcium phosphate deposits on its surface in BHI broth, like MTA and Dycal can, and increased numbers of adherent bacteria were still killed by the antibacterial activity of the material.

The developed composite resin is cured by light irradiation for 60 s, and therefore, the curing time for the new composite material containing the light-curable resin is reduced to approximately 1 minute compared to 202 minutes for MTA [44], which improves the handling characteristics of the material. The curing time for the developed material is also somewhat shorter than that of Dycal (2.5–3.5 min) [45]. Although the pulp-capping effects of QAS antibacterial resin and PC have been confirmed by previous studies [29,30,39], that of the novel material combining both materials still remains to be verified in a further *in vivo* study. Furthermore, the present study only evaluated the antibacterial activity in an *in vitro* study, and additional important parameters for pulp capping materials such as the physical properties, chemical properties, bioactivity and biological properties, need to be evaluated in further experiments. Therefore, continued assessments of the physicochemical and biological properties of the material developed in this study are in progress.

Conclusion

The results of the study indicate that, compared to commercial and negative control materials, the novel light-cured pulp capping composite containing PC as a filler and MAE-DB monomer as an antibacterial agent greatly inhibited *S. mutans* biofilm formation on its surface. Conversely, none of the tested materials had any effect on *S. mutans* growth in the culture medium surrounding the material samples. This contact-inhibition activity of the new pulp capping composite was shown to persist through 6 months of aging in water, suggesting that the developed material holds great promise for application as an antibacterial resin in pulp capping treatment.

Acknowledgments

The Endodontics Department and Microbiology Department of the Fourth Military Medical University are gratefully acknowledged for providing equipment and technical support. We are very grateful to Esstech (Essington, PA) and Aalborg (Anqing, China) for donating materials.

Author Contributions

Conceived and designed the experiments: JHC. Performed the experiments: YWY LH YD WZ JHB JJW. Analyzed the data: YWY HCZ LL JG. Contributed reagents/materials/analysis tools: JHC JG. Wrote the paper: YWY LH YD.

References

1. Hayashi M, Fujitani M, Yamaki C, Momoi Y (2011) Ways of enhancing pulp preservation by stepwise excavation–a systematic review. J Dent 39: 95–107.
2. Al-Hiyasat AS, Barrieshi-Nusair KM, Al-Omari MA (2006) The radiographic outcomes of direct pulp-capping procedures performed by dental students: a retrospective study. J Am Dent Assoc 137: 1699–1705.
3. Momoi Y, Hayashi M, Fujitani M, Fukushima M, Imazato S, et al. (2012) Clinical guidelines for treating caries in adults following a minimal intervention policy—evidence and consensus based report. J Dent 40: 95–105.
4. Bjorndal L, Reit C, Bruun G, Markvart M, Kjaeldgaard M, et al. (2010) Treatment of deep caries lesions in adults: randomized clinical trials comparing stepwise vs. direct complete excavation, and direct pulp capping vs. partial pulpotomy. Eur J Oral Sci 118: 290–297.
5. Bergenholtz G (2001) Factors in pulpal repair after oral exposure. Adv Dent Res 15: 84.
6. Tran XV, Gorin C, Willig C, Baroukh B, Pellat B, et al. (2012) Effect of a calcium-silicate-based restorative cement on pulp repair. J Dent Res 91: 1166–1171.
7. Simon SR, Berdal A, Cooper PR, Lumley PJ, Tomson PL, et al. (2011) Dentin-pulp complex regeneration: from lab to clinic. Adv Dent Res 23: 340–345.
8. Dammaschke T, Stratmann U, Wolff P, Sagheri D, Schafer E (2010) Direct pulp capping with mineral trioxide aggregate: an immunohistologic comparison with calcium hydroxide in rodents. J Endod 36: 814–819.
9. Yasuda Y, Kamaguchi A, Saito T (2008) In vitro evaluation of the antimicrobial activity of a new resin-based endodontic sealer against endodontic pathogens. J Oral Sci 50: 309–313.
10. Miyagak DC, de Carvalho EM, Robazza CR, Chavasco JK, Levorato GL (2006) In vitro evaluation of the antimicrobial activity of endodontic sealers. Braz Oral Res 20: 303–306.
11. Estrela C, Bammann LL, Estrela CR, Silva RS, Pecora JD (2000) Antimicrobial and chemical study of MTA, Portland cement, calcium hydroxide paste, Sealapex and Dycal. Braz Dent J 11: 3–9.
12. Imazato S, Torii M, Tsuchitani Y, McCabe JF, Russell RR (1994) Incorporation of bacterial inhibitor into resin composite. J Dent Res 73: 1437–1443.
13. Imazato S, Imai T, Russell RR, Torii M, Ebisu S (1998) Antibacterial activity of cured dental resin incorporating the antibacterial monomer MDPB and an adhesion-promoting monomer. J Biomed Mater Res 39: 511–515.
14. Xiao YH, Chen JH, Fang M, Xing XD, Wang H, et al. (2008) Antibacterial effects of three experimental quaternary ammonium salt (QAS) monomers on bacteria associated with oral infections. J Oral Sci 50: 323–327.
15. Xiao YH, Ma S, Chen JH, Chai ZG, Li F, et al. (2009) Antibacterial activity and bonding ability of an adhesive incorporating an antibacterial monomer DMAE-CB. J Biomed Mater Res B Appl Biomater 90: 813–817.
16. Li F, Chai ZG, Sun MN, Wang F, Ma S, et al. (2009) Anti-biofilm effect of dental adhesive with cationic monomer. J Dent Res 88: 372–376.
17. Li F, Chen J, Chai Z, Zhang L, Xiao Y, et al. (2009) Effects of a dental adhesive incorporating antibacterial monomer on the growth, adherence and membrane integrity of Streptococcus mutans. J Dent 37: 289–296.
18. Li F, Weir MD, Fouad AF, Xu HH (2014) Effect of salivary pellicle on antibacterial activity of novel antibacterial dental adhesives using a dental plaque microcosm biofilm model. Dent Mater 30: 182–191.
19. Cheng L, Weir MD, Xu HH, Antonucci JM, Kraigsley AM, et al. (2012) Antibacterial amorphous calcium phosphate nanocomposites with a quaternary ammonium dimethacrylate and silver nanoparticles. Dent Mater 28: 561–572.
20. Huang L, Xiao YH, Xing XD, Li F, Ma S, et al. (2011) Antibacterial activity and cytotoxicity of two novel cross-linking antibacterial monomers on oral pathogens. Arch Oral Biol 56: 367–373.
21. Huang L, Sun X, Xiao YH, Dong Y, Tong ZC, et al. (2012) Antibacterial effect of a resin incorporating a novel polymerizable quaternary ammonium salt MAE-DB against Streptococcus mutans. J Biomed Mater Res B Appl Biomater 100: 1353–1358.
22. Shayegan A, Petein M, Vanden Abbeele A (2009) The use of beta-tricalcium phosphate, white MTA, white Portland cement and calcium hydroxide for direct pulp capping of primary pig teeth. Dent Traumatol 25: 413–419.
23. Torabinejad M, Parirokh M (2010) Mineral trioxide aggregate: a comprehensive literature review–part II: leakage and biocompatibility investigations. J Endod 36: 190–202.
24. Asgary S, Eghbal MJ, Parirokh M, Ghoddusi J, Kheirieh S, et al. (2009) Comparison of mineral trioxide aggregate's composition with Portland cements and a new endodontic cement. J Endod 35: 243–250.
25. Aguilar FG, Roberti Garcia LF, Panzeri Pires-de-Souza FC (2012) Biocompatibility of new calcium aluminate cement (EndoBinder). J Endod 38: 367–371.
26. Song JS, Mante FK, Romanow WJ, Kim S (2006) Chemical analysis of powder and set forms of Portland cement, gray ProRoot MTA, white ProRoot MTA, and gray MTA-Angelus. Oral Surg Oral Med Oral Pathol Oral Radiol Endod 102: 809–815.
27. Borges AH, Pedro FL, Miranda CE, Semenoff-Segundo A, Pecora JD, et al. (2010) Comparative study of physico-chemical properties of MTA-based and Portland cements. Acta Odontol Latinoam 23: 175–181.
28. Camilleri J, Montesin FE, Di Silvio L, Pitt Ford TR (2005) The chemical constitution and biocompatibility of accelerated Portland cement for endodontic use. Int Endod J 38: 834–842.
29. Saidon J, He J, Zhu Q, Safavi K, Spangberg LS (2003) Cell and tissue reactions to mineral trioxide aggregate and Portland cement. Oral Surg Oral Med Oral Pathol Oral Radiol Endod 95: 483–489.
30. Min KS, Kim HI, Park HJ, Pi SH, Hong CU, et al. (2007) Human pulp cells response to Portland cement in vitro. J Endod 33: 163–166.
31. Li F, Weir MD, Chen J, Xu HH (2013) Comparison of quaternary ammonium-containing with nano-silver-containing adhesive in antibacterial properties and cytotoxicity. Dent Mater 29: 450–461.
32. Accorinte Mde L, Holland R, Reis A, Bortoluzzi MC, Murata SS, et al. (2008) Evaluation of mineral trioxide aggregate and calcium hydroxide cement as pulp-capping agents in human teeth. J Endod 34: 1–6.
33. Bjorndal L, Mjor IA (2001) Pulp-dentin biology in restorative dentistry. Part 4: Dental caries–characteristics of lesions and pulpal reactions. Quintessence Int 32: 717–736.
34. Formosa LM, Mallia B, Camilleri J (2013) The chemical properties of light- and chemical-curing composites with mineral trioxide aggregate filler. Dent Mater 29: e11–19.
35. Gandolfi MG, Taddei P, Siboni F, Modena E, Ciapetti G, et al. (2011) Development of the foremost light-curable calcium-silicate MTA cement as root-end in oral surgery. Chemical-physical properties, bioactivity and biological behavior. Dent Mater 27: e134–157.
36. Gandolfi MG, Taddei P, Siboni F, Modena E, De Stefano ED, et al. (2011) Biomimetic remineralization of human dentin using promising innovative calcium-silicate hybrid "smart" materials. Dent Mater 27: 1055–1069.
37. Cox CF, Hafez AA, Akimoto N, Otsuki M, Suzuki S, et al. (1998) Biocompatibility of primer, adhesive and resin composite systems on non-exposed and exposed pulps of non-human primate teeth. Am J Dent 11 Spec No: S55–63.
38. Kitasako Y, Inokoshi S, Tagami J (1999) Effects of direct resin pulp capping techniques on short-term response of mechanically exposed pulps. J Dent 27: 257–263.
39. Tziafas D, Koliniotou-Koumpia E, Tziafa C, Papadimitriou S (2007) Effects of a new antibacterial adhesive on the repair capacity of the pulp-dentine complex in infected teeth. Int Endod J 40: 58–66.
40. Antonucci JM, Zeiger DN, Tang K, Lin-Gibson S, Fowler BO, et al. (2012) Synthesis and characterization of dimethacrylates containing quaternary ammonium functionalities for dental applications. Dent Mater 28: 219–228.
41. Cheng L, Weir MD, Xu HH, Kraigsley AM, Lin NJ, et al. (2012) Antibacterial and physical properties of calcium-phosphate and calcium-fluoride nanocomposites with chlorhexidine. Dent Mater 28: 573–583.
42. Beyth N, Yudovin-Farber I, Bahir R, Domb AJ, Weiss EI (2006) Antibacterial activity of dental composites containing quaternary ammonium polyethylenimine nanoparticles against Streptococcus mutans. Biomaterials 27: 3995–4002.
43. Namba N, Yoshida Y, Nagaoka N, Takashima S, Matsuura-Yoshimoto K, et al. (2009) Antibacterial effect of bactericide immobilized in resin matrix. Dent Mater 25: 424–430.
44. Ber BS, Hatton JF, Stewart GP (2007) Chemical modification of ProRoot MTA to improve handling characteristics and decrease setting time. J Endod 33: 1231–1234.
45. Shen Q, Sun J, Wu J, Liu C, Chen F (2010) An in vitro investigation of the mechanical-chemical and biological properties of calcium phosphate/calcium silicate/bismutite cement for dental pulp capping. J Biomed Mater Res B Appl Biomater 94: 141–148.

19

A Sialoreceptor Binding Motif in the *Mycoplasma synoviae* Adhesin VlhA

Meghan May[1]*, Dylan W. Dunne[2], Daniel R. Brown[3]

1 Department of Biomedical Sciences, College of Osteopathic Medicine, University of New England, Biddeford, Maine, United States of America, **2** Department of Biological Sciences, Jess and Mildred Fisher College of Science and Mathematics, Towson University, Towson, Maryland, United States of America, **3** Department of Infectious Diseases and Pathology, College of Veterinary Medicine, University of Florida, Gainesville, Florida, United States of America

Abstract

Mycoplasma synoviae depends on its adhesin VlhA to mediate cytadherence to sialylated host cell receptors. Allelic variants of VlhA arise through recombination between an assemblage of promoterless *vlhA* pseudogenes and a single transcription promoter site, creating lineages of *M. synoviae* that each express a different *vlhA* allele. The predicted full-length VlhA sequences adjacent to the promoter of nine lineages of *M. synoviae* varying in avidity of cytadherence were aligned with that of the reference strain MS53 and with a 60-a.a. hemagglutinating VlhA C-terminal fragment from a Tunisian lineage of strain WVU1853[T]. Seven different sequence variants of an imperfectly conserved, single-copy, 12-a.a. candidate cytadherence motif were evident amid the flanking variable residues of the 11 total sequences examined. The motif was predicted to adopt a short hairpin structure in a low-complexity region near the C-terminus of VlhA. Biotinylated synthetic oligopeptides representing four selected variants of the 12-a.a. motif, with the whole synthesized 60-a.a. fragment as a positive control, differed ($P<0.01$) in the extent they bound to chicken erythrocyte membranes. All bound to a greater extent ($P<0.01$) than scrambled or irrelevant VlhA domain negative control peptides did. Experimentally introduced branched-chain amino acid (BCAA) substitutions Val3Ile and Leu7Ile did not significantly alter binding, whereas fold-destabilizing substitutions Thr4Gly and Ala9Gly tended to reduce it ($P<0.05$). Binding was also reduced to background levels ($P<0.01$) when the peptides were exposed to desialylated membranes, or were pre-saturated with free sialic acid before exposure to untreated membranes. From this evidence we conclude that the motif P-X-(BCAA)-X-F-X-(BCAA)-X-A-K-X-G binds sialic acid and likely mediates VlhA-dependent *M. synoviae* attachment to host cells. This conserved mechanism retains the potential for fine-scale rheostasis in binding avidity, which could be a general characteristic of pathogens that depend on analogous systems of antigenically variable adhesins. The motif may be useful to identify previously unrecognized adhesins.

Editor: Mitchell F. Balish, Miami University, United States of America

Funding: This work was supported by the Robert M. Fisher Foundation (MM). The funder had no role in study design, data collection and analysis, decision to publish, or preparation of the manuscript.

Competing Interests: The authors have declared that no competing interests exist.

* Email: mmay3@une.edu

Introduction

The bacterial pathogen *Mycoplasma synoviae* is associated with a broad spectrum of clinical manifestations ranging from inapparent infection to systemic disease of poultry. Infection is most commonly associated with inflammatory lesions of the joints, respiratory and/or reproductive tract and results in reduced feed conversion and poor egg quality. Less commonly, *M. synoviae* can be found infecting additional tissues in galliform birds (*e.g.* spleen, liver, central nervous system, skeletal muscle, and eye) [1–4] and respiratory tissues or synovial membranes of distantly related avian species such as ducks, geese, pigeons, and sparrows [5].

Attachment to sialylated receptors on host cells is mediated by the *M. synoviae* variable lipoprotein hemagglutinin VlhA [6–7]. Previous analyses indicated that the *vlhA* gene family has been laterally transferred between *M. synoviae* and *Mycoplasma gallisepticum* possibly during coinfection of a shared avian host [8–9]. In *M. synoviae*, antigenic variants of this adhesin result from

unidirectional recombination between a single expression site and a large reservoir of *vlhA* pseudogenes [10]. In contrast, altered expression in *M. gallisepticum* stems from the expansion and contraction of a poly-GAA repeat upstream of the promoters of each copy of *vlhA* [11]. The selective pressure of specific host immune responses to these antigens is thought to drive diversity in *vlhA* allele expression [10–13]. Despite the critical importance of cytadherence to the establishment and maintenance of infection, discrete VlhA types were demonstrated to have significantly different avidities for host cell binding, which can be quantified by agglutination of erythrocytes [14]. *M. synoviae*'s capacity for cytadherence maps surprisingly to a hypervariable C-terminal domain of VlhA called MSPA [15–16]. The precise means of attachment and how this capacity is retained despite such extensive sequence polymorphism and allele switching are not known. We sought to identify and characterize the specific motif that mediates adhesion of VlhA proteins to host cells.

PHM Residue	1	2	3	4	5	6	7	8	9	10	11	12
Strains/lineages												
F10-2-AS and K4907	P	K	V	T	F	D	V	A	Q	K	E	G
FMT	P	K	V	T	F	N	L	A	A	K	E	G
K5016	P	K	V	T	F	T	V	T	A	K	N	G
K5395	P	T	V	T	F	N	L	A	A	K	E	G
MS53	P	K	V	T	F	N	L	T	P	K	E	G
MS117, MS173, MS178	P	T	V	T	F	T	V	A	A	K	D	G
WVU1853^T/Florida and Tunisia	P	K	V	T	F	T	V	E	A	K	P	G
Preliminary consensus:	P	X	V	T	F	X	(B)	X	X	K	X	G
Site-directed Mutants												
FMT T4G	P	K	V	G	F	N	L	A	A	K	E	G
FMT A9G	P	K	V	T	F	N	L	A	G	K	E	G
FMT V3I	P	K	I	T	F	N	L	A	A	K	E	G
FMT L7I	P	K	V	T	F	N	I	A	A	K	E	G
FinalPHM:	P	X	(B)	T	F	X	(B)	X	A	K	X	G
FMT - scrambled:	L	A	F	G	A	V	K	K	T	P	E	N
Irrelevant peptide:	P	N	A	V	F	V	Q	Q	M	K	D	D

Figure 1. Aligned PHM and control peptide sequences. The putative hemagglutination motif (PHM) was deduced by aligning the adhesin protein VlhA allele present at the expression site of ten specimens of *M. synoviae* with a 60-a.a. hemagglutinating VlhA C-terminal fragment from the Tunisian lineage of strain WVU1853[T], then inspecting the alignment for contiguous residues inferred to be under stabilizing selection. Peptides representing five variants of the PHM, including strains having a>20-fold range in quantitative hemagglutination phenotypes [14], were synthesized. Directed mutations were introduced at selected residues relative to the PHM from strain FMT, which had only one difference (Thr6Asn) from the most common amino acid at each residue. The mutations Val3Ile and Leu7Ile were predicted to be inconsequential, while Thr4Gly and Ala9Gly were predicted to affect PHM structure and/or function. Negative control peptides used in erythrocyte membrane-binding assays are also shown. Functionally non-synonymous differences relative to the most common amino acid at each residue are shaded in black, synonymous differences are shaded in gray, and identical residues are not shaded. (B) = branched chain amino acid.

Materials and Methods

Identification and Structural Modeling of the Putative Hemagglutination Motif (PHM)

The predicted full-length VlhA sequences adjacent to the single transcription promoter of nine lineages of *M. synoviae* varying in avidity of cytadherence (F10-2AS, FMT, K4907, K5016, K5395, MS117, MS173, MS178, and a>30X-passaged Florida lineage of strain WVU1853[T]) [14] were aligned with that of the reference strain MS53 [8] and with a 60-a.a. hemagglutinating VlhA C-terminal fragment from a ca. 12X-passaged Tunisian lineage of strain WVU1853[T] [15] by using ClustalΩ [17]. The multiple alignment was manually inspected for conserved motifs, evident as contiguous residues inferred to be under stabilizing selection (ω<1) by using Bayesian models of sequence evolution in the Selecton v2.4 software suite [18]. The secondary structures of full-length VlhA, MSPA and its C-terminal 60 residues, and of the putative hemagglutination motifs (PHMs) described were modeled using the Phyre2 suite of template-directed and *ab initio* protein structure prediction algorithms (http://www.sbg.bio.ic.ac.uk/phyre2) [19]. The effects of individual amino acid substitutions on peptide structural stability were predicted by applying the Site Directed Mutator algorithm (http://mordred.bioc.cam.ac.uk/~sdm/sdm.php) [20] to the.pdb files generated by Phyre2. Substitutions having stability scores ($\Delta\Delta G$) between −0.5 and 0.5 were predicted to be neutral, whereas those <−2 or>2 were predicted to be highly destabilizing. The potential to bind sialic acid (KEGG Compound C00270; PubChem.sdf 445063) or any other ligand in the KEGG Compound database was predicted by applying the eFindSite ligand binding site prediction algorithm (http://brylinski.cct.lsu.edu/) [21–22] also to the.pdb files generated by Phyre2.

Quantitative Binding of PHM Peptides

Twelve-a.a. peptides representing five variants of the PHM from strains FMT, K5016, K5395, MS53 and WVU1853[T], plus the whole 60-a.a. hemagglutinating fragment of the Tunisian lineage of strain WVU1853[T], were synthesized, biotinylated and lyophilized (Biomatik, Wilmington, DE). Purity of each lyophilized preparation was confirmed by HPLC to be 90–92% full-length peptide. Those strains were chosen because FMT, K5016, K5395 and the Florida lineage of WVU1853[T] spanned a>20-fold range in quantitative hemagglutination phenotypes, and the entire *vlhA* locus sequence of the reference strain MS53 has been published. [8,14]. Peptides having single directed mutations introduced at the conserved residues 3 or 4, or non-conserved residues 7 or 9, were also synthesized using the strain FMT motif PKVTFNLAAKEG as a parent. FMT was chosen as the parent motif because it had only one difference (Thr6Asn) from the most commonly observed amino acid at each residue (Figure 1). The functionally synonymous substitutions Val3Ile and Leu7Ile (BLOSUM62 [23] scores>0) were predicted to be inconsequential, while non-synonymous Thr4Gly and Ala9Gly (BLOSUM62 scores ≤0) were predicted to affect PHM structure and/or function.

The capacity of the peptides to bind to native or desialylated chicken erythrocyte membranes was assessed quantitatively in an ELISA format. Microtiter plates were coated with 5% v/v suspensions of chicken erythrocytes (Lampire Biologicals,

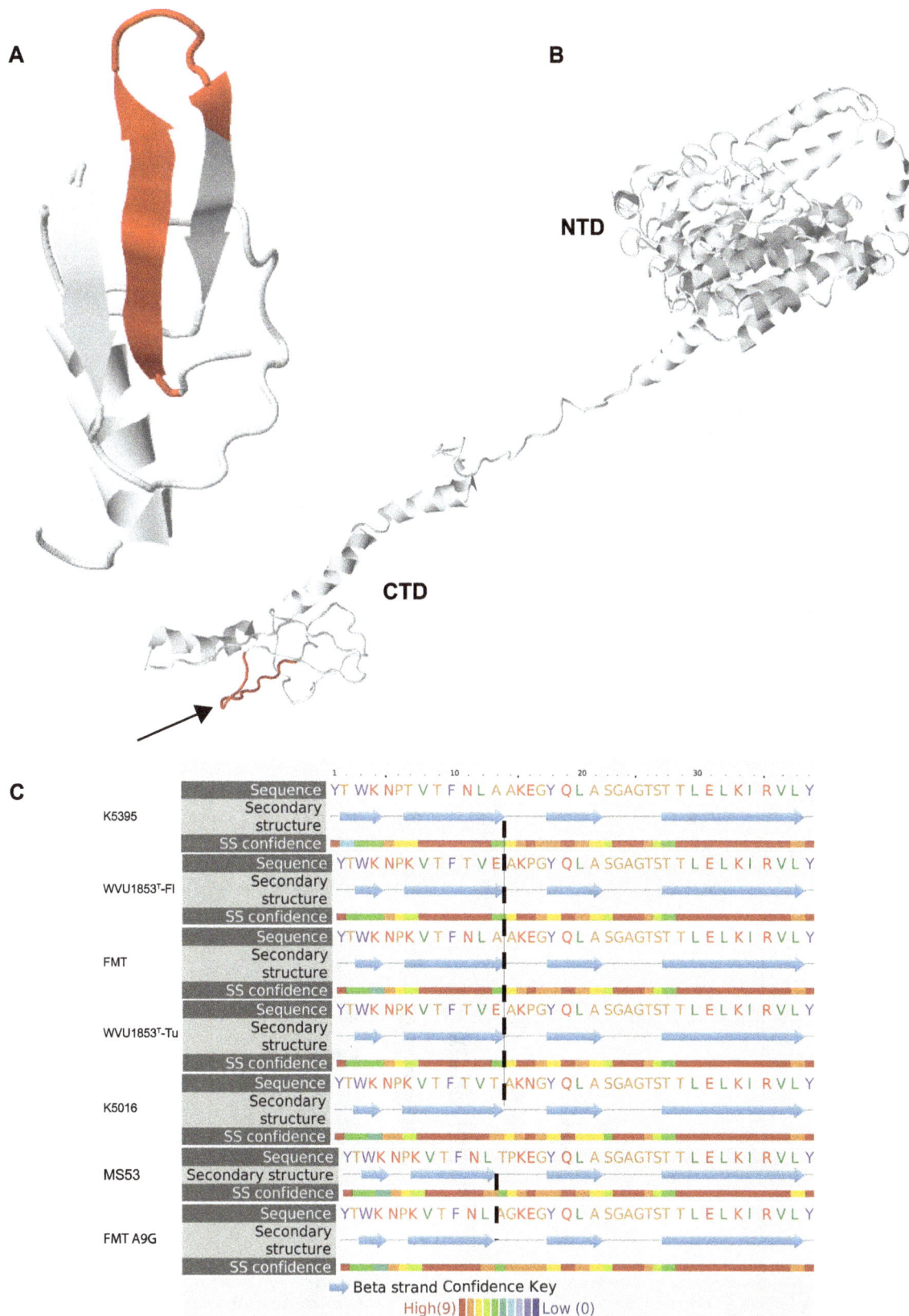

Figure 2. PHM structural predictions. (**A**) The putative hemagglutination motif (PHM; red) was predicted to adopt a hairpin structure of two anti-parallel β strands separated by a short disordered loop. (**B**) The motif (red, indicated by arrow) mapped to a low-complexity region near the carboxyterminal domain (CTD) of the *M. synoviae* adhesin protein VlhA cleavage product MSPA, shown here in the structure predicted for the Tunisian lineage of strain WVU1853[T]. The N-terminal domain (NTD) of MSPA was predicted to have much greater 3-dimensional complexity. (**C**) The length of the disordered loop was predicted to be longer in PHM peptides that bound to avian erythrocyte membranes (representing Florida and

Tunisian lineages of strain WVU1853[T] and strains FMT, K5016 and K5395) than in the reduced-binding peptide mutant FMT-Ala9Gly and the non-binding peptide representing strain MS53.

Pipersville, PA) diluted 1:3 in 0.5 M sodium bicarbonate lysis buffer, pH 10.0, to a total volume of 300 µL per well. Desialylated membranes were prepared by pre-treatment of the erythrocytes with 10 U/ml of sialidase purified from *Clostridium perfringens* (Sigma-Aldrich, St. Louis, MO) for 1 hr at 37°C. Following coating for 12 hr at 4°C, cellular debris including hemoglobin was removed by washing each well 3× with 300 µL of PBS, pH 7.4, and sealed plates were blocked 1 hr at 37°C with 300 µL per well of 5% v/v fetal bovine serum in PBS.

After washing the membrane-coated and blocked wells 3× with 300 µL of PBS, 50 µg of biotinylated peptide solubilized in 50 µL of water was added to each of duplicate wells and allowed to bind for 1 hr at 37°C. After washing each well 3× with 300 µL of PBS, bound peptides were detected using horseradish peroxidase-conjugated streptavidin (2 µg/mL, Sigma-Aldrich, St. Louis, MO) and the chromogenic substrate 3,3',5,5'-tetramethylbenzidine (Thermo Fisher Scientific, Waltham, MA) with an acid stop followed by spectrophotometric analysis ($\lambda = 450$ nm). The hemagglutinating 60-mer of the Tunisian lineage of strain WVU1853[T] served as the positive control peptide, and negative controls were a scrambled version of the PHM from strain FMT (LAFGAVKKTPEN) and an irrelevant peptide (PNAVFVQQMKDD) from a distant site in the expressed VlhA of the Florida lineage of strain WVU1853[T] (GenBank AEA01932.1). The effect of pre-saturation with ligand was tested by first incubating the peptides in 250 mg/ml N-acetylneuraminic acid (Sigma-Aldrich, St. Louis, MO) in water without pH adjustment at a peptide: ligand molar ratio of $1:2 \times 10^4$ for 1 hr at 37°C.

Statistical Procedures

The effect of peptide sequence on extent of adherence to membranes ($n = 3$ independent replications of each treatment combination, with duplicate measurements of each peptide within replicate) was analyzed by ANOVA, with Tukey-Kramer Honestly Significant Difference (HSD) post-hoc comparisons used to group the means when the main effect was significant ($P < 0.05$ or less). The effects of membrane pre-treatment with sialidase and peptide pre-saturation with sialic acid were analyzed by ANOVA, with HSD or Dunnett's post-hoc comparisons to the corresponding native specimens when the main effect was significant. Statistical analyses were performed using Origin 9 (OriginLab, Northampton, MA) software.

Motif Distribution in *M. synoviae* and *M. gallisepticum*

M. synoviae strain MS53 *vlhA* pseudogene sequences and *M. gallisepticum* strains R, F, WI01, NY01, NC06, CA06, VA94, NC95, NC08, and NC96 were obtained from GenBank (accession numbers NC_007294.1, NC_004829.2, NC_017503.1, NC_018410.1, NC_018409.1, NC_018411.1, NC_018412.1, NC_018406.1, NC_018407.1, NC_018413.1, and NC_018408.1, respectively). Occurrences of PHM-encoding sequences were totaled and normalized to the total length of *vlhA*-encoding sequence in each strain. Each member of the *vlhA* pseudogene reservoir of *M. synoviae* strain MS53 was used to construct a neighbor-joining tree (bootstrap n = 100) using ClustalW2 [24]. The designated outgroup was *vlhA* 4.02 from *M. gallisepticum*.

Results

Identification of the PHM

When the full-length expressed VlhA protein MSPA sequences of nine strains of *M. synoviae* that vary in avidity of cytadherence were aligned with MSPA of the reference strain MS53 [8] and a 60-a.a. hemagglutinating peptide derived from the C-terminus of MSPA expressed by the Tunisian lineage of strain WVU1853[T] [15], an imperfectly conserved 12-a.a. motif was evident in all sequences (Figure 1). A total of seven different PHM sequence variants were evident among the 11 total sequences aligned.

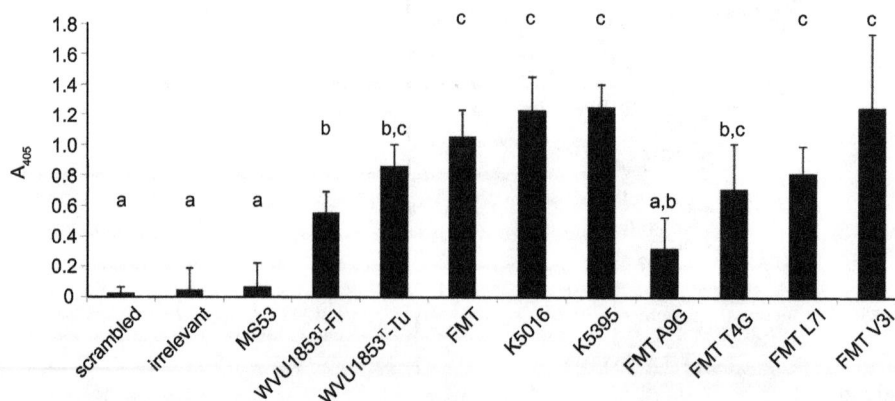

Figure 3. Erythrocyte membrane binding by PHM peptides. Bars depict mean ± standard error of the amount of synthetic peptide bound to avian erythrocyte membranes in an ELISA format (n = 3 independent replicates, with duplicate measurements of each peptide within replicate). The peptides represented variants of the putative hemagglutination motif (PHM) at the VlhA expression site of *M. synoviae* strains MS53, WVU1853[T] (Florida and Tunisian lineages), FMT, K5016 and K5395, which spanned a >20-fold range in quantitative hemagglutination phenotypes [14]. The positive control was the Tunisian lineage of strain WVU1853[T], and negative controls were scrambled strain FMT peptide and an irrelevant peptide from a distant site in VlhA from the Florida lineage of strain WVU1853[T]. Different letters above the bar indicate means that differ (P<0.05 or less) by Tukey-Kramer Honestly Significant Difference test. As predicted, the directed substitution Ala9Gly significantly reduced binding versus the parent peptide from strain FMT, and Thr4Gly tended to reduce binding, whereas Val3Ile and Leu7Ile did not significantly alter binding.

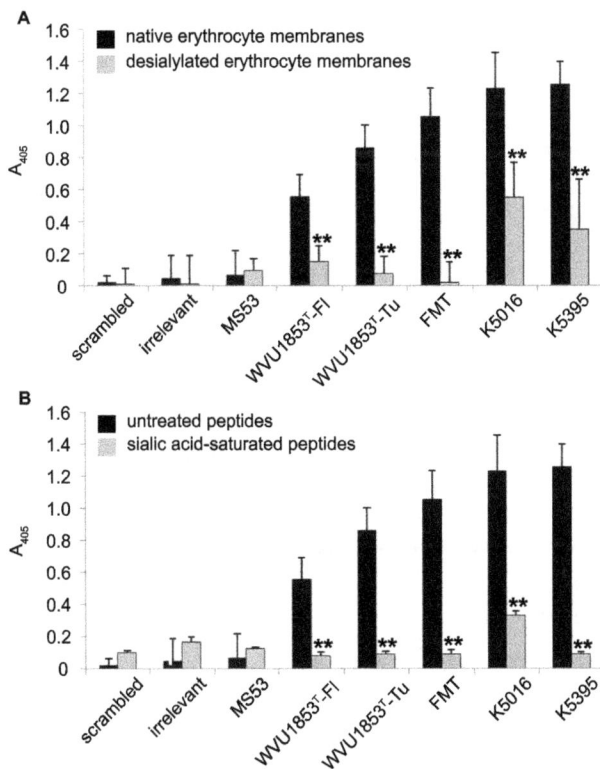

Figure 4. Effects of sialylation and desialylation on PHM peptide binding. Bars depict mean ± standard error of the amount of synthetic peptide bound to avian erythrocyte membranes in an ELISA format (n = 3 independent replicates, with duplicate measurements of each peptide within replicate). (**A**) Desialylation of erythrocyte membranes significantly reduced PHM peptide binding relative to native membranes (** = P<0.01) for all strains of M. synoviae except MS53, which bound to native or desialylated erythrocyte membranes at background levels. (**B**) Presaturation of PHM peptides with free sialic acid before exposure to native erythrocyte membranes significantly reduced binding relative to untreated peptides (** = P<0.01) for all strains except MS53, on which sialic acid had no effect.

Strains FMT, K5016, K5395 and MS53 all had unique PHM sequences; the sequences in strains F10-2-AS and K4907 were identical; the sequences in Florida and Tunisian lineages of WVU1853[T] were identical; and the sequences in Argentine strains MS117, MS173 and MS178 were all identical. Six of twelve residues in the PHM were perfectly conserved across strains, two (residues 6 and 7) were conserved in polarity and hydrophobicity, respectively, and four were variable. Polar Asn_6 or Asp_6 were invariably paired with Leu_7, while Thr_6 was invariably paired with Val_7. The motif was predicted to adopt a short hairpin secondary structure of two anti-parallel beta strands, separated by a disordered loop of four or five residues, in a region of low structural complexity (regional structure prediction confidence < 70%) near the C-terminus of MSPA (Figure 2a, b). Fifty-three percent of residues in the full-length VlhA were modeled at >90% confidence [19], with the regions of greatest confidence being similar to the streptococcal adhesin emb (99.8% confidence) and the staphylococcal extracellular matrix-binding protein ebhA (99.4%). The degree of structural complexity in the C-terminus of MSPA was otherwise too low for the algorithms to predict binding of any specific ligand.

Synthetic biotinylated peptides representing the full-length 60-a.a. hemagglutinating fragment and four strain variants of its candidate 12-a.a. cytadherence motif (Figure 1) bound to chicken erythrocyte membranes in an ELISA format and could be detected by probing with horseradish peroxidase-conjugated streptavidin. Four of the peptides bound to membranes to a significantly greater extent (P<0.05) than scrambled or irrelevant control peptides did, but a peptide representing the corresponding motif from strain MS53 did not bind to membranes to any extent greater than background (Figure 3). Single neutral substitutions (predicted $\Delta\Delta G = -0.25$) experimentally introduced at conserved residue 3 (Val3Ile) or non-conserved residue 7 (Leu7Ile) did not alter binding to membranes with respect to the extent of binding by the parent motif of strain FMT, whereas the experimental destabilizing substitution Thr4Gly (predicted $\Delta\Delta G = -2.31$) tended to reduce binding (Figure 3). The motif of strain MS53 differs naturally from all others by Ala9Pro (BLOSUM62 $= -1$; predicted $\Delta\Delta G = -2.22$), and the even more destabilizing substitution Ala9Gly (predicted $\Delta\Delta G = -3.88$) nearly abolished binding when introduced into the parent motif of strain FMT (P< 0.05; Figure 3). These effects correlated with a predicted change in length of the disordered loop in the hairpin secondary structure of the motif (Figure 2c).

Binding of the peptides to desialylated membranes was significantly reduced (P<0.01) relative to untreated membranes for all peptides except those representing strain MS53 and the scrambled and irrelevant controls (Figure 4a). When pre-incubated with free sialic acid, all peptides except the one representing strain MS53 and the scrambled and irrelevant controls had significantly diminished (P<0.01) capacity for membrane binding (Figure 4b). From this evidence we conclude that the composite amino acid motif P-X-(BCAA)-X-F-X-(BCAA)-X-A-K-X-G binds sialic acid and likely mediates VlhA-dependent M. synoviae attachment to sialylated receptors on the surface of avian erythrocytes.

PHM Distribution among M. synoviae Strain MS53 vlhA Pseudogenes and Mycoplasma gallisepticum vlhA Homologs

Candidate PHM sequences occurred in 45 of the 70 putative vlhA pseudogenes of M. synoviae strain MS53 [8], 39% of the time with no deviation from the consensus among the alleles expressed by the strains examined, 20% with a single deviation, and 17% with two deviations from consensus. Phylogenetic clustering of vlhA pseudogenes containing intact copies of the PHM did not correlate with their syntenic order in the strain MS53 genome (Figure S1). The PHM occurred at least 18-fold more frequently in strain MS53 (0.65 motifs/kb of vlhA sequence) than in the genomes of any of 10 strains of M. gallisepticum (0.014–0.037 motifs/kb of vlhA sequence), a species known to employ a different primary cytadherence mechanism [25] (Figure 5a). The rate of occurrence of imperfect PHMs was comparable between the two species (Figure 5b).

Discussion

One of the defining moments of many infections is the attachment of a disease-causing agent to its host. Understanding how the parasitic bacterial species M. synoviae colonizes a host cell's surface is paramount to understanding how to prevent infection. It is known that the protein family VlhA is responsible for attachment by M. synoviae, but the functional motifs of the adhesin and the molecular basis for rheostasis in binding avidity have not been characterized. Proteins in this family from multiple strains of M. synoviae have been identified as having a role in the attachment to host blood cells [7,15]. Khiari et al. [15] mapped

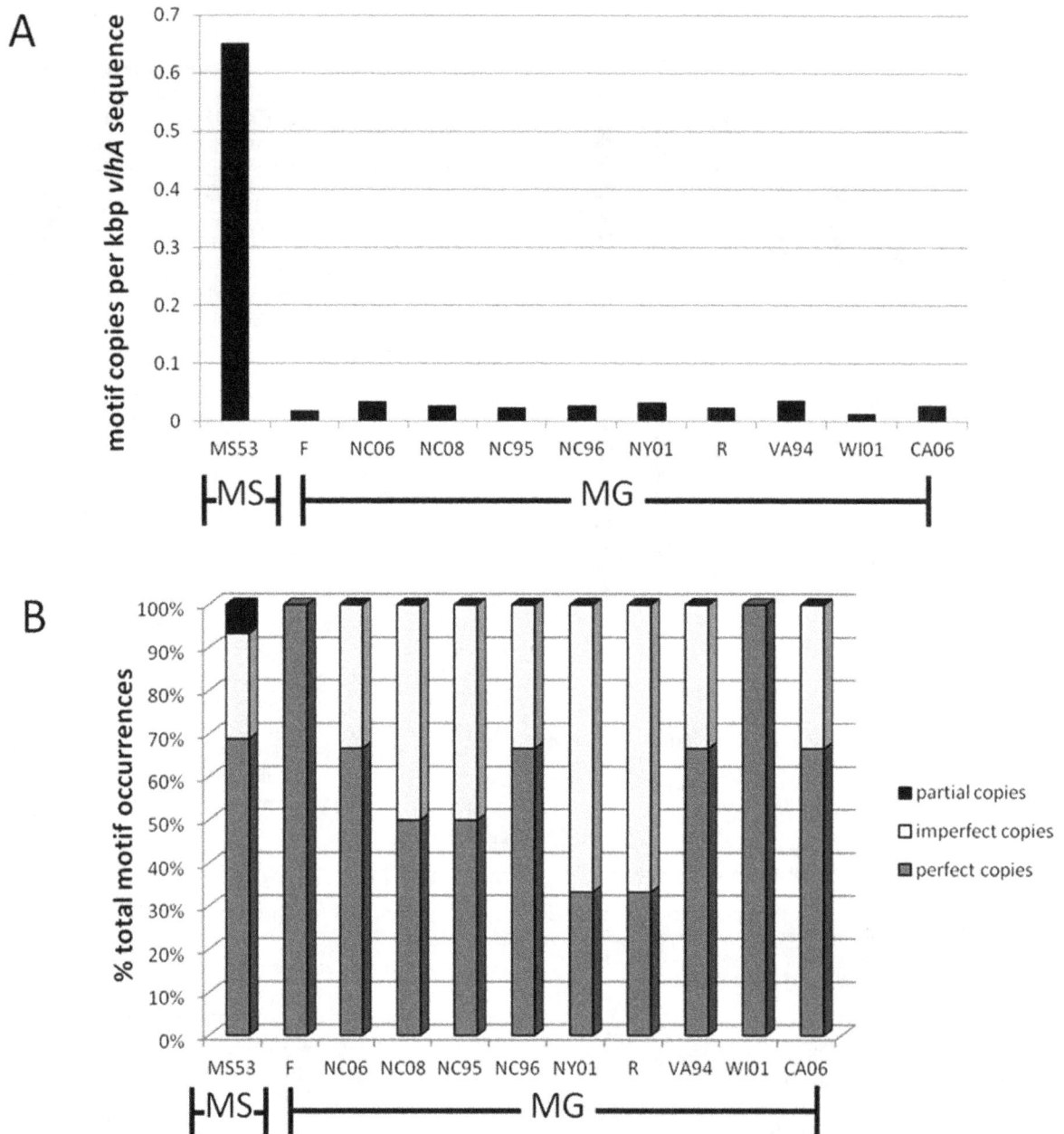

Figure 5. PHM distribution in vlhA genes and pseudogenes in M. synoviae and M. gallisepticum. (A) PHM-encoding sequence as a function of total kbp of vlhA sequence is elevated 18-fold in M. synoviae (MS) reference strain MS53, the only strain for which the entire vlhA locus sequence has been published, relative to 10 fully-sequenced strains of M. gallisepticum (MG), evidence that it is far more common in M. synoviae. **(B)** The relative proportions of perfect and imperfect PHM copies were comparable between strains of M. synoviae and M. gallisepticum.

the capacity for attachment to the carboxyterminus of VlhA, and we utilized that finding to identify a specific motif sufficient to mediate VlhA binding to sialylated host cells.

Sequence conservation across adherent strains enabled the identification of a 12-residue putative hemagglutination motif that could be characterized further. This motif was predicted to have remarkably little structural complexity, in contrast to the complex topology of sialic acid ligand-binding domains of other microbes [26–27]. While residues at PHM positions 3 and 7 were conserved, substitution with similar residues having BLOSUM62 scores>0 did not alter function. The conserved Thr residue at position 4 could be changed to the dissimilar residue Gly (BLOSUM62

$= -3$) without loss of function. It is thus likely that the binding mechanism will tolerate synonymous substitutions at positions 3 and 7, and nonsynonymous substitutions at position 4. Residue 9 was a conserved Ala in all adherent strains. Strain MS53, which has an unknown attachment phenotype but is an attenuated strain, had the nonsynonmymous substitution Ala9Pro (BLOSUM62 $= -1$). Changing the strain FMT peptide to Gly_9 (BLOSUM62 $= 0$) significantly diminished binding, and the strain MS53 peptide was non-adherent. Taken together, these results indicate that Ala_9 is critical to PHM domain function. Our results indicate that the composite amino acid motif P-X-(BCAA)-X-F-X-(BCAA)-X-A-K-X-G mediates MSPA binding to avian erythrocytes. The potential to accommodate all amino acids with BLOSUM62 scores>0 at PHM positions 3 and 7 (i.e., Ala, Met, Thr and Met, Phe, respectively) rather than restricting the parameters to branched-chain amino acids (Ile, Leu, Val) merits further analysis.

Previous studies indicated that whole M. synoviae cells interact with sialylated host cell receptors in order to facilitate attachment. Extrapolation from PHM peptide-binding to whole cell attachment necessarily requires demonstration of peptide-sialic acid interactions. Desialylation of avian erythrocytes prior to antigen preparation resulted in significant losses of binding capacity for all PHM peptides except the scrambled and irrelevant controls and strain MS53, for which desialylation had no effect on binding. In a reciprocal experiment, pre-adsorption of peptides with free sialic acid prior to exposure to intact erythrocyte antigen similarly diminished binding capacity for all PHM peptides except the scrambled and irrelevant controls and strain MS53. These results indicate a specific interaction between sialic acid and the PHM and support the hypothesis that the PHM domain mediates attachment of whole M. synoviae cells to host sialoreceptors.

The occurrance of PHM domains was not uniform among the pseudogenes of M. synoviae strain MS53, the only strain for which the entire pseudogene reservoir has been sequenced [8]. A majority (69%) of pseudogenes had perfect or near-perfect PHMs, while 31% had no discernible PHMs. To provide some context for the distribution of PHM domains in the sample of VlhA sequences existing within M. synoviae strain MS53, we examined the frequency and distribution in an alternative sample of VlhA sequences that exist distributed across multiple strains of M. gallisepticum. In contrast to the 45 copies in M. synoviae strain MS53, sequenced M. gallisepticum strains ranged from having just a single copy of vlhA encoding a PHM domain (strains WI01 and F) up to a maximum of only 3 copies (strains R, NY01, NC06, CA06, VA94, NC95, and NC96). Normalization to the total amount of vlhA sequence within species confirmed that M. synoviae has a greatly elevated instance of PHM-encoding

sequence relative to M. gallisepticum, and that the low frequency of PHM is consistent across strains of M. gallisepticum. The multiple independent cytadherence mechanisms of M. gallisepticum [28–32] may allow the decay of PHM domains within VlhA proteins, while selective pressure to retain the functional motif in the homologous proteins in M. synoviae is substantially greater due to the absence of other mechanisms of cytadherence.

This work describes a novel functional motif associated with adherence to sialic acid, and its distribution across vlhA pseudogenes. This very specific protein fragment pattern may be a target to design novel drug therapies or vaccines to alleviate or prevent infection due to M. synoviae as well as other pathogens that use similar mechanisms to attach to their hosts, and allows for the identification of currently unrecognized microbial adhesins targeting sialoreceptors.

Supporting Information

Figure S1   Distribution and relatedness of PHM-encoding pseudogenes. PHM-encoding pseudogenes (shaded) did not cluster together as a separate group from non-encoding pseudogenes. Relatedness of pseudogenes did not reflect gene synteny.

Acknowledgments

We thank Edan Tulman (University of Connecticut) for helpful discussions regarding vlhA loci in M. gallisepticum. This work was supported by the Robert M. Fisher Foundation (MM).

Author Contributions

Conceived and designed the experiments: MM DRB. Performed the experiments: DD MM. Analyzed the data: MM DRB. Contributed reagents/materials/analysis tools: MM. Wrote the paper: MM DRB.

References

1. Stipkovits L, Kempf I (1996) Mycoplasmoses in poultry. Rev Sci Tech. 15(4): 1495–525.

2. Sentíes-Cué G, Shivaprasad HL, Chin RP (2005) Systemic Mycoplasma synoviae infection in broiler chickens. Avian Pathol. 34(2): 137–42.

3. Chin RP, Meteyer CU, Yamamoto R, Shivaprasad HL, Klein PN (1991) Isolation of Mycoplasma synoviae from the brains of commercial meat turkeys with meningeal vasculitis. Avian Dis. 35(3): 631–7.

4. Lockaby SB, Hoerr FJ, Lauerman LH, Kleven SH (1998) Pathogenicity of Mycoplasma synoviae in broiler chickens. Vet Pathol. 35(3): 178–90.

5. Brown DR, May M, Bradbury JM, Balish MF, Calcutt MJ, et al. (2010) Genus I. Mycoplasma. In: Krieg NR, Ludwig W, Brown DR, Whitman WB, Hedlund BP, Paster BJ, Staley JT, et al., editors. Bergey's Manual of Systematic Bacteriology Volume 4. Springer, Inc.: New York, NY.

6. Manchee R, Taylor-Robinson D (1969) Utilization of neuraminic acid receptors by mycoplasmas. J Bacteriol. 98(3): 914–9.

7. Noormohammadi A, Markham P, Duffy M, Whithear K, Browning G (1998) Multigene families encoding the major hemagglutinins in phylogenetically distinct mycoplasmas. Infect Immun. 66(7): 3470–5.

8. Vasconcelos A, Ferreira H, Bizarro C, Bonatto S, Carvalho M, et al. (2005) Swine and poultry pathogens: the complete genome sequences of two strains of Mycoplasma hyopneumoniae and a strain of Mycoplasma synoviae. J Bacteriol. 187(16): 5568–77.

9. Szczepanek SM, Tulman ER, Gorton TS, Liao X, Lu Z, et al. (2010) Comparative genomic analyses of attenuated strains of Mycoplasma gallisepticum. Infect Immun. 78(4): 1760–71.

10. Noormohammadi A, Markham P, Kanci A, Whithear K, Browning G (2000) A novel mechanism for control of antigenic variation in the haemagglutinin gene family of mycoplasma synoviae. Mol Microbiol. 35(4): 911–23.

11. Glew MD, Baseggio N, Markham PF, Browning GF, Walker ID (1998) Expression of the pMGA genes of Mycoplasma gallisepticum is controlled by variation in the GAA trinucleotide repeat lengths within the 5' noncoding regions. Infect Immun. 66(12): 5833–41.

12. Citti C, Browning GF, Rosengarten R (2005) Phenotypic diversity and cell invasion in host subversion by pathogenic mycoplasmas. In: Blanchard A,

Browning GF, editors. Mycoplasmas Molecular Biology Pathogenicity and Strategies for Control. Horizon Bioscience: Norfolk, UK.

13. Zimmerman C-U (2014) Current insights into phase and antigenic variation in mycoplasmas. In: Browning GF, Citti C, editors. Mollicutes Molecular Biology and Pathogenesis. Caister Academic Press: Norfolk, UK.

14. May M, Brown DR. (2011) Diversity of expressed vlhA adhesin sequences and intermediate hemagglutination phenotypes in Mycoplasma synoviae. J Bacteriol. 193(9): 2116–21.

15. Khiari AB, Guériri I, Mohammed RB, Mardassi BB (2010) Characterization of a variant vlhA gene of Mycoplasma synoviae, strain WVU 1853, with a highly divergent haemagglutinin region. BMC Microbiol. 10: 6. [doi: 1471-2180-10-6 [pii] 10.1186/1471-2180-10-6].

16. Noormohammadi A, Markham P, Whithear K, Walker I, Gurevich V, et al. (1997) Mycoplasma synoviae has two distinct phase-variable major membrane antigens, one of which is a putative hemagglutinin. Infect Immun. 65(7): 2542–7.

17. Sievers F, Wilm A, Dineen D, Gibson TJ, Karplus K, et al. (2011) Fast, scalable generation of high-quality protein multiple sequence alignments using Clustal Omega. Mol Syst Biol. 7: 539. [doi: 10.1038/msb.2011.75.]

18. May M, Brown DR (2009) Diversifying and stabilizing selection of sialidase and N-acetylneuraminate catabolism in Mycoplasma synoviae. J Bacteriol. 191(11): 3588–93.

19. Kelley LA, Sternberg MJ (2009) Protein structure prediction on the Web: a case study using the Phyre server. Nat Protoc. 4(3): 363–71.

20. Worth CL, Preissner R, Blundell TL (2011). SDM–a server for predicting effects of mutations on protein stability and malfunction. Nucleic Acids Res. 39(Web Server issue): W215–22. [doi: 10.1093/nar/gkr363].

21. Brylinski M, Feinstein WP (2013) eFindSite: improved prediction of ligand binding sites in protein models using meta-threading, machine learning and auxiliary ligands. J Comput Aided Mol Des. 27(6): 551–67.

22. Feinstein W, Brylinski M (2014) eFindSite: Enhanced fingerprint-based virtual screening against predicted ligand binding sites in protein models. Mol Inform. 33(2): 15 [doi: 10.1002/minf.201300143]

23. Henikoff S, Henikoff JG (1992) Amino acid substitution matrices from protein blocks. Proc Natl Acad Sci U S A. 89(22): 10915–9.

24. Larkin MA, Blackshields G, Brown NP, Chenna R, McGettigan PA, et al. (2007) Clustal W and Clustal X version 2.0. Bioinformatics. 23(21): 2947–8.

25. Papazisi L, Frasca S, Gladd M, Liao X, Yogev D, Geary SJ (2002) GapA and CrmA coexpression is essential for *Mycoplasma gallisepticum* cytadherence and virulence. Infect Immun. 70(12): 6839–45.

26. Tharakaraman K, Jayaraman A, Raman R, Viswanathan K, Stebbins NW, et al. (2013) Glycan receptor binding of the influenza A virus H7N9 hemagglutinin. Cell. 153(7): 1486–93.

27. Pang SS, Nguyen ST, Perry AJ, Day CJ, Panjikar S, et al. (2014) The three-dimensional structure of the extracellular adhesion domain of the sialic acid-binding adhesin SabA from *Helicobacter pylori*. J Biol Chem. 289(10): 6332–40.

28. Boguslavsky S, Menaker D, Lysnyansky I, Liu T, Levisohn S, et al. (2000) Molecular characterization of the *Mycoplasma gallisepticum* pvpA gene which encodes a putative variable cytadhesin protein. Infect Immun. 68(7): 3956–64.

29. Forsyth MH, Tourtellotte ME, Geary SJ (1992) Localization of an immuno-dominant 64 kDa lipoprotein (LP 64) in the membrane of *Mycoplasma gallisepticum* and its role in cytadherence. Mol Microbiol. 6(15): 2099–106.

30. Goh MS, Gorton TS, Forsyth MH, Troy KE, Geary SJ (1998) Molecular and biochemical analysis of a 105 kDa *Mycoplasma gallisepticum* cytadhesin (GapA). Microbiology. 144 (11): 2971–8.

31. Jenkins C, Geary SJ, Gladd M, Djordjevic SP (2007) The *Mycoplasma gallisepticum* OsmC-like protein MG1142 resides on the cell surface and binds heparin. Microbiology. 153(5): 1455–63.

32. May M, Papazisi L, Gorton TS, Geary SJ (2006) Identification of fibronectin-binding proteins in *Mycoplasma gallisepticum* strain R. Infect Immun. 74(3): 1777–85.

20

Space-Dependent Formation of Central Pair Microtubules and Their Interactions with Radial Spokes

Yuki Nakazawa[1¤a], Tetsuro Ariyoshi[1¤b], Akira Noga[1], Ritsu Kamiya[1,2], Masafumi Hirono[1*]

1 Department of Biological Sciences, Graduate School of Science, University of Tokyo, Bunkyo-ku, Tokyo, Japan, 2 Department of Life Science, Faculty of Science, Gakushuin University, Toshima-ku, Tokyo, Japan

Abstract

Cilia and flagella contain nine outer doublet microtubules and a pair of central microtubules. The central pair of microtubules (CP) is important for cilia/flagella beating, as clearly shown by primary ciliary dyskinesia resulting from the loss of the CP. The CP is thought to regulate axonemal dyneins through interaction with radial spokes (RSs). However, the nature of the CP-RS interaction is poorly understood. Here we examine the appearance of CPs in the axonemes of a *Chlamydomonas* mutant, *bld12*, which produces axonemes with 8 to 11 outer-doublets. Most of its 8-doublet axonemes lack CPs. However, in the double mutant of *bld12* and *pf14*, a mutant lacking the RS, most 8-doublet axonemes contain the CP. Thus formation of the CP apparently depends on the internal space limited by the outer doublets and RSs. In 10- or 11-doublet axonemes, only 3–5 RSs are attached to the CP and the doublet arrangement is distorted most likely because the RSs attached to the CP pull the outer doublets toward the axonemal center. The CP orientation in the axonemes varies in double mutants formed between *bld12* and mutants lacking particular CP projections. The mutant *bld12* thus provides the first direct and visual information about the CP-RS interaction, as well as about the mechanism of CP formation.

Editor: Takashi Toda, Cancer Research UK London Research Institute, United Kingdom

Funding: This work was supported by JSPS KAKENHI Grant Numbers 23657046, 24370079, and 25113503. The funders had no role in study design, data collection and analysis, decision to publish, or preparation of the manuscript.

Competing Interests: The authors have declared that no competing interests exist.

* Email: hirono@bs.s.u-tokyo.ac.jp

¤a Current address: Center for Neuroscience, University of California Davis, Davis, California, United States of America
¤b Current address: Department of Neurobiology, Graduate School of Medicine, University of Tokyo, Bunkyo-ku, Tokyo, Japan

Introduction

Motile cilia and flagella are ancient organelles present in various eukaryotic organisms including humans. Defects in structure or motility of cilia and flagella result in a category of diseases called primary ciliary dyskinesia [1,2]. The axoneme of motile cilia and flagella has a strikingly conserved structure consisting of nine outer doublet microtubules and two central-pair (CP) microtubules. These microtubules have various types of projections, such as outer and inner dynein arms, radial spokes (RSs), CP projections and nexin links, which directly or indirectly interact with each other [3–8]. The interactions between these projections must play fundamental roles for maintaining the cylindrical arrangement of the outer doublets, for producing the motive force, and for converting the force to ciliary/flagellar bending waves [9–11]. However, the nature of interaction between these structures is not well understood.

The ninefold symmetrical arrangement of the outer doublets is determined by the centriole (basal body), which has a cylindrical shape consisting of nine short triplet microtubules. Axonemal doublet microtubules assemble onto the A- and B-tubules of each triplet microtubule through a junction called the transition zone [11]. In contrast to the outer doublets assembling on clearly defined template structures, the assembly process of the CP is largely unknown. In *Chlamydomonas* flagella, the distal end of the CP is capped with a plate while the proximal end does not attach to any recognizable structure [12,13]. In *Tetrahymena* cilia, the distal end is also capped with a plate, while the proximal end of one CP microtubule is covered with the axosome, an amorphous structure observed in ciliate axonemes, and the other CP microtubule is apparently associated with no distinct structures [14,15]. Neither of the ends appears to function as the nucleation site for CP assembly [16–20]. However, what nucleates the CP assembly and what determines the CP number in the axoneme are not known.

The CP and RS appear to function as a regulatory system for ciliary and flagellar motility [21–25]. For generation of axonemal beating, a subset of the axonemal dyneins must be activated at a specific phase of beating, and the region of the activated dynein must move along the axoneme as the bending propagates toward the tip. Although the mechanism of this dynein regulation is yet to be elucidated, the CP/RS system clearly plays a crucial role in coordinating dynein activities [9,10,23]. Several lines of evidence suggest that, in some protists including *Chlamydomonas* and *Paramecium*, the CP assumes a twisted conformation and rotates within the axoneme once per beating cycle. The signal of CP orientation is most likely transmitted to the dynein arms through the RS [9,10,26–28]. The CP-RS interaction probably involves mechanical force since the RS is the structure that keeps the CP at the center of the axoneme [24]. In accordance with the probable mechanical nature of interaction, a recent study has suggested that the RS pushes the CP in beating axonemes [29]. However, the evidence is rather indirect and thus whether the RS pushes or pulls the CP appears to need further studies.

We previously reported a *Chlamydomonas* mutant, *bld12*, that has severe defects in a subcentriolar structure termed the cartwheel [30]. The cartwheel, consisting of a central hub and nine spokes, is located at the proximal end of the centriole as a stack of several layers [31,32]. The mutant *bld12* lacks the central part of the cartwheel due to a null mutation in the gene coding for SAS-6, a component of the cartwheel [30]. X-ray crystallography and biochemical analyses showed that SAS-6 forms a dimer having two globular heads and a coiled coil tail, and the dimers assemble into a ring through their hydrophobic interaction between the heads [33,34].

Lack of the cartwheel in *bld12* causes severe defects in the centriole assembly: ~80% of the centrioles observed are split into fragments and only ~20% are assembled in the cylindrical structure. Interestingly, the number of the triplets varies from seven to eleven in the cylindrical centrioles. As a consequence, flagellar axonemes produced in a small fraction of *bld12* cells contain variable numbers of outer doublet microtubules ranging from eight to eleven [30]. In this study, we investigated the effects of the variation in the outer doublet number on the appearance of the CP within the axoneme. The results revealed the importance of the spatial factor for the formation of the CP, and furthermore, provided evidence for the presence of attractive force between the CP and RS.

Material and Methods

Strains

Chlamydomonas strains CC124 (wild-type), *pf6*, *pf14*, and *pf16* were obtained from the *Chlamydomonas* Resource center, and *cpc1* from Dr. Mitchell of State University of New York Upstate Medical University [35]. The mutant *bld12* (*bld12-1*) was previously isolated in our laboratory (Nakazawa et al., 2007). The double and triple mutants, *bld12pf6*, *bld12pf14*, *bld12pf16*, *bld12cpc1*, *bld12cpc1pf6*, *pf14pf6*, *pf14pf16*, and *pf14cpc1*, were produced by genetic crosses [36]. Cells were grown in Tris-acetate-phosphate (TAP) media [37] with aeration on a 12 h/12 h light/dark cycle, or under constant illumination with agitation.

Preparation of flagellar axonemes

Flagella were isolated from *bld12-1* or *bld12-1cw92* cells by the dibucaine method of Witman [38]. Detached flagella were collected by centrifugation at 10,000×g, overlaid on a sucrose cushion (25% sucrose, 20 mM HEPES pH 7.4) and centrifuged at 1,000×g for 10 min at 4°C. Flagella at the boundary between the upper phase and the sucrose solution were collected, and demembranated by treatment with 0.5% Nonidet P-40 in HMDE solution (30 mM HEPES, 5 mM MgCl$_2$, 1 mM dithiothreitol (DTT), 1 mM EGTA, pH 7.4) [38]. The axonemes collected by centrifugation were resuspended with HMDE solution. For analysis of flagellar length, isolated flagella were collected and their lengths were measured using ImageJ software.

Electron microscopy

Axonemes were prefixed with 2% glutaraldehyde and 1% tannic acids in 50 mM phosphate buffer pH 7.2 for 1–2 h at 0°C, and postfixed with 1% OsO$_4$ in the phosphate buffer for 1 h. The samples were dehydrated by passing through graded concentrations of alcohol solutions, and embedded in EPON 812. Thin sections (50–70 nm) were stained with aqueous uranyl acetate and Reynold's lead citrate, and observed with a JEM1010 electron microscope. Images were obtained using a film-base camera or MegaView III digital camera (JEOL, Tokhyo). For analysis of axonemal cross sections, images were chosen on the basis of clear appearance of all outer doublets, to ensure that the axoneme under examination was cut almost normal to the axoneme axis. The diameter and inner-doublet spacing were measured using ImageJ software. Correlation between the diameter and the doublet number was analyzed using a linear regression model. Group difference in inter-doublet spacing was analyzed using ANOVA. Statistical significance for the test was set at P<0.05.

Analysis of CP-RS association and helical properties of the CP

For analyses of the CP-RS association, cross section images of 10- or 11-doublet axonemes of *bld12*, *bld12pf6*, *bld12cpc1*, *bld12cpc1pf6*, and *bld12pf16* were chosen and collected based on the clarity of the C1 and C2 microtubules. In the case of *bld12pf16*, images of the 9-doublet axonemes were used for the analysis in addition to the 10- or 11-doublet axonemes. Each image was oriented with the dynein arms projecting clockwise. The CP surface in the image, which was approximated to a circle, was equally divided into 12 sectors. The spokeheads attached to or detached from the CP were identified by visual inspection. The center of the consecutive spokeheads attached to the CP was defined as the spokehead-interaction site. Distribution of the interaction sites on the CP surface was represented by polar histograms.

In the CP orientation analysis in cross section images of *pf14*, *pf14pf6*, *pf14cpc1*, *pf14pf16* axonemes, distribution of the sector closest to the outer doublet wall was represented by polar histograms. Statistical significance of the difference in distribution was evaluated by the χ^2 test. Statistical significance was set at P<0.05.

Results

Abnormal axonemes in *bld12*

As we previously reported, ~10% of *bld12* cells produced one or two flagella when cell walls were removed by treatment with autolysin [30]. Under this condition, about 8% of cells were uniflagellated and 2% biflagellated. Of the biflagellated cells, about 50% had flagella of unequal lengths. The flagellar length was variable but always shorter than that of wild type (Figure 1A). The *bld12* flagella displayed a variety of motility phenotypes, ranging from complete paralysis to sporadic twitching to almost normal beating.

Electron microscopy showed that the percentage of 8-, 9-, 10-, and 11-doublet axonemes was, respectively, ~5%, ~90%, ~5%, and ~0.1% (Figure 1B) [30]. While the diameter of the axoneme increased with the doublet number, the space between the adjacent doublets was constant (Figure 1C), suggesting that the inter-doublet structures such as the inner- and outer-dynein arms and the nexin links were not distorted in the abnormal axonemes. In the images of these axonemes, we noticed two remarkable features that are not seen in normal axonemes: ~95% of the 8-doublet axonemes had no CP microtubules; and, in 10- or 11-doublet axonemes, the circular arrangement of the doublets was distorted because only three to five RSs were attached to the CP. Similarly distorted axonemes were also observed in intact flagella of this mutant (Yuki Nakazawa, unpublished observation).

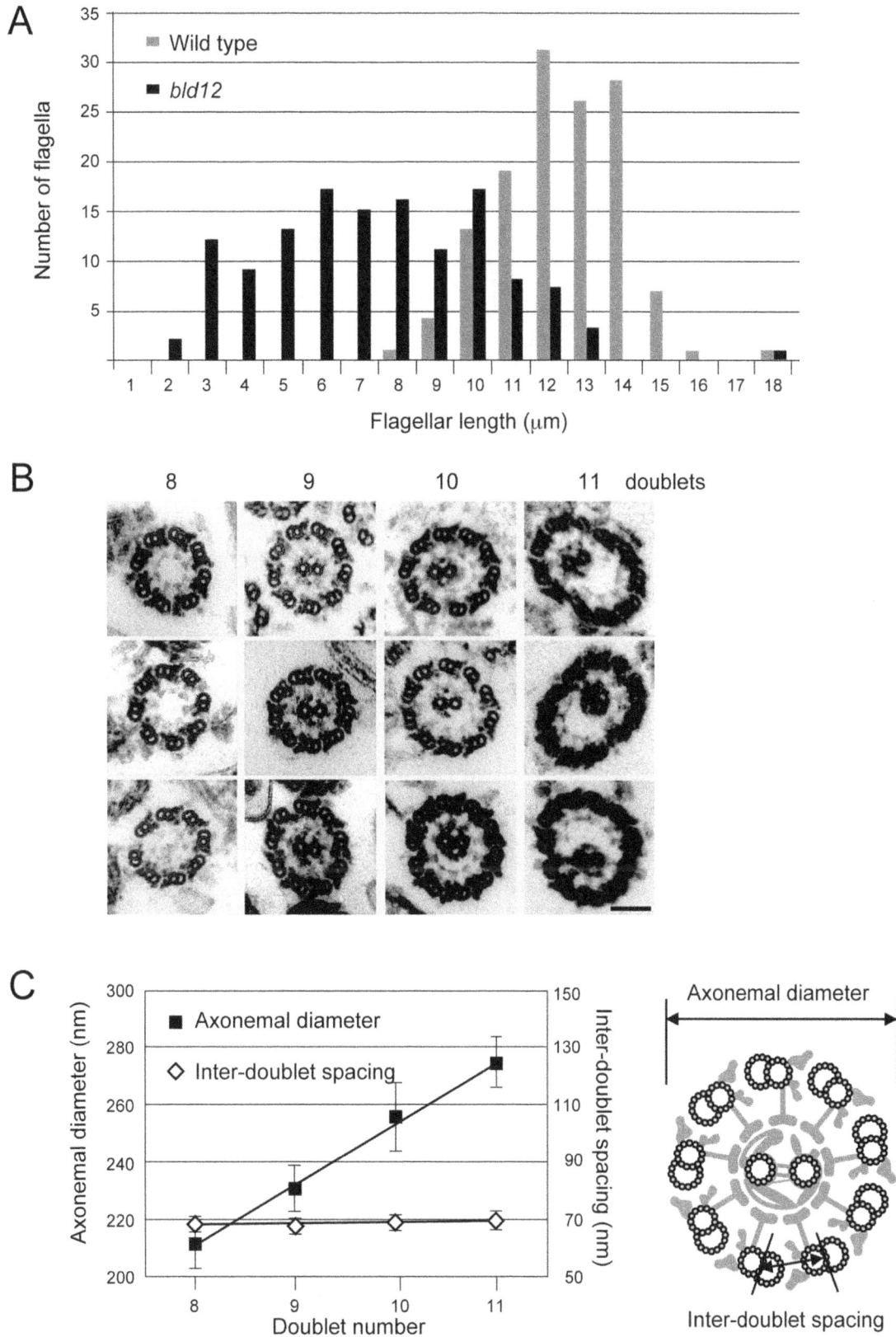

Figure 1. Abnormal features of *bld12* axonemes. (A) Length distributions of wild type (gray) and *bld12* (black) flagella. (B) Axonemes with 8, 9, 10, and 11 outer doublet microtubules. Bar, 100 nm. (C) Diameter and inter-doublet spacing in axonemes with various doublet numbers. Diameter increases with the doublet numbers, while the inter-doublet spacing is constant. The linear regression slope of the diameter is 22.14±1.69 nm ($R^2 = 0.900$). The difference between the inter-doublet spacing is not significant (ANOVA, $P = 0.59$).

Effects of RS removal on CP formation

Absence of the CP in most of the 8-doublet axonemes led us to assume that these flagella do not have enough room to accommodate a CP in the central area. To test this hypothesis, we produced the double mutant of *bld12* and *pf14,* a mutant that lacks the RSs and has a larger internal space [39]. This mutant, *bld12pf14,* also produced axonemes with a variable number of doublets, ranging from 7 to 11 (Figure 2). As expected, all of the 8-doublet axonemes and even 7-doublet axonemes had the CP (Figure 2, Table 1). In addition, ~5% of the 9-doublet axonemes or ~74% of the 10-doublet axonemes contained three or four central microtubules (Figure 2, Table 1). In *pf14* also, axonemes with three central microtubules were observed although the occurrence was rare (~0.7% in 2000 cross-section images). These observations indicate that the formation of the CP depends on the size of the space limited by the RSs and the outer doublet microtubules.

Spokeheads preferentially bind to distinct sites on the CP surface

Cross section images of the 10- or 11-doublet axonemes of *bld12* showed that five or six doublets were tethered by RSs to the CP to form a semicircle of a normal diameter. The rest of the doublets, not tethered to the CP due to the detachment of RSs, are arranged in another semicircle bulging outward (Figure 1B). Whether an RS was attached or detached could be easily judged from the position of the bulky spokehead. The doublet arrangement was distorted at the junctions of the two semicircles. In contrast, no such distortion was observed in the RS-lacking axonemes of *bld12pf14* (Figure 2). These observations suggest that the RS binds to the CP to help the doublets align in a circular arrangement of a constant diameter, and that the CP-RS binding is strong enough to distort the arrangement in 10- or 11-doublet axonemes.

To investigate whether this RS binding occurs on particular regions on the CP surface, we examined cross-section images of 10- or 11-doublet axonemes for the possible location on the CP where the spokeheads preferentially attach. We divided the CP image in cross section into 12 sectors, and scored the frequency of each sector to locate at the center of the group of CP-associated RSs (Figure 3, A–C). An analysis of 56 cross-section images revealed that the CP surface had two preferred sites for association with the RS: one near the C1a and the other near the C1b projection. These two preferred sites must bind to the spokeheads more strongly than the other regions of the CP.

Previous studies showed that the *Chlamydomonas* CP, when released from the axoneme, forms a helical complex; when contained in the axoneme, it must be forced to assume a straight form with a 360 degree twist per the length of the flagellum [40,41]. This tendency of the CP to assume a helix might bias the distribution of the apparent spoke-interaction sites in the 10- or 11-doublet axonemes (Figure 3, B and C). To address this possibility, we examined the helical tendency of the CP in the spoke-less *pf14* axonemes, in which the CP should assume a small-amplitude helical form facing its outer surface to the outer doublet wall (Figure 3, D and E). An analysis of 80 cross sections of *pf14* axonemes indicated that the C1 microtubule is located outer side of the CP helix, i.e. closest to the doublet wall. This distribution pattern is clearly different from the pattern in the axonemes of *bld12,* which showed two preferred orientation regions (Figure 3, B–E). We therefore concluded that the helical tendency of the CP did not mask the CP orientation resulting from its interaction with RSs.

Removal of CP projections identifies multiple weak CP-spoke association sites

We next examined CP-RS interactions in double mutants between *bld12* and each one of four mutants that lack specific CP projections (Figure 4). The CP mutants used were *pf6* lacking the prominent projection C1a [42]; *cpc1* lacking C1b and C2b [35]; *pf6cpc1* double mutant lacking these three projections; and *pf16* lacking the C1 microtubule [42]. Interestingly, the spokeheads preferentially bound to the C1b–C2b region when the C1a projection was absent, whereas they bound to the C1a–C2a region when C1b and C2b were absent. In *pf6cpc1* or *pf16,* the spokeheads tended to bind to the C1d and C2c regions, or to the C2b region, where binding was only infrequently observed in *bld12* axonemes (Figure 4, A and B). As controls, we also analyzed the axonemal images of double mutants *pf14pf6, pf14cpc1,* and *pf14pf16* because we were particularly concerned that the mutant CPs might have varied tendencies to assume helical forms, which might affect the results of our analysis. The histograms obtained in the control experiments showed that the helical tendency was not largely affected by the *pf6, cpc1,* or *pf16* mutations (Figure 4C). For example, the *pf14pf16* CP, which lacks the C1 microtubule, displayed the same tendency as the *pf14* CP (Figure 3D); that is, the side of the C2 that would be positioned closest to the C1 in the wild type CP was still positioned on the outside of the helix. These results suggest that, although different regions of the CP surface differ in the spokehead binding affinity, almost the whole area of the CP surface can bind to the spokeheads.

Table 1. The number of axonemes with particular numbers of outer doublets and central microtubules observed in *bld12pf14.*

	# of central microtubules per axoneme		
# of outer doublets per axoneme	0 or 1	2	3 or 4
7	1	2	0
8	0	465	0
9	15	801	42
10	3	5	23
11	0	1	1

(Total number of axonemes counted, 1,369).

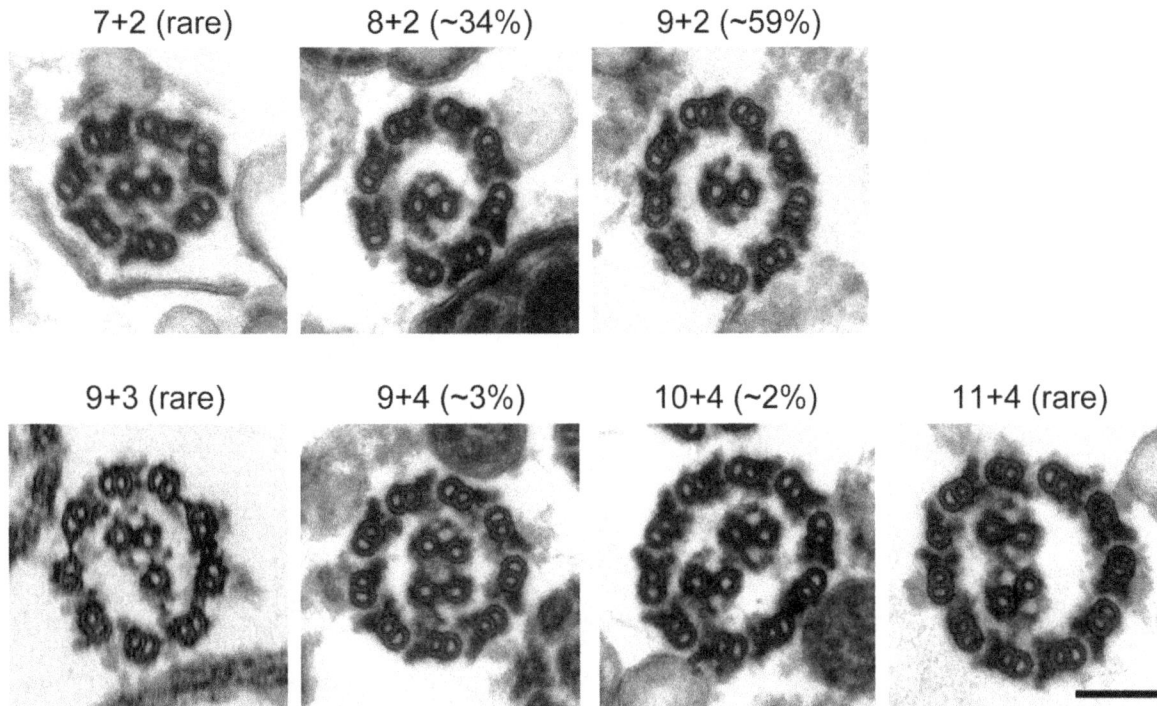

Figure 2. CPs in RS-lacking axonemes. Cross section images of *bld12pf14* axonemes with 7 to 10 outer doublets are shown. The frequency of each pattern in the observed images is indicated in the parenthesis (n = 1,369, Table 1). All CPs have the normal polarity in the axoneme as judged by their CP projections. Bar, 100 nm.

Discussion

Formation of the CP

We showed that the CP did not assemble in 8-doublet axonemes of *bld12*, but assembled in those of *bld12pf14* lacking the RS. Furthermore, while 9-doublet axonemes of *bld12pf14* only rarely contained two CPs, its 10-doublet axonemes frequently contained two CPs (Figure 2, Table 1). Formation of two CPs has previously been observed in flagella and cilia that lack the RS: for example, flagella of *Chlamydomonas pf14* and *pf14pf6cpc1* mutants [20,23]; and nodal cilia of rabbit [43]. These observations suggest that CP assembly depends on the internal space of the axoneme. Our results, demonstrating a clear correlation between the CP assembly and the space size, lend strong support to the previous proposal (Table 1).

The distal end of the CP in *Chlamydomonas* flagella and *Tetrahymena* cilia is capped by a plate-like structure that is indirectly attached to the membrane through a spherical bead. In contrast, the proximal end of the CP, located near the transition zone, is associated with no detectable structure [12,13]. In some *Chlamydomonas* mutants with defects in the transition zone, as well as in isolated tracheal epithelial cells, the CP microtubules grow proximally into the centriole, suggesting that the proximal end is not the site of nucleation for CP growth [16,17]. Lechtreck et al. (2013) reported that, when gametes of a *Chlamydomonas* mutant lacking the CP are mated with wild type gametes, the mutant flagella in the fused cells start to produce the CP at the middle portion of the axoneme. This observation indicates that no organizing center is required for the CP assembly at either end of the flagellum. These authors also observed that the RS-lacking *pf14* axonemes contain two CPs (four microtubules) with correct polarity. Together with their findings, our observation of the extra CP in *bld12pf14* implies that the space size within the axoneme is

an important factor that directs CP formation in the axoneme. Possibly, when the CP precursors are present in the axoneme, the CP may form spontaneously without interacting with any template or RSs. In such a case, only the available space and the amount of the precursors may limit the number of the CPs produced.

Nature of CP-RS interaction for the regulation of flagellar motility

A minor population of *bld12* axonemes having 10 or 11 outer doubles exhibited distortion in the circular arrangement of the outer doublets. This observation clearly indicates that the RS binds to the CP and the binding exerts force. Previous studies have provided substantial evidence that the CP and RS form a signal transduction pathway that modulates dynein activity through phosphorylation of a specific subunit of inner-arm dynein [44–46]. The RS is likely to chemically and mechanically control the dynein activity based on the interaction between the CP projections and the spokeheads [9,21]. However, whether or not the RS exerts mechanical force on the outer doublet has been unknown. Our present study is the first to show that RSs actually pull the outer doublet microtubules toward the axonemal center. Such a mechanical force may be at the center of the regulatory function of the CP/RS.

The CP-RS interaction should be transient and the binding strength is weak enough to allow such an interaction. This is because the *Chlamydomonas* CP rotates within the axoneme [40,41] and the RS slides over the CP for a certain distance as the axoneme propagates bending waves [47]. In this study, however, we showed that the interaction is still strong enough to distort the circular arrangement of the outer doublets. This finding prompts us to speculate that the mechanical force transmitted by the RS could change the relative position of the doublet microtubules and

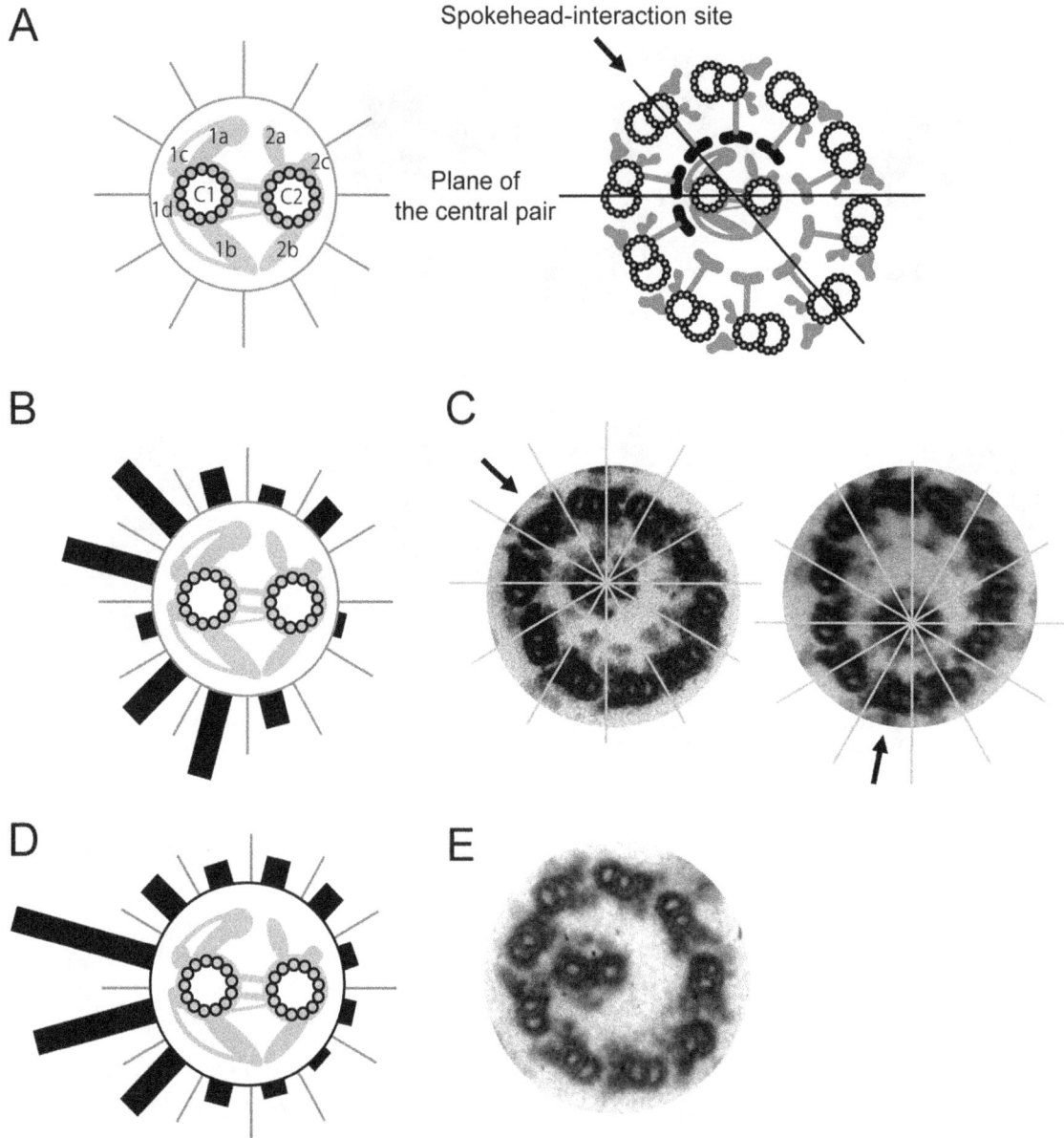

Figure 3. Frequency of CP-RS association in different sectors of the CPs. (A) The CP cross section was divided into 12 sectors by six lines including the one connecting the centers of the two central microtubules. The center of the consecutive spokeheads in contact with the CP was defined as the spokehead-interaction site. Major CP projections are designated [35]. (B) A polar histogram representing the distribution of the spokehead-interaction sites on the CP in cross section images of 10- or 11-doublet axonemes of *bld12* (n = 56). The length of each bar represents relative frequency to locate at the spokehead-interaction site. (C) Axoneme images that correspond to the two peaks in the histogram in (B). (D) Distribution of the CP region in contact with the wall of outer doublets in *pf14* axonemes (n = 80). The histogram suggests that the C1 microtubule is located on the outer surface on the helical CP. (E) A cross section image that represents the peak distribution in the histogram in (D). The differences between the distributions of *bld12* (B) and *pf14* (D) are significant (χ^2 test, p<0.05).

dyneins, and thereby transiently activate or inactivate the dyneins located in a particular region of the axoneme. A change in the dynein-microtubule positioning has also been postulated in the geometric clutch model of Lindemann [48]. The location of dynein molecules activated or inactivated by the CP/RS should propagate along the axoneme as the twisted CP rotates. In the cilia and flagella of multicellular organisms and some unicellular organisms, the CP neither assumes a helical shape nor rotates in the axoneme although sharing the structure and components with the *Chlamydomonas* CP [23,49,50]; in those organisms, the stationary CP determines the plane of axonemal beating possibly

by activating dynein molecules on a particular side of the axoneme [51,52].

Our image analyses of mutant axonemes suggest that, while most area of the CP surface can bind to the RS, the two major projections on the C1 microtubule bind stronger than the other projections on the C2 microtubule. This asymmetric distribution of the RS binding affinities on the CP surface must be important for the signal generation by the rotating CP. Although the present study does not provide information as to whether the stronger binding promotes or inhibits dynein-driven microtubule sliding, previous studies suggest that the C1 microtubule or its projections

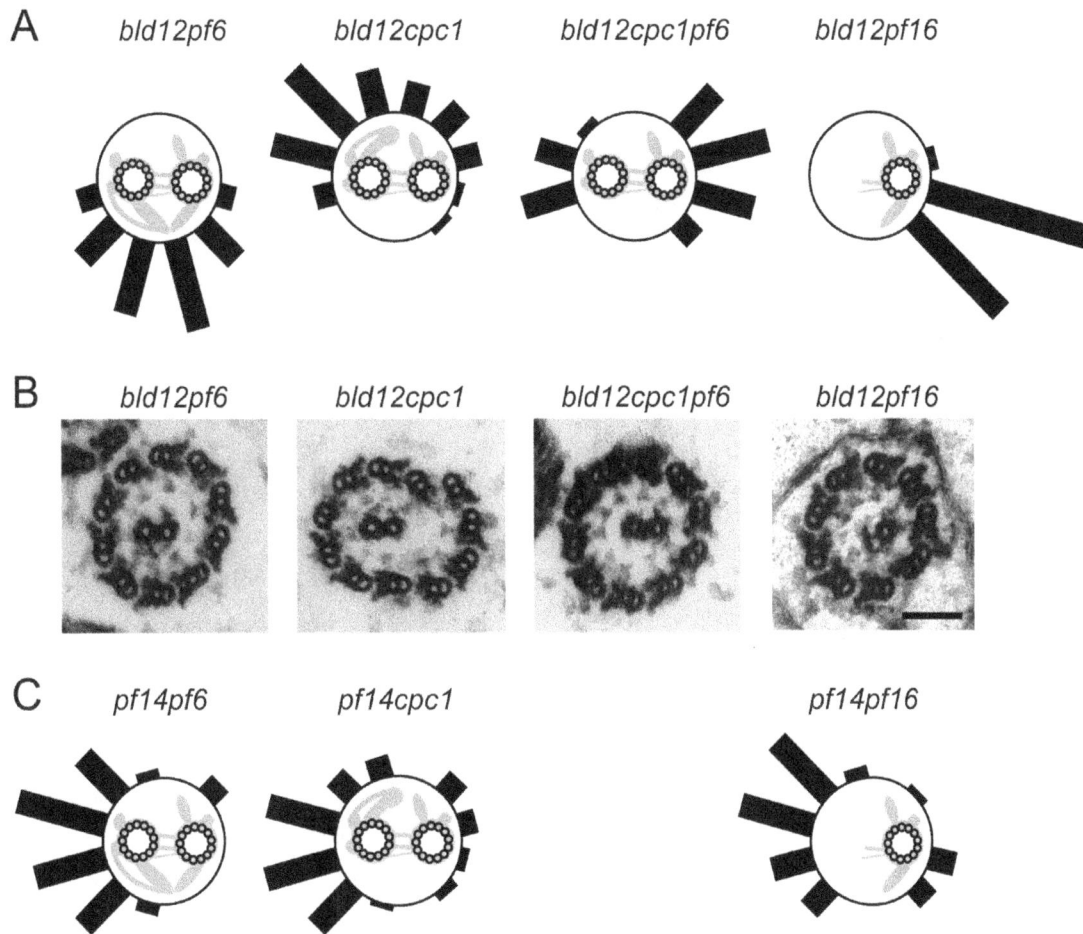

Figure 4. Removal of CP projections manifests weak CP-spokehead association. (A) Distributions of the spokehead-interaction sites on the CP in cross section images of 10- or 11-doublet axonemes of *bld12pf6*, *bld12cpc1*, *bld12cpc1pf6*, and *bld12pf16* (n = 59, 49, 40, and 41), which lack specific CP projections. (B) Cross section images that represent the peaks in the histograms in (A). (C) Distributions of contact sites on the CP surface with the outer doublet wall in cross section images of *pf14pf6*, *pf14cpc1*, and *pf14pf16* (n = 33, 69, and 46). The differences between the distributions of *bld12pf6* and *pf14pf6*; between *bld12cpc1* and *pf14cpc1*, and between *bld12pf16* and *pf14pf16*, are significant (χ^2 test, p<0.05 for each of the three pairs). These distribution patterns are similar to the pattern in *pf14*.

enhances microtubule sliding in axonemes [46,53–55]. Thus the doublet-pulling by stronger RS binding to the C1 surface may activate dyneins and promote microtubule sliding.

Our results are in apparent contradiction with a non-specific CP-RS interaction model recently proposed by Oda et al. (2014). These authors showed that the flagella of the *pf6* mutant lacking the C1a projection recovered motility if any one of three protein tags (hemagglutinin, biotin carboxyl carrier protein, and green fluorescent protein) was attached on top of the spokehead. Because the extent of motility recovery increased in the order of the size of the tag, they proposed that the added tag elongated the RS and compensated the loss of the C1a projection, possibly by enabling RSs to collide with the CP. The proposed physical CP-RS interaction must be non-specific since the three protein tags used are structurally unrelated. In contrast, our analysis of 10- or 11-doublet axoneme images showed that the C1a projection, together with the C1b projection, preferentially associates with the RSs among all the CP projections, favoring the view that the C1a projection pulls the outer doublet in a fairly specific manner. Both their results and our results must reflect some aspects of the CP-RS interaction, but their relationship is not understood. The

molecular mechanism of the CP-RS interaction remains one of the most interesting problems in cilia/flagella motility studies.

An obvious question regarding the present study is whether or not the axonemes with aberrant numbers of outer doublets are motile. We may imagine that 8-doublet flagella are non-motile because they lack the CP, like the flagella of non-motile mutants such as *pf18* and *pf19* [56–58]. However, it is difficult to predict whether or not axonemes with 10 or 11 doublets can display some motility. Development of techniques that determine the number of axonemal microtubules under the microscope, or those that permit constant production of flagella with 10–11 doublets, may well provide answers. We can hope that the answer will provide a strong clue as to why motile cilia and flagella almost always contain nine outer doublets.

Acknowledgments

We thank Dr. David Mitchell for providing the *Chlamydomonas* mutant strain *cpc1*.

Author Contributions

Conceived and designed the experiments: YN RK MH. Performed the experiments: YN TA AN RK MH. Analyzed the data: YN TA AN RK MH. Contributed to the writing of the manuscript: YN RK MH.

References

1. Badano JL, Mitsuma N, Beales PL, Katsanis N (2006) The ciliopathies: an emerging class of human genetic disorders. Annual review of genomics and human genetics 7: 125–148.
2. Onoufriadis A, Shoemark A, Schmidts M, Patel M, Jimenez G, et al. (2014) Targeted NGS gene panel identifies mutations in RSPH1 causing primary ciliary dyskinesia and a common mechanism for ciliary central pair agenesis due to radial spoke defects. Human molecular genetics 23: 3362–3374.
3. Bui KH, Yagi T, Yamamoto R, Kamiya R, Ishikawa T (2012) Polarity and asymmetry in the arrangement of dynein and related structures in the Chlamydomonas axoneme. The Journal of cell biology 198: 913–925.
4. Bui KH, Sakakibara H, Movassagh T, Oiwa K, Ishikawa T (2008) Molecular architecture of inner dynein arms in situ in Chlamydomonas reinhardtii flagella. The Journal of cell biology 183: 923–932.
5. Pigino G, Bui KH, Maheshwari A, Lupetti P, Diener D, et al. (2011) Cryoelectron tomography of radial spokes in cilia and flagella. The Journal of cell biology 195: 673–687.
6. Nicastro D, Schwartz C, Pierson J, Gaudette R, Porter ME, et al. (2006) The molecular architecture of axonemes revealed by cryoelectron tomography. Science 313: 944–948.
7. Heuser T, Raytchev M, Krell J, Porter ME, Nicastro D (2009) The dynein regulatory complex is the nexin link and a major regulatory node in cilia and flagella. The Journal of cell biology 187: 921–933.
8. Carbajal-Gonzalez BI, Heuser T, Fu X, Lin J, Smith BW, et al. (2013) Conserved structural motifs in the central pair complex of eukaryotic flagella. Cytoskeleton (Hoboken, NJ) 70: 101–120.
9. Smith EF, Yang P (2004) The radial spokes and central apparatus: mechano-chemical transducers that regulate flagellar motility. Cell motility and the cytoskeleton 57: 8–17.
10. Kamiya R (2002) Functional diversity of axonemal dyneins as studied in Chlamydomonas mutants. International review of cytology 219: 115–155.
11. Mizuno N, Taschner M, Engel BD, Lorentzen E (2012) Structural studies of ciliary components. Journal of molecular biology 422: 163–180.
12. Dentler WL, Rosenbaum JL (1977) Flagellar elongation and shortening in Chlamydomonas. III. structures attached to the tips of flagellar microtubules and their relationship to the directionality of flagellar microtubule assembly. The Journal of cell biology 74: 747–759.
13. Dentler WL (1980) Structures linking the tips of ciliary and flagellar microtubules to the membrane. Journal of cell science 42: 207–220.
14. Allen RD (1969) The morphogenesis of basal bodies and accessory structures of the cortex of the ciliated protozoan Tetrahymena pyriformis. The Journal of cell biology 40: 716–733.
15. Dute R, Kung C (1978) Ultrastructure of the proximal region of somatic cilia in Paramecium tetraurelia. The Journal of cell biology 78: 451–464.
16. Dentler WL, LeCluyse EL (1982) The effects of structures attached to the tips of tracheal ciliary microtubules on the nucleation of microtubule assembly in vitro. Progress in clinical and biological research 80: 13–18.
17. Jarvik JW, Suhan JP (1991) The role of the flagellar transition region: Inferences from the analysis of a Chlamydomonas mutant with defective transition region structures. Journal of cell science 99: 731–740.
18. Silflow CD, Liu B, LaVoie M, Richardson EA, Palevitz BA (1999) Gamma-tubulin in Chlamydomonas: characterization of the gene and localization of the gene product in cells. Cell motility and the cytoskeleton 42: 285–297.
19. McKean PG, Baines A, Vaughan S, Gull K (2003) Gamma-tubulin functions in the nucleation of a discrete subset of microtubules in the eukaryotic flagellum. Current biology: CB 13: 598–602.
20. Lechtreck KF, Gould TJ, Witman GB (2013) Flagellar central pair assembly in Chlamydomonas reinhardtii. Cilia 2: 15.
21. Wirschell M, Nicastro D, Porter ME, Sale WS (2009) Chapter 9 - The Regulation of Axonemal Bending. In: Harris EH, Stern DB, Witman GB, editors. The Chlamydomonas Sourcebook (Second Edition). London: Academic Press. 253–282.
22. Yang P, Smith EF (2009) Chapter 7 - The Flagellar Radial Spokes. In: Harris EH, Stern DB, Witman GB, editors. The Chlamydomonas Sourcebook (Second Edition). London: Academic Press. 209–234.
23. Mitchell DR (2009) Chapter 8 - The Flagellar Central Pair Apparatus. In: Harris EH, Stern DB, Witman GB, editors. The Chlamydomonas Sourcebook (Second Edition). London: Academic Press. 235–252.
24. Witman GB, Plummer J, Sander G (1978) Chlamydomonas flagellar mutants lacking radial spokes and central tubules. Structure, composition, and function of specific axonemal components. The Journal of cell biology 76: 729–747.
25. Lechtreck KF, Delmotte P, Robinson ML, Sanderson MJ, Witman GB (2008) Mutations in Hydin impair ciliary motility in mice. The Journal of cell biology 180: 633–643.
26. Huang B, Ramanis Z, Luck DJ (1982) Suppressor mutations in Chlamydomonas reveal a regulatory mechanism for Flagellar function. Cell 28: 115–124.

27. Piperno G, Mead K, Shestak W (1992) The inner dynein arms I2 interact with a "dynein regulatory complex" in Chlamydomonas flagella. The Journal of cell biology 118: 1455–1463.
28. Omoto CK, Kung C (1979) The pair of central tubules rotates during ciliary beat in Paramecium. Nature 279: 532–534.
29. Oda T, Yanagisawa H, Yagi T, Kikkawa M (2014) Mechanosignaling between central apparatus and radial spokes controls axonemal dynein activity. The Journal of cell biology 204: 807–819.
30. Nakazawa Y, Hiraki M, Kamiya R, Hirono M (2007) SAS-6 is a cartwheel protein that establishes the 9-fold symmetry of the centriole. Current biology: CB 17: 2169–2174.
31. Gibbons IR, Grimstone AV (1960) On flagellar structure in certain flagellates. The Journal of biophysical and biochemical cytology 7: 697–716.
32. Cavalier-Smith T (1974) Basal body and flagellar development during the vegetative cell cycle and the sexual cycle of Chlamydomonas reinhardii. Journal of cell science 16: 529–556.
33. Kitagawa D, Vakonakis I, Olieric N, Hilbert M, Keller D, et al. (2011) Structural basis of the 9-fold symmetry of centrioles. Cell 144: 364–375.
34. van Breugel M, Hirono M, Andreeva A, Yanagisawa HA, Yamaguchi S, et al. (2011) Structures of SAS-6 suggest its organization in centrioles. Science (New York, NY) 331: 1196–1199.
35. Mitchell DR, Sale WS (1999) Characterization of a Chlamydomonas insertional mutant that disrupts flagellar central pair microtubule-associated structures. The Journal of cell biology 144: 293–304.
36. Dutcher SK (1995) Mating and tetrad analysis in Chlamydomonas reinhardtii. Methods in cell biology 47: 531–540.
37. Gorman DS, Levine RP (1965) Cytochrome f and plastocyanin: their sequence in the photosynthetic electron transport chain of Chlamydomonas reinhardi. Proceedings of the National Academy of Sciences of the United States of America 54: 1665–1669.
38. Witman GB (1986) [28] Isolation of Chlamydomonas flagella and flagellar axonemes. In: Richard BV, editor. Methods in enzymology: Academic Press. 280–290.
39. Piperno G, Huang B, Luck DJ (1977) Two-dimensional analysis of flagellar proteins from wild-type and paralyzed mutants of Chlamydomonas reinhardtii. Proceedings of the National Academy of Sciences of the United States of America 74: 1600–1604.
40. Kamiya R (1982) Extrusion and Rotation of the central-pair microtubules in detergent-treated Chlamydomonas flagella. Progress in clinical and biological research 80: 169–173.
41. Mitchell DR, Nakatsugawa M (2004) Bend propagation drives central pair rotation in Chlamydomonas reinhardtii flagella. The Journal of cell biology 166: 709–715.
42. Dutcher SK, Huang B, Luck DJ (1984) Genetic dissection of the central pair microtubules of the flagella of Chlamydomonas reinhardtii. The Journal of cell biology 98: 229–236.
43. Feistel K, Blum M (2006) Three types of cilia including a novel 9+4 axoneme on the notochordal plate of the rabbit embryo. Developmental dynamics: an official publication of the American Association of Anatomists 235: 3348–3358.
44. Habermacher G, Sale WS (1995) Regulation of dynein-driven microtubule sliding by an axonemal kinase and phosphatase in Chlamydomonas flagella. Cell motility and the cytoskeleton 32: 106–109.
45. Yang P, Sale WS (2000) Casein kinase I is anchored on axonemal doublet microtubules and regulates flagellar dynein phosphorylation and activity. The Journal of biological chemistry 275: 18905–18912.
46. Smith EF (2002) Regulation of flagellar dynein by the axonemal central apparatus. Cell motility and the cytoskeleton 52: 33–42.
47. Warner FD, Satir P (1974) The structural basis of ciliary bend formation. Radial spoke positional changes accompanying microtubule sliding. The Journal of cell biology 63: 35–63.
48. Lindemann CB, Kanous KS (1995) "Geometric clutch" hypothesis of axonemal function: key issues and testable predictions. Cell motility and the cytoskeleton 31: 1–8.
49. Tamm SL, Tamm S (1981) Ciliary reversal without rotation of axonemal structures in ctenophore comb plates. The Journal of cell biology 89: 495–509.
50. Gadelha C, Wickstead B, McKean PG, Gull K (2006) Basal body and flagellum mutants reveal a rotational constraint of the central pair microtubules in the axonemes of trypanosomes. Journal of cell science 119: 2405–2413.
51. Gibbons IR (1961) The relationship between the fine structure and direction of beat in gill cilia of a lamellibranch mollusc. The Journal of biophysical and biochemical cytology 11: 179–205.
52. Yoshimura M, Shingyoji C (1999) Effects of the central pair apparatus on microtubule sliding velocity in sea urchin sperm flagella. Cell structure and function 24: 43–54.

53. Brown JM, Dipetrillo CG, Smith EF, Witman GB (2012) A FAP46 mutant provides new insights into the function and assembly of the C1d complex of the ciliary central apparatus. Journal of cell science 125: 3904–3913.

54. Smith EF (2002) Regulation of flagellar dynein by calcium and a role for an axonemal calmodulin and calmodulin-dependent kinase. Molecular biology of the cell 13: 3303–3313.

55. Wargo MJ, Smith EF (2003) Asymmetry of the central apparatus defines the location of active microtubule sliding in *Chlamydomonas* flagella. Proceedings of the National Academy of Sciences of the United States of America 100: 137–142.

56. Randall J, Warr JR, Hopkins JM, McVittie A (1964) A SINGLE-GENE MUTATION OF *CHLAMYDOMONAS REINHARDII* AFFECTING MOTILITY: A GENETIC AND ELECTRON MICROSCOPE STUDY. Nature 203: 912–914.

57. Randall JT, Cavalier-Smith T, McVittie A, Warr JR, Hopkins JM (1967) Developmental and control processes in the basal bodies and flagella of *Chlamydomonas reinhardtii*. Dev Biol suppl 1: 43–83.

58. Warr JR, McVittie A, Randall JT, Hopkins JM (1966) Genetic control of flagellar structure in *Chlamydomonas reinhardtii*. Genet Res 7: 335–351.

21

Host, Pathogen, and Environmental Characteristics Predict White-Nose Syndrome Mortality in Captive Little Brown Myotis (*Myotis lucifugus*)

Joseph S. Johnson[1], DeeAnn M. Reeder[1], James W. McMichael III[1], Melissa B. Meierhofer[1], Daniel W. F. Stern[1], Shayne S. Lumadue[1], Lauren E. Sigler[1], Harrison D. Winters[1], Megan E. Vodzak[1], Allen Kurta[2], Joseph A. Kath[3], Kenneth A. Field[1]*

1 Department of Biology, Bucknell University, Lewisburg, Pennsylvania, United States of America, 2 Department of Biology, Eastern Michigan University, Ypsilanti, Michigan, United States of America, 3 Illinois Department of Natural Resources, Springfield, Illinois, United States of America

Abstract

An estimated 5.7 million or more bats died in North America between 2006 and 2012 due to infection with the fungus *Pseudogymnoascus destructans* (*Pd*) that causes white-nose syndrome (WNS) during hibernation. The behavioral and physiological changes associated with hibernation leave bats vulnerable to WNS, but the persistence of bats within the contaminated regions of North America suggests that survival might vary predictably among individuals or in relation to environmental conditions. To investigate variables influencing WNS mortality, we conducted a captive study of 147 little brown myotis (*Myotis lucifugus*) inoculated with 0, 500, 5 000, 50 000, or 500 000 *Pd* conidia and hibernated for five months at either 4 or 10°C. We found that female bats were significantly more likely to survive hibernation, as were bats hibernated at 4°C, and bats with greater body condition at the start of hibernation. Although all bats inoculated with *Pd* exhibited shorter torpor bouts compared to controls, a characteristic of WNS, only bats inoculated with 500 conidia had significantly lower survival odds compared to controls. These data show that host and environmental characteristics are significant predictors of WNS mortality, and that exposure to up to 500 conidia is sufficient to cause a fatal infection. These results also illustrate a need to quantify dynamics of *Pd* exposure in free-ranging bats, as dynamics of WNS produced in captive studies inoculating bats with several hundred thousand conidia may differ from those in the wild.

Editor: Michelle L. Baker, CSIRO, Australia

Funding: Funding for this project was provided by the United States Fish and Wildlife Service grant F12AP01210 (DMR and KAF) and the Woodtiger Foundation (DMR). The funders had no role in study design, data collection and analysis, decision to publish, or preparation of the manuscript.

Competing Interests: The authors have declared that no competing interests exist.

* Email: kfield@bucknell.edu

Introduction

White-nose syndrome (WNS) is a fungal disease affecting hibernating bats, causing the death of an estimated 5.7–6.7 million bats since its initial discovery in North America in 2006 (USFWS 2012). WNS is caused by the cold-adapted fungus *Pseudogymnoascus destructans* (*Pd*), which invades the dermis and epidermis of bats during hibernation [1–3]. Bats are vulnerable to *Pd* because their immune system is suppressed along with nearly all physiological processes during hibernation [4–6], and because the cold temperature and high humidity typical in many bat hibernacula represent ideal conditions for *Pd* growth [7–9]. *Pd* was first identified as *Geomyces destructans* in 2009 [10], and was reclassified to the genus *Pseudogymnoascus* in 2013 [11]. The fungus is not native to North America, and is believed to have been introduced from Europe [12,13]. Although large-scale mortality of bats has never been documented in Europe, 90% mortality of bats occurs in North American hibernacula after *Pd* is introduced [14,15]. Such high mortality rates have led to predictions of regional extinctions, although not all bat species appear equally affected by the disease [15,16].

The little brown myotis (*Myotis lucifugus*) is among the species most heavily impacted by WNS [15]. Little brown myotis infected with *Pd* arouse more frequently from hibernation than unaffected bats, resulting in exhaustion of fat reserves needed to survive the winter [13,17]. Although the trigger for this increase in arousals has yet to be confirmed, there is evidence that hypotonic dehydration of infected bats may influence arousal behaviors [18]. Thus, it is likely that little brown myotis are particularly vulnerable to *Pd* because they have naturally high rates of evaporative water loss during winter, and because their small size (<10 g) limits their total fat reserves upon entering hibernation and the number of periodic arousals that can be sustained during winter [19–21].

Despite the high mortality rates of little brown myotis inhabiting *Pd*-contaminated hibernacula, summer maternity colonies of little brown myotis still persist within the contaminated region of North America. The persistence of these maternity colonies, which are typically composed of female bats that return to familiar roosts each year after hibernation [22–24], suggests that some bats have survived several winters of exposure to *Pd*. Although it is possible that these bats over-winter in hibernacula still unexposed to *Pd* or survived due to random factors, it is more likely that some bats have naturally higher survival rates when exposed to the fungus. This survival may result from immunological resistance to *Pd*, differences in physiology (e.g., larger body size or lower rate of evaporative water loss), or behavioral ecology (e.g., use of hibernaculum microclimates less favorable for fungal growth) that result in higher resilience. Studies of the little brown myotis demonstrate individual variability in both winter ecology and physiology, including differences in torpor patterns and energy use, and selection of microclimates within a hibernaculum, representing possible foundations for variation in survival [25–27].

If individual variation in ecology and physiology relate to WNS mortality, then mortality should vary predictably. For example, male little brown myotis utilize their winter energy reserves more rapidly than females [26,27], potentially making them more vulnerable to the further depletion of fat reserves when exposed to *Pd*. Mortality may also vary among bats inhabiting different hibernacula or areas within a hibernaculum that have different microclimates. Higher population declines have been observed among populations of little brown myotis inhabiting warmer hibernacula [28], a trend possibly linked to the growth rate of *Pd*, which peaks between 12.5 and 15.8°C and declines at warmer and colder temperatures [8]. Temperatures inside hibernacula of little brown myotis are typically below 10°C but range widely, including environmental conditions with varying suitability for *Pd* growth [7,29]. Because temperature also affects the winter torpor behaviors and energy expenditure of bats [30,31], variables such as the temperature and individual behavior and physiology are likely to have interacting effects on fungal growth and WNS mortality.

Thus, understanding whether or not some bats are better able to survive WNS requires an understanding of the interaction of the environment, host, and pathogen, a concept presented in the disease triangle [9,32,33]. However, our current understanding of WNS mortality lacks such context because laboratory studies investigating the disease are conducted under a single environmental condition, typically exposing bats to 500 000 *Pd* conidia and hibernating them at 7°C, without consideration for individual variation in survival [3,13].

Our purpose was to examine WNS mortality and survival in a captive population of little brown myotis in the context of this disease triangle. We hypothesized that the number of *Pd* conidia bats are initially exposed to affects fungal load at the end of hibernation, duration of torpor bouts, and mortality. We further hypothesized that bats hibernating at warm temperatures have higher *Pd* loads at the end of hibernation, exhibit shorter torpor bouts, and experience greater mortality. Finally, we hypothesized that mortality would be inversely related to body condition at the onset of hibernation, and that mortality would be greatest among males.

Results

Of the 147 little brown myotis, 69 (47%) survived the five-month captive hibernation study. Three emaciated bats died within 1 week of removal from hibernation and were not considered to have survived hibernation. Logistic regression analysis showed that mortality among little brown myotis was not random ($\chi^2 = 65.9$, df = 6, $P<0.001$), with *Pd* inoculation ($P=0.037$), temperature ($P<0.001$), sex ($P=0.024$), and pre-hibernation body condition ($P<0.001$) significantly influencing mortality (Fig. 1). Mortality in control groups consisted mostly (89%) of males (3 of 3 bats in the 4°C control; 5 of 6 bats in the

10°C control), all of which had body condition indices at the onset of hibernation that were below the median. Among inoculation treatments, only bats exposed to 500 *Pd* conidia had regression coefficients significantly different from 0, meaning it was the only inoculation treatment with mortality odds significantly greater than the control groups (Table 1). Mortality odds for males were significantly greater than females, greater for bats hibernating at 10°C compared to 4°C, and initial body condition had a negative, linear effect on mortality (Table 1). Average body condition of females (0.224±0.02 s.d.) was significantly higher ($t = -6.0$, df 145, $P<0.001$) than males (0.206±0.02) at the onset of hibernation, but the correlation between body condition and sex in the logistic regression model (0.255) was not great enough to merit removal of either variable. Thus, host and environmental variables helped predict mortality, with males, bats of both sexes with low body condition, and bats hibernating at warmer temperatures less likely to survive.

Differences in torpor patterns among treatment groups resembled differences in mortality (Fig. 2). Skin temperature data were successfully collected for 113 bats (77%). An analysis of variance (ANOVA) showed that average torpor bout duration varied with *Pd* inoculation ($F_{4, 102}$ 23.0, $P<0.001$) and hibernation temperature ($F_{1, 102} = 115$, $P<0.001$). Bats exhibited longer torpor bouts at 4°C, and bats in the control groups had longer torpor bouts than bats in each inoculation treatment ($P<0.05$; Fig. 2). Additionally, bats inoculated with 500 conidia also exhibited shorter torpor bouts than bats inoculated with 500 000 conidia ($P<0.05$). There was a significant interaction between *Pd* inoculation and temperature ($F_{4, 102} = 6.6$, $P<0.001$) reflecting a larger decrease in average torpor bout duration associated with *Pd* inoculation at 10°C (Fig. 2). There was no significant effect of sex on torpor bout duration ($F_{1, 103} = 3.5$, $P = 0.063$), although the observed power was low (0.46), due to low sample sizes and high variance for each sex within groups. Comparison of control males and females (Fig. 3) revealed longer torpor bouts among females hibernating at 4°C ($t = -1.984$, df = 12, $P = 0.04$) but not 10°C ($t = -1.302$, df =9, $P=0.11$). Overall, we found that *Pd* inoculation, especially at low doses, resulted in an increase in arousals from hibernation.

Greater amounts of *Pd* detected on wing swabs were not associated with higher mortality or more frequent arousals from torpor. *Pd* was not detected by quantitative polymerase chain reaction (qPCR) on any bats ($n = 147$) upon arrival at our facility, and was not detected on any control bat at the end of hibernation ($n = 29$). *Pd* loads detected on inoculated bats at the end of hibernation were highly variable (Fig. 4). We found no difference in the median fungal loads detected on bats in each inoculation treatment between temperatures. Within each temperature, however, median fungal loads varied significantly among inoculation groups (4°C: $H = 35.4$, $P<0.001$; 10°C: $H = 22.6$, $P< 0.001$; Fig. 4). At both temperatures, significantly less *Pd* was detected on bats inoculated with 500 conidia compared to all other treatment groups. At 4°C, less *Pd* was detected on bats inoculated with 5 000 conidia compared to bats inoculated with 500 000.

Inoculated bats surviving the hibernation experiment showed marked declines in *Pd* loads within several weeks of removal from hibernation. Forty-five inoculated bats were swabbed both upon removal from hibernation as well as 19 days later. *Pd* was detected on neither date for 8 bats (18%), and *Pd* loads declined to zero in 16 bats (35%). The remaining 21 bats (47%) exhibited a 15-fold median decrease in *Pd* load, with a median load of 11 487 genomic equivalents (range = 292–640 000) at the end of hibernation, followed by a median load of 735 genomic equivalents (range: 10–18 611) 19 days later.

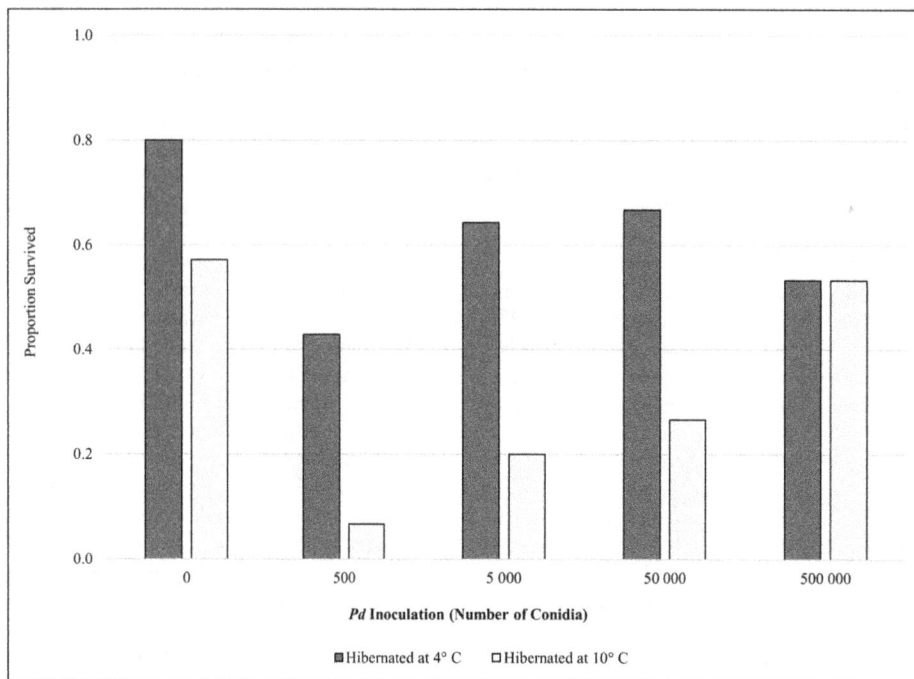

Figure 1. Comparison of survival rates (percent) for little brown myotis (*Myotis lucifugus*) inoculated with different doses of *Pseudogymnoascus destructans* (*Pd*) conidia and hibernated for five months at either 4 or 10°C.

Discussion

We found that WNS mortality is influenced by the level of *Pd* exposure, characteristics of the host, and the environment, and that several variables have interacting effects. As we predicted, differences in torpor patterns mirrored differences in mortality, but contrary to our expectations, bats inoculated with the lowest *Pd* dose experienced the greatest mortality rate and shortest torpor bouts during our study. Together, these data provide important insights on WNS survivors and have several implications for the possibility of long-term survival of little brown myotis in eastern North America.

Bats with greater body condition, indicative of greater fat reserves [34], were more likely to survive our experiment. Because

bats rely on the metabolism of fat to arouse from torpor and sustain brief periods of euthermy during hibernation [4], fat reserves limit the number of times a bat can arouse [35]. Thus, bats with greater body condition at the onset of hibernation can sustain more arousals during the course of a winter, making them better suited to surviving the increased frequency of arousals associated with WNS [13,17]. In a study of free-ranging bats affected by WNS, however, Reeder and colleagues [17] found no relationship between date of death and body condition. This discrepancy likely results from the confounding effects of other pertinent variables influencing the disease. As our results show, WNS mortality is driven by the interaction of variables pertaining to not only the host, but also the pathogen and the environment.

Table 1. Logistic regression analysis of little brown myotis (*Myotis lucifugus*) survival when experimentally inoculated with *Pseudogymnoascus destructans* under varying conditions.

Variable	W	P-value	Odds Ratio	95% Confidence Interval
Body Condition	15.6	<0.001	0.57	0.43–0.75
Temperature				
10°C	15.6	<0.001	5.8	2.4–14.0
Sex				
Male	5.1	0.024	2.8	1.1–6.8
Pd inoculation				
500 spores	8.4	0.004	9.1	2.0–40.4
5 000 spores	1.5	0.217	2.4	0.61–8.8
50 000 spores	0.2	0.663	1.3	0.36–5.1
500 000 spores	0.3	0.609	1.4	0.38–5.6

For categorical variables, results are given in respect to a reference condition of 10°C, female, and inoculated with no fungal spores.

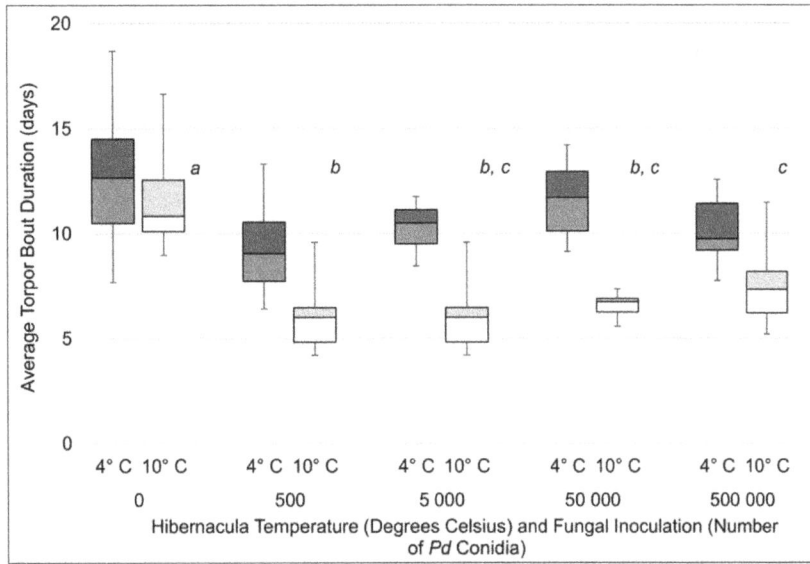

Figure 2. Average duration of torpor bouts (days) for little brown myotis (*Myotis lucifugus*) inoculated with different doses of *Pseudogymnoascus destructans* (*Pd*) conidia and hibernated for five months at either 4 or 10°C. Within each temperature, treatments not sharing common superscript letters were significantly different (*P*<0.05). All doses differed between temperatures (*P*<0.05).

Understanding how our results, obtained under carefully controlled conditions, compare to survival and mortality of wild populations requires the incorporation of all these variables, and provides a foundation for hypotheses related to the persistence of little brown myotis in the WNS-affected region of North America.

Females also exhibited greater survival probability in our study. Female little brown myotis are frequently documented with greater mass or body condition compared to males in the late fall or early winter [17,27,36,37], a difference also present in our captive sample. Although females in our study had greater body condition than males at the onset of hibernation, we did not find a large correlation between sex and body condition in our survival analysis, demonstrating that while large body condition contributes to survival in females, there are other sex-based differences contributing to variation in mortality. Jonasson and Willis found that hibernating little brown myotis females have less pronounced declines in body mass over winter compared to males [26,27], but were unable to attribute this to differences in torpor patterns during hibernation between the two sexes. We were also unable to detect differences in mean torpor bout duration between males and females in our overall analysis, although statistical power was low. A limited comparison of torpor bouts between males and

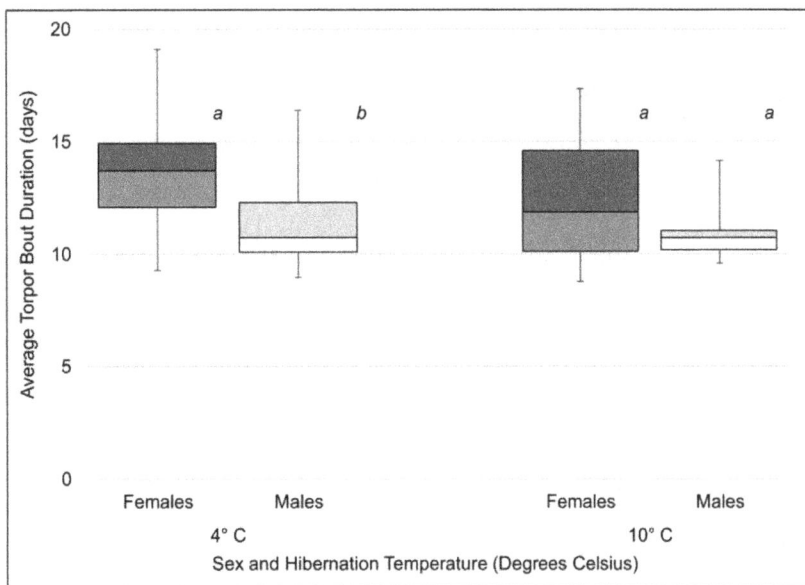

Figure 3. Average duration of torpor bouts (days) of male and female little brown myotis (*Myotis lucifugus*) not inoculated with *Pseudogymnoascus destructans* (*Pd*) and hibernated at either 4 or 10°C. Within each temperature, sexes not sharing common superscript letters were significantly different (*P*<0.05).

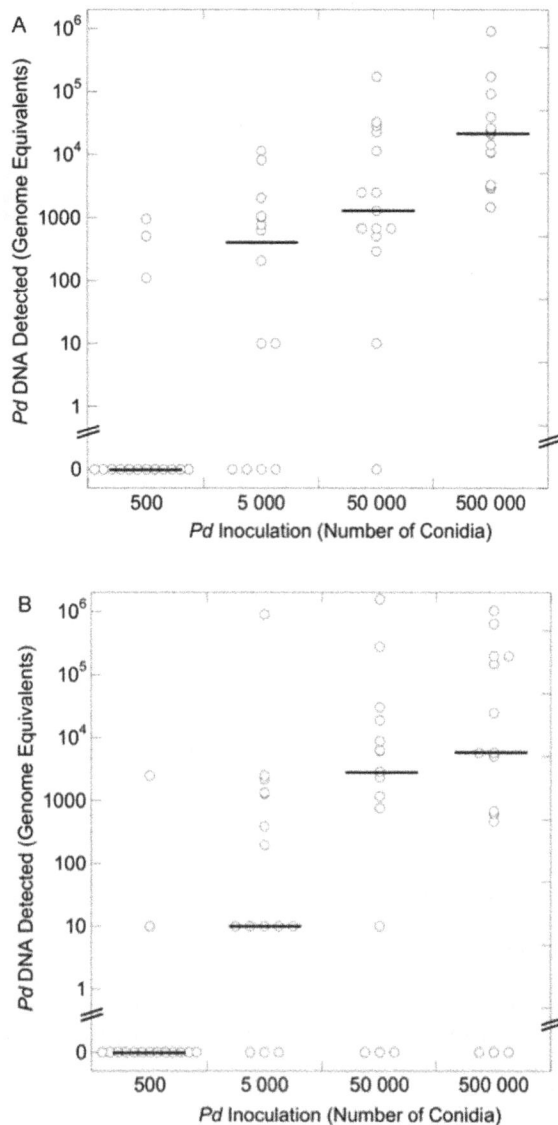

Figure 4. *Pseudogymnoascus destructans* (*Pd*) **DNA detected at the end of hibernation on little brown myotis (*Myotis lucifugus*) inoculated with varying doses of *Pd* conidia and hibernated for five months at either 4°C (A) or 10°C (B).** Individual observations are represented with open circles and medians represented by horizontal lines. At both temperatures, significantly less *Pd* was detected on bats in the 500 conidia group than on bats in other treatment groups. At 4°C, less *Pd* was detected on bats inoculated with 5 000 conidia compared to bats inoculated with 500 000.

frequent arousals while females arouse less frequently to emerge from hibernation with the fat stores necessary for ovulation [27,38–42]. While some data support this hypothesis [27], the large variation observed in winter torpor behavior provides evidence that each sex exhibits diversity in their torpor behaviors [27]. Furthermore, arousal from hibernation and energy savings while torpid are not only determined by sex. Frequency of arousals and torpid metabolic rates decrease with temperature, resulting in greater energy savings [30,31]. Boyles and colleagues suggested that both sexes of little brown myotis select microclimates within caves for hibernation based upon their body condition, i.e. bats with less fat hibernating in colder regions to conserve energy [25]. Thus, torpor patterns in free-ranging little brown myotis are influenced by the interaction of numerous variables, including sex, body condition, and environmental conditions.

Because colder temperatures are conducive to greater energy savings for bats (provided ambient temperature remains above the hypothalamic set-point) [4,31] and are associated with slower fungal growth [8], we predicted that WNS mortality would be greater at higher temperatures. This was supported by our mortality and torpor duration results, the latter of which found a significant interaction between *Pd* inoculation and temperature, and is consistent with population declines observed in little brown myotis hibernacula, where warmer hibernacula exhibited the largest declines [28]. Similarly, we hypothesized fungal loads would be greater at 10°C, but contrary to our expectations, we did not detect differences in *Pd* loads between temperatures. Thus, *Pd* loads appear to be poor indicators of the severity of infection and WNS, as both mortality and frequency of arousals from hibernation increased at 10°C. It is important to note, however, that because 90% relative humidity was maintained in environmental chambers at both temperatures, the absolute humidity of the air was approximately 40% greater at 10°C. This difference in absolute humidity between temperatures could potentially result in different progressions of WNS, resulting in differences in the rates of evaporative water loss in bats or fungal invasion of the skin [43]. Thus, the role of absolute humidity was unclear from our experiment. Regardless, the high variability in *Pd* loads detected at both temperatures highlights the variability in *Pd* growth on bats relative to growth patterns in culture. In free-ranging bats exposed to more variable initial *Pd* exposures than those used in our experiment, and inhabiting hibernacula with conditions that can fluctuate throughout the winter, change in fungal loads are likely to be even more variable. In addition to being more variable, exposure to *Pd* in free-ranging bats is likely to occur repeatedly during the winter, as bats move about within and among hibernacula. These dynamics of *Pd* spread are poorly understood, however, and more research in this area is needed.

Also contrary to our prediction, we observed the greatest mortality and shortest torpor bouts in bats inoculated with the least concentrated solution of *Pd* conidia. This paradoxical result could be explained if lower concentrations of *Pd* grow differently than *Pd* at high densities. We hypothesize that *Pd* germination is inversely related to the density of conidia, resulting in more rapid fungal invasion and mortality in bats inoculated with 500 conidia. Density-dependent growth has been documented in many fungal species and is known as autoinhibition or self-inhibition, a process that can be mediated by volatile organic compounds produced by multiple genera of fungi [44,45]. In the first study to show inoculation of bats with *Pd* causes WNS, Lorch and colleagues [3,13] noted that their captive study was not long enough to result in mortality despite histological evidence of infection among inoculated bats. Density-dependent growth of *Pd* may explain why mortality did not occur within the time period of their study,

females in control groups did reveal differences in torpor behaviors, however, potentially explaining why females had higher survival rates than males. This was true of inoculated as well as control bats; 83% (*n* = 5) of the mortality observed among control bats hibernated at 10°C consisted of males with body condition below that of any female in the group.

Differences in winter body condition and torpor behaviors between male and female little brown myotis are believed to be related to the reproductive biology of the species. Because copulation occurs throughout fall and winter, and ovulation occurs in spring after emergence from hibernation, several have argued that males benefit from a winter torpor strategy favoring which inoculated bats with 500 000 conidia. Our mortality data are only suggestive of self-inhibition in *Pd*, however, and research documenting germination at varying concentrations of conidia is needed to directly address this hypothesis. Such research, along with studies documenting natural exposure dynamics among free-ranging bats, are needed to better inform captive studies of WNS, which typically inoculate bats with 500 000 conidia [3,13]. Inoculations resulting in mortality patterns that differ from wild populations may produce misleading insights into WNS.

It is notable that we were often unable to detect *Pd* DNA on swabs from bats inoculated with 500 conidia. This demonstrates that the number of *Pd* conidia did not exponentially increase on bats in this treatment. We hypothesize that bats have some ability to control the fungal infection at this level of exposure. Although the mechanism of control is uncertain, the increased frequency of periodic arousals observed in these treatments likely plays some role. Arousals provide opportunities for euthermic rest, grooming [46], and immune upregulation, although the brevity of periodic arousals in bats compared to other hibernating mammals likely limits potential immune responses to *Pd* [47–49]. As previously discussed, however, the number of arousals bats can energetically sustain are limited, and the frequent arousals in bats inoculated with 500 conidia resulted in high mortality despite an ability to control the fungus. Furthermore, *Pd* always remained on some bats within the 500 conidia treatment groups, serving as vectors for continued *Pd* exposure within this hibernation chamber.

Mortality in the remaining inoculation treatments was not significantly greater than controls in our model. This lack of difference was driven by the low mortality observed in the remaining inoculation treatments hibernated at 4°C (33–47%) and relatively high mortality in the 10°C control group (43%). The mortalities in the 10°C control group ($n = 6$) are well explained by the logistic regression model. Mortalities in this group were primarily (83%; $n = 5$) males with body condition indices at the onset of hibernation that were below the median body condition. It is well documented that lower temperatures are more energetically favorable for hibernating bats [4,31], a conclusion supported by our own data. Thus, it is not surprising that we observed high mortality among male bats, which aroused more frequently from hibernation, with low fat reserves when placed in an energetically unfavorable environment. Mortality among bats with low body condition in both control groups may also result from placing bats in environmental conditions that differ from their native hibernacula. The 4° and 10°C environmental chambers represented temperatures that are colder and warmer, respectively, than both of the hibernacula we sampled in Illinois and Michigan. Research with captive big brown bats found that although hibernating bats conform to temperatures inside of environmental chambers, torpid metabolic rates are influenced by the temperature regime bats were accustomed to in their native hibernacula [50]. As a result, conditions inside both chambers may be more energetically stressful than can be predicted based upon temperature alone. Regardless of why some mortality occurred in the control groups, the lack of difference in mortality between the control and treatment groups exposed to >500 conidia should not be interpreted to mean that bats in these treatments did not have WNS. To the contrary, bats in all inoculation treatment groups at both temperatures exhibited significantly shorter torpor bouts than controls, a key sign of WNS [13,17]. The reduction in torpor bout length demonstrates that bats in all inoculation treatments developed one of the hallmarks of WNS and would exhaust their energy reserves in a longer hibernation period, unlike bats in the control group that had normal torpor bout lengths, but our results show this mortality would occur after bats exposed to a smaller

number of conidia early during hibernation. It is important to note that we did not use histopathologic criteria to confirm WNS in bats in our experiment, and assumed that *Pd* inoculation was the cause of the increased frequency of arousals and increased mortality compared to control bats, an assumption that is strongly supported by recent research [3,13].

At the northern edge of their range, little brown myotis are reported to hibernate for two months longer than the duration of our experiment [37]. Thus, the ability of free-ranging bats to survive exposure to *Pd* must be considered in the context of winter duration and hibernaculum temperature. Our model predicts that little brown myotis with greater body condition indices inhabiting the regions of North America where the hibernation period lasts approximately 5 months will be able to persist in *Pd*-contaminated hibernacula, provided bats have access to cold roosting microclimates. Although the maximum winter duration little brown myotis can survive with *Pd* is uncertain, hibernacula temperatures below those included our study may confer even greater survival benefits.

Variables relating to the environment, host, and pathogen interact to produce disease [32]. Our study presents WNS survival and mortality within the context of the disease triangle, showing that little brown myotis females, and individuals of both sexes with higher body condition, are more resilient to *Pd*, and that cold hibernacula further increase individual odds of survival. These results suggest a scenario in which little brown myotis may continue to persist in the affected region of North America, with selection favoring individuals with large fat reserves and preference for cold hibernation sites. Because our study was conducted with naïve individuals under controlled conditions, however, additional research on survival in free-ranging populations, and the possible role of the immune system in pathology or resistance, are needed to better understand the fate of little brown myotis and other cave-hibernating species in eastern North America.

Materials and Methods

Animal Collection

This study was carried out on non-endangered animals in strict accordance with the recommendations in the Guide for the Care and Use of Laboratory Animals of the National Institutes of Health. All methods were approved by the Institutional Animal Care and Use Committee at Bucknell University (protocol number DMR-016). Animals were collected at Blackball Mine in Utica, Illinois, USA by state wildlife officials (including JAK with Illinois Department of Natural Resources) on non-endangered bats; thus numbered permits were not required or issued. Animals were collected at Iron Mountain Iron Mine in Vulcan, Michigan, USA under Scientific Collector' Permit SC 1475 from the Michigan Department of Natural Resources to DMR. In accordance with the permit and with state wildlife policies, research was either conducted on state land or on private property, with the explicit permission of private landowners.

We collected 147 hibernating little brown myotis (70 male; 77 female) from Blackball Mine and Iron Mountain Mine on 2–3 November, 2013. Bats were placed in individual cloth bags and transported to Bucknell University in Pennsylvania inside a portable refrigeration unit (Dometic Ltd., Bedfordview, South Africa) set to an internal temperature of 4°C. We determined the sex, weight, and right forearm length of each bat upon arrival at the laboratory and determined the body condition of each bat by dividing the mass by the forearm length [34]. Although the hibernacula that bats were collected from were believed to be unexposed to *Pd*, we swabbed the wings and muzzle of each bat with a sterile cotton swab to collect any *Pd* cells in order to verify that bats had not been exposed. Both wings were swabbed five times on both the dorsal and ventral sides. All bats were fitted with modified iButton temperature dataloggers (Embedded Data Systems, Lawrenceburg, KY, USA), programmed to record skin temperature (T_{sk}) at 30-min intervals [17].

Fungal Inoculation and Hibernation

Bats were placed into treatment groups representing the number of Pd conidia that bats were to be inoculated with prior to being placed into hibernation: 0 (control), 500, 5 000, 50 000, or 500 000 conidia. The Pd culture was derived from an isolate from an infected little brown myotis in Pennsylvania in 2010. Conidia were enumerated using a hemocytometer and 0.25% Trypan Blue staining, and viability of spores was confirmed by culture on Sabouraud agar plates. Each group was replicated once at 4 and once at 10°C. To the extent possible, we randomly selected an equal number of males and females from each hibernaculum to be placed into each treatment group ($n = 14$–15 bats per treatment). Once separated into treatment groups, each bat was either sham inoculated (controls) with 50 μL phosphate buffered saline with 0.05% Tween-20 (PBST) or inoculated with the appropriate number of Pd conidia suspended in 50 μL PBST. The solution was pipetted onto the ventral surface of one wing below the wrist, and distributed along the wing by gentle manipulation of the wing. Similar captive inoculation methods have been clearly demonstrated to cause WNS in recent studies [3,13]. Bats in each treatment group were housed together in open air aluminum cages (Zoo Med Laboratories Inc., San Luis Obispo, CA, USA), provided with *ad lib* water, and placed into environmental chambers set to maintain a constant temperature (4 or 10°C) and ≥90% relative humidity. Control and inoculated treatment groups were housed in separate environmental chambers. Within the chamber housing inoculated treatment cages, individual cages were not in contact with one another, preventing any contact among bats, and, therefore, Pd transmission among cages. A study with a similar design found that Pd transmission did not occur between cages of inoculated and un-inoculated bats when cages were separated within the same environmental chamber [3]. Temperature and relative humidity dataloggers (TransiTempII, MadgeTech, Warner, New Hampshire, USA) were placed inside each chamber to confirm environmental conditions. To avoid disturbance and unnatural arousals from hibernation, chambers were only opened once per month to provide fresh water and remove any moribund bats.

All bats were swabbed a second time following removal from hibernation to estimate Pd loads. Bats were left in hibernation for ca. 5 months (148 d), after which dataloggers were removed from all bats, and surviving individuals were placed in an indoor flight cage where they were hand-fed gut-loaded mealworms until able to self-feed. Conditions inside the flight cage were maintained at approximately 21°C and 60% relative humidity. Inoculated bats surviving hibernation were swabbed a third time 19 d after the end of hibernation to determine the change in Pd loads after bats were removed from an environment favorable for the growth of the fungus. Because surviving bats were not euthanized in this experiment, and moribund bats were only removed once per month, typically several days or weeks after mortality, no tissues were available for a histological confirmation of WNS [2].

Quantifying Fungal DNA

We used qPCR to determine Pd loads on bats prior to and following emergence from hibernation. To prepare for genomic DNA (gDNA) extraction, swabs were incubated at 37°C for 30 minutes in Tris-EDTA buffer (10 mM Tris, 1 mM EDTA;

Amresco, Solon, Ohio, USA) containing 20 U/ml Lyticase (Sigma-Aldrich, St. Louis, Missouri, USA) and 30 mM Dithiothreitol (Sigma-Aldrich). Genomic DNA was extracted using the QIAamp DNA Micro kit (Qiagen Inc., Valencia, California, USA), following the manufacturer's instructions, including the addition of 1 μg carrier RNA.

We used a Taqman 5' endonuclease assay targeting the IGS region of the rRNA complex to detect Pd gDNA extracted from swabs. Primers were synthesized by Integrated DNA Technologies (Coralville, IA, USA) using the sequences [51]: forward primer nu-IGS-0169-5'Gd: 5'– TGC CTC TCC GCC ATT AGT G –3'; reverse primer nu-IGS-0235-3'–Gd: 5'– ACC ACC GGC TCG CTA GGT A –3'; and probe nu-IGS-0182/0204-Gd: 5'– (FAM) CGT TAC AGC TTG CTC GGG CTG CC (BHQ-1) –3'. Each 25 μL PCR reaction contained 12.5 μL Bio-Rad 2× Supermix (Hercules, California, USA), 10.5 μL sample elution, 1 μL of each primer (0.4 μM), and 1 μL probe (0.2 μM). Reactions were performed using a Bio-Rad iCycler starting with an initial 3 min incubation at 95°C, followed by 40 cycles of 30 s at 95°C and 30 s at 60°C. As a control, unused swabs known to be negative for Pd, and swabs exposed to a known quantity of Pd, were included on each plate, as well as no-template control. In order to quantify gDNA on swabs, we created standards by spiking swabs with 10 or 10 000 conidia. Standards were purified in parallel with each batch of samples and run in triplicate on each PCR plate.

The cycle threshold (Ct) was determined using the thresholds set by the data analysis software (iCycler iQ version 3.0a). The average Ct for a swab spiked with 10 Pd cells was 34.8 ($n = 34$), and the assay was found to be linear at all Ct values lower than this. A Ct of 38.1 was calculated to represent 1 conidia and was used as the limit of detection for the assay. All swabs with non-exponential fluorescence increases or with a Ct between 34.8 and 38.1 Ct were considered ambiguous and reanalyzed to confirm that the amount of Pd detected was greater than 1 conidia. Swabs with ambiguous results in two analyses were not considered ($n = 3$). All samples with final Ct values between 34.8 and 38.1 were considered to have a Pd load of ≤10 genomic equivalents, while samples with Ct values greater than 38.1 were considered Pd-negative. The amplification efficiency of the PCR reaction was calculated to be ~100% based on the slope of the standard curve. Swabs with a Ct less than 34.8 were used to calculate the number of conidia present according to the formula: $10\ 000 \times 2^{Ct(exp)-Ct(ss)}$, where *exp* is a swab sample from a bat and *ss* is a swab spiked with 10 000 conidia.

Data Analysis

Mortality and survival were analyzed using a binary (logit function) logistic regression model including temperature (categorical), Pd inoculation (categorical), sex (categorical), and body condition (scale) as dependent variables. To aid interpretation of results, body condition values (range: 0.161–0.269) were multiplied by 100 prior to inclusion in the model. We did not include the state of origin (Michigan and Illinois) as a variable because within each sex, preliminary analyses found that the body condition did not differ between states (two-tailed t-tests, $P > 0.05$) and because mortality rates were similar (Illinois: 46%; Michigan: 48%). T_{sk} data recorded at 30-min intervals were analyzed to characterize the torpor behavior of bats during hibernation. For each bat, we determined the date and time for each arousal from torpor and calculated the average duration of torpor bouts. Bats were considered aroused from torpor when T_{sk} was ≥20°C for ≥1 reading, or when T_{sk} ≥15°C for ≥2 readings. We compared the average duration of torpor bouts with an ANOVA using temperature, inoculation, sex, and a temperature-inoculation interaction as main effects to test our hypotheses regarding torpor behaviors. Due to low statistical power for the variable sex, we also conducted two 1-tailed t-tests comparing the duration of torpor bouts of males to females at each temperature. Data for torpor bout duration were transformed by calculating the natural logarithm prior to analysis to meet statistical assumptions. Pd loading data could not be transformed to meet assumptions of normality and were compared among inoculation treatments using separate Kruskal-Wallis tests for each hibernation temperature. Means comparisons were made using a Wilcoxon test for each pair of treatments, with a sequential Bonferroni-Holm correction [52]. Pd loads were compared between temperatures using a Wilcoxon test for each inoculation treatment. Control groups were not included in Pd load analyses, because Pd was never detected on control animals. All tests used a significance threshold of 0.05 and

Fisher' Least Significant Difference to compare means where appropriate.

Acknowledgments

We would like to thank M Pucciarello, K DeRuff, S Reeder, B Rogers, MH Schwartz, C Seery, and M Hayes of Bucknell University for their invaluable assistance with this research. We are also grateful to the animal care staff at Bucknell, especially C Rhone, G Long, and M Gavitt, for help caring for captive bats.

Author Contributions

Conceived and designed the experiments: DMR KAF. Performed the experiments: JSJ DMR JWM MBM DWFS SSL LES HDW MEV KAF. Analyzed the data: JSJ DMR JWM KAF. Contributed reagents/materials/analysis tools: AK JAK. Wrote the paper: JSJ DMR KAF.

References

1. Blehert DS, Hicks AC, Behr M, Meteyer CU, Berlowski-Zier BM, et al. (2009) Bat white-nose syndrome: an emerging fungal pathogen? Science 323: 227–227.

2. Meteyer CU, Buckles EL, Blehert DS, Hicks AC, Green DE, et al. (2009) Histopathologic criteria to confirm white-nose syndrome in bats. Journal of Veterinary Diagnostic Investigation 21: 411–414.

3. Lorch JM, Meteyer CU, Behr MJ, Boyles JG, Cryan PM, et al. (2011) Experimental infection of bats with Geomyces destructans causes white-nose syndrome. Nature 480: 376–378.

4. Carey HV, Andrews MT, Martin SL (2003) Mammalian hibernation: cellular and molecular responses to depressed metabolism and low temperature. Physiol Rev 83: 1153–1181.

5. Bouma HR, Carey HV, Kroese FG (2010a) Hibernation: the immune system at rest. J Leukoc Biol 88: 619–624.

6. Storey KB, Heldmaier G, Rider MH (2010) Mammalian hibernation: Physiology, cell signaling, and gene controls on metabolic rate depression. In: Anonymous Dormancy and Resistance in Harsh Environments.: Springer. pp.227–252.

7. Webb PI, Speakman JR, Racey PA (1996) How hot is a hibernaculum? A review of the temperatures at which bats hibernate. Can J Zool 74: 761–765.

8. Verant ML, Boyles JG, Waldrep W Jr, Wibbelt G, Blehert DS (2012) Temperature-dependent growth of Geomyces destructans, the fungus that causes bat white-nose syndrome. PLoS One 7: e46280.

9. Reeder DM, Moore MS (2013) White-nose syndrome: A deadly emerging infectious disease of hibernating bats. In: Adams RA, Pedersen SC, editors. Bat Evolution, Ecology, and Conservation. USA: Springer New York. pp.413–434.

10. Gargas A, Trest M, Christensen M, Volk TJ, Blehert D (2009) Geomyces destructans sp. nov., associated with bat white-nose syndrome. Mycotaxon 108: 147–154.

11. Minnis AM, Lindner DL (2013) Phylogenetic evaluation of Geomyces and allies reveals no close relatives of Pseudogymnoascus destructans, comb. nov., in bat hibernacula of eastern North America. Fungal Biology 117: 638–649.

12. Puechmaille SJ, Verdeyroux P, Fuller H, Gouilh MA, Bekaert M, et al. (2010) White-nose syndrome fungus (Geomyces destructans) in bat, France. Emerging Infectious Diseases 16: 290.

13. Warnecke L, Turner JM, Bollinger TK, Lorch JM, Misra V, et al. (2012) Inoculation of bats with European Geomyces destructans supports the novel pathogen hypothesis for the origin of white-nose syndrome. Proc Natl Acad Sci U S A 109: 6999–7003.

14. Puechmaille SJ, Frick WF, Kunz TH, Racey PA, Voigt CC, et al. (2011) White-nose syndrome: is this emerging disease a threat to European bats? Trends Ecol Evol 26: 570–576.

15. Turner GG, Reeder DM, Coleman JTH (2011) A five year assessment of mortality and geographic spread of white-nose syndrome in North American bats and a look to the future. Bat Research News 52: 13–27.

16. Frick WF, Pollock JF, Hicks AC, Langwig KE, Reynolds DS, et al. (2010) An emerging disease causes regional population collapse of a common North American bat species. Science 329: 679–682.

17. Reeder DM, Frank CL, Turner GG, Meteyer CU, Kurta A, et al. (2012) Frequent arousal from hibernation linked to severity of infection and mortality in bats with white-nose syndrome. PLoS One 7: e38920.

18. Cryan PM, Meteyer CU, Blehert DS, Lorch JM, Reeder DM, et al. (2013) Electrolyte depletion in white-nose syndrome bats. J Wildl Dis 49: 398–402.

19. Thomas DW, Dorais M, Bergeron J (1990) Winter energy budgets and cost of arousals for hibernating little brown bats, Myotis lucifugus. J Mammal 71: 475–479.

20. Thomas DW, Cloutier D (1992) Evaporative water loss by hibernating little brown bats, Myotis lucifugus. Physiol Zool: 443–456.

21. Cryan PM, Meteyer CU, Boyles JG, Blehert DS (2010) Wing pathology of white-nose syndrome in bats suggests life-threatening disruption of physiology. BMC Biol 8: 135-7007-8-135.

22. Humphrey SR, Cope JB (1976) Population ecology of the little brown bat, Myotis lucifugus, in Indiana and north-central Kentucky. American Society of Mammalogists.

23. Dixon MD (2011) Population genetic structure and natal philopatry in the widespread North American bat Myotis lucifugus. J Mammal 92: 1343–1351.

24. Norquay KJ, Martinez-Nuñez F, Dubois JE, Monson KM, Willis CK (2013) Long-distance movements of little brown bats (Myotis lucifugus). J Mammal 94: 506–515.

25. Boyles JG, Dunbar MB, Storm JJ, Brack V (2007) Energy availability influences microclimate selection of hibernating bats. J Exp Biol 210: 4345–4350.

26. Jonasson KA, Willis CK (2011) Changes in body condition of hibernating bats support the thrifty female hypothesis and predict consequences for populations with white-nose syndrome. PLoS One 6: e21061.

27. Jonasson KA, Willis CK (2012) Hibernation energetics of free-ranging little brown bats. J Exp Biol 215: 2141–2149.

28. Langwig KE, Frick WF, Bried JT, Hicks AC, Kunz TH, et al. (2012) Sociality, density-dependence and microclimates determine the persistence of populations suffering from a novel fungal disease, white-nose syndrome. Ecol Lett 15: 1050–1057.

29. Humphries MM, Thomas DW, Speakman JR (2002) Climate-mediated energetic constraints on the distribution of hibernating mammals. Nature 418: 313–316.

30. Twente JW, Twente J, Brack V Jr (1985) The duration of the period of hibernation of three species of vespertilionid bats. II. Laboratory studies. Can J Zool 63: 2955–2961.

31. Geiser F (2004) Metabolic rate and body temperature reduction during hibernation and daily torpor. Annu Rev Physiol 66: 239–274.

32. van der Plank JE (1963) Plant diseases: epidemics and control. New York and London: Academic Press.

33. Blehert DS (2012) Fungal disease and the developing story of bat white-nose syndrome. PLoS Pathog 8: e1002779.

34. Speakman J, Racey P (1986) The influence of body condition on sexual development of male brown long-eared bats (Plecotus auritus) in the wild. J Zool 210: 515–525.

35. Thomas DW, Cloutier D, Gagne D (1990) Arrhythmic breathing, apnea and non-steady state oxygen uptake in hibernating little brown bats (Myotis lucifugus). J Exp Biol 149: 395–406.

36. Fenton MB, Brockett M (1970) Population studies of Myotis lucifugus: (Chiroptera: Vespertilionidae) in ontario. In: Anonymous: Royal Ontario Museum. pp.1–34.

37. Fenton MB, Barclay RM (1980) Myotis lucifugus. Mammalian Species 142: 1–8.

38. Gustafson A, Shemesh M (1976) Changes in plasma testosterone levels during the annual reproductive cycle of the hibernating bat, Myotis lucifugus lucifugus with a survey of plasma testosterone levels in adult male vertebrates. Biol Reprod 15: 9–24.

39. Thomas DW, Fenton MB, Barclay RM (1979) Social behavior of the little brown bat, Myotis lucifugus. Behav Ecol Sociobiol 6: 129–136.

40. Buchanan GD (1987) Timing of ovulation and early embryonic development in Myotis lucifugus (Chiroptera: Vespertilionidae) from northern central Ontario. Am J Anat 178: 335–340.

41. Humphries MM, Thomas DW, Kramer DL (2003) The role of energy availability in mammalian hibernation: a cost-benefit approach. Physiological and Biochemical Zoology 76: 165–179.

42. Kunz TH, Wrazen JA, Burnett C (1998) Changes in body mass and fat reserves in pre-hibernating little brown bats (Myotis lucifugus). Ecoscience 5: 8–17.

43. Willis CK, Menzies AK, Boyles JG, Wojciechowski MS (2011) Evaporative water loss is a plausible explanation for mortality of bats from white-nose syndrome. Integr Comp Biol 51: 364–373.

44. Chitarra GS, Abee T, Rombouts FM, Posthumus MA, Dijksterhuis J (2004) Germination of Penicillium paneum conidia is regulated by 1-octen-3-ol, a volatile self-inhibitor. Appl Environ Microbiol 70: 2823–2829.

45. Herrero-Garcia E, Garzia A, Cordobés S, Espeso EA, Ugalde U (2011) 8-Carbon oxylipins inhibit germination and growth, and stimulate aerial conidiation in *Aspergillus nidulans*. Fungal Biology 115: 393–400.

46. Brownlee-Bouboulis SA, Reeder DM (2013) White-nose syndrome-affected little brown myotis (*Myotis lucifugus*) increase grooming and other active behaviors during arousals from hibernation. J Wildl Dis 49: 850–859.

47. Moore MS, Reichard JD, Murtha TD, Nabhan ML, Pian RE, et al. (2013) Hibernating little brown myotis (*Myotis lucifugus*) show variable immunological responses to white-nose syndrome. PloS One 8: e58976.

48. Prendergast BJ, Freeman DA, Zucker I, Nelson RJ (2002) Periodic arousal from hibernation is necessary for initiation of immune responses in ground squirrels. Am J Physiol Regul Integr Comp Physiol 282: R1054–62.

49. Luis AD, Hudson PJ (2006) Hibernation patterns in mammals: a role for bacterial growth? Funct Ecol 20: 471–477.

50. Dunbar MB, Brigham RM (2010) Thermoregulatory variation among populations of bats along a latitudinal gradient. Journal of Comparative Physiology B 180: 885–893.

51. Muller LK, Lorch JM, Lindner DL, O'Connor M, Gargas A, et al. (2013) Bat white-nose syndrome: a real-time TaqMan polymerase chain reaction test targeting the intergenic spacer region of *Geomyces destructans*. Mycologia 105: 253–259.

52. Holm S (1979) A simple sequentially rejective multiple test procedure. Scandinavian Journal of Statistics: 65–70.

22

Fluoride-Tolerant Mutants of *Aspergillus niger* Show Enhanced Phosphate Solubilization Capacity

Ubiana de Cássia Silva[1], Gilberto de Oliveira Mendes[2,3], Nina Morena R. M. Silva[1], Josiane Leal Duarte[1], Ivo Ribeiro Silva[2,4], Marcos Rogério Tótola[1,4], Maurício Dutra Costa[1,4]*

1 Departamento de Microbiologia, Universidade Federal de Viçosa, Viçosa, MG, Brasil, **2** Departamento de Solos, Universidade Federal de Viçosa, Viçosa, MG, Brasil, **3** Instituto de Ciências Agrárias, Universidade Federal de Uberlândia, Monte Carmelo, MG, Brasil, **4** Pesquisador Bolsista do Conselho Nacional de Desenvolvimento Científico e Tecnológico (CNPq), Brasília, DF, Brasil

Abstract

P-solubilizing microorganisms are a promising alternative for a sustainable use of P against a backdrop of depletion of high-grade rock phosphates (RPs). Nevertheless, toxic elements present in RPs, such as fluorine, can negatively affect microbial solubilization. Thus, this study aimed at selecting *Aspergillus niger* mutants efficient at P solubilization in the presence of fluoride (F^-). The mutants were obtained by exposition of conidia to UV light followed by screening in a medium supplemented with $Ca_3(PO_4)_2$ and F^-. The mutant FS1-555 showed the highest solubilization in the presence of F^-, releasing approximately 70% of the P contained in $Ca_3(PO_4)_2$, a value 1.7 times higher than that obtained for the wild type (WT). The mutant FS1-331 showed improved ability of solubilizing fluorapatites, increasing the solubilization of Araxá, Catalão, and Patos RPs by 1.7, 1.6, and 2.5 times that of the WT, respectively. These mutants also grew better in the presence of F^-, indicating that mutagenesis allowed the acquisition of F^- tolerance. Higher production of oxalic acid by FS1-331 correlated with its improved capacity for RP solubilization. This mutant represents a significant improvement and possess a high potential for application in solubilization systems with fluoride-rich phosphate sources.

Editor: Franck Chauvat, CEA-Saclay, France

Funding: Funding provided by Conselho Nacional de Desenvolvimento Científico e Tecnológico (CNPq) (www.cnpq.br), scholarships to UCS, NMRMS, IRS, MRT, and MDC. Coordenação de Aperfeiçoamento de Pessoal de Nível Superior (Capes) (www.capes.gov.br), scholarship to GOM. Fundação de Amparo à Pesquisa do Estado de Minas Gerais (FAPEMIG) (www.fapemig.br), grant GAG-APQ-00712-12. The funders had no role in study design, data collection and analysis, decision to publish, or preparation of the manuscript.

Competing Interests: The authors have declared that no competing interests exist.

* Email: mdcosta@ufv.br

Introduction

Phosphate fertilizers are used intensively in agriculture for improving crop production. The use of low-reactivity rock phosphates (RPs), such as igneous RPs, combined with P-solubilizing microorganisms (PSM) has been shown to be an alternative for a sustainable use of P [1–4]. Several bacteria and fungi in the soil are able to solubilize P and to participate in the biogeochemical cycling of this element [5]. The ability of PSM to solubilize P is mainly associated with the release of metabolites with chelating or complexing properties, such as organic acids [3,6]. The release of H^+ during NH_4^+ assimilation and other metabolic processes that trigger H^+ excretion are also reported as mechanisms of P solubilization [3,7].

Aspergillus niger is a PSM with high P solubilization activity due to its capacity of medium acidification and production of organic acids with high metal complexation activity [3]. *Aspergillus niger* has been shown to solubilize either synthetic or natural apatites, i.e. RPs [3,8–10]. Nevertheless, the chemical characteristics of RPs can interfere with the production of organic acids by PSM [4,11,12] and elements released during the solubilization may be toxic to microbial metabolism [4,13]. Recently, it was demonstrated that F^- released from fluorapatite strongly inhibits RP solubilization by *A. niger* [4]. Fluoride also decreased fungal growth and the production of citric acid. Numerous other cellular processes can be negatively affected by F^-, such as ion transport, secretion, endocytosis, gene expression, and, especially, enzymatic activity [14].

Given the ubiquitous distribution of F^- in RPs, most microbial RP solubilization systems studied so far have probably been operated under suboptimal conditions [4] and, thus, strategies to overcome the toxic effects of F^- on P solubilization must be developed. Such strategies might involve: i) solubilization systems in which only the microbial metabolites, and not the microorganism, are put into contact with RP [15]; ii) addition of adsorbents to remove F^- released during RP solubilization [16]; iii) isolation of strains naturally tolerant to F^- from environmental samples; and iv) mutagenesis of PSM to obtain mutants tolerant to F^-.

Mutagenesis can be done by different strategies, such as the genetic engineering for the introduction of new information into the genome or deletion of chromosomal regions, induction of random mutations with physical and chemical mutagens, and manipulation of the sexual and parasexual cycles [17]. UV light is used for the genetic improvement of fungi [18–20]. The irradiation with UV light induces the formation of lesions on the DNA. The most common damages are the formation of cyclobutane pyrimidine dimers and the pyrimidine(6-4)pyrimidone photoproducts [21]. In the last few years, mutants of *Aspergillus tubingensis* [22] and *Penicillium rugulosum* [12] with increased P solubilization ability were obtained by random mutation with UV light. Thus, the objective of this work was to select *A. niger* mutants with increased P solubilization capacity in the presence of F^-. Additionally, the effect of mutagenesis on fungal processes involved in P solubilization was also evaluated, namely the production of organic acids and medium acidification.

Materials and Methods

Microorganism and cultivation conditions

The strain *A. niger* FS1 was obtained from the Collection of Phosphate Solubilizing Fungi, Microbiology Department, Institute of Biotechnology Applied to Agriculture (BIOAGRO), Universidade Federal de Viçosa, Viçosa, MG, Brazil. Batch fermentations were performed in 125-mL Erlenmeyer flasks with 50 mL of the National Botanical Research Institute's phosphate growth medium (NBRIP) [23] [5 g $Ca_3(PO_4)_2$, 10 g glucose, 5 g $MgCl_2.6H_2O$, 0.25 g $MgSO_4.7H_2O$, 0.2 g KCl, 0.1 g $(NH_4)_2SO_4$, 1 L deionized water]. Variations of this medium were obtained by replacing the P source or by supplementation with F^-, as specified in each experiment. The medium pH was adjusted to 7.0 before the addition of the P sources. *Aspergillus niger* inoculum was added to flasks at the concentration of 10^6 conidia from a suspension prepared in 0.1% (v/v) Tween 80. The flasks were incubated on an orbital shaker for 60 h at 32°C and 160 rpm [4]. Uninoculated flasks were used as controls.

Mutagenesis of *A. niger* and screening of mutants for P solubilization in the presence of fluoride

Ten milliliters of a suspension of 10^6 conidia mL^{-1}, prepared in 0.1% (v/v) Tween 80, were irradiated for 16 min using a 13.8-W UV lamp (Mineralight) aiming at, approximately, 10% of survival. After UV irradiation, the conidia were spread onto Petri dishes (90×15 mm) containing potato dextrose agar (PDA) and incubated at 28°C for 48 h. The *A. niger* colonies obtained on PDA were then screened on solid NBRIP medium supplemented with NaF at 50 mg F^- L^{-1} (NBRIP-F). This concentration corresponds to the F^- amount that would be released from Araxá RP (3 g L^{-1}) if all the F^- was made soluble [4]. The Petri dishes were incubated at 28°C for four days.

Colonies showing clear solubilization halos were transferred in triplicate to Petri dishes with NBRIP-F medium and incubated for six days at 28°C. After that, the diameters of the P solubilization halos of the mutants and the wild type (WT) were measured and compared. P-solubilizing ability was confirmed by quantification of soluble P (see analytical methods below) in liquid medium. Batch fermentations were done in liquid NBRIP-F medium and NBRIP medium with 3 g L^{-1} of Araxá RP (Table 1) as the only P source instead of $Ca_3(PO_4)_2$. Inoculum preparation and cultivation conditions were as described above. Monosporic purification, successive cultivation on PDA and on NBRIP-F media was performed to ensure mitotic stability [24] of the *A. niger* mutants that showed the desired phenotype.

Characterization of *A. niger* mutants

Mutants showing higher or lower P-solubilizing ability than the WT were selected for further studies. The mutants were cultivated in NBRIP containing a soluble P source (1 g L^{-1} K_2HPO_4) instead of $Ca_3(PO_4)_2$, without or with F^- (50 mg L^{-1}), to allow the evaluation of the mutagenesis effects on the production of organic acids and fungal growth. The flasks were incubated on an orbital shaker for 60 h at 32°C and 160 rpm. The experiments were

performed in triplicate following a completely randomized design (CRD).

The effect of mutagenesis on P solubilization in the presence of F^- was also investigated. For this, the selected mutants were grown in liquid NBRIP-F medium and NBRIP with Araxá RP (3 g L^{-1}) as the only P source. NBRIP medium without F^- was used as a control. The experiment was performed in triplicate under a CRD. At the end of incubation, solubilized P, dry biomass, pH, titratable acidity, and the concentration of organic acids were analyzed (see analytical methods below).

Effect of fluoride on RP solubilization by *A. niger* mutants

The mutants and the WT were grown on NBRIP medium with Araxá RP as the only P source and supplemented with increasing concentrations of F^-, ranging from 0 to 50 mg L^{-1} at intervals of 5 mg L^{-1}. At the end of the experiment, solubilized P, fungal biomass, $Y_{P/B}$ [P/biomass yield = Solubilized P (mg)/dry biomass (g), in 50 mL of medium] [3], and pH were determined. Treatments were arranged in a CRD with three replications at the central point. The results were submitted to regression analyses.

Solubilization of different P sources by *A. niger* mutants

The mutants with the highest P solubilization activity were also tested with other P sources besides Araxá RP. Pure P sources, namely $AlPO_4$ and $FePO_4$, were added to NBRIP medium [without $Ca_3(PO_4)_2$] at an equivalent concentration of 1 g P L^{-1}. The RPs evaluated were Catalão, Patos de Minas, and Itafós (Table 1). Due to the high variability in the P and F content, all the P-bearing rocks were added at 3 g L^{-1}. At the end of the experiment, soluble P and fungal biomass were determined (see analytical methods below). The experiment was conducted under a CRD with three replications.

Analytical methods

After incubation, the spent media were centrifuged at 5,000 *g* for 20 min and filtered through quantitative filter paper (8-μm pores). The filtrate was used to determine solubilized P, pH, titratable acidity, and organic acids. Solubilized P was quantified spectrophotometrically by an ascorbic acid method [25]. Titratable acidity was measured by titrating 5 mL of the culture filtrate to pH 7 with 0.1 M NaOH using bromothymol blue as a pH indicator. The fungal biomass retained on the filter paper was collected, dried in an oven at 70°C to constant weight, and incinerated at 500°C for 8 h. Biomass yield was determined by subtracting the weight of the residue left after incineration from the weight of the dried fungal mycelium. This method avoids the overestimation of fungal biomass by the adherence of phosphate particles on the mycelium [24].

Based on previous results for the isolate *A. niger* FS1, organic acid analyses were focused on citric, gluconic, and oxalic acids [3]. For this, the culture filtrate was further passed through 0.22-μm nylon filters. Citric and gluconic acids were determined by ultraperformance liquid chromatography-tandem mass spectrometry (UPLC/MS/MS) using a UPLC Agilent 1290 Series coupled to a 6400 Triple Quadrupole mass spectrometer. Chromatographic separations were carried out using an Agilent ZORBAX Eclipse Plus C18 column (1.8 μm, 2.1 mm×50 mm). The column temperature was controlled at 35°C. An isocratic flow of 97% water and 3% acetonitrile at a flow rate of 0.45 mL min^{-1} was used. The sample injection volume was 10 or 20 μL according to the acid concentration in each sample. The eluate from UPLC was introduced into MS through an APCI source in negative mode. The acids were identified in an analysis time of 1.5 min using

Table 1. Phosphorus and fluorine content and particle size of rock phosphates (RP).

RP	P (g kg^{-1})			F	Particle size
	Total	2% CA[a]	NAC[b]	(g kg^{-1})	(μm)
Araxá	139	19	5	16	<75
Catalão	162	21	4	22	<75
Itafós	39	8	-	9	<600
Patos de Minas	144	15	-	26	<75

[a]Soluble in 2% citric acid.
[b]Soluble in neutral ammonium citrate.

multiple reaction monitoring (MRM). The transitions 191.18 to 110.90 and 195.15 to 128.93 were monitored to citric and gluconic acids, respectively, according to standards (Sigma Chemical Co. St. Louis, MO). For oxalic acid analyses, to ensure the solubilization of calcium oxalate precipitates formed during the solubilization process, the culture supernatants were acidified with 37% HCl to a pH value of approximately 0.5 before filtration [4]. The compound was determined using an Ultimate Dionex 3000 HPLC equipped with a refraction index (RI) detector. The chromatographic separation was carried out in a Rezex ROA-Organic Acid H$^+$ (8%) column (8 μm, 300 mm×7.8 mm) with sample injection volume of 20 μL and analysis time of 15 min. The mobile phase corresponded to sulfuric acid (5 mmol L^{-1}) with a flow rate of 0.7 mL min^{-1}. Oxalic acid was quantified by reference to the peak areas obtained with appropriate standards (Merck, Germany).

Statistical analyses

The data were subjected to ANOVA and multiple mean comparisons were performed through the Tukey or Scott-Knott tests ($P<0.05$) using the statistical software Minitab 16 and Sisvar.

Results

Mutagenesis of A. niger

Twenty-nine mutants showing higher and lower solubilization halos than the WT were obtained in NBRIP-F medium after exposure to UV light (data not shown). These mutants were tested in liquid medium to quantify the solubilization of Araxá RP and Ca$_3$(PO$_4$)$_2$ supplemented with F$^-$ (Table 2). Based on these data, three strains were selected for further studies according to the following criteria: FS1-555 showed the highest increase in P solubilization from Ca$_3$(PO$_4$)$_2$ with F$^-$ (67%); FS1-331 showed the highest increase in P solubilization from Araxá RP (64%) and FS1-375 showed decreased solubilized P for both sources. The mutants FS1-55 and FS1-42 also showed reduced ability to solubilize P, but were not chosen for further analyses because of their substantial growth decrease (Table 2).

Characterization of A. niger mutants

Mutagenesis and F$^-$ altered the profile of organic acid production by the strains (Fig. 1a, b). Only the mutants FS1-331 and FS1-555 produced oxalic acid in the medium with K$_2$HPO$_4$ (Fig. 1a). Citric and gluconic acids were produced by all the strains (Fig. 1a). The addition of F$^-$ inhibited the production of citric and oxalic acids (Fig. 1b). F$^-$ also decreased fungal growth, however, the mutants were more tolerant to it than the WT (Fig. 1c).

The mutant FS1-331 showed the highest value of solubilized P when grown in the presence of Araxá RP, solubilizing 70% more P

than the WT (Table 3). The FS1-555 also increased the solubilization of Araxá RP by 15% compared to the WT. The titratable acidity was higher in the treatments inoculated with the mutants FS1-331 and FS1-555 on Araxá RP. In the medium with Ca$_3$(PO$_4$)$_2$+F$^-$, the mutant FS1-555 was the most effective at P solubilization. Moreover, in the media inoculated with this mutant, the lowest pH values were observed. In the medium with Ca$_3$(PO$_4$)$_2$, no difference was observed between the values of solubilized P by the mutants FS1-331 and FS1-555 and the WT. A significant decrease in P-solubilization and higher pH values were observed for the mutant FS1-375 in the three media evaluated. The biomass of FS1-375 was also lower than that of the WT for all the P sources tested.

The production of the citric and oxalic acids differed among the mutants and the WT (Table 4). The mutants FS1-331 and FS1-555 were the only ones that produced oxalic acid in the presence of Araxá RP. All the mutants produced less citric acid than the WT, while the production of gluconic acid was similar for all strains. Only gluconic and citric acids were detected in the medium with Ca$_3$(PO$_4$)$_2$+F$^-$, and the mutant FS1-555 produced the highest amount of gluconic acid. The mutant FS1-375 produced the lowest amounts of organic acids in the three media evaluated.

Effect of fluoride on RP solubilization by A. niger mutants

The selected positive mutants for Araxá RP solubilization (FS1-331 and FS1-555) were grown in media containing increasing F$^-$ concentrations to simulate the release of F$^-$ from Araxá RP and its effect on P solubilization by the mutants compared to the WT. The mutant FS1-331 showed higher capacity of P solubilization at low F$^-$ doses, solubilizing up to 90% more P than the WT (Fig. 2a). The mutant FS1-555 was the most efficient P solubilizer at higher F$^-$ doses. In general, biomass production of the mutants was lower than that of the WT (Fig. 2b). The biomass production of FS1-331 was less affected by increasing F$^-$ dose (Fig. 2b) and its Y$_{P/B}$ was reduced at higher F$^-$ doses (Fig. 2c).

Higher pH values were observed in response to increasing F$^-$ doses in the growth medium inoculated with FS1-331. For FS1-555, the Y$_{P/B}$ was not affected, but the Y$_{P/B}$ was higher than that of the WT for all doses evaluated (data not shown).

Solubilization of different P sources by A. niger mutants

The mutant FS1-555 solubilized the highest amounts of P from pure P sources, namely AlPO$_4$ and FePO$_4$ (Fig. 3a). When compared to the WT, this mutant increased the concentration of solubilized P from AlPO$_4$ and FePO$_4$ 1.7 and 3.7 times, respectively. In the media with the RPs, FS1-331 showed the highest P-solubilizing ability, increasing the solubilization of Catalão RP by 55% and that of Patos de Minas RP by 150% in

Table 2. Solubilized P and biomass produced by *Aspergillus niger* FS1 mutants compared to the wild type.

Strains	NBRIP-F			NBRIP with Araxá RP		
	Soluble P (mg L^{-1})	I or D (%)[a]	Dry biomass (mg flask^{-1})	Soluble P (mg L^{-1})	I or D (%)	Dry biomass (mg flask^{-1})
WT	346.1 b	-	42.7 b	61.1 d	-	45.6 b
FS1-555	554.8 a	67	34.9 c	85.3 c	42	28.7 e
FS1-326	487.9 a	45	8.0 g	40.4 e	−41	30.0 e
FS1-307	458.5 a	36	45.1 b	69.2 d	14	36.2 c
FS1-512	435.4 b	29	21.4 f	75.7 c	25	26.6 f
FS1-506	429.5 b	23	30.5 d	49.4 d	−15	25.5 f
FS1-261	424.5 b	21	46.3 b	67.2 c	8	26.5 f
FS1-408	416.8 b	62	5.9 g	86.2 c	50	21.2 f
FS1-440	406.9 b	19	39.1 c	101.2 b	51	44.5 b
FS1-442	403.9 b	16	36.6 c	78.1 b	22	23.3 f
FS1-270	400.9 b	48	8.6 g	104.1 a	55	29.8 e
FS1-347	395.8 b	13	9.9 g	37.3 e	−30	13.5 g
FS1-250	392.7 b	9	51.6 a	77.2 c	30	46.4 a
FS1-331	391.2 b	12	24.2 f	110.9 a	64	17.2 g
FS1-262	372.6 b	5	46.9 b	69.9 d	16	48.3 a
FS1-98	370.6 b	7	42.6 b	50.7 d	−13	33.5 c
FS1-8	368.1 b	6	44.7 b	74.6 c	17	33.3 c
FS1-22	366.6 b	6	42.7 b	43.3 e	−23	33.0 d
FS1-406	360.8 b	4	22.0 f	67.4 d	10	23.3 f
FS1-164	339.9 c	−2	27.0 e	53.3 d	−12	38.4 c
FS1-537	334.9 c	−4	34.8 c	96.3 b	61	41.9 b
FS1-48	303.0 c	−11	23.3 f	75.0 c	22	35.3 c
FS1-166	287.1 c	−15	30.7 d	54.7 d	−10	47.3 a
FS1-28	285.8 c	−15	28.5 e	57.9 d	−5	34.7 c
FS1-41	284.7 c	−15	37.0 c	55.9 d	−8	43.7 b
FS1-123	242.5 c	−33	47.6 b	55.1 d	−10	49.8 a
FS1-110	212.5 c	−25	52.4 a	58.1 d	−6	40.6 b
FS1-375	196.2 c	−48	20.1 f	30.8 e	−52	43.6 b
FS1-55	120.8 d	−56	18.0 f	14.3 f	−72	7.7 h
FS1-42	59.5 d	−71	1.0 h	33.4 e	−43	12.7 g

The experiments were carried out in liquid NBRIP medium supplemented with F^{-} at 50 mg L^{-1} (NBRIP-F) or NBRIP with 3 g L^{-1} of Araxá RP as the only P source. Flasks were incubated for 60 h at 32°C and 160 rpm.
[a]The percent increase or decrease of solubilized P of the mutants was calculated based on the solubilized P by the WT in the media NBRIP-F and NBRIP with RP, respectively 346 and 61 mg L^{-1}. P (%) = (mMUT−mWT)/mWT×100; mMUT: mean solubilized P for the mutant; mWT: mean solubilized P for the wild type. Means followed by the same letter are not significantly different according to the Scott Knott test ($P<0.05$).

comparison to the WT. However, none of the mutants was more efficient than the WT in the solubilization of Itafós RP. Finally, the mutants produced similar amounts of biomass for all P sources (Fig. 3b).

Discussion

Mutagenesis using UV light allowed the generation of strains with increased P-solubilizing ability in the presence of F^{-}. The most prominent phenotypic difference between the mutants and the WT was the profile of organic acids produced (Fig. 1, Table 4). Organic acids are effective agents in mobilizing P from RPs or soil particles due to their capacity to form chelates with cations linked to P in poorly soluble forms [26,27]. However, the type of organic acids produced in a microbial solubilization system is of great importance, given that the effectiveness of an organic

acid as a chelating agent is highly dependent on the chemical structure, type, and position of the carboxyl and hydroxyl groups in the molecule [27,28].

The mutants FS1-331 and FS1-555 were the only ones that produced detectable quantities of oxalic acid in the medium with Araxá RP (Table 4). The capacity to produce this acid under such conditions is probably one of the features that confer the superiority of these mutants over the WT at RP solubilization, since the production of gluconic and citric acids by the mutants was not higher than that of the WT. Previous works with the starting strain FS1 have already suggested the importance of oxalic acid for P solubilization [3,16]. The lack of oxalic acid production by the WT in the present work is probably a consequence of the short incubation time adopted, which, in turn, highlights the efficiency of the mutants in producing this acid. Oxalic acid was

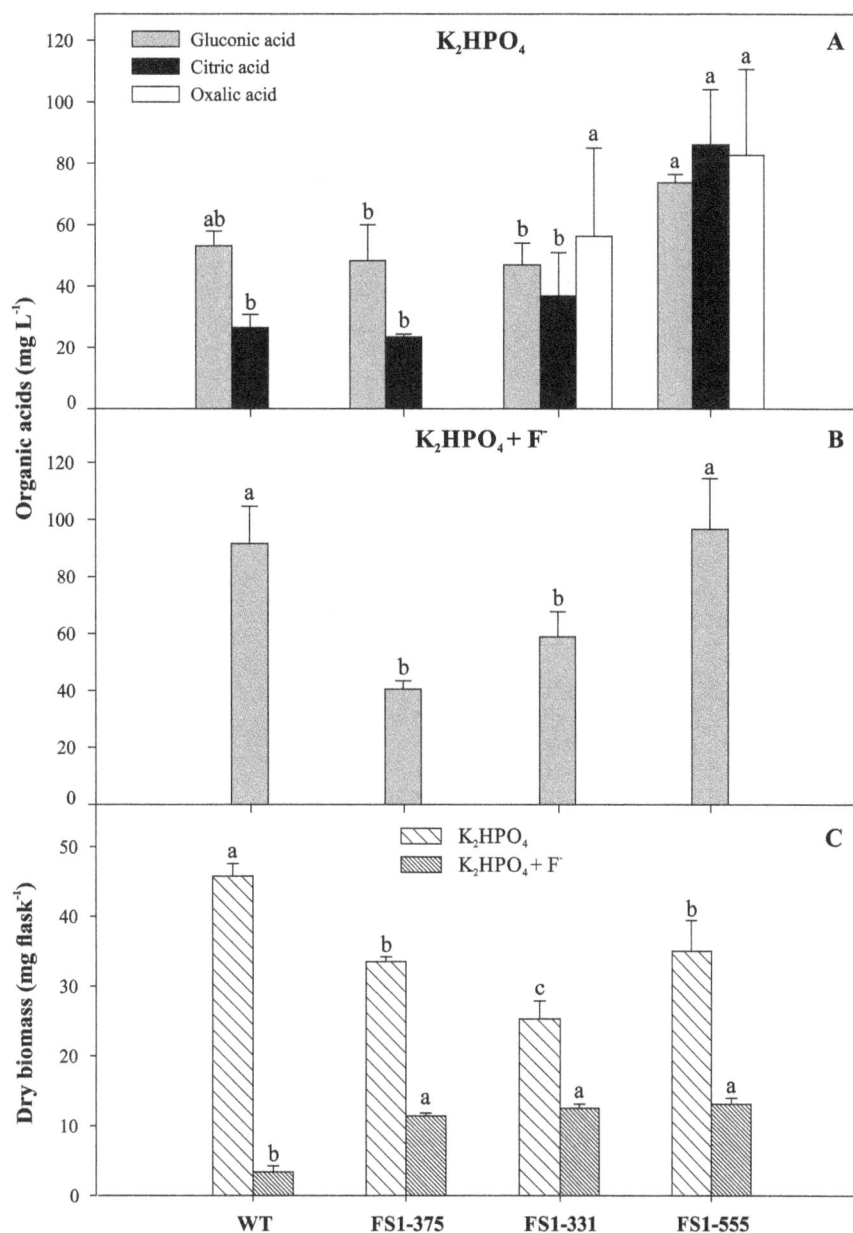

Figure 1. Organic acids produced by *Aspergillus niger* FS1 mutants and the wild type grown on NBRIP medium with K$_2$HPO$_4$ (1 g L^{-1}) as the P source (A) and supplemented with fluoride (50 mg L^{-1}) (B); and fungal dry biomass produced in both conditions (C). The experiment was incubated for 60 h at 32°C and 160 rpm. For each cultivation condition, columns with the same letter are not significantly different by the Tukey's test ($P<0.05$). Error bars represent the mean standard deviation (n = 3).

reported as one of the most effective organic acid in releasing P from RPs [27]. The chemical structure of oxalic acid (C$_2$H$_2$O$_4$) is formed by the linkage of two carboxyl groups. The proximity of these carboxyl groups increases its chelation ability [29]. Additionally, oxalate has a high tendency to precipitate with Ca^{2+}, favoring the solubilization of apatite RPs [28]. Nonetheless, there must be other factors besides oxalic acid production that are related to the superiority of the mutants. Depending on the chemical composition of RPs, another organic acid, e.g. citric acid, can be more effective in solubilizing P [27]. Moreover, chemical elements released from RP can modulate the metabolism of each fungal strain [30], which could explain the differences between mutants in solubilizing different P sources (Fig. 3a). Finally,

increased tolerance to F$^-$ is another feature that probably improved the performance of the mutants.

As expected, mutagenesis changed the response of the fungi to F$^-$. All mutants grew more than the WT in the medium with soluble P supplemented with F$^-$ (Fig. 1c). These data suggest that mutagenesis allowed the isolation of mutants that were more tolerant to F$^-$, considering that decreases in fungal growth are one of the major effects of this ion [4,31]. However, in the medium with Ca$_3$(PO$_4$)$_2$+ F$^-$ the mutants grew less than the WT (Table 3). The Ca^{2+} ions released from Ca$_3$(PO$_4$)$_2$ can react with F$^-$ to form a low-solubility complex (CaF$_2$) [32] which may partially alleviate F$^-$ toxicity. This would permit higher growth of the WT, as already observed under low F$^-$ (Fig. 1c, 2b). The negative mutant FS1-375 showed significant decreases in P solubilization in the

Table 3. Solubilized P, dry biomass, pH and titratable acidity in NBRIP medium with Araxá RP (3 g L^{-1}) as the only P source, $Ca_3(PO_4)_2$ (5 g L^{-1}) + F^- (50 mg L^{-1}), or $Ca_3(PO_4)_2$ (5 g L^{-1}) after the cultivation of *Aspergillus niger* FS1 mutants and the wild type for 60 h at 32°C and 160 rpm.

Strains	Solubilized P (mg L^{-1})	Dry biomass (mg flask^{-1})	pH	Titratable acidity (mmol H^+ L^{-1})
Araxá RP				
WT[a]	60.70 c	41.63 a	2.91 b	2.5 bc
FS1-375[b]	33.32 d	34.17 b	3.27 a	0.8 c
FS1-331	102.75 a	26.43 c	2.79 b	6.4 a
FS1-555	69.66 b	28.73 c	2.77 b	6.0 ab
$Ca_3(PO_4)_2$+ F^-				
WT	382.81 b	35.87 a	3.49 a	13.0 ab
FS1-375	199.58 c	16.07 d	3.46 a	4.7 b
FS1-331	372.25 b	20.73 c	3.20 ab	12.5 ab
FS1-555	558.90 a	29 b	3.07 b	17.1 a
$Ca_3(PO_4)_2$				
WT	766.31 a	51.17 a	2.86 b	25.2 a
FS1-375	458.68 b	23.43 c	3.45 a	13.9 a
FS1-331	694.54 a	32.63 b	3.08 b	16.7 a
FS1-555	744.82 a	52.63 a	3.47 a	14.8 a

Means followed by the same letter are not significantly different according to the Tukey's test ($P<0.05$).
[a]WT: *Aspergillus niger* FS1 wild type.
[b]Mutant with a significant decrease in P solubilization potential (negative mutant).

presence of all the P sources tested (Table 3). This result can be due to a decreased organic acid production, especially citric and oxalic acids (Fig. 1, Table 4), suggesting that this mutant was the most sensitive to F^-. The production of citric acid during RP solubilization is almost completely inhibited by F^- [4]. Moreover, F^- has antimicrobial action and can alter numerous cellular

processes, such as respiration, metabolism, ion transport, secretion, endocytosis, and gene expression [14,33].

The mutants FS1-331 and FS1-555 were more effective than the WT at solubilizing Araxá RP even at increased F^- doses (Fig. 2a). However, P solubilization by FS1-331 decreased sharply with increasing F^- doses. At higher doses, this mutant solubilized

Table 4. Organic acids (mg L^{-1}) produced by *Aspergillus niger* FS1 mutants and the wild type in NBRIP medium with Araxá RP (3 g L^{-1}) as the only P source, $Ca_3(PO_4)_2$ (5 g L^{-1}) + F^- (50 mg L^{-1}), or $Ca_3(PO_4)_2$ (5 g L^{-1}) after 60 h of incubation at 32°C and 160 rpm.

Strains	Gluconic acid	Citric acid	Oxalic acid
Araxá RP			
WT	287 a	164 a	nd
FS1-375	190 a	100 c	nd
FS1-331	225 a	126 b	27 a
FS1-555	240 a	135 b	22 b
$Ca_3(PO_4)_2$+ F^-			
WT	1855 b	455 b	nd
FS1-375	1281 c	103 c	nd
FS1-331	1013 c	711 a	nd
FS1-555	2482 a	250 bc	nd
$Ca_3(PO_4)_2$			
WT	4676 c	1114 a	nd
FS1-375	5930 b	197 c	nd
FS1-331	7655 a	629 b	nd
FS1-555	6382 ab	618 b	nd

Means followed by the same letter are not significantly different according to the Tukey's test ($P<0.05$).
nd: not detected.

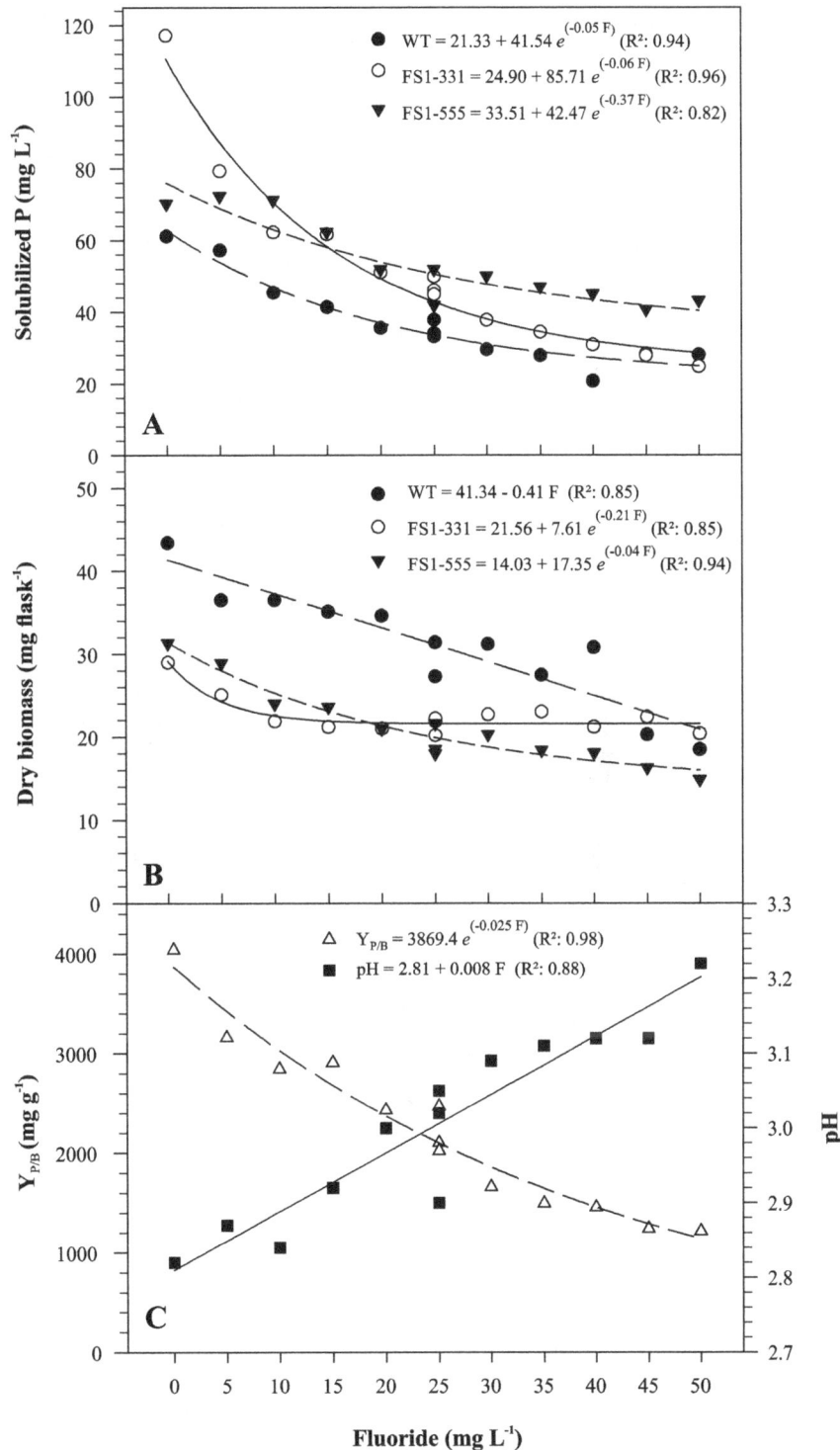

Figure 2. Effect of fluoride on rock phosphate solubilization by *Aspergillus niger* FS1 mutants and the wild type. (A) Solubilized P; (B) dry biomass; (C) P/biomass yield ($Y_{P/B}$ = mg solubilized P per g of biomass) and medium pH for the mutant FS1-331. The strains were grown in NBRIP for 60 h at 32°C and 160 rpm. All regression coefficients are significant as determined by *t* test ($P < 0.01$).

less than FS1-555, indicating that the latter is more tolerant to F⁻. This can be also observed in the medium with $Ca_3(PO_4)_2$+ F⁻, where the FS1-555 solubilized more P than the FS1-331 and the WT (Table 3). In the case of the FS1-555 and the WT, the decreases in P solubilization can be associated to the toxic effects of F⁻ on fungal growth (Fig. 2b). However, above 10 mg L⁻¹ of F⁻

there was no further decrease in biomass production by FS1-331, while the $Y_{P/B}$ of this mutant decreased with increasing F⁻ doses (Fig. 2c). These data show that the biomass became less efficient at P solubilization probably because of deleterious effects of F⁻ on metabolic processes involved in P solubilization, i.e. production of organic acids (Table 4) and release of H⁺ resulting from cellular

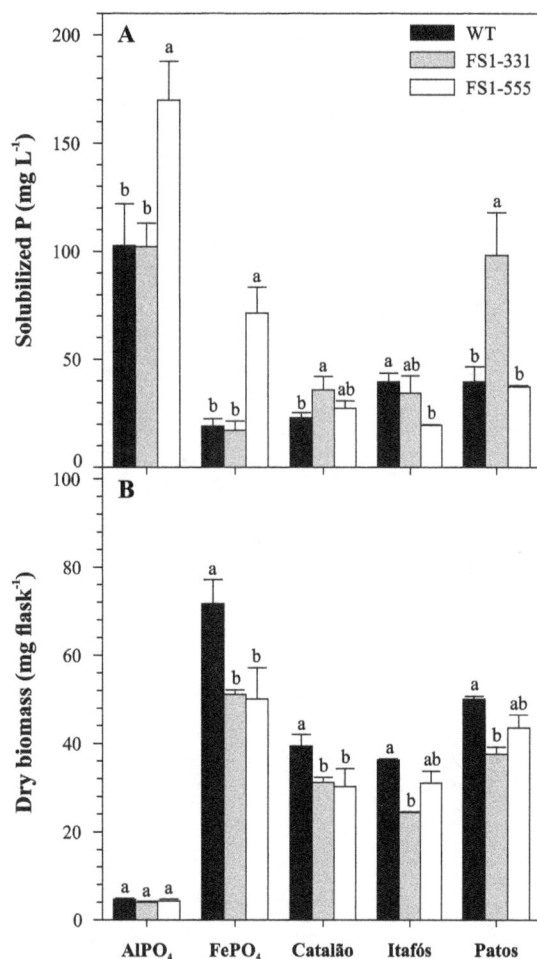

Figure 3. Solubilization of different P sources (A) and biomass produced (B) by *Aspergillus niger* **FS1 mutants and the wild type grown in NBRIP during 60 h at 32°C and 160 rpm.** For each P source, columns with the same letter are not significantly different by the Tukey's test ($P<0.05$). Error bars represent the mean standard deviation (n = 3).

respiration and/or NH_4^+ assimilation [7], as evidenced by the higher pH observed at higher F^- doses (Fig. 2c).

In the medium containing the soluble P sources, higher amounts of oxalic acid were detected for FS1-331 and FS1-555 (Fig. 1a). Oxalic acid accumulation by *A. niger* is stimulated by the addition of P into the medium [34]. This characteristic is very interesting for microbial solubilization systems based on batch culture, where the concentration of soluble P increases along the time. Conversely, the synthesis of citric and oxalic acids in all treatments with soluble P supplemented with F^- was inhibited (Fig. 1b). Fluoride inhibits the glycolytic pathway and the Krebs cycle

through binding to the active center of the enzymes of these pathways [14]. Enolase is inhibited by F^- [31] and, thus, in the medium with F^-, a decrease in the pyruvate pool, an important precursor for the synthesis of citric and oxalic acids [35], may have inhibited the production of these acids. However, in the medium with $Ca_3(PO_4)_2$ the addition of F^- had less effect on citric acid production (Table 4). As discussed above, F^- toxicity can be alleviated by formation of CaF_2.

Gluconic acid was produced under all experimental conditions and presented little variation among the strains (Fig. 1, Table 4). In general, the presence of F^- is not inhibitory for gluconic acid production [4]. Gluconic acid has been found in various solubilization systems and contributes mainly with protons for the solubilization reaction [11,36]. It seems that the increased production of gluconic acid in the medium with $Ca_3(PO_4)_2$ supplemented with F^- (Table 4) was the reason for the lower pH and, consequently, the higher levels of solubilized P (Table 3) observed for FS1-555.

When the mutants were tested in different P sources, FS1-331 was more effective at solubilizing RPs (Catalão and Patos de Minas), while FS1-555 was more effective in the media with pure synthetic sources ($AlPO_4$ and $FePO_4$) (Fig. 3). FS1-555 possesses important characteristics for P solubilization, such as high production of citric and oxalic acids (Fig. 1a). However, this mutant seems to be more sensitive to other elements released from RP. Further studies should be done to elucidate this point.

In this work, *A. niger* mutants with improved P-solubilizing activity and higher tolerance to F^- were obtained using UV light-induced mutagenesis. These mutants also presented increased production of oxalic acid. Given the effectiveness of oxalic acid to solubilize apatite RPs and that most RPs are rich in F^-, the mutants obtained, especially FS1-331, represent a significant improvement and possess a high potential for application in solubilization systems with fluoride-rich phosphate sources. Direct inoculation of these mutants in the soil-plant environment is also a prospect. However, some factors should be studied to accomplish this, such as the competition ability of the mutants against the indigenous community and the selection of vehicles for inoculation of fungal propagules into the soil.

Acknowledgments

The authors are grateful to the Núcleo de Análises de Biomoléculas UFV/Brazil (NUBIOMOL) for the facilities to analyze the organic acids. Thanks are also due to Hilário Cuquetto Mantovani for his assistance in oxalic acid analyses.

Author Contributions

Conceived and designed the experiments: UCS GOM IRS MRT MDC. Performed the experiments: UCS NMRMS JLD. Analyzed the data: UCS GOM NMRMS JLD MDC. Contributed reagents/materials/analysis tools: IRS MRT MDC. Contributed to the writing of the manuscript: UCS GOM MDC.

References

1. Farhat MB, Farhat A, Bejar W, Kammoun R, Bouchaala K, et al. (2009) Characterization of the mineral phosphate solubilizing activity of *Serratia marcescens* CTM 50650 isolated from the phosphate mine of Gafsa. Arch Microbiol 191: 815–824.

2. Vassilev N, Medina A, Azcon R, Vassileva M (2006) Microbial solubilization of rock phosphate on media containing agro-industrial wastes and effect of the resulting products on plant growth and P uptake. Plant Soil 287: 77–84.

3. Mendes GO, Freitas ALM, Pereira OL, Silva IR, Vassilev NB, et al. (2014) Mechanisms of phosphate solubilization by fungal isolates when exposed to different P sources. Ann Microbiol 64: 239–249.

4. Mendes GO, Vassilev NB, Bonduki VHA, Silva IR, Ribeiro JI, et al. (2013) Inhibition of *Aspergillus niger* phosphate solubilization by fluoride released from rock phosphate. Appl Environ Microbiol 79: 4906–4913.

5. Richardson AE, Simpson RJ (2011) Soil microorganisms mediating phosphorus availability. Plant Physiol 156: 989–996.

6. Vassilev N, Mendes G, Costa M, Vassileva M (2014) Biotechnological tools for enhancing microbial solubilization of insoluble inorganic phosphates. Geomicrobiol J 31: 751–763.

7. Illmer P, Schinner F (1995) Solubilization of inorganic calcium phosphates - Solubilization mechanisms. Soil Biol Biochem 27: 257–263.

8. Vassilev N, Baca MT, Vassileva M, Franco I, Azcon R (1995) Rock phosphate solubilization by *Aspergillus niger* grown on sugar-beet waste medium. Appl Microbiol Biotechnol 44: 546–549.

9. Xiao C, Zhang H, Fang Y, Chi R (2013) Evaluation for rock phosphate solubilization in fermentation and soil–plant system using a stress-tolerant phosphate-solubilizing *Aspergillus niger* WHAK1. Appl Biochem Biotechnol 169: 123–133.

10. Chuang CC, Kuo YL, Chao CC, Chao WL (2007) Solubilization of inorganic phosphates and plant growth promotion by *Aspergillus niger*. Biol Fertility Soils 43: 575–584.

11. Schneider KD, van Straaten P, de Orduna RM, Glasauer S, Trevors J, et al. (2010) Comparing phosphorus mobilization strategies using *Aspergillus niger* for the mineral dissolution of three phosphate rocks. J Appl Microbiol 108: 366–374.

12. Reyes I, Baziramakenga R, Bernier L, Antoun H (2001) Solubilization of phosphate rocks and minerals by a wild-type strain and two UV-induced mutants of *Penicillium rugulosum*. Soil Biol Biochem 33: 1741–1747.

13. Banik S, Dey B (1982) Available phosphate content of an alluvial soil as influenced by inoculation of some isolated phosphate-solubilizing micro-organisms. Plant Soil 69: 353–364.

14. Barbier O, Arreola-Mendoza L, Del Razo LM (2010) Molecular mechanisms of fluoride toxicity. Chem-Biol Interact 188: 319–333.

15. Goldstein AH, Rogers RD, Mead G (1993) Mining by microbe. Nat Biotechnol 11: 1250–1254.

16. Mendes GO, Zafra DL, Vassilev NB, da Silva IR, Ribeiro JI, et al. (2014) Biochar enhances *Aspergillus niger* rock phosphate solubilization by increasing organic acid production and alleviating fluoride toxicity. Appl Environ Microbiol 80: 3081–3085.

17. Nevalainen KMH (2001) Strain improvement in filamentous fungi-an overview. In: George GK, Dilip KA, editors. Applied Mycology and Biotechnology. Elsevier. 289–304.

18. Hao XC, Yu XB, Yan ZL (2006) Optimization of the medium for the production of cellulase by the mutant *Trichoderma reesei* WX-112 using response surface methodology. Food Technol Biotechnol 44: 89–94.

19. Lotfy WA, Ghanem KM, El-Helow ER (2007) Citric acid production by a novel *Aspergillus niger* isolate: I. Mutagenesis and cost reduction studies. Bioresour Technol 98: 3464–3469.

20. Maresma BG, Castillo BG, Fernandez RC, da Silva ES, Maiorano AE, et al. (2010) Mutagenesis of *Aspergillus oryzae* IPT-301 to improve the production of β- fructofuranosidase. Braz J Microbiol 41: 186–195.

21. Pfeifer GP, You Y-H, Besaratinia A (2005) Mutations induced by ultraviolet light. Mutat Res 571: 19–31.

22. Achal V, Savant VV, Reddy MS (2007) Phosphate solubilization by a wild type strain and UV-induced mutants of *Aspergillus tubingensis*. Soil Biol Biochem 39: 695–699.

23. Nautiyal CS (1999) An efficient microbiological growth medium for screening phosphate solubilizing microorganisms. FEMS Microbiol Lett 170: 265–270.

24. Reyes I, Bernier L, Simard RR, Tanguay P, Antoun H (1999) Characteristics of phosphate solubilization by an isolate of a tropical *Penicillium rugulosum* and two UV-induced mutants. FEMS Microbiol Ecol 28: 291–295.

25. Braga JM, Defelipo BV (1974) Determinação espectrofotométrica de fósforo em extratos de solo e material vegetal. R Ceres 21: 73–85.

26. Bolan NS, Naidu R, Mahimairaja S, Baskaran S (1994) Influence of low-molecular-weight organic acids on the solubilization of phosphates. Biol Fertility Soils 18: 311–319.

27. Kpomblekou-A K, Tabatabai MA (1994) Effect of organic acids on release of phosphorus from phosphate rocks. Soil Sci 158: 442–453.

28. Jones DL (1998) Organic acids in the rhizosphere – a critical review. Plant Soil 205: 25–44.

29. Razzaghe-Karim M, Robert M (1975) Altération des micas et géochimie de l'aluminium: rôle de la configuration de la molécule organique sur l'aptitude á la complexation. CR AcSc Paris t 280: 2.645–642.648.

30. Gadd GM (1993) Interactions of fungi with toxic metals. New Phytol 124: 25–60.

31. Agrawal PK, Bhatt CS, Viswanathan L (1983) Effect of some metabolic inhibitors on citric acid production by *Aspergillus niger*. Enzyme Microb Technol 5: 373–376.

32. Aigueperse J, Mollard P, Devilliers D, Chemla M, Faron R, et al. (2000) Fluorine compounds, Inorganic. Ullmann's Encyclopedia of Industrial Chemistry. Weinheim: Wiley-VCH Verlag GmbH & Co. KGaA. 397–441.

33. Marquis RE, Clock SA, Mota-Meira M (2003) Fluoride and organic weak acids as modulators of microbial physiology. FEMS Microbiol Rev 26: 493–510.

34. Kubicek CP, Schreferl-Kunar G, Wohrer W, Rohr M (1988) Evidence for a cytoplasmic pathway of oxalate biosynthesis in *Aspergillus niger*. Appl Environ Microbiol 54: 633–637.

35. Magnuson JK, Lasure LL (2004) Organic acid production by filamentous fungi. In: Tkacz JS, Lange L, editors. Advances in fungal biotechnology for industry, agriculture, and medicine. New York: Kluwer Academic/Plenum Publishers. 307–340.

36. Lin T-F, Huang H-I, Shen F-T, Young C-C (2006) The protons of gluconic acid are the major factor responsible for the dissolution of tricalcium phosphate by *Burkholderia cepacia* CC-Al74. Bioresour Technol 97: 957–960.

23

Survey of Surface Proteins from the Pathogenic *Mycoplasma hyopneumoniae* Strain 7448 Using a Biotin Cell Surface Labeling Approach

Luciano Antonio Reolon[2,3¤], Carolina Lumertz Martello[1], Irene Silveira Schrank[2,3,4], Henrique Bunselmeyer Ferreira[1,3,4]*

1 Laboratório de Genômica Estrutural e Funcional, Centro de Biotecnologia, UFRGS, Porto Alegre, RS, Brazil, **2** Laboratório de microrganismos diazotróficos, Centro de Biotecnologia, UFRGS, Porto Alegre, RS, Brazil, **3** Programa de Pós-Graduação em Biologia Celular e Molecular, Centro de Biotecnologia, Universidade Federal do Rio Grande do Sul (UFRGS), Porto Alegre, RS, Brazil, **4** Departamento de Biologia Molecular e Biotecnologia, Instituto de Biociências, UFRGS, RS, Brazil

Abstract

The characterization of the repertoire of proteins exposed on the cell surface by *Mycoplasma hyopneumoniae* (*M. hyopneumoniae*), the etiological agent of enzootic pneumonia in pigs, is critical to understand physiological processes associated with bacterial infection capacity, survival and pathogenesis. Previous *in silico* studies predicted that about a third of the genes in the *M. hyopneumoniae* genome code for surface proteins, but so far, just a few of them have experimental confirmation of their expression and surface localization. In this work, *M. hyopneumoniae* surface proteins were labeled in intact cells with biotin, and affinity-captured biotin-labeled proteins were identified by a gel-based liquid chromatography-tandem mass spectrometry approach. A total of 20 gel slices were separately analyzed by mass spectrometry, resulting in 165 protein identifications corresponding to 59 different protein species. The identified surface exposed proteins better defined the set of *M. hyopneumoniae* proteins exposed to the host and added confidence to *in silico* predictions. Several proteins potentially related to pathogenesis, were identified, including known adhesins and also hypothetical proteins with adhesin-like topologies, consisting of a transmembrane helix and a large tail exposed at the cell surface. The results provided a better picture of the *M. hyopneumoniae* cell surface that will help in the understanding of processes important for bacterial pathogenesis. Considering the experimental demonstration of surface exposure, adhesion-like topology predictions and absence of orthologs in the closely related, non-pathogenic species *Mycoplasma flocculare*, several proteins could be proposed as potential targets for the development of drugs, vaccines and/or immunodiagnostic tests for enzootic pneumonia.

Editor: Wei Wang, Henan Agricultural Univerisity, China

Funding: This work was funding from MCT/CNPq and FAPERGS. The funder had no role in study design, data collection and analysis, decision to publish, or preparation of the manuscript.

Competing Interests: The authors have declared that no competing interests exist.

* Email: Henrique@cbiot.ufrgs.br

¤ Current address: Centro Universitário Ritter dos Reis, Porto Alegre, RS, Brazil

Introduction

Mycoplasmas belong to the class Mollicutes and are among the smallest free-living organisms capable of self-replication. Evolutionarily related to Gram-positive bacteria, mycoplasmas have undergone extensive genome reduction, which led to simplification or loss of some metabolic pathways and structural cell components. They are unable to synthesize peptidoglycans or their precursors, and therefore, present no cell wall. The mycoplasma cell membrane is formed from proteins, phospholipids and cholesterol; cholesterol is an essential nutrient to bacterial growth and is responsible for membrane rigidness and stability [1,2].

M. hyopneumoniae is the etiological agent of enzootic pneumonia (EP), a chronic disease characterized by a dry and nonproductive cough, most evident when pigs are roused, retarded growth, and inefficient food conversion [3]. In pigs, *M. hyopneumoniae* is found to be attached to the cilia of the tracheal epithelial cells, causing a reduction in ciliary action [4,5], and predisposing the swine to infection by other pathogens, such as *Pasteurella multocida* [6] and porcine reproductive and respiratory syndrome virus (PRRSV) [7].

M. hyopneumoniae adhesion to the swine tracheal epithelial cells is essential to disease establishment, and the characterization of adhesion-mediating molecules has been the focus of most studies on the bacterial mechanisms of virulence and pathogenesis. Bacterial adhesive capability is related to several proteins, such as the well-described P97 adhesin [8,9,10] and adhesin-like proteins, such as P216 [11], P159 [12], P102 [13], P146 [14] and P116 [15]. It has been suggested, however, that several other proteins that so far remain uncharacterized are involved in the *M. hyopneumoniae* cell adhesion process [12].

Membrane (integral or associated) proteins are directly exposed on the cell surface and play key roles in cell adhesion, and evasion and/or modulation of the host immune system, events which are important for environmental, bacterial and host cell interactions [16,17]. The identification of membrane proteins represents a great challenge, especially due to their mainly hydrophobic nature and selective loss during purification, which especially occurs in the precipitation and solubilization steps.

Several methods have been applied to the experimental identification of mycoplasma membrane proteins, especially methods involving selective solubilization the use of detergents, such as Triton X100 [18,19] and Triton X114 [20]. These methods yield enriched membrane protein fractions, including lipoproteins, but do not completely avoid contamination with cytosolic and ribosomal proteins [21]. Selective labeling using hydrophilic and membrane-impermeable reagents, such as Sulfo-NHS-Biotin, is an alternative way to reduce this contamination and allow the recovery of a more specific fraction of surface exposed proteins [22].

Here, we describe a proteomic approach, based on intact cell surface labeling and labeled protein purification, coupled to gel-based liquid chromatography-tandem mass spectrometry (GeLC-MS/MS), to identify M. hyopneumoniae surface exposed proteins. This strategy allowed to identify surface proteins possibly involved in pathogenesis, including some previously annotated as hypothetical proteins. Moreover, a comparative analysis was carried with the closely related non-pathogenic species Mycoplasma flocculare (M. flocculare), to verify which of the identified surface proteins are found only in the pathogenic counterpart. The potential of some of the identified M. hyopneumoniae surface exposed proteins as novel targets for the development of vaccines, diagnostic tests and therapeutic drugs is discussed.

Methods

Bacterial strain and culture conditions

The Mycoplasma hyopneumoniae pathogenic strain 7448 was isolated from an infected swine from Lindóia do Sul (Santa Catarina, Brazil), and cultured in Friis medium as previously described [23].

Biotin labeling and affinity recovery of labeled M. hyopneumoniae strain 7448 proteins

A cell pellet from 100 ml of fresh M. hyopneumoniae 7448 culture was collected by centrifugation at 3360 x g for 15 min. The pellet was washed three times with cold phosphate-buffered saline (PBS; pH 7.2) and resuspended in the same buffer with the addition of 1 mg of sulfosuccinimidyl biotin (EZ-Link Sulfo-NHS-Biotin; Pierce, USA)/ml, in a final concentration of 2 mM of biotin reagent, according to the manufacturer's instructions. The labeling reaction was performed as previously described [22] for 30 min at 4°C, and the residual sulfo-NHS-biotin was quenched by adding glycine to a final concentration of 100 mM. To remove all unspecific and inactivated Sulfo-NHS-Biotin, the cell suspension was washed twice with 100 mM glycine (final concentration) in PBS. Cells were then lysed by five rounds of sonication (30 s, 20 kHz, 1 min interval between rounds) and labeled proteins were recovered by affinity chromatography in a Monomeric Avidin Resin (Pierce, USA), with gravity flow, using phosphate-buffered saline (PBS, pH 7.0) + 1% Triton X-100 as an equilibration and wash buffer and the same buffer with the addition of biotin to a final concentration of 2 mM was used to block biotin non-reversible binding sites and to elute the bound biotinylated molecule. All steps were performed in the presence of protease

inhibitors (Sigma-Aldrich, USA). Chromatography was monitored by measuring the absorbance at 280 nm. As a control, a cell pellet from 100 ml of another fresh M. hyopneumoniae 7448 culture was collected by centrifugation at 3360×g for 15 min, washed three times with cold phosphate-buffered saline (PBS; pH 7.2) and lysed by sonication (as described above) prior to biotin labeling of total proteins. After this, labeled proteins were recovered and treated as described above for protein labeling of intact cells. Crude protein extracts were produced in the same way, excluding biotin labeling steps.

Protein samples were concentrated, and salts and detergents were removed by a chloroform/methanol precipitation step and freeze dried until use. All the above procedures of cell culture, and protein labeling and affinity purification were performed in triplicate.

Electrophoretic prefractionation followed by liquid chromatography tandem mass spectrometry of M. hyopneumoniae biotin-labeled proteins

A gel-based liquid chromatography-tandem mass spectrometry (GeLC-MS/MS) [24] approach, using protein prefractionation by sodium dodecyl sulfate-polyacrylamide gel electrophoresis (SDS-PAGE) and in-gel digestion (IGD), followed by liquid chromatography–tandem mass spectrometry (LC-MS/MS), was used for the identification of biotin-labeled proteins of M. hyopneumoniae strain 7448. The workflow for GeLC-MS/MS was performed as follows. Freeze-dried protein samples (obtained as described in section 2.2) were resuspended in PBS (pH 7.0) and protein concentration was determined using the Qubit Protein Assay Kit (Invitrogen, USA), according to the manufacturer's instructions. Samples corresponding to 15 μg of biotin-labeled proteins in PBS, after addition of urea to a final concentration of 8 M, were fractionated by SDS-PAGE on 4–10% Mini-PROTEAN tetra cell precast gels at 80 mV using Tris-SDS running buffer, and stained with Coomassie Brilliant-Blue G250 (Sigma-Aldrich, USA). For each protein sample resolved by SDS-PAGE the corresponding gel lane was divided into 20 slices of similar size (~15 mm^2 in area), which were manually excised from the gel and individually processed as follows. They were initially submitted to three washes with 400 μl of 50% acetonitrile and 50 mM ammonium bicarbonate pH 8.0 for 15 min, followed by one washing step with 400 μl of acetonitrile. Gel slices were then incubated with 200 μl of 10 mM DTT in 50 mM ammonium bicarbonate at room temperature for 60 min for reduction, and proteins were subsequently alkylated by incubation with 50 mM iodoacetamide in 50 mM ammonium bicarbonate for 45 min in the dark at room temperature. Gel slices were dried in a CentriVap centrifuge (Labconco, USA). For IGD, gel slices were covered with a trypsin solution (Promega, USA) (20 μg in 1 mL of ammonium bicarbonate 50 mM) and samples were incubated overnight at 37°C. Extraction of the resulting peptides from gel slices was carried out by two successive 1 h incubations with 50 μl of 50% acetonitrile and trifluoroacetic acid (TFA).

Peptides resulting from IGD were separated in a Nanoease C18 (75 μm ID) capillary column by elution with a water/acetonitrile 0.1% formic acid gradient in a capillary liquid chromatography system (Waters, Milford, US). Liquid chromatography was coupled online with an electrospray ionization (ESI) quadrupole time-of-flight (Q-TOF) Ultima API mass spectrometer (Micromass, UK). Data were acquired in data-dependent mode (DDA), and multiple charged peptide ions (+2 and +3) were automatically mass selected and dissociated in MS/MS experiments. Typical LC and ESI conditions included a flow of 200 nL/min, a nanoflow capillary voltage of 3.5 kV, a block temperature of 100°C, and a cone voltage of 100 V. For each replicate, three independent LC-MS/MS measurements were performed. Mass spectrometry was performed in the Unidade de Química de Proteínas e Espectrometria de Massas (Uniprote-MS), Centro de Biotecnologia, UFRGS (Porto Alegre, Brazil).

Protein identification based on peptide MS/MS data was performed using Mascot software (Matrix Science, UK). All tandem mass spectra were searched against a database generated via an *in silico* digest of all proteins encoded by the *M. hyopneumoniae* 7448 genome with the following search parameters: trypsin was used as the cutting enzyme, mass tolerance for the monoisotopic peptide window was set to ±0.2 Da, the MS/MS tolerance window was set to ±0.2 Da, one missed cleavage was allowed, and carbamidomethyl and oxidized methionine were chosen as variable modifications.

In silico analyses of protein identified by GeLC-MS/MS

Cluster of Orthologous in Genomes (COG) annotations were assigned based on sequence similarity searches of CDS entries from *M. hyopneumoniae* strain 7448 (http://www.ncbi.nlm.nih.gov/genome, NC_007332) against the COG annotated proteins database (http://www.ncbi.nlm.nih.gov/COG) [25]. Lipoprotein (LP) *in silico* prediction was performed using LIPOPREDICT (http://www.lipopredict.cdac.in/) and LIPO CBU (http://services.cbu.uib.no/tools/lipo). Ortholog analysis was performed as previously described [26].

Results

GeLC-MS/MS analysis of *M. hyopneumoniae* cell surface labeled proteins

Biotin-labeled *M. hyopneumoniae* strain 7448 proteins were resolved by SDS-PAGE, and proteins from gel slices (Figure S1, lane 2 and 3) were subjected to nanoHPLC-nanoESI-Q-TOF-MS/MS identification. MS/MS analyses were performed in replicates (technical replicates) for each of the three biological samples, with virtually the same results. The GeLC-MS/MS identification of labeled intact cell (LIC) samples, resulted in a total of 165 protein identifications (Table S1), corresponding to 59 different protein species (Table S2), which represent approximately 10% of the total proteins encoded by the genome of *M. hyopneumoniae* strain 7448. In control samples from labeled lysed cells (LLC), 96 different proteins species were identified, including several typical intracellular proteins, such as ribosomal proteins, not identified in samples from intact cells (Table S3).

Considering the repertoire of 59 surface exposed proteins detected by GeLC-MS/MS of labeled intact cell (LIC) samples, 34 were identified from two or more gel slices, including well-known adhesion related proteins, such as P76, P97, P102, P146 and P216, along with 19 (32%) hypothetical proteins. Thirty five of these proteins (58.3%) were previously predicted as surface protein [26]. We also performed an *in silico* prediction of *M. hyopneumoniae* 7448 lipoproteins (LPs) (Table S4), and, from the total of 25 predicted LPs (summing those predicted by LIPOPREDICT and LIPO CBU tools), 16 (64%) were detected in our survey of surface exposed proteins. Among the MS-detected surface proteins predicted *in silico* as such or as LPs, there are well-known adhesion related proteins, such as P76, P97, P102, P146 and P216, 46 K surface antigen precursor, ABC transporter xylose-binding lipoprotein, and prolipoprotein p65, along with 16 hypothetical proteins.

A total of 21 proteins out of the 59 identified as surface exposed by GeLC-MS/MS of labeled intact cell (LIC) samples (35%) were not predicted *in silico* as such [26] or as lipoproteins (our results)

(Table S2). This group includes proteins traditionally involved in intracellular processes. An additional analysis of these proteins reveled that at least for some of them, like pyruvate dehydrogenase subunits, glyceraldehyde 3-phosphate dehydrogenase (GAPDH), L-lactate dehydrogenase, elongation factor Tu (Ef-Tu), molecular chaperone DnaK, and NADH-dependent flavin oxidoreductase, there are previous evidences in the literature (see Discussion) suggesting their surface localization in mycoplasmas or other pathogenic bacteria.

Functional classification of *M. hyopneumoniae* surface proteins from LIC samples identified by GeLC-MS/MS

In an attempt to infer potential physiological/functional features, the functional classification of the 59 proteins experimentally identified as surface exposed was performed based on the COG [25] (Figure 1, Table S2). According to COG, more than half of the GeLC-MS/MS identified *M. hyopneumoniae* surface proteins (35 out of 59; 58%) from LIC samples were classified as having unknown function, a class which included adhesion-related (6 proteins) and hypothetical proteins (19 proteins). Energy production and conversion was the second well-represented class (9 out of 59; 15%), followed by nucleotide transport and metabolism (5 out of 59; 8%), carbohydrate transport and metabolism (3 out of 59; 5%), general function prediction (3 out of 59; 5%), posttranslational modification/protein turnover/chaperones (2 out of 59; 2%), inorganic ion transport and metabolism (1 out of 59; 1.6%), translation, ribosomal structure and biogenesis (1 out of 59; 1.6%) and translation (1 out of 59; 1.6%).

Topology predictions of *M. hyopneumoniae* surface proteins from LIC samples identified by GeLC-MS/MS

In silico topology predictions were performed for the 59 proteins experimentally identified as surface exposed. These predictions (Table S2) showed that 19 of them had transmembrane domains (TM proteins), including the adhesion related proteins P76, P97, P102, P146 and P216, the transporter potassium uptake protein, and the cell division protein. Interestingly, other 10 of the 19 putative transmembrane proteins also presented adhesion-like predicted topologies, consisting of a transmembrane helix and a large tail exposed at the cell surface. These proteins are protein MHP7448_0372, 46 K surface antigen precursor, periplasmic sugar-binding protein, and 7 hypothetical proteins (MHP7448_0138, MHP7448_0352, MHP7448_0373, MHP7448_0467, MHP7448_0468, MHP7448_0629, and MHP7448_0661).

Considering the experimental demonstration of surface exposure, adhesin-like topology predictions and absence of orthologue in the closely related, non-pathogenic species *M. flocculare*, 12 of the proteins identified by GeLC-MS/MS in LIC samples could be selected as potential targets for the development of drugs, vaccines and/or immunodiagnostic tests (Table 1). These include proteins previously related to pathogenicity in bacteria (NADH-dependent flavin oxidoreductase, and 46 K surface antigen precursor), and several previously uncharacterized *M. hyopneumoniae* hypothetical proteins.

Discussion

Cell surface proteins are directly exposed to the environment and can mediate important pathogen-host interactions, such as adhesion, cell signaling and immune modulation, all of which are relevant for pathogenesis. For *M. hyopneumoniae*, it is predicted that about a third of the genes in its genome code for surface

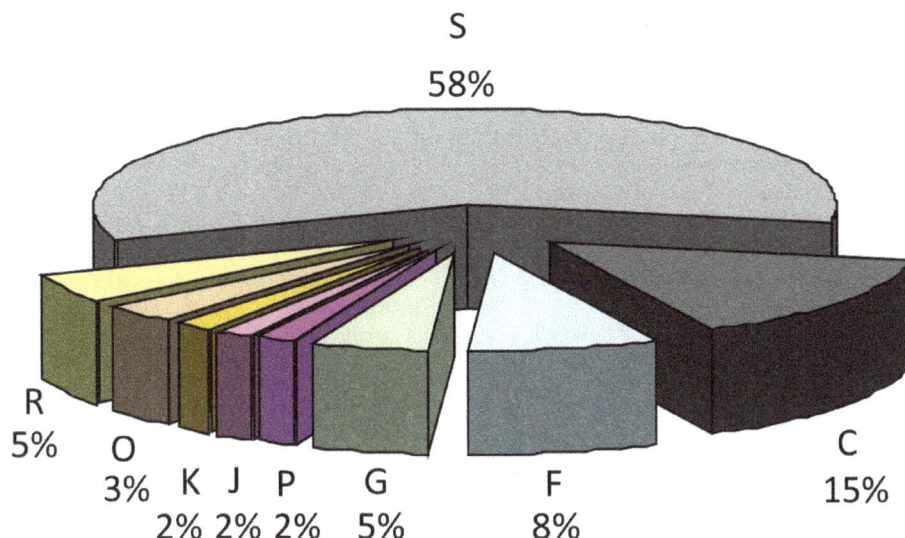

Figure 1. Functional analysis of *M. hyopneumoniae* surface proteins identified by GeLC-MS/MS in LIC samples. Percentages of proteins predicted in each functional category are indicated in the sectors of the circle. COG functional classes are as follows. Major class information storage and processes: (J) Translation, ribosomal structure and biogenesis, (K) Transcription; Major class cellular processes: (O) Post-translational modification, protein turnover, and chaperones, (P) Inorganic ion transport and metabolism; Major class Metabolism: (C) Energy production and conversion, (G) Carbohydrate transport and metabolism, (F) Nucleotide transport and metabolism; Major class poorly characterized: (R) General function prediction only, (S) Function unknown.

proteins [26], but so far, just a few of them have experimental confirmation of their expression and surface localization. Here, we employed a biotin cell surface labeling approach followed by avidin affinity recovery of labeled proteins coupled to identification by GeLC-MS/MS to survey the exposed surface proteins from *M. hyopneumoniae*. Experimental demonstration of surface exposure, adhesion-like topology predictions and absence of orthologs in the closely related, non-pathogenic *M. flocculare*

allowed to propose several proteins as candidates for the development of drugs, vaccines and/or immunodiagnostic tests for enzootic pneumonia.

Biotinylation has already been described as useful for the specific labeling of exposed tails of surface proteins [21,22]. The GeLC-MS/MS workflow was chosen for the qualitative identification of biotin-labeled *M. hyopneumoniae* surface proteins due to some advantages for hydrophobic protein analysis, including

Table 1. Adhesin topology prediction and additional criteria for the selection of the identified surface exposed proteins as potential targets for the development of drugs, vaccines and/or immunodiagnostic tests.

Locus Tag[1]	Protein product[2]	Protein name	Adhesin-like topology[3]	Absence in *M. flocculare*[4]	Studies suggested a surface localization[5]
MHP7448_0088	YP_287488.1	Hypothetical protein MHP7448_0088		•	
MHP7448_0138	YP_287535.1	Hypothetical protein MHP7448_0138	•		
MHP7448_0234	YP_287631.1	Periplasmic sugar-binding protein	•	•	[46]
MHP7448_0309	YP_287705.1	NADH-dependent flavin oxidoreductase		•	[48,49,50]
MHP7448_0324	YP_287719.1	Hypothetical protein MHP7448_0324		•	
MHP7448_0333	YP_287728.1	Hypothetical protein MHP7448_0333		•	
MHP7448_0352	YP_287746.1	Hypothetical protein MHP7448_0352	•		[29]
MHP7448_0372	YP_287766.1	Protein MHP7448_0372	•		
MHP7448_0373	YP_287767.1	Hypothetical protein MHP7448_0373	•		[30]
MHP7448_0513	YP_287902.1	46 K surface antigen precursor	•		
MHP7448_0629	YP_288014.1	Hypothetical protein MHP7448_0629	•		
MHP7448_0661	YP_288046.1	Hypothetical protein MHP7448_0661	•		

[1]Locus tag as defined for *M. hyopneumoniae* strain 7448 (http://www.ncbi.nlm.nih.gov/genome, NC_007332).
[2]Protein product according to NCBI database (http://www.ncbi.nlm.nih.gov).
[3]Topology prediction similar to well known adhesins (Table S2).
[4]No ortholog found closely related, non-pathogenic species *M. flocculare*.
[5]Published literature.

complete solubilization by SDS and removal of detergent and salts [24]. The biotinylation-GeLC-MS/MS combined approach successfully enriched samples and identified *M. hyopneumoniae* surface proteins, with no ribosomal protein recovery and with the identification of few classical cytoplasm proteins (Table S2). Besides, a large proportion of adhesion related proteins and LPs was selectively recovered and identified from LIC samples, in contrast with labeled lysed cell (LLC) control samples (cells lysed prior to labeling), from which proteins involved in a broad range of cellular functions, localized in both cytoplasm and membrane were identified (Table S3).

The adherence of *M. hyopneumoniae* to tracheal host cells is a crucial step for colonization and establishment of disease in infected pigs [1,3,16]. Adhesins play key roles in the process of pathogen binding to a host cell, although cell adhesion is a multifactorial process that involves surface proteins from both bacterial and host cells [27]. Well-known *M. hyopneumoniae* adhesins, namely P97, P97-like, P216, P102, and P76 proteins, which have been associated with binding to porcine cilia, heparin, fibronectin and plasminogen [8,9,10,11,13], had their surface localization confirmed by GeLC-MS/MS identification in LIC samples (Table S2). Other well represented proteins in our proteomic survey were LPs, among which were ABC transporter xylose-binding lipoprotein, and P60-like lipoprotein. ABC transporter xylose-binding lipoprotein is similar to the outer membrane lipoprotein P48 of some *Mycoplasma* species [28], described as an immunomodulatory protein required for intracellular invasion. The P60-like lipoprotein ortholog of *Mycoplasma hominis* was previously described [29] as a surface protein associated with P80 protein, whose *M. hyopneumoniae* ortholog (MHP7448_352) was also identified in our survey.

The GeLC-MS/MS protein identification in LIC samples also revealed the presence of 19 different hypothetical proteins in the *M. hyopneumoniae* cell surface. This corresponds to 6.3% of the total of 298 *M. hyopneumoniae* 7448 CDS products annotated as hypothetical proteins [26]. The products of two of these CDSs (MHP7448_0373 and MHP7448_0662) have been recently associated with cell adhesion. The MHP7448_0373 hypothetical protein was described as a heparin binding protein in porcine cilia [30], and the MHP7448_0662 hypothetical protein, which is a paralog of P102 was described as an adhesion protein able to bind fibronectin extracellular compounds [31] overexpressed in pathogenic strains [32]. Furthermore, our topology analysis showed that five of the 19 identified hypothetical proteins have adhesin-like topologies (Table S2).

Some proteins traditionally related to cellular processes that occur mainly in the cytoplasm were also identified in the *M. hyopneumoniae* cell surface. At least some of them, namely pyruvate dehydrogenase, GAPDH, L-lactate dehydrogenase, Ef-Tu and DnaK, have also been described as bacterial surface components, with involvement in bacterium-host interaction and pathogenicity. Pyruvate dehydrogenase, an immunogenic *M. hyopneumoniae* protein [33,34], was described as a surface protein involved in the bacterial binding to the host extracellular matrix in *Mycoplasma pneumoniae* [35]. GAPDH, usually located in the cytoplasm and known to play a central role as a glycolytic enzyme, has been identified on the surface of pathogenic bacteria and was related to pathogenic processes [36], such as adhesion and host matrix binding [37], immunomodulation and immune evasion [38]. L-lactate dehydrogenase, also known as P36, was described as immunogenic for pigs infected with *M. hyopneumoniae* [39]. The translational factor Ef-Tu was also identified as an immunogenic cell surface protein in mycoplasmas [18]. The molecular chaperone DnaK has been shown to be surface

accessible in *M. hyopneumoniae* [40] and was suggested as a surface or secreted protein of several other pathogenic bacteria, as *Bacillus anthracis* [41] and *Mycobacterium tuberculosis* [42].

The identification of the same protein species from two or more gel slices was taken as further evidence of proteolytic post-translational processing. This was observed for 34 of the 59 different protein species identified by GeLC-MS/MS in LIC samples, including adhesins and hypothetical proteins (Table S1). Proteolytic post-translational processing of *M. hyopneumoniae* proteins has been previously shown [43,44], and an alternative explanation based on protein degradation during sample processing was discarded in our survey due to the use of broad range protease inhibitors.

Nearly two thirds (39 out of 59) of the GeLC-MS/MS identified *M. hyopneumoniae* 7448 proteins in LIC samples was predicted *in silico* as surface proteins, considering the previous work by Siqueira *et al.* [26] and our complementary LP prediction. The extensive overlap between our proteomic approach and the *in silico* predictions is another indicator of the success of the selective surface biotin labeling, increasing the confidence of our results. Our proteomic results also contributed to better understanding of the *M. hyopneumoniae* 7448 repertoire of surface exposed proteins by also detecting, as discussed above, annotated (16) and hypothetical (5) proteins previously not predicted as bacterial surface components (Table S2). Considering sensitivity limitations of both labeling and MS detection [21], the identified proteins can be considered to be among the more abundant protein species on the *M. hyopneumoniae* 7448 cell surface.

The pathogenic *M. hyopneumoniae* and the non-pathogenic *M. flocculare* cohabit the swine respiratory tract and both species adhere to the cilia of tracheal epithelial cells, but just *M. hyopneumoniae* causes tissue damage [45]. Considering the extensive genetic similarity between *M. hyopneumoniae* and *M. flocculare* [26], we also surveyed the *M. flocculare* genome for orthologs of the *M. hyopneumoniae* 7448 proteins identified in LIC samples. Although *M. hyopneumoniae* 7448 and *M. flocculare* share nearly 90% of their predicted repertoire of surface proteins [26], 5 out of the 59 (8%) proteins identified in LIC samples are not found in *M. flocculare*. The *M. hyopneumoniae* proteins not shared with *M. flocculare* are three hypothetical proteins (MHP7448_0088, MHP7448_0324 and MHP7448_0333); a periplasmic sugar-binding protein that serves as primary receptor for diverse solutes in transport system, chemotaxis and signaling [46]; and a NAD-dependent flavin oxidoreductase. Interestingly, the involvement of oxidoreductase enzymes in pathogenicity has been described [47,48,49], and recent studies demonstrated that the *Mycobacterium avium* subsp. *paratuberculosis* NADH- flavin oxidoreductase from is essential for invasion of epithelial cells [50].

Besides the absence in *M. flocculare*, another aspect that calls attention to the periplasmic sugar-binding protein and to some other surface proteins identified in *M. hyopneumoniae* LIC samples was their adhesin-like topology. These criteria, along with previous studies of possible involvement of ortholog proteins in pathogen-host interactions, allowed us to propose at least 12 proteins (Table 1) as novel and promising targets for future studies.

Conclusion

The continuous development of immunization strategies, diagnosis tests and new drugs and antibiotics for disease treatment is essential for the control and prevention of EP [51]. Besides, some aspects of *M. hyopneumoniae* pathogenicity still need further clarification, including cell adhesion and immunomodulation [2,5,13]. Bacterial cell surface proteins play key roles in several cellular events important for bacterial growth, colonization and host inception. The selective labeling of surface proteins and their identification by GeLC-MS/MS described here provided new insights about cell surface protein composition in *M. hyopneumoniae* and allowed to point out novel and promising targets for the development of vaccines, diagnostic tests and therapeutic drugs. Further insights on *M. hyopneumoniae* pathogenicity are expected when the approaches standardized here were coupled to quantitative proteomic strategies for the comparison of the surface protein content of pathogenic and non-pathogenic *M. hyopneumoniae* strains.

Supporting Information

Figure S1 Biotin labeling and affinity capture of M. hyopneumoniae proteins from labeled lysed cell (LLC) and labeled intact cell (LIC) samples. SDS-PAGE 10%, and stained with Coomassie Brilliant Blue. Lane M – marker, Precision Plus prestained protein standards (Bio-Rad). Lane 1– crude protein extracts from M. hyopneumoniae 7448 (15 µg). Lane 2– avidin affinity capture of proteins from M. hyopneumoniae LLC samples (15 µg). Lane 3 - avidin affinity capture of proteins from M. hyopneumoniae LIC samples (15 µg). Bands 1 to 20 were subjected to Nano-LC/MS/MS analysis.

Table S1 *Proteins identified by GeLC-MS/MS in labeled intact cell (LIC) samples*.

Table S2 Different protein species identified by GeLC-MS/MS in labeled intact cell (LIC) samples.

Table S3 Different protein species identified by GeLC-MS/MS in control samples from labeled lysed cells (LLC).

Table S4 Lipoprotein predictions.

Author Contributions

Conceived and designed the experiments: LAR ISS HBF. Performed the experiments: LAR CLM ISS HBF. Analyzed the data: LAR CLM ISS HBF. Contributed to the writing of the manuscript: LAR ISS HBF.

References

1. Fadiel A, Eichenbaum KD, El SN, Epperson B (2007) Mycoplasma genomics: tailoring the genome for minimal life requirements through reductive evolution. Front Biosci 12: 2020–2028. 2207 [pii].
2. Razin S, Yogev D, Naot Y (1998) Molecular biology and pathogenicity of mycoplasmas. Microbiol Mol Biol Rev 62: 1094–1156.
3. Maes D, Verdonck M, Deluyker H, de KA (1996) Enzootic pneumonia in pigs. Vet Q 18: 104–109. 10.1080/01652176.1996.9694628 [doi].
4. DeBey MC, Ross RF (1994) Ciliostasis and loss of cilia induced by Mycoplasma hyopneumoniae in porcine tracheal organ cultures. Infect Immun 62: 5312–5318.
5. Zielinski GC, Ross RF (1993) Adherence of Mycoplasma hyopneumoniae to porcine ciliated respiratory tract cells. Am J Vet Res 54: 1262–1269.
6. Amass SF, Clark LK, van Alstine WG, Bowersock TL, Murphy DA, et al. (1994) Interaction of Mycoplasma hyopneumoniae and Pasteurella multocida infections in swine. J Am Vet Med Assoc 204: 102–107.
7. Thacker EL, Halbur PG, Ross RF, Thanawongnuwech R, Thacker BJ (1999) Mycoplasma hyopneumoniae potentiation of porcine reproductive and respiratory syndrome virus-induced pneumonia. J Clin Microbiol 37: 620–627.
8. Zhang Q, Young TF, Ross RF (1994) Microtiter plate adherence assay and receptor analogs for Mycoplasma hyopneumoniae. Infect Immun 62: 1616–1622.
9. Djordjevic SP, Cordwell SJ, Djordjevic MA, Wilton J, Minion FC (2004) Proteolytic processing of the Mycoplasma hyopneumoniae cilium adhesin. Infect Immun 72: 2791–2802.
10. Deutscher AT, Jenkins C, Minion FC, Seymour LM, Padula MP, et al. (2010) Repeat regions R1 and R2 in the P97 paralogue Mhp271 of Mycoplasma hyopneumoniae bind heparin, fibronectin and porcine cilia. Mol Microbiol 78: 444–458. 10.1111/j.1365-2958.2010.07345.x [doi].
11. Wilton J, Jenkins C, Cordwell SJ, Falconer L, Minion FC, et al. (2009) Mhp493 (P216) is a proteolytically processed, cilium and heparin binding protein of Mycoplasma hyopneumoniae. Mol Microbiol 71: 566–582. MMI6546 [pii];10.1111/j.1365-2958.2008.06546.x [doi].
12. Burnett TA, Dinkla K, Rohde M, Chhatwal GS, Uphoff C, et al. (2006) P159 is a proteolytically processed, surface adhesin of Mycoplasma hyopneumoniae: defined domains of P159 bind heparin and promote adherence to eukaryote cells. Mol Microbiol 60: 669–686. MMI5139 [pii];10.1111/j.1365-2958.2006.05139.x [doi].
13. Seymour LM, Jenkins C, Deutscher AT, Raymond BB, Padula MP, et al. (2012) Mhp182 (P102) binds fibronectin and contributes to the recruitment of plasmin(ogen) to the Mycoplasma hyopneumoniae cell surface. Cell Microbiol 14: 81–94. 10.1111/j.1462-5822.2011.01702.x [doi].
14. Bogema DR, Deutscher AT, Woolley LK, Seymour LM, Raymond BB, et al. (2012) Characterization of cleavage events in the multifunctional cilium adhesin Mhp684 (P146) reveals a mechanism by which Mycoplasma hyopneumoniae regulates surface topography. MBio 3. mBio.00282-11 [pii];10.1128/mBio.00282-11 [doi].
15. Seymour LM, Deutscher AT, Jenkins C, Kuit TA, Falconer L, et al. (2010) A processed multidomain mycoplasma hyopneumoniae adhesin binds fibronectin, plasminogen, and swine respiratory cilia. J Biol Chem 285: 33971–33978. M110.104463 [pii];10.1074/jbc.M110.104463 [doi].
16. Rottem S (2003) Interaction of mycoplasmas with host cells. Physiol Rev 83: 417–432. 10.1152/physrev.00030.2002 [doi].
17. You XX, Zeng YH, Wu YM (2006) Interactions between mycoplasma lipid-associated membrane proteins and the host cells. J Zhejiang Univ Sci B 7: 342–350. 10.1631/jzus.2006.B0342 [doi].
18. Cacciotto C, Addis MF, Pagnozzi D, Chessa B, Coradduzza E, et al. (2010) The liposoluble proteome of Mycoplasma agalactiae: an insight into the minimal protein complement of a bacterial membrane. BMC Microbiol 10: 225. 1471-2180-10-225 [pii];10.1186/1471-2180-10-225 [doi].
19. Corona L, Cillara G, Tola S (2013) Proteomic approach for identification of immunogenic proteins of Mycoplasma mycoides subsp. capri. Vet Microbiol 167: 434–439. S0378-1135(13)00442-2 [pii];10.1016/j.vetmic.2013.08.024 [doi].
20. Krasteva I, Liljander A, Fischer A, Smith DG, Inglis NF, et al. (2014) Characterization of the in vitro core surface proteome of Mycoplasma mycoides subsp. mycoides, the causative agent of contagious bovine pleuropneumonia. Vet Microbiol 168: 116–123. S0378-1135(13)00498-7 [pii];10.1016/j.vetmic.2013.10.025 [doi].
21. Cordwell SJ (2006) Technologies for bacterial surface proteomics. Curr Opin Microbiol 9: 320–329. S1369-5274(06)00057-9 [pii];10.1016/j.mib.2006.04.008 [doi].
22. Cullen PA, Xu X, Matsunaga J, Sanchez Y, Ko AI, et al. (2005) Surfaceome of Leptospira spp. Infect Immun 73: 4853–4863. 73/8/4853 [pii];10.1128/IAI.73.8.4853-4863.2005 [doi].
23. Vasconcelos AT, Ferreira HB, Bizarro CV, Bonatto SL, Carvalho MO, et al. (2005) Swine and poultry pathogens: the complete genome sequences of two strains of Mycoplasma hyopneumoniae and a strain of Mycoplasma synoviae. J Bacteriol 187: 5568–5577. 187/16/5568 [pii];10.1128/JB.187.16.5568-5577.2005 [doi].
24. Piersma SR, Warmoes MO, de Wit M, de Reus I, Knol JC, et al. (2013) Whole gel processing procedure for GeLC-MS/MS based proteomics. Proteome Sci 11: 17. 1477-5956-11-17 [pii];10.1186/1477-5956-11-17 [doi].
25. Tatusov RL, Fedorova ND, Jackson JD, Jacobs AR, Kiryutin B, et al. (2003) The COG database: an updated version includes eukaryotes. BMC Bioinformatics 4: 41. 10.1186/1471-2105-4-41 [doi];1471-2105-4-41 [pii].
26. Siqueira FM, Thompson CE, Virginio VG, Gonchoroski T, Reolon L, et al. (2013) New insights on the biology of swine respiratory tract mycoplasmas from a comparative genome analysis. BMC Genomics 14: 175. 1471-2164-14-175 [pii];10.1186/1471-2164-14-175 [doi].
27. Krause DC (1996) Mycoplasma pneumoniae cytadherence: unravelling the tie that binds. Mol Microbiol 20: 247–253.
28. Hall RE, Kestler DP, Agarwal S, Goldstein KM (1999) Expression of the monocytic differentiation/activation factor P48 in Mycoplasma species. Microb Pathog 27: 145–153. 10.1006/mpat.1999.0293 [doi];S0882-4010(99)90293-0 [pii].
29. Hopfe M, Hoffmann R, Henrich B (2004) P80, the HinT interacting membrane protein, is a secreted antigen of Mycoplasma hominis. BMC Microbiol 4: 46. 1471-2180-4-46 [pii];10.1186/1471-2180-4-46 [doi].
30. Deutscher AT, Tacchi JL, Minion FC, Padula MP, Crossett B, et al. (2012) Mycoplasma hyopneumoniae Surface proteins Mhp385 and Mhp384 bind host cilia and glycosaminoglycans and are endoproteolytically processed by proteases that recognize different cleavage motifs. J Proteome Res 11: 1924–1936. 10.1021/pr201115v [doi].

31. Bogema DR, Scott NE, Padula MP, Tacchi JL, Raymond BB, et al. (2011) Sequence TTKF downward arrow QE defines the site of proteolytic cleavage in Mhp683 protein, a novel glycosaminoglycan and cilium adhesin of Mycoplasma hyopneumoniae. J Biol Chem 286: 41217–41229. M111.226084 [pii];10.1074/jbc.M111.226084 [doi].

32. Pinto PM, Klein CS, Zaha A, Ferreira HB (2009) Comparative proteomic analysis of pathogenic and non-pathogenic strains from the swine pathogen Mycoplasma hyopneumoniae. Proteome Sci 7: 45. 1477-5956-7-45 [pii];10.1186/1477-5956-7-45 [doi].

33. Pinto PM, Chemale G, de Castro LA, Costa AP, Kich JD, et al. (2007) Proteomic survey of the pathogenic Mycoplasma hyopneumoniae strain 7448 and identification of novel post-translationally modified and antigenic proteins. Vet Microbiol 121: 83–93. S0378-1135(06)00472-X [pii];10.1016/j.vetmic.2006.11.018 [doi].

34. Matic JN, Wilton JL, Towers RJ, Scarman AL, Minion FC, et al. (2003) The pyruvate dehydrogenase complex of Mycoplasma hyopneumoniae contains a novel lipoyl domain arrangement. Gene 319: 99–106. S0378111903007984 [pii].

35. Thomas C, Jacobs E, Dumke R (2013) Characterization of pyruvate dehydrogenase subunit B and enolase as plasminogen-binding proteins in Mycoplasma pneumoniae. Microbiology 159: 352–365. mic.0.061184-0 [pii];10.1099/mic.0.061184-0 [doi].

36. Seidler NW (2013) GAPDH, as a virulence factor. Adv Exp Med Biol 985: 149–178. 10.1007/978-94-007-4716-6_5 [doi].

37. Alvarez RA, Blaylock MW, Baseman JB (2003) Surface localized glyceraldehyde-3-phosphate dehydrogenase of Mycoplasma genitalium binds mucin. Mol Microbiol 48: 1417–1425. 3518 [pii].

38. Terao Y, Yamaguchi M, Hamada S, Kawabata S (2006) Multifunctional glyceraldehyde-3-phosphate dehydrogenase of Streptococcus pyogenes is essential for evasion from neutrophils. J Biol Chem 281: 14215–14223. M513408200 [pii];10.1074/jbc.M513408200 [doi].

39. Frey J, Haldimann A, Kobisch M, Nicolet J (1994) Immune response against the L-lactate dehydrogenase of Mycoplasma hyopneumoniae in enzootic pneumonia of swine. Microb Pathog 17: 313–322. S0882-4010(84)71077-1 [pii];10.1006/mpat.1994.1077 [doi].

40. Chou SY, Chung TL, Chen RJ, Ro LH, Tsui PI, et al. (1997) Molecular cloning and analysis of a HSP (heat shock protein)-like 42 kDa antigen gene of Mycoplasma hyopneumoniae. Biochem Mol Biol Int 41: 821–831.

41. Chitlaru T, Gat O, Grosfeld H, Inbar I, Gozlan Y, et al. (2007) Identification of in vivo-expressed immunogenic proteins by serological proteome analysis of the Bacillus anthracis secretome. Infect Immun 75: 2841–2852. IAI.02029-06 [pii];10.1128/IAI.02029-06 [doi].

42. Xolalpa W, Vallecillo AJ, Lara M, Mendoza-Hernandez G, Comini M, et al. (2007) Identification of novel bacterial plasminogen-binding proteins in the human pathogen Mycobacterium tuberculosis. Proteomics 7: 3332–3341. 10.1002/pmic.200600876 [doi].

43. Raymond BB, Tacchi JL, Jarocki VM, Minion FC, Padula MP, et al. (2013) P159 from Mycoplasma hyopneumoniae binds porcine cilia and heparin and is cleaved in a manner akin to ectodomain shedding. J Proteome Res 12: 5891–5903. 10.1021/pr400903s [doi].

44. Tacchi JL, Raymond BB, Jarocki VM, Berry IJ, Padula MP, et al. (2014) Cilium adhesin P216 (MHJ_0493) is a target of ectodomain shedding and aminopeptidase activity on the surface of Mycoplasma hyopneumoniae. J Proteome Res 13: 2920–2930. 10.1021/pr500087c [doi].

45. Kobisch M, Friis NF (1996) Swine mycoplasmoses. Rev Sci Tech 15: 1569–1605.

46. Mowbray SL, Cole LB (1992) 1.7 A X-ray structure of the periplasmic ribose receptor from Escherichia coli. J Mol Biol 225: 155–175. 0022-2836(92)91033-L [pii].

47. Yu J, Kroll JS (1999) DsbA: a protein-folding catalyst contributing to bacterial virulence. Microbes Infect 1: 1221–1228. S1286-4579(99)00239-7 [pii].

48. Jenkins C, Geary SJ, Gladd M, Djordjevic SP (2007) The Mycoplasma gallisepticum OsmC-like protein MG1142 resides on the cell surface and binds heparin. Microbiology 153: 1455–1463. 153/5/1455 [pii];10.1099/mic.0.2006/004937-0 [doi].

49. Jenkins C, Samudrala R, Geary SJ, Djordjevic SP (2008) Structural and functional characterization of an organic hydroperoxide resistance protein from Mycoplasma gallisepticum. J Bacteriol 190: 2206–2216. JB.01685-07 [pii];10.1128/JB.01685-07 [doi].

50. Alonso-Hearn M, Patel D, Danelishvili L, Meunier-Goddik L, Bermudez LE (2008) The Mycobacterium avium subsp. paratuberculosis MAP3464 gene encodes an oxidoreductase involved in invasion of bovine epithelial cells through the activation of host cell Cdc42. Infect Immun 76: 170–178. IAI.01913-06 [pii];10.1128/IAI.01913-06 [doi].

51. Maes D, Segales J, Meyns T, Sibila M, Pieters M, et al. (2008) Control of Mycoplasma hyopneumoniae infections in pigs. Vet Microbiol 126: 297–309. S0378-1135(07)00450-6 [pii];10.1016/j.vetmic.2007.09.008 [doi].

24

Contribution of S-Layer Proteins to the Mosquitocidal Activity of *Lysinibacillus sphaericus*

Mariana Claudia Allievi, María Mercedes Palomino, Mariano Prado Acosta, Leonardo Lanati, Sandra Mónica Ruzal*, Carmen Sánchez-Rivas

Departamento de Química Biológica, Facultad de Ciencias Exactas y Naturales, Universidad de Buenos Aires, IQUIBICEN-CONICET, Buenos Aires, Argentina

Abstract

Lysinibacillus sphaericus strains belonging the antigenic group H5a5b produce spores with larvicidal activity against larvae of *Culex* mosquitoes. C7, a new isolated strain, which presents similar biochemical characteristics and Bin toxins in their spores as the reference strain 2362, was, however, more active against larvae of *Culex* mosquitoes. The contribution of the surface layer protein (S-layer) to this behaviour was envisaged since this envelope protein has been implicated in the pathogenicity of several bacilli, and we had previously reported its association to spores. Microscopic observation by immunofluorescence detection with anti S-layer antibody in the spores confirms their attachment. S-layers and BinA and BinB toxins formed high molecular weight multimers in spores as shown by SDS-PAGE and western blot detection. Purified S-layer from both *L. sphaericus* C7 and 2362 strain cultures was by itself toxic against *Culex sp* larvae, however, that from C7 strain was also toxic against *Aedes aegypti*. Synergistic effect between purified S-layer and spore-crystal preparations was observed against *Culex* sp. and *Aedes aegypti* larvae. This effect was more evident with the C7 strain. *In silico* analyses of the S-layer sequence suggest the presence of chitin-binding and hemolytic domains. Both biochemical characteristics were detected for both S-layers strains that must justify their contribution to pathogenicity.

Editor: Guido Favia, University of Camerino, Italy

Funding: Support was provided by Consejo Nacional de Investigaciones Científicas y Técnicas CONICET PIP0229 and Agencia Nacional de Promoción Científica y Tecnológica ANPCyT PICT-2012-0789. The funders had no role in study design, data collection and analysis, decision to publish, or preparation of the manuscript.

Competing Interests: The authors have declared that no competing interests exist.

* Email: sandra@qb.fcen.uba.ar

Introduction

Lysinibacillus sphaericus, together with *Bacillus thuringiensis* var. *israelensis*, represents the best ecological insecticide against mosquitoes and an environmental friendly alternative to chemical insecticides. These Gram-positive bacteria synthesize spores together with crystal-proteins, Cry and/or Cyt toxins, both being a very stable bioinsecticide [1]. In *B. thuringiensis* var. *israelensis*, crystal Cry and Cyt proteins are involved in the recognition of the insect target, the disruption of the membrane and finally the hemolysis. This last activity has been shown to be related to the low appearance of resistant mosquitoes. Moreover, other less stable toxins are produced during the cells vegetative growth: Vip for *Bacillus thuringiensis* [2] [3] or Mtx for *L. sphaericus* [4].

Lysinibacillus sphaericus, formerly *Bacillus sphaericus*, was renamed due to the presence of lysine and aspartic acid in the composition of their peptidoglycan [5]. These are a heterogeneous group of gram positive sporulating *Bacillus* some of which are entomopathogenic against mosquito larvae [4]. Hybridization studies of their DNA lead to classify them in 5 groups (I to V), but the most toxic strains belong to the homology group IIA [6] [7] and flagellar serotype H5a5b being 2362 the reference strain. These bacteria present particular metabolic traits: although they do not use hexoses or pentoses as carbon sources [8], they are able to use the amino-sugar N-acetylglucosamine, the monomer of chitin, and posses an active PTS transporter (Phosphoenolpyruvate phosphotransferase system) essentially implicated in its utilization [9] [10]. However, in contrast to *B. thuringiensis* strains [11], no chitinase activity has been detected in these bacteria.

L. sphaericus spores present an important exosporium allowing spore and crystals to remain firmly associated [12]. These crystalline inclusions are composed by two proteins named BinA and BinB, which can form dimer and/or associate in mixed proportions [13]. During the vegetative growth phase *L. sphaericus* strains produce several toxic proteins named Mtx1, 2, 3 [14] [15]. Besides being very efficient in synergic experiments with BinA-BinB, Mtx proteins are not synthesized during the sporulation phase and are degraded by proteases synthesized during this period. In fact, recombinants containing the cloned *mtx1* gene under a *bin* promoter allow Mtx1 synthesis during sporulation, but again the protein was rapidly degraded while sporulation proceeds [16].

While *B. thuringiensis* var. *israelensis* spore-crystal preparations are highly active against *Aedes, Culex, Anopheles* and *Simulium* [1] [4] those from *L. sphaericus* are essentially active against *Culex* and *Anopheles* species. This complementarity in behavior and targets has been exploited by using mixed preparations and recombinants containing the cloned toxic genes from *L. sphaericus* [17] [18] [19] into *B. thuringiensis*. However, *B. thuringiensis* is reported to be highly sensitive to the presence of chemical and metal contaminants, while *L. sphaericus* shows a better persistence in contaminated ponds [4]. Also several reports have shown the ability of this bacterium to survive [20] [21] and bioabsorb metal at concentrations otherwise toxic [22] and this property is linked to the presence of their S-layer envelope.

Moreover, it is worthwhile mentioning that the S-layer from several bacteria have also been implicated in their pathogenicity; this is especially so for *Bacillus* species as *B. cereus*, *B. anthracis* [23] [24], *Paenibacillus alvei* [25], different strain-variants from *B. thuringiensis* [26] [27] [28] and new isolated from *L. sphaericus* [29]. Altogether these properties have increased the interest in *L. sphaericus* strains and the drive to find new isolates of this species.

The reference strain 2362 is endowed with a high molecular weight S-layer (120 kDa) [30] that was also present in spore preparations [22]. Sequence based analysis performed for this protein allowed to predict the presence of hemolytic and chitin binding domains that might contribute to their entomopathogenicity. Such properties are associated with the pathogenicity of the Cry proteins from *B. thuringiensis* [31]. This led us to investigate the insecticidal properties of the S-layer from two *L. sphaericus* strains: the reference strain 2362 and C7 (a new isolate for which a higher insecticidal activity of the mixture of spore-crystals preparations has been observed). The mosquitocidal activity against *Culex* and *Aedes* larvae of spore-crystals and purified S-layers was assayed. Hemolytic activity and chitin binding properties of the S-layer were analyzed to characterize their influence on the mosquitocidal activity of *Lysinibacillus sphaericus*.

Materials and Methods

Bacterial strains, antibodies and media

L. sphaericus 2362 was obtained from Institut Pasteur. C7 strain was isolated from a crocodile lagoon in Cuba. Flagellar antisera (provided by A. Delécluse, Institut Pasteur) and phages typing (provided by A. Yousten) characterizations allowed us to classify them in the same group as 2362 (H5a, H5b). In addition, the antibody against the S-layer from 2362 (provided by L. Lewis) [30] recognizes the S-layer from the C7 strain. The anti-BinA or BinB were kindly provided by JF. Charles (Institut Pasteur). Strains were grown in LB (Luria Bertrani) or NYSM (Nutrient Yeast Extract Salt Medium) medium [32] as indicated in the text in aerated conditions at 32°C. For spores preparations growth was on solid NYSM medium at 32°C and spore-crystals were collected after 3 day incubation.

Phage typing assay

The phage-typing of the isolated strain was performed as described previously and the phages used were provided by A. A. Yousten [33]. Bacteria were grown on NYSM agar at 30°C. Phages 1A and 2 were propagated on strain *Bacillus sphaericus* SSII-1; phages 12 and SST were propagated on strain Kellen K; and phages 4 and 5 were propagated on strain 1593. For typing, bacteria were seeded into NYSM soft agar overlays and 20 μl of diluted phage suspension was spotted onto the surface. Plates were incubated at 30°C for 18 h. The production of individual plaques at the same dilution or at a 10-fold more concentrated dilution than those producing individual plaques on the normal phage propagating strain was noted as a positive result. Patterns of test results were compared with those for standard strains 1593 and 2362 (phage group 3), SSII-1 (phage group 2), and Kellen K (phage group 1).

Spores-crystal preparations

Strains were grown on solid NYSM medium and incubated 2–3 days at 32°C. Plates were then washed with 10 ml of 1 M NaCl, washed four times with milliQ H_2O and suspended in 10 mM sodium acetate pH 4.5 that prevents germination [34]. Dry-pellets containing a mixture of spores-crystals were stored at −20°C.

Cleaning off S-layer protein from spores

Spores-crystals from 3-day cultures were scrapped and washed with 10 ml of 1 M NaCl, then in deionized water, in 0.5 mM EDTA which eliminates the S-layer, and finally four times in deionized water [22]. These spores were used to evaluate synergy with purified S-layer proteins.

S-layer purification from cultures

Lysinibacillus sphaericus 2362 and C7 strains were grown in LB medium. Cells from 100 ml exponential cultures (OD_{600} = 1) were harvested and washed once with PBS. The S-layer proteins were extracted from the cells by cationic substitution by resuspending in 10 ml of 6 M LiCl, vortexing and incubating for 30 min at room temperature. After centrifugation at 15,000×g for 15 min, the supernatant was collected and dialyzed against 10 mM $CaCl_2$ overnight at 4°C. After centrifugation (10,000×g for 20 min), the pellets containing the S-layers were resuspended in deionized water and stored at −20°C [22].

Protein analysis

Spores proteins were alkali-solubilized (30 min at 37°C in 0.05 N NaOH) [17]. After alkaline extraction the preparations were centrifuged at 12,000×g 5 min to produce Pellet (P) and supernatant (S). The spores and S-layers preparations were heated at 90°C 5 min in loading buffer (10% glycerol, 4% SDS, 4 M urea, 2% β-mercaptoethanol, and 0.05% bromophenol blue) and subjected to electrophoresis in 12.5% SDS-PAGE. Gels were stained with Coomassie Brilliant Blue. The same amount of material was loaded in each well (10^7 CFU of spores and 0.5 μg of S-layer protein).

For western blot analysis, gels were electrotransferred with a Semi-dry Blotter (Amersham Biosciences) to PVDF membranes (Macherey-Nagel, Germany) soaked with polyclonal anti-rabbit antibodies against S-layer or BinA or BinB (diluted 1:2000 for S-layer and 1:50000 for BinA and BinB), and visualized with biotin-conjugated anti-rabbit followed by streptavidin-HRP conjugate (Pierce). Chemiluminescence was detected with luminol substrate (ECL from Sigma). Images were obtained with a Fuji LAS1000 digitalizer.

Detection of S-layer protein in spores by immunofluorescence

An immunofluorescence assay was performed in order to detect the presence of S-layer protein probably associated to the exosporium as described in *B. anthracis* [35] [36]. Spores from 3-day cultures were scrapped, washed with 10 ml of 1 M NaCl and resuspended in 10 mM sodium acetate pH 4.5 that prevents germination [34]. After centrifugation and double washing with PBS, the pellets were resuspended in PBS. A 20 μl (10^8 CFU) spore sample was then mixed with 0.5 μl antibody against S-layer and incubated 1 h at 25°C at a low shaking speed. The sample was then centrifuged, washed twice with PBS, resuspended in 100 μl PBS and subjected to the second antibody. 1.5 μl of antibody [Alexa fluor 647 goat anti-rabbit IgG (H+L) (Invitrogen)] was added and the sample was incubated for 30 min at 25°C. Spores were then pelleted, washed twice with PBS, and resuspended in

50 μl PBS. A drop was applied to a glass slide and overlaid with a coverslip. Immunofluorescence was observed with fluorescence microscopy on an Epifluorescense LED Axio Scope A1 model Microphot microscope (Carl Zeiss), N-Achroplan 100x/1.25 Oil PH3, filters mCherry FS64HE. The images were taken with a EOS T3 1100D digital camera (Canon) (10,1 Mega Pixel CMOS, 3.888×2.592 pixel eff, lens 18–55 mm).

Larvicidal assays

Experiments were carried with autochthonous seasonal species of *Culex sp* and *Aedes aegypti* larvae supplied by "Grupo de estudios de Mosquito" of the University of Buenos Aires [37]. Eggs were spread in dechlorinated water, fed with yeast extract, maintained at 28°C with daily illumination. After 4 days the majority of larvae reached 2–3rd instar development and were ready to use.

Petri dishes (10 cm diameter) containing 20 ml dechlorinated water and 20 larvae (2nd and 3rd instar) were inoculated with different concentrations of spores, S-layer protein or both as indicated in Tables 1 and 2 and were incubated at room temperature [38]. Larval mortality was evaluated 24 and 48 hours after exposure and compared to controls. The concentrations giving 50% mortality (LC50) was expressed as number of spores (CFU/ml) or S-layer proteins (μg/ml) after subtracting the mortality of controls. Assays were performed in duplicate and repeated three times. A representative experiment is shown.

Binding of S-layers to chitin compounds

To determine if the S-layer proteins recognize the chitin present in the cuticulum of insects, the binding specificity of these proteins for insoluble polysaccharides was used as follows: 50 μg of purified proteins were mixed with 10 μg of polysaccharide in a total volume of 500 μl PBS and incubated for 16 h at 32°C. The samples were centrifuged at 3,000×g for 3 min to separate free protein from substrate bound protein. Protein concentrations in the supernatant were determined using Bradford (BioRad). The percentage of adsorbed protein was calculated by substracting the free protein amount from the total protein in the sample. Polysaccharides were also incubated with bovine serum albumin, BSA, as negative control.

The polysaccharides used for this experiment were powder chitin, crab chitosan and fish chitosan resuspended in H$_2$O deionised and sonicated (to solubilise particles). These assays were performed at least six or more times as independent experiments in duplicate each time.

Hemolytic assays

Hemolysis is a way to evaluate membrane perturbation, a feature present in entomopathogenic toxins [31]. % Hemolysis was determined as previously described [39]. Briefly, sheep red blood cells were separated by centrifugation (1,000×g for 5 min) and resuspended to 1% with PBS. The final volume of the reaction mixture was 1 ml containing 0.5 ml of washed blood cells and various concentrations of S-layer protein (0.5–150 μg) in the same buffer and incubated at 37°C for 30 min. After centrifugation at 1,200×g for 5 min hemolysis was quantified measuring the absorbance of the supernatant at 405 nm. Positive control was 100% hemolysis after incubation of the same volume of sheep red blood cells with dechlorinated H$_2$O. Negative controls were red blood cells incubated with buffer. As a specificity control, antibodies (1:200 dilution) against S-layer were incubated for one hour along with the S-layer protein. Subsequently, the mixture was tested for hemolysis. These assays were performed

five or more times as independent experiments in duplicate each time.

Statistical Analysis

Statistical significance was evaluated by the Mann Whitney-U test for nonparametric data by Infostat software [40]. P<0.05 was considered to be statistically significant.

Results

Spore-crystals preparations contain S-layers

In a search for *Lysinibacillus sphaericus* strains with enhanced insecticidal activity, one particular isolate C7 was obtained. The C7 strain belongs to the same phage group 3 as the 2362 control. Its spore-crystals preparations were at least 100-fold more active than 2362 against *Culex pipiens* 2nd and 3rd instars larvae (LC$_{50}$ as μg total protein/assay were 1500±340 and 15±2 for 2362 and C7 respectively). Moreover, the strain had the same metabolic features as the reference strain 2362: growth with N-acetylglucosamine but not with glucose as carbon source, antigenic and phage-typing characteristics, presence of Mtx and Bin in spores. Since no clear difference in the yield of Bin toxins present in spores was observed, we undertook the analysis of the S-layer content of spores. We had previously observed that S-layers remain associated to spores [22], but these proteins could be a remnant of the vegetative envelope, or a constituent of the spore itself. For this purpose two approaches were used: Western Blot and immunofluorescence. Spore-crystal preparations were analyzed and detected using Western Blot. In this condition several high molecular weight bands (>120 kDa) are usually observed. This is generally attributed to associations between the toxic components BinA and BinB [13]. Western blot analyses with antibodies against Bin toxins and S-layer protein reveal the latters presence in the high molecular weight band (HMW up to 130 kDa), confirming that the S-layer and Bin components remain closely associated (Fig. 1A and B). Furthermore, the more toxic strain, C7, produced spores containing higher amounts of S-layer protein, which was also recognized by the specific antibody from 2362.

An immunofluorescence assay was used to directly detect the presence of S-layer protein in spores in association with the exosporium as has been described in *B. antracis* [35] [36] S-layer protein outside the spores, probably in association with the exosporium, was detected through direct observation of S-layer protein in non permeabilized spores (Fig. 2).

Larvicidal activity of S-Layer proteins from vegetative cultures and spores

S-layer extracted by LiCl and analyzed by SDS-PAGE and Western Blot showed a band running as 120–130 kDa (Fig. 1C and D). These purified S-layers proteins from C7 and 2362 strains were used to feed *Culex sp*. larvae as described in Materials and Methods. The dose response mortalities were obtained. The S-layer from C7 was as efficient as that of 2362 against *Culex* larvae (LC$_{50}$ 2.2 and 2.0 μg/ml, respectively) (Table 1). Surprisingly, when assayed against *Aedes aegypti*, only that of C7 showed a substantial activity (10-fold more toxic C7 than 2362) (Table 1). This led us to investigate possible interactions of the S-layers and spores in the larvicidal activity.

Synergy of spores and S-layers in the larvicidal activity

For this purpose individual batches of spores-preparations from 2362 and C7 strains were analyzed in order to determine their activity in the same batch of *Culex sp*. and *Aedes aegypti* larvae.

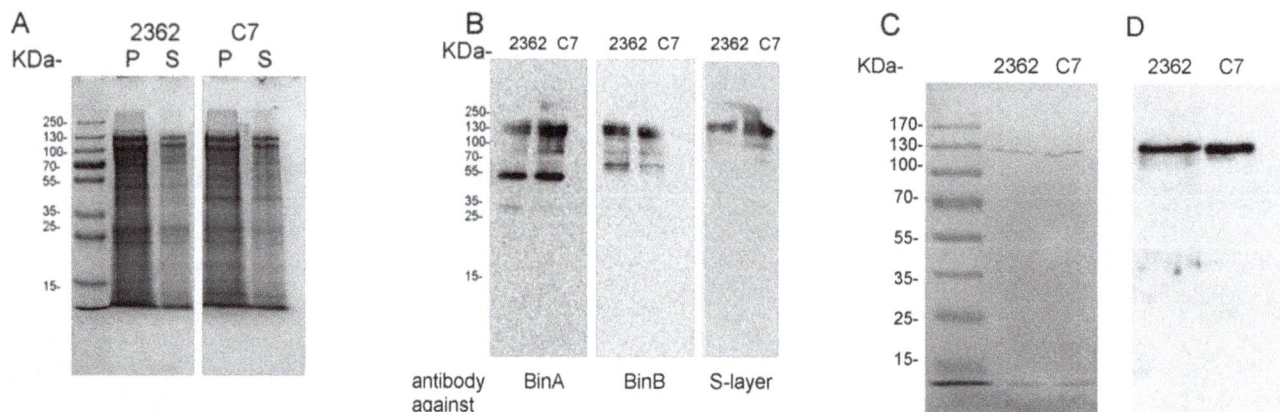

Figure 1. Detection of S-layers in spore-preparations and vegetative cultures. Pellet (P) and supernatant (S) fractions of spore-preparations from 2362 and C7 strains, obtained from the alkaline treatment described in Materials and Methods, were subjected to SDS-PAGE 12.5% (A) and Western Blot analysis for detection with specific antibodies against BinA or BinB or S-layer proteins (B). Purified S-layers from 2362 and C7 strains were obtained from vegetative cultures as described in Materials and Methods. 6 µg of each preparation were subjected to SDS-PAGE 12.5% electrophoresis (C) and analyses by Western Blot with specific antibody against the 2362 strain's S-layer (D).

This caution is necessary since batches of mosquitoes vary between seasons and origin.

In this condition we determined the LC_{50} for each preparation (spores and S-layers) and strain (2362 and C7). In this assay (Table 1), spores from C7 were 2 and 25 times more toxic against *Culex* and *Aedes* respectively than those from strain 2362.

In order to check for synergistic effects S-layer was removed from spores with EDTA [22]. Using concentrations of spores and S-layers giving mortality below 50%, a synergistic evaluation between them was assayed (Table 2). As shown, the mixtures (spores <u>and</u> S-layers), at sub-lethal concentrations, had 2 to 3 times higher activity than each component on its own, indicating that a synergistic effect between spores and S-layers took place.

These findings demonstrate the importance of the S-layer protein present in spores preparations in the larvicidal activity of these bacteria.

Sequence based analysis for putative functionalities that could support larvicidal activity in the S-layer protein

In order to characterize if a biochemical support for *Lysinibacillus sphaericus* pathogenicity is provided by it S-layer, a search of possible domains present in this protein was performed. Using different programs that are described in the Material S1 (SMART, EMBOSS Matcher Pairwise Sequence Alignment, and Clustal-O) the sequences of SlpC proteins were analyzed.

AAA50256.1, surface layer protein of *Lysinibacillus sphaericus* 2362, was used. When we analyzed the SMART database to predict functionality and physical distribution, we found a similar arrangement of surface layer homology domains (SLH) and Internal Repeats as that observed for the chitinase ChiW (BAM67143.1) of *Paenibacillus sp.* FPU-7 (Fig. S1A) [41]. Homologue architecture was observed between the two proteins: three SLH domains and two Internal Repeats. A local similarity to GH 18 chitinase domains was found for the internal repeats (Fig. S1B and S1C) using EMBOSS Matcher program, which identifies local similarities in two input sequences using a rigorous algorithm and allows for the modification of default substitution scoring matrices (BLOSUM) for sequence alignment between distantly related proteins. Using Clustal-O, a global alignment tool for multiple proteins sequences, homology with ACA38715.1 hemolysin-type calcium-binding domain-containing protein (874 aa, from the complete sequence of *Lysinibacillus sphaericus* C3–41 accession number NC_010382), was obtained with a high score alignment (Fig. S1D). The predicted functionalities were then investigated using *in vitro* assays for chitin binding and hemolytic activity in these S-layers preparations.

Chitin binding assays of S-layers

An *in vitro* assay was performed in order to check whether the pathogenicity of these S-layers involved the recognition of the

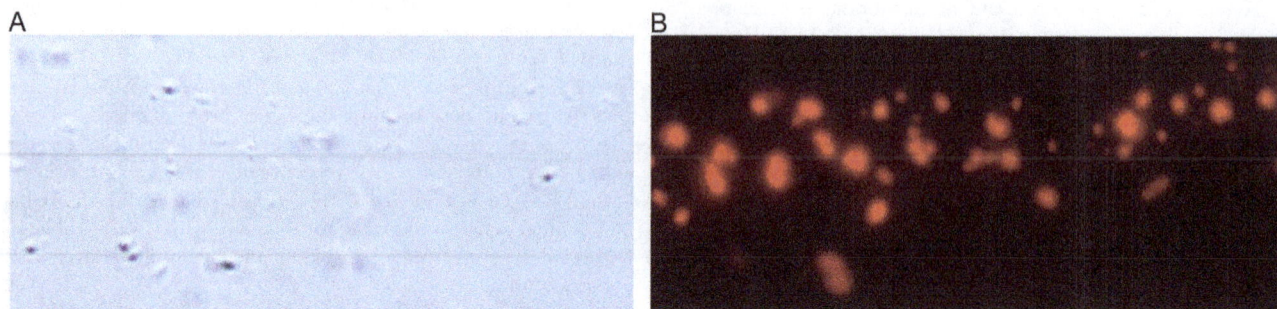

Figure 2. Detection of S-layers in spores by immunofluorescence. Spores preparations from C7 (A and B) were cleaned as described in Materials and Methods. A: Microscopic white light observation. B: Fluorescent (652 nm excitation and 668 nm emission) observation of the S-layer in the same preparation.

Table 1. Larvicidal activity of S-layers and spores from 2362 and C7 strains against *Culex sp* and *Aedes aegypti* larvae.

	Culex sp.	*Aedes aegypti*
S-layers from	**LC50 (μg/ml)**	**LC50 (μg/ml)**
2362	2.2 (0.2)	>80
C7	2.0 (0.2)	8.0 (0.5)
Spores from	**LC50 (cfu/ml)**	**LC50 (cfu/ml)**
2362	1.00×10^3 (0.02×10^3)	1.20×10^6 (0.06×10^6)
C7	0.50×10^3 (0.02×10^3)	5.00×10^4 (0.25×10^4)

S-layer and spores preparations were added with 20 *Culex sp.* or *Aedes aegypti* larvae as indicated in Material and Methods. After 48 h lethality was evaluated. Assays were performed with duplicate and repeated three times and a representative experiment is shown. In parenthesis, standard deviation.

chitin present in the cuticulum of their target. In a first assay, SDS-PAGE analysis of S-layers was performed in the presence or absence of colloidal chitin. Mobility retardation was observed for the S-layer but not for the bovine serum albumin (BSA) protein used as control (data not shown). To quantify the chitin binding property, a more sensitive assay was performed mixing purified S-layer proteins with insoluble chitosan, crab chitosan or colloidal chitin as described in Materials and Methods. As shown in Figure 3, the binding of S-layer proteins by these compounds featured similar efficacy, while this was not the case for BSA. Besides, the presence of N-acetylglucosamine (NAG), the monomer of chitin, inhibited the binding thus ensuring the specificity of the association between S-layers and chitin. Although binding to chitin was observed, we could not detect chitinase activity in the S-layer for these substrates.

Hemolytic activity of S-layers

We assayed hemolytic activity with purified S-layer preparation to determine the amount of hemoglobin release by the lyses of sheep red blood cells to obtain a direct quantitative measurement (see Materials and Methods). As shown in Figure 4, unlike BSA, the protein used as control, both S-layer proteins had hemolytic activity. We verified the specificity of the hemolysis by neutralizing the S-layer with specific antibodies that decreased the hemolytic activity to a 50%. We found that hemolytic activity was inhibited when N-acetylglucosamine was added, pointing to sugar recognition (data not shown).

Discussion

The pathogenic role of the S-layer has already been shown for several *Bacillus* species like *B. cereus* and *B. anthracis* [23] [24], *Paenibacillus alvei* [25], different strains from *B. thuringiensis* [26] [27] [28] and *Lysinibacillus sphaericus* [29]. However, no

function has been assigned to it other than its role as an adhesion factor for *B. anthracis* [23].

In the present work we confirm our previous observation [22] that spores from *L. sphaericus* retain S-layer proteins. Western Blot analyses of spore preparations from 2362 and C7 strains revealed associations of S-layer and Bin proteins (Fig. 1B). Immunofluorescence was used to visualize the S-layer (Fig. 2) surrounding spores. Since Bin proteins are deposited on the exosporium, we suspect that S-layer proteins are also associated to this structure. In fact, in *B. anthracis* the analysis of the exosporium shows the presence of several envelope proteins, including S-layers [35] [36] [42]. On the other hand, unlike the Mtx toxins present during the exponential growth, the S-layers were not degraded [16] by the proteolytic activity of the sporulation process, suggesting that their presence in spores was not fortuitous and that their location on the exosporium must contribute to this.

Moreover, the S-layer protein present in vegetative cultures (in the absence of spores), was mosquitocidal by itself as was the S-layer protein from new isolates reported by Lozano *et al* [29]. Concerning the larvicidal activity, it can be remarked that the S-layer from C7 was more active than that from 2362 against mosquitoes, in particular against *Aedes aegypti*, a species poorly sensitive to *L. sphaericus* toxins (LC50 was 10–times lower as shown in Table 1). We also reported a synergistic effect between cleaned spores (devoid of S-layer by EDTA treatment) and S-layers. When both are present, their activity is higher than that of the individual preparations (Tables 2).

The *slpC* gene from 2362 is 100% homologous to several entomopathogenic strains from the same antigenic group and especially to C3-41, a completely sequenced strain [43]. This led us to investigate functionalities within the protein sequence that could account for pathogenicity. Sequence based analysis reported in Material S1 revealed two possible biochemical domains: a chitin

Table 2. Synergy between spores and S-layers against *Culex sp.* Larvae.

	2362	C7
Mortality with Spores	32%	20%
Mortality with S-layer	2%	5%
Mortality with both	62%	85%

S-layer and spores preparations were added separately or mixed at sub-lethal concentrations with 20 *Culex sp.* larvae as indicated in Material and Methods. After 24 h lethality was evaluated. Only the % of died larvae are reported. The concentration of spores or S-layer protein individually or together per experiment were 500 CFU/ml and 0.25 μg for 2362 strain and 50 CFU/ml and 0.6 μg for C7 strain.

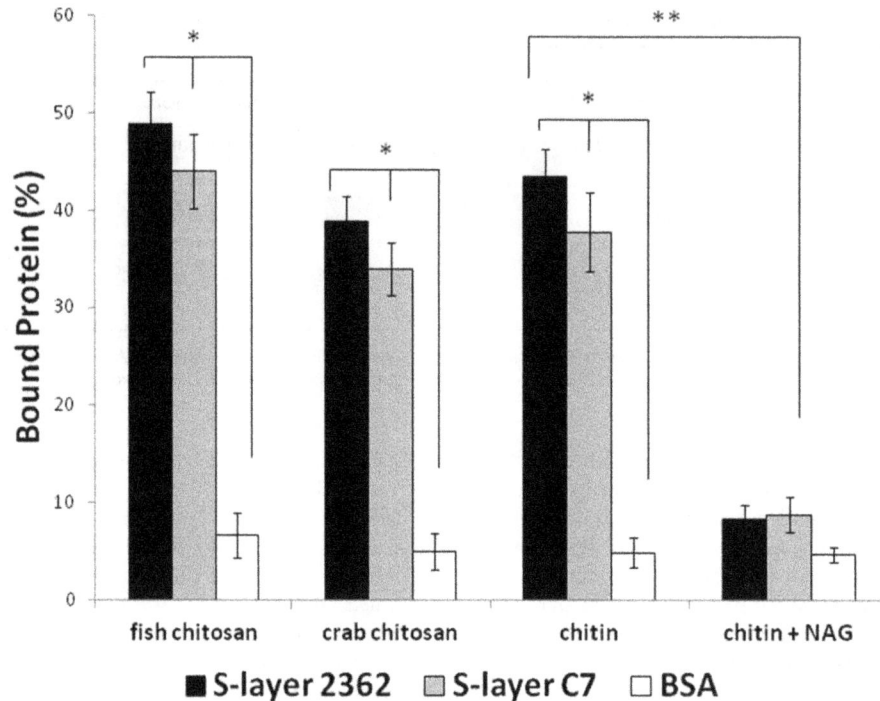

Figure 3. Binding assays of S-layer to chitin derivatives. Insoluble preparations of chitosan, crab chitosan, chitin (10 µg) were vigorously mixed with 50 µg of the different proteins and allow standing for 16 h at 32°C: S-layers from either 2362 or C7 strains, Bovine serum albumin (BSA). Free protein was determined and % bound calculated. Six independent experiments with duplicate samples were performed. Bars show the mean ± SD. Mann Whitney-U test was used to determine statistically significant differences between S-layers proteins and BSA control protein. *, P<0.05. With colloidal chitin, N-acetylglucosamine (NAG) (25 mM) was also added to verify binding inhibition. Five independent experiments with duplicate samples were performed. Mann Whitney-U test was used to determine statistically significant differences with and without NAG. **, P<0.05.

binding domain and a hemolytic domain that may contribute to the entomopathogenicity.

The chitin binding capacity (Fig. 3) and hemolytic activity (Fig. 4) were found to be present in the S-layer protein preparations and are both indicative of pathogenicity. Proteins with chitin-binding domains and chitinase activity have been described as pathogenicity factors in other species [44] [45]. In fact, the introduction of a chitinase gene from *B. thuringiensis* in a

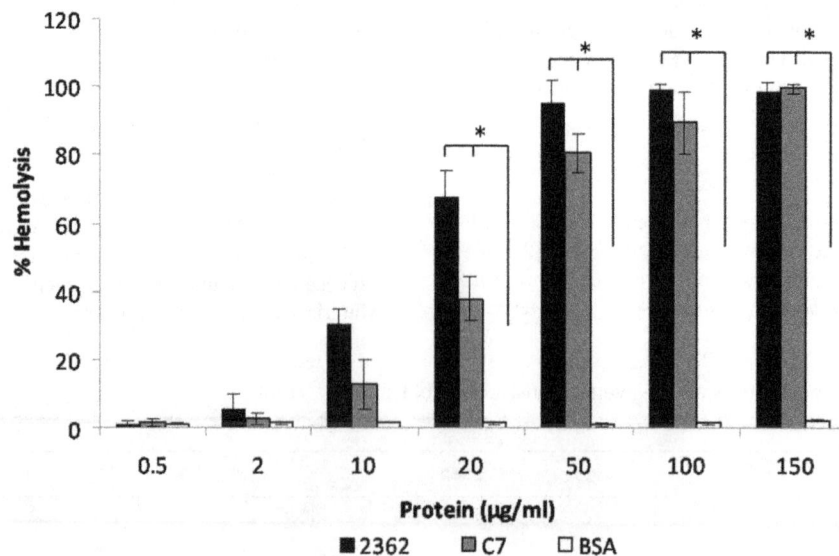

Figure 4. Hemolytic activity from S-layer of *L. sphaericus* 2362 and C7. Hemolysis was analyzed with 1% sheep red blood cells suspension in PBS. Bovine serum albumin (BSA) protein was used for unspecific effect. Six independent experiments with duplicate samples were performed. Bars show the mean ± SD. Mann Whitney-U test was used to determine statistically significant differences between S-layers proteins and BSA control protein. *, P<0.05.

L. sphaericus strain increases its insecticidal activity [46]. The presence of a ubiquitous chitin-binding domain in an external protein would be a welcome feature that would facilitate its binding to chitin containing substrates such as insects, thus expanding the number of susceptible mosquitoes species. A chitin binding domain in the S-layer protein would favor the interaction and attachment to insects as an anchor for the bacterium to its potential host. It is worthwhile to remark that chitinases and chitin-binding proteins have been implicated as virulence factors in several species [45]. Since chitinases are enzymes with a large spectrum of substrates and not easy to characterize, we have not been able to detect any chitinase activity in *L. sphaericus* cultures nor in their S-layer using the substrates reported in Fig. 3 (insoluble chitosan, crab chitosan or colloidal chitin). Although the S-layer protein sequence analysis showed a putative chitinase active site within the Glyco_18 domain, the carboxi-terminal cleft responsible for the catalytic activity seems to be absent (Fig. S1). However, it is worthwhile stressing that *L. sphaericus*, a species known for its failure to transport and use glucose (hexoses) as carbon source, has an active PTS transport system and functions for the use of N-acetylglucosamine (NAG), the monomer of chitin [9] [10]. Since *L. sphaericus* share the same habitat as fungi able to degrade chitin and deliver NAG, we may speculate that an efficient PTS-NAG would ensure the survival of these bacteria in the same environment while the presence of the S-layer protein with its chitin binding capacity would favor their interaction and also the attachment to insects, a step necessary for the development of pathogenicity. In fact, the presence of NAG interferes with chitin binding as shown in Figure 3.

Furthermore, the S-layer protein was shown to have hemolytic activity. Also as for chitin binding NAG was found to interfere with hemolytic activity (not shown). Hemolysis is attributed to membrane distortions caused by proteins interacting with lipids. This capacity should help in the pathogenic effect within the larvae, probably contributing to bring down the number of resistant mosquitoes, in the same way the Cyt proteins act in *B. thuringiensis* strains.

B. thuringiensis Cyt and Cry toxins showed a synergistic pattern against *Aedes* larvae [47] [31] similar to our observations with S-layers and spores containing Bins (Table 2).

The presence of S-layer in spore preparations has been reported for different *B. thuringiensis* strains [26] [27] [28] but their contribution to pathogenicity has not been confirmed. In such reports, the S-layer proteins were shown to be present as an inclusion inside the spore-crystals cytoplasm and were highly unstable, thus very difficult to isolate and characterize. Since we suspect that in *L. sphaericus* the S-layer is stabilized probably due to its location in the exosporium, as is the case in *B. anthracis*, it would be worthwhile to analyze it by cryo-EM [42].

Several strains belonging to the same antigenic group as 2362 presented S-layer proteins which cross-react with the antibody used in this work [30]; we also observed this cross-reactivity with that of the C7 strain, which implies they share a great homology. The reported sequences of the *slpC* gene from 2362 and C3-41 were identical. Moreover, the analysis of peptides obtained by trypsin digestion and MALDI-TOF analysis showed that the peptides sequences of S-layer from 2362 and C7 were identical [48]. We wonder why the larvicidal activity was so variable among these strains presenting similar SlpC proteins. We cannot account differences between strains to chitin binding or hemolysis activities since the *in vitro* assay using mimetic substrates (chitin compounds and sheep red blood cells) might not be exactly identical to their target in mosquitoes larvae gut.

The absence of larvicidal activity of the *slpC* gene from C3-41 cloned into *E. coli* [49], would suggest that post-translational modifications of the S-layer taking place in the original host, might be necessary to ensure pathogenicity. A relationship between pathogenicity and S-layer post-translational modifications has been reported for other species [50] [51] [52].

In conclusion, we observed that the S-layer of *L. sphaericus* play several functions. In fact their presence during an osmotic stress is essential [48]. Besides, the S-layers, either from vegetative cultures or associated to spores, were larvicidal against mosquitoes; however that of C7 strain was more active and presented a wider spectrum of activity. The S-layer proteins high molecular weight could indicate that they might result from an assembly of several functional domains in a new protein. Also, multiple *slp* genes have been annotated for the C3-41 strain, suggesting the importance of this protein for this species.

Concerning the bioinsecticide activity, this is the first report of the presence of both toxins (BinA-B) and S-layer protein in spores of *Lysinibacillus sphaericus* strains, where both contribute to its pathogenicity. Together with the high capacity for metal biosorption of these spores [22], which must explain their higher survival in contaminated ponds [20], strains with a wider bioinsecticide spectrum such as C7, would be an interesting alternative to *B. thuringiensis* formulations.

The analysis of the post-translational modifications of the S-layer proteins, which may contribute to the proteins' mosquito-cidal activity and specificity, would be the aim of our future investigation.

Supporting Information

Figure S1   Sequence-based analysis. S1A: Comparison of the structural disposition obtained with SMART SEARCH (Simple Modular Architecture Research Tool) for gb|AAA50256.1| surface layer protein [*Lysinibacillus sphaericus*] and dbj|BAM67143.1| chitinase [*Paenibacillus* sp. FPU-7]. **S1B and S1C:** EMBOSS Matcher Pairwise Sequence Alignment between gb|AAA50256.1| surface layer protein [*Lysinibacillus sphaericus*] and dbj|BAM67143.1| chitinase [*Paenibacillus* sp. FPU-7]. B) for the predicted Pfam SLH domains, C) for both predicted Internal Repeat with Glyco_18_1 domains. Symbols: "*" identical aminoacid, ":" indicates group similarity, "." indicates low group similarity. Amino Acid Notation according to IUPAC-IUB-CBN. N° is for number of residues and position within proteins. Grey boxes are SLH domains. **S1D:** CLUSTAL O (1.2.0) sequence alignment between AAA50256.1 surface layer protein [*Lysinibacillus sphaericus* 2362, 1176 aa] and ACA38715.1 hemolysin-type calcium-binding domain-containing protein, *Lysinibacillus sphaericus*, C3-41, 874 aa): score 494.1 bits. Symbols: "*" identical aminoacid, ":" indicates group similarity, "." indicates low group similarity. Amino Acid Notation according to IUPAC-IUB-CBN. N° is for number of residues and position within proteins. Grey boxes are SLH domains in AAA50256.1 only.

Material S1   Sequence-based analysis. Free access sites were used to predict protein structure and function for the Surface layer protein AAA50256 [*Lysinibacillus sphaericus* 2362] and compared to possible orthologous proteins. URL links: **SMART**, (http://smart.embl-heidelberg.de/) Simple modular architecture research tool for Comparison of the structural disposition obtained with SMART SEARCH (Simple Modular Architecture Research Tool). **EMBOSS** Matcher Pairwise Sequence Alignment ebi?tool=emboss_matcher&context=protein) identifies local similarities in two input sequences using a rigorous algorithm based on Bill Pearson's lalign application. It enables to modify the default substitution scoring matrices (BLOSUM) for sequence alignment between distantly related proteins. **Clustal-O**, Global alignment tool (http://www.ebi.ac.uk/Tools/msa/clustalo/), Clustal Omega is a multiple sequence alignment program for proteins. It produces biologically meaningful multiple sequence alignments of divergent sequences.

Acknowledgments

We would like to specially thank: Linn Lewis, Jean François Charles, Armelle Deléscluse, Allan Yousten for providing us with the antibodies and phages necessary for identification of strains and proteins analyzed in this work. We also wish to thank Liliana Rondón for her help in the immunofluorescence microscopy examinations of spores with their S-layer and Silvia Strauss for careful language revision.

Author Contributions

Conceived and designed the experiments: MCA MMP MPA SMR CSR. Performed the experiments: MCA MMP MPA LL SMR. Analyzed the data: MCA MMP MPA SMR CSR. Contributed reagents/materials/analysis tools: MCA MMP MPA LL SMR CSR. Contributed to the writing of the manuscript: MCA MMP MPA SMR CSR.

References

1. Lacey LA (2007) *Bacillus thuringiensis* serovariety *israelensis* and *Bacillus sphaericus* for mosquito control. J Am Mosq Control Assoc. 23(2 Suppl): 133–63.

2. Estruch JJ, Warren GW, Mullins MA, Nye GJ, Craig JA, et al. (1996) Vip3A, a novel *Bacillus thuringiensis* vegetative insecticidal protein with a wide spectrum of activities against lepidopteran insects. Proc Natl Acad Sci U S A. 93(11): 5389–94.

3. van Frankenhuyzen KJ (2013) Cross-order and cross-phylum activity of *Bacillus thuringiensis* pesticidal proteins. Invertebr Pathol. 114(1): 76–85.

4. Berry C (2012). The bacterium, *Lysinibacillus sphaericus*, as an insect pathogen. J Invertebr Pathol 109: 1–10.

5. Ahmed I, Yokota A, Yamazoe A, Fujiwara T (2007) Proposal of *Lysinibacillus boronitolerans* gen. nov. sp. nov., and transfer of *Bacillus fusiformis* to *Lysinibacillus fusiformis* comb. nov. and *Bacillus sphaericus* to *Lysinibacillus sphaericus* comb. nov. Int. J. Syst. Evol. Microbiol. 57, 1117–1125.

6. Krych VK, Johnson JL, Yousten AA (1980) Deoxyribonucleic acid homologies among strains of *Bacillus sphaericus*. Int J Syst Bacteriol 30: 476–484.

7. Rippere KE, Johnson JL, Yousten AA (1997) DNA similarities among mosquito-pathogenic and nonpathogenic strains of *Bacillus sphaericus*. Int J Syst Bacteriol 47: 214–216.

8. Russell BL, Jelley SA, Yousten AA (1989) Carbohydrate metabolism in the mosquito pathogen *Bacillus sphaericus* 2362. Appl Environ Microbiol. 55(2): 294–7.

9. Alice AF, Pérez-Martínez G, Sánchez-Rivas C (2002) Existence of a true phosphofructokinase in *Bacillus sphaericus*: cloning and sequencing of the *pfk* gene. Appl Environ Microbiol. 68: 6410–5.

10. Alice AF, Pérez-Martínez G, Sánchez-Rivas C (2003) Phosphoenolpyruvate phosphotransferase system and N-acetylglucosamine metabolism in *Bacillus sphaericus*. Microbiol. 149: 1687–98.

11. Gomez-Ramírez M, Rojas-Avelizapa LI, Cruz-Camarillo R (2001) The chitinase of *Bacillus thuringiensis*, p. 273–282. In R. A. A. Muzzarelli ed.), Chitin enzymology. Atec Edizioni, Atec, Italy.

12. Yousten AA, Davidson EW (1982) Ultrastructural Analysis of Spores and Parasporal Crystals Formed by *Bacillus sphaericus* 2297. Environ Microbiol. 44(6): 1449–55.

13. Smith AW, Cámara-Artigas A, Brune DC, Allen JP (2005) Implications of high-molecular-weight oligomers of the binary toxin from *Bacillus sphaericus*. J Invertebr Pathol. 88: 27–33.

14. Thanabalu T, Hindley J, Jackson-Yap J, Berry C (1991) Cloning, sequencing and expression of a gene encoding a 100-kilodalton mosquitocidal toxin from *Bacillus sphaericus* SSII-1. J. Bacteriol. 173: 2776–2785.

15. Charles JF, Nielson-LeRoux C, Delécluse A (1996) *Bacillus sphaericus* toxins: molecular biology and mode of action. Annu Rev Entomol. 41: 451–72.

16. Yang Y, Wang L, Gaviria A, Yuan Z, Berry C (2007) Proteolytic stability of insecticidal toxins expressed in recombinant bacilli. Appl Environ Microbiol 73: 218–25.

17. Thiéry I, Hamon S, Delécluse A, Orduz S (1998) The introduction into *Bacillus sphaericus* of the *Bacillus thuringiensis* subsp. *medellin* Cyt1Ab1 gene results in higher susceptibility of resistant mosquito larva populations to *B. sphaericus*. Appl Environ Microbiol. 64: 3910–6.

18. Wirth MC, Yang Y, Walton WE, Federici BA, Berry C (2007) Mtx toxins synergize *Bacillus sphaericus* and Cry11Aa against susceptible and insecticide-resistant *Culex quinquefasciatus* larvae. Appl Environ Microbiol. 73: 6066–71.

19. Chenniappan K, Ayyadurai N (2012) Synergistic activity of Cyt1A from *Bacillus thuringiensis* subsp. *israelensis* with *Bacillus sphaericus* B101 H5a5b against *Bacillus sphaericus* B101 H5a5b-resistant strains of *Anopheles stephensi Liston* (Diptera: Culicidae). Parasitol Res 110: 381–388.

20. Merroun ML, Raff J, Rossberg A, Hennig C, Reich T, et al (2005) Complexation of uranium by cells and S-layer sheets of *Bacillus sphaericus* JG-A12. Appl. Environ. Microbiol. 71: 5532–5543.

21. Pollmann K, Raff J, Merroun M, Fahmy K, Selenska-Pobell S (2006) Metal binding by bacteria from uranium mining waste piles and its technological applications. Biotechnology Advances 24: 58–68.

22. Allievi MC, Sabbione F, Prado-Acosta M, Palomino MM, Ruzal SM, et al. (2011) Metal Biosorption by Surface-Layer Proteins from *Bacillus* Species. Microbiol. Biotechnol. 21: 147–153.

23. Kern J, Schneewind O (2010) BslA, the S-layer adhesin of *B. anthracis*, is a virulence factor for anthrax pathogenesis. Mol Microbiol. 75: 324–32.

24. Wang YT, Oh SY, Hendrickx AP, Lunderberg JM, Schneewind O (2013) *Bacillus cereus* G9241 S-layer assembly contributes to the pathogenesis of anthrax-like disease in mice. J. Bacteriol. 195: 596–605.

25. Janesch B, Messner P, Schäffer C (2013) Are the SLH-Domains Essential for Cell Surface display and glycosylation of the S-Layer protein from *Paenibacillus alvei* CCM 2051T? J. Bacteriol. 195: 596–605.

26. Peña G, Miranda-Rios J, de la Riva G, Pardo-López L, Soberón M, et al. (2006) A *Bacillus thuringiensis* S-layer protein involved in toxicity against *Epilachna varivestis* (Coleoptera: Coccinellidae). Appl Environ Microbiol 72: 353–360.

27. Guo G, Zhang L, Zhou Z, Ma Q, Liu J, et al. (2008) A new group of parasporal inclusions encoded by the S-layer gene of *Bacillus thuringiensis*. FEMS Microbiol Lett 282: 1–7.

28. Zhou Z, Peng D, Zheng J, Guo G, Tian L, et al. (2011) Two groups of S-layer proteins, SLP1s and SLP2s, in *Bacillus thuringiensis* co-exist in the S-layer and in parasporal inclusions. BMB Rep. 44: 323–8.

29. Lozano LC, Ayala JA, Dussán J (2011) *Lysinibacillus sphaericus* S-layer protein toxicity against *Culex quinquefasciatus*. Biotechnol Lett 33: 2037–2041.

30. Lewis LO, Yousten AA, Murray RGE (1987) Characterization of the surface protein layers of the mosquito-pathogenic strains of *Bacillus sphaericus* J. Bacteriol 169: 72–79.

31. López-Diaz JA, Cantón PE, Gill SS, Soberón M, Bravo A (2013) Oligomerization is a key step in Cyt1Aa membrane insertion and toxicity but not necessary to synergize Cry11Aa toxicity in *Aedes aegypti* larvae. Environ Microbiol. 15: 3030–3039.

32. Myers PS, Yousten AA (1980) Localization of a mosquito-larval toxin of *Bacillus sphaericus* 1593. Appl. Environ. Microbiol. 39: 1205–1211.

33. Yousten AA (1984) *Bacillus sphaericus*: microbiological factors related to its potential as a mosquito larvicide. Adv Biotechnol Processes. 1984; 3: 315–43.

34. Travers RS, Martin PA, Reichelderfer CF (1987) Selective process for efficient isolation of soil *Bacillus* spp. Appl Environ Microbiol. 53(6): 1263–1266.

35. Redmond C, Baillie LW, Hibbs S, Moir AJ, Moir A (2004) Identification of proteins in the exosporium of *Bacillus anthracis*. Microbiology. 150(Pt 2): 355–63.

36. Kailas L, Terry C, Abbott N, Taylor R, Mullin N, et al. (2011) Surface architecture of endospores of the *Bacillus cereus/anthracis/thuringiensis* family at the subnanometer scale. Proc Natl Acad Sci U S A.108(38): 16014–9.

37. Fischer S, Schweigmann N (2004) *Culex* mosquitoes in temporary urban rain pools: seasonal dynamics and relation to environmental variables. J Vector Ecol. 29(2): 365–73.

38. Thiery I, Delécluse A, Tamayo MC, Orduz S (1997) Identification of a gene for Cyt1A-like hemolysin from *Bacillus thuringiensis* subsp. *medellin* and expression in a crystal-negative *B. thuringiensis* strain. Appl Environ Microbiol. 63(2): 468–73.

39. Rodriguez-Almazan C, Ruiz de Escudero I, Cantón PE, Muñoz-Garay C, Pérez C, et al. (2011) The amino- and carboxyl-terminal fragments of the *Bacillus thuringensis* Cyt1Aa toxin have differential roles in toxin oligomerization and pore formation. Biochem. 50: 388–96.

40. Balzarini MG, Gonzalez L, Tablada M, Casanoves F, Di Rienzo JA, et al. (2008) Infostat. Manual del Usuario, Editorial Brujas, Córdoba, Argentina.

41. Itoh T, Hibi T, Fujii Y, Sugimoto I, Fujiwara A, et al. (2013) Cooperative degradation of chitin by extracellular and cell surface-expressed chitinases from *Paenibacillus* sp. strain FPU-7. Appl Environ Microbiol. 79: 7482–90.

42. Rodenburg CM, McPherson SA, Turnbough CL Jr, Dokland T (2014) Cryo-EM analysis of the organization of BclA and BxpB in the *Bacillus anthracis* exosporium. J Struct Biol. 186(1): 181–7.

43. Hu X, Fan W, Han B, Liu H, Zheng D, et al. (2008) Complete genome sequence of the mosquitocidal bacterium *Bacillus sphaericus* C3–41 and comparison with those of closely related *Bacillus* species. J. Bacteriol 190: 2892–2902.

44. Sampson MN, Gooday GW (1998) Involvement of chitinases of *Bacillus thuringiensis* during pathogenesis in insects. Microbiology 144: 2189–2194.

45. Frederiksen RF, Paspaliari DK, Larsen T, Storgaard BG, Larsen MH, et al. (2013) Bacterial chitinases and chitin-binding proteins as virulence factors. Microbiol. 159: 833–47.

46. Cai Y, Ya J, Hu X, Han B, Yuan Z (2007) Improving the insecticidal activity against resistant *Culex quinquefasciatus* mosquitoes by expression of chitinase gene *chiAC* in *Bacillus sphaericus*. Appl Environ Microbiol. 73: 7744–7746.

47. Wirth MC, Delécluse A, Walton WE (2001) Cyt1Ab1 and Cyt2Ba1 from *Bacillus thuringiensis* subsp. *medellin* and *B. thuringiensis* subsp. *israelensis* synergize *Bacillus sphaericus* against *Aedes aegypti* and resistant *Culex quinquefasciatus* (Diptera: Culicidae). Appl Environ Microbiol. 67: 3280–4.

48. Allievi MC (2012) S-layer de *Bacillus sphaericus:* caracterización, regulación, análisis funcional y aplicaciones. Thesis. University of Buenos Aires. Available: http://bldigital.bl.fcen.uba.ar/Download/Tesis/Tesis_5143_Allievi.pdf Accessed 28 June 2012.

49. Hu X, Li J, Hansen BM, Yuan Z (2008) Phylogenetic analysis and heterologous expression of surface layer protein SlpC of *Bacillus sphaericus* C3–41. Biosci Biotechnol Biochem 72: 1257–1263.

50. Forsberg LS, Choudhury B, Leoff C, Marston CK, Hoffmaster AR, et al. (2011) Secondary cell wall polysaccharides from *Bacillus cereus* strains G9241, 03BB87 and 03BB102 causing fatal pneumonia share similar glycosyl structures with the polysaccharides from *Bacillus anthracis*. Glycobiol. 21: 934–948.

51. Balomenou S, Fouet A, Tzanodaskalaki M, Couture-Tosi E, Bouriotis V, et al. (2013) Distinct functions of polysaccharide deacetylases in cell shape, neutral polysaccharide synthesis and virulence of *Bacillus anthracis*. Mol Microbiol. 87: 867–83.

52. Friedlander A, Quinn CP, Kannenberg EL, Carlson RW (2012) Localization and structural analysis of a conserved pyruvylated epitope in *Bacillus anthracis* secondary cell wall polysaccharides and characterization of the galactose-deficient wall polysaccharide from avirulent *B. anthracis* CDC 684. Glycobiol. 22: 1103–1117.

Permissions

All chapters in this book were first published in PLOS ONE, by The Public Library of Science; hereby published with permission under the Creative Commons Attribution License or equivalent. Every chapter published in this book has been scrutinized by our experts. Their significance has been extensively debated. The topics covered herein carry significant findings which will fuel the growth of the discipline. They may even be implemented as practical applications or may be referred to as a beginning point for another development.

The contributors of this book come from diverse backgrounds, making this book a truly international effort. This book will bring forth new frontiers with its revolutionizing research information and detailed analysis of the nascent developments around the world.

We would like to thank all the contributing authors for lending their expertise to make the book truly unique. They have played a crucial role in the development of this book. Without their invaluable contributions this book wouldn't have been possible. They have made vital efforts to compile up to date information on the varied aspects of this subject to make this book a valuable addition to the collection of many professionals and students.

This book was conceptualized with the vision of imparting up-to-date information and advanced data in this field. To ensure the same, a matchless editorial board was set up. Every individual on the board went through rigorous rounds of assessment to prove their worth. After which they invested a large part of their time researching and compiling the most relevant data for our readers.

The editorial board has been involved in producing this book since its inception. They have spent rigorous hours researching and exploring the diverse topics which have resulted in the successful publishing of this book. They have passed on their knowledge of decades through this book. To expedite this challenging task, the publisher supported the team at every step. A small team of assistant editors was also appointed to further simplify the editing procedure and attain best results for the readers.

Apart from the editorial board, the designing team has also invested a significant amount of their time in understanding the subject and creating the most relevant covers. They scrutinized every image to scout for the most suitable representation of the subject and create an appropriate cover for the book.

The publishing team has been an ardent support to the editorial, designing and production team. Their endless efforts to recruit the best for this project, has resulted in the accomplishment of this book. They are a veteran in the field of academics and their pool of knowledge is as vast as their experience in printing. Their expertise and guidance has proved useful at every step. Their uncompromising quality standards have made this book an exceptional effort. Their encouragement from time to time has been an inspiration for everyone.

The publisher and the editorial board hope that this book will prove to be a valuable piece of knowledge for researchers, students, practitioners and scholars across the globe.

List of Contributors

Salai Madhumathi Parkunan
Departments of Microbiology and Immunology, University of Oklahoma Health Sciences Center, Oklahoma City, Oklahoma, United States of America

Roger Astley
Department of Ophthalmology, University of Oklahoma Health Sciences Center, Oklahoma City, Oklahoma, United States of America

Michelle C. Callegan
Departments of Microbiology and Immunology, University of Oklahoma Health Sciences Center, Oklahoma City, Oklahoma, United States of America
Department of Ophthalmology, University of Oklahoma Health Sciences Center, Oklahoma City, Oklahoma, United States of America
Dean A. McGee Eye Institute, Oklahoma City, Oklahoma, United States of America

Anella Saggese, Veronica Scamardella, Teja Sirec, Giuseppina Cangiano, Rachele Isticato, Ezio Ricca and Loredana Baccigalupi
Department of Biology, Federico II University of Naples, Naples, Italy

Francesca Pane and Angela Amoresano
Department of Chemistry, Federico II University of Naples, Naples, Italy

Laura A. Huppert and Briana M. Burton
Department of Molecular and Cellular Biology, Harvard University, Cambridge, Massachusetts, United States of America

Talia L. Ramsdell, Michael R. Chase and Sarah M. Fortune
Department of Immunology and Infectious Diseases, Harvard School of Public Health, Boston, Massachusetts, United States of America

David A. Sarracino
Thermo Fisher Scientific, BRIMS Unit, Cambridge, Massachusetts, United States of America

Dongfeng Ning, Alin Song, Fenliang Fan and Zhaojun Li
Ministry of Agriculture Key Laboratory of Crop Nutrition and Fertilization, Institute of Agricultural Resources and Regional Planning, Chinese Academy of Agricultural Sciences, Beijing, China

Yongchao Liang
Ministry of Agriculture Key Laboratory of Crop Nutrition and Fertilization, Institute of Agricultural Resources and Regional Planning, Chinese Academy of Agricultural Sciences, Beijing, China
Ministry of Education Key Laboratory of Environment Remediation and Ecological Health, College of Environmental and Resource Sciences, Zhejiang University, Hangzhou, China

Lovleen Tina Joshi and Les Baillie
Cardiff School of Pharmacy and Pharmaceutical Sciences, Cardiff University, Cardiff, United Kingdom

Buddha L. Mali and Chris D. Geddes
Institute of Fluorescence, University of Maryland, Baltimore County, Baltimore, Maryland, United States of America

Xueshu Xie and Roman A. Zubarev
Division of Physiological Chemistry I, Department of Medical Biochemistry and Biophysics, Karolinska Institutet, Stockholm, Sweden

Viduthalai R. Regina and Duncan S. Sutherland
Interdisciplinary Nanoscience Center (iNANO), Aarhus C, Denmark

Arcot R. Lokanathan
Interdisciplinary Nanoscience Center (iNANO), Aarhus C, Denmark
Department of Forest Products Technology, Aalto University, Aalto, Finland

Jakub J. Modrzyński
Interdisciplinary Nanoscience Center (iNANO), Aarhus C, Denmark
Department of Plant and Environmental Sciences, University of Copenhagen, Frederiksberg C, Denmark

Rikke L. Meyer
Interdisciplinary Nanoscience Center (iNANO), Aarhus C, Denmark
Department of Bioscience, Aarhus University, Aarhus C, Denmark

R. Hemamalini and Sunil Khare
Enzyme and Microbial Biochemistry Lab, Department of Chemistry, Indian Institute of Technology, Delhi, New Delhi, India

Alexandre Courtiol and Karin Müller
Leibniz Institute for Zoo and Wildlife Research, Berlin, Germany

Stephanie Speck
Leibniz Institute for Zoo and Wildlife Research, Berlin, Germany
Institute of Animal Hygiene and Veterinary Public Health, University of Leipzig, Leipzig, Germany

Christof Junkes and Margitta Dathe
Leibniz Institute of Molecular Pharmacology, Berlin, Germany

Martin Schulze
Institute for Reproduction of Farm Animals Schoenow e. V., Bernau, Germany

Zhiming Ouyang, Jianli Zhou and Michael V. Norgard
Department of Microbiology, University of Texas Southwestern Medical Center, Dallas, Texas, United States of America

Mara S. Roset, Juan M. Spera, Leonardo Minatel, Juliana Cassataro and Gabriel Briones
Instituto de Investigaciones Biotecnológicas "Rodolfo Ugalde" - Instituto Tecnológico de Chascomús (IIB-INTECH), Universidad Nacional de San Martín (UNSAM), Consejo Nacional de Investigaciones Científicas y Técnicas (CONICET), Buenos Aires, Argentina

Andrés E. Ibañez and Guillermo H. Giambartolomei
Laboratorio de Inmunogenética, INIGEM-CONICET, Hospital de Clínicas "José de San Martín," Facultad de Medicina, Universidad de Buenos Aires (UBA), Buenos Aires, Argentina

Job Alves de Souza Filho and Sergio C. Oliveira
Department of Biochemistry and Immunology, Institute of Biological Sciences, Federal University of Minas Gerais, Belo Horizonte, Minas Gerais, Brazil

Xiao-Yan Yang, Liang Zhang, Nan Li, Junlong Han, Jing Zhang, Xuesong Sun and Qing-Yu He
Key Laboratory of Functional Protein Research of Guangdong Higher Education Institutes, Institute of Life and Health Engineering, College of Life Science and Technology, Jinan University, Guangzhou, China

Bin Sun
School of Pharmaceutical Sciences, Southern Medical University, Guangzhou, China

Jie Xu and Zhen Shi
State Key Laboratory of Tropical Oceanography, South China Sea Institute of Oceanology, Chinese Academy of Sciences, Guangzhou, China

Mingming Sun, Paul J. Harrison and Hongbin Liu
Division of Life Sciences, The Hong Kong University of Science and Technology, Clear Water Bay, Kowloon, Hong Kong

Kasumi Ishida, Junji Matsuo, Tomohiro Yamazaki and Hiroyuki Yamaguchi
Department of Medical Laboratory Science, Faculty of Health Sciences, Hokkaido University, Sapporo, Hokkaido, Japan

Tsuyoshi Sekizuka, Fumihiko Takeuchi and Makoto Kuroda
Pathogen Genomics Center, National Institute of Infectious Diseases, Tokyo, Japan

Kyoko Hayashida and Chihiro Sugimoto
Research Center for Zoonosis Control, Hokkaido University, Sapporo, Hokkaido, Japan

Shinji Nakamura
Division of Biomedical Imaging Research, Juntendo University Graduate School of Medicine, Tokyo, Japan

Mitsutaka Yoshida and Kaori Takahashi
Division of Ultrastructural Research, Juntendo University Graduate School of Medicine, Tokyo, Japan

Hiroki Nagai
Research Institute for Microbial Diseases, Osaka University, Osaka, Japan

Noriko Ido
Iwate Prefecture Central Livestock Hygiene Service Center, Iwate, Japan

Ken-ichi Lee, Masahiro Kusumoto and Taketoshi Iwata
Bacterial and Parasitic Disease Research Division, National Institute of Animal Health, National Agriculture and Food Research Organization, Ibaraki, Japan

Kaori Iwabuchi
Research Institute for Environmental Sciences and Public Health of Iwate Prefecture, Iwate, Japan,

Hidemasa Izumiya and Makoto Ohnishi
Department of Bacteriology, National Institute of Infectious Diseases, Tokyo, Japan

Ikuo Uchida
Hokkaido Research Station, National Institute of Animal Health, National Agriculture and Food Research Organization, Hokkaido, Japan

Masato Akiba
Bacterial and Parasitic Disease Research Division, National Institute of Animal Health, National
Graduate School of Life and Environmental Sciences, Osaka Prefecture University, Osaka, Japan

Iztok Dogsa, Ziva Marsetic and Ines Mandic-Mulec
Department of Food Science and Technology, Biotechnical Faculty, University of Ljubljana, Ljubljana, Slovenia

Sanjarbek Hudaiberdiev and Kumari Sonal Choudhary
Group of Protein Structure and Bioinformatics, International Centre for Genetic Engineering and Biotechnology, Trieste, Italy

Roberto Vera and Sándor Pongor
Group of Protein Structure and Bioinformatics, International Centre for Genetic Engineering and Biotechnology, Trieste, Italy
Faculty of Information Technology and Bionics, Pázmány Péter Catholic University, Budapest, Hungary

Michelle M. Nerandzic
Research Service, Louis Stokes Cleveland Veterans Affairs Medical Center, Cleveland, Ohio, United States of America

Curtis J. Donskey
Research Service, Louis Stokes Cleveland Veterans Affairs Medical Center, Cleveland, Ohio, United States of America
Geriatric Research, Education and Clinical Center, Cleveland Veterans Affairs Medical Center, Cleveland, Ohio, United States of America

Christopher W. Fisher
STERIS Corporation, Healthcare Group, Mentor, Ohio, United States of America

Yanwei Yang, Yan Dong, Wei Zhou, Jinghao Ban, Jingjing Wei, Yan Liu, Jing Gao and Jihua Chen
State Key Laboratory of Military Stomatology, Department of Prosthodontics, School of Stomatology, Fourth Military Medical University, Xi'an, China

Li Huang
State Key Laboratory of Military Stomatology, Department of General Dentistry and Emergency, School of Stomatology, Fourth Military Medical University, Xi'an, China

Hongchen Zhang
Department of Clinical Nursing, School of Nursing, Fourth Military Medical University, Xi'an, China

Meghan May
Department of Biomedical Sciences, College of Osteopathic Medicine, University of New England, Biddeford, Maine, United States of America
Dylan W. Dunne
Department of Biological Sciences, Jess and Mildred Fisher College of Science and Mathematics, Towson University, Towson, Maryland, United States of America

Daniel R. Brown
Department of Infectious Diseases and Pathology, College of Veterinary Medicine, University of Florida, Gainesville, Florida, United States of America

Akira Noga and Masafumi Hirono
Department of Biological Sciences, Graduate School of Science, University of Tokyo, Bunkyo-ku, Tokyo, Japan

Yuki Nakazawa
Department of Biological Sciences, Graduate School of Science, University of Tokyo, Bunkyo-ku, Tokyo, Japan
Center for Neuroscience, University of California Davis, Davis, California, United States of America

Tetsuro Ariyoshi
Department of Biological Sciences, Graduate School of Science, University of Tokyo, Bunkyo-ku, Tokyo, Japan
Department of Neurobiology, Graduate School of Medicine, University of Tokyo, Bunkyo-ku, Tokyo, Japan

Ritsu Kamiya
Department of Biological Sciences, Graduate School of Science, University of Tokyo, Bunkyo-ku, Tokyo, Japan
Department of Life Science, Faculty of Science, Gakushuin University, Toshima-ku, Tokyo, Japan

Joseph S. Johnson, DeeAnn M. Reeder, James W. McMichael III, Melissa B. Meierhofer, Daniel W. F. Stern, Shayne S. Lumadue, Lauren E. Sigler, Harrison D. Winters, Megan E. Vodzak and Kenneth A. Field
Department of Biology, Bucknell University, Lewisburg, Pennsylvania, United States of America

Allen Kurta
Department of Biology, Eastern Michigan University, Ypsilanti, Michigan, United States of America

Joseph A. Kath
Illinois Department of Natural Resources, Springfield, Illinois, United States of America

Ubiana de Cássia Silva, Nina Morena R. M. Silva and Josiane Leal Duarte
Departamento de Microbiologia, Universidade Federal de Viçosa, Vic‚osa, MG, Brasil

Gilberto de Oliveira Mendes
Departamento de Solos, Universidade Federal de Viçosa, Viçosa, MG, Brasil
Instituto de Ciências Agrárias, Universidade Federal de Uberlândia, Monte Carmelo, MG, Brasil

Ivo Ribeiro Silva
Departamento de Solos, Universidade Federal de Viçosa, Viçosa, MG, Brasil
Pesquisador Bolsista do Conselho Nacional de Desenvolvimento Científico e Tecnológico (CNPq), Brasília, DF, Brasil

Marcos Rogério Tótola and Maurício Dutra Costa
Departamento de Microbiologia, Universidade Federal de Viçosa, Vic‚osa, MG, Brasil
Pesquisador Bolsista do Conselho Nacional de Desenvolvimento Científico e Tecnológico (CNPq), Brasília, DF, Brasil

Carolina Lumertz Martello
Laboratório de Genômica Estrutural e Funcional, Centro de Biotecnologia, UFRGS, Porto Alegre, RS, Brazil

Luciano Antonio Reolon
Laboratório de microrganismos diazotróficos, Centro de Biotecnologia, UFRGS, Porto Alegre, RS, Brazil

Programa de Pós-Graduação em Biologia Celular e Molecular, Centro de Biotecnologia, Universidade Federal do Rio Grande do Sul (UFRGS), Porto Alegre, RS, Brazil
Centro Universitário Ritter dos Reis, Porto Alegre, RS, Brazil

Irene Silveira Schrank
Laboratório de microrganismos diazotróficos, Centro de Biotecnologia, UFRGS, Porto Alegre, RS, Brazil
Programa de Pós-Graduação em Biologia Celular e Molecular, Centro de Biotecnologia, Universidade Federal do Rio Grande do Sul (UFRGS), Porto Alegre, RS, Brazil
Departamento de Biologia Molecular e Biotecnologia, Instituto de Biociências, UFRGS, RS, Brazil

Henrique Bunselmeyer Ferreira
Laboratório de Genômica Estrutural e Funcional, Centro de Biotecnologia, UFRGS, Porto Alegre, RS, Brazil
Programa de Pós-Graduação em Biologia Celular e Molecular, Centro de Biotecnologia, Universidade Federal do Rio Grande do Sul (UFRGS), Porto Alegre, RS, Brazil
Departamento de Biologia Molecular e Biotecnologia, Instituto de Biociências, UFRGS, RS, Brazil

Mariana Claudia Allievi, María Mercedes Palomino, Mariano Prado Acosta, Leonardo Lanati, Sandra Mónica Ruzal and Carmen Sánchez-Rivas
Departamento de Química Biológica, Facultad de Ciencias Exactas y Naturales, Universidad de Buenos Aires, IQUIBICEN-CONICET, Buenos Aires, Argentina

Index